Focus on Bacterial Biofilms

Edited by Theerthankar Das

Published in London, United Kingdom

Focus on Bacterial Biofilms
http://dx.doi.org/10.5772/intechopen.96843
Edited by Theerthankar Das

Contributors
Dayamoy Mondal, Bettina Wollanke, Hartmut Gerhards, Ajay Kumar Oli, Palaksha K. Kanive Javaregowda, Apoorva Jain, Chandrakanth R. Kelmani, Sankar Veintramuthu, Selliamman Ravi Mahipriya, Deepak Dwivedi, Trishla Sehgal, Galina Satchanska, Norzawani Jaffar, Theerthankar Das, Brandon C. Young, Zeuko'O Menkem Elisabeth, Roger Bayston, Zhen Jia, Meenakshi Sharma, Pragati Yadav, Deepika Tripathi, Lakshmi Singh, Dewi F. Suniarti, Ria Puspitawati, Rezon Yanuar, Ranny R. Herdiantoputri, Tarik Aanniz, Safae El Mazouri, Wissal Bakri, Mouna Ouadghiri, Ilham Kandoussi, Azeddine Ibrahimi, Hajar Wakrim, Lahcen Belyamani, Saulo Henrique Rodrigues, Mariana Ottaiano Gonçalves, Lavinia Cipriano, Camila Cristina de Foggi, Marcelo Assis, Elson Longo, Cristina Paiva de Sousa, Andréa Cristina Bogas, Evandro Leite de Souza, Cynthia Amaning Danquah, Prince Amankwah Baffour Minkah, Theresah A. Agana, Phanankosi Moyo, Michael Tetteh, Isaiah Osei Duah Junior, Kofi Bonsu Amankwah, Samuel Owusu Somuah, Vinesh J. Maharaj, Michael Ofori, Salma Kloula Ben Ghorbel, Abdelwaheb Chatti, Rim Werheni

First published in London, United Kingdom, 2022 by IntechOpen
IntechOpen is the global imprint of INTECHOPEN LIMITED, registered in England and Wales, registration number: 11086078, 5 Princes Gate Court, London, SW7 2QJ, United Kingdom

British Library Cataloguing-in-Publication Data
A catalogue record for this book is available from the British Library

Additional hard and PDF copies can be obtained from orders@intechopen.com

Focus on Bacterial Biofilms
Edited by Theerthankar Das
p. cm.
Print ISBN 978-1-80355-795-3
Online ISBN 978-1-80355-796-0
eBook (PDF) ISBN 978-1-80355-797-7

For EU product safety concerns:
IN TECH d.o.o., Prolaz Marije Krucifikse Kozulić 3, 51000 Rijeka, Croatia,
info@intechopen.com or visit our website at intechopen.com.

Meet the editor

Dr. Theerthankar Das is a senior research fellow at the University of Sydney, Australia, with expertise in bacterial virulence factors, biofilm formation, and antimicrobial drug discovery. He has been awarded research funding/grants from the Australian Government totaling more than 4.5 million AUD. To date, Dr. Das has authored and co-authored forty-two publications in journals and eight book chapters. He is an academic editor of books and a reviewer for many high-impact scientific journals. He currently supervises Ph.D., master's, and honors students and is involved in academic teachings. Dr. Das' research has initiated the establishment of strong collaborations with industry, hospital, and research institutes.

Contents

Preface

Biofilms and their allied infections have an enormous negative influence on multiple sectors, including human health, by increasing morbidity and mortality rate, hospital admission, and associated treatment cost. Disease in livestock and plants and bacterial contamination leads to billions of dollars in losses for food, meat, dairy, and agriculture industries. Bacterial biofilms also damage the environment by contaminating water bodies via corroding pipelines (water, oil, and gas), ship hulls, and medical equipment. Bacterial biofilms have a detrimental impact on the global economy, accounting for billions of dollars in losses annually. It is paramount for everybody, including scientists, medical professionals, health care workers, and the public, to learn about this recurring issue and do everything possible to mitigate the damaging impacts of biofilms.

Focus on Bacterial Biofilms covers a wide array of subjects relevant to bacterial biofilms, focusing on the fundamentals of bacterial biofilms, the mechanism of biofilm formation, biofilm-associated infections, and allied catastrophic loss to both human health and economics. In addition, this book also addresses bacterial virulence factors, quorum sensing in bacteria, antimicrobial resistance in bacteria, and strategies to develop new antibacterial agents. The principal objective is to provide readers with a clear and comprehensive overview of biofilm formation and its detrimental impacts. At the end of the book, some chapters also highlight the application of bacteria for beneficial and industrial applications.

Research on bacterial biofilms has attracted interest for many decades, as evidenced by the thousands of journals, conferences, and projects cultivated in this field. In recent years, published research papers, conferences, and opinions from expert scientists, clinicians, and healthcare workers have undoubtedly enhanced the scientific basis for bacterial biofilms' pros and cons.

To this end, I would like to express my appreciation to all the scientists and researchers from different research institutes and universities around the world who put forth an enormous effort and contributed their chapters to the completion of this book. I am also thankful to IntechOpen for the opportunity to serve as editor of this project.

Theerthankar Das, Ph.D.
Infection, Immunity and Inflammation Theme,
Sydney Institute for Infectious Diseases,
Charles Perkins Centre,
School of Medical Sciences,
The University of Sydney,
Sydney, Australia

Section 1

Introduction

Chapter 1

Introductory Chapter: Highlighting Pros and Cons of Bacterial Biofilms

Theerthankar Das and Brandon C. Young

1. Introduction

Bacteria are unicellular microorganisms that belong to the classification of prokaryotes (other prokaryotic organisms are Archaea) that lack a membrane-bound nucleus (genetic material DNA is present in the cytoplasm) and other organelles such as mitochondria [1, 2]. Bacteria are generally classified in three shapes: rod, sphere/cocci and spiral. They can divide/multiply/grow and metabolise either in the presence of oxygen (aerobic) or in the absence of oxygen (anaerobic). In addition, they can be facultative and survive under aerobic, anoxic (low oxygen) or anaerobic conditions. Bacteria get energy by making adenosine triphosphate (ATP) through glycolysis, pyruvate oxidation, citric acid cycle and the electron transport chain in the presence of oxygen (aerobic respiration), whereas in the absence of oxygen, ATP is produced via fermentation of glycolysis derived products (anaerobic respiration) [1, 3, 4]. In microbiology, bacteria are differentiated into two groups: (i) Gram-positive and (ii) Gram-negative depending upon bacterial ability to retain Gram stain or crystal violet stain. Gram-positive bacteria have thick peptidoglycan cell walls that strongly bind and retain crystal violet stain. In contrast, Gram-negative bacteria have a thinner peptidoglycan cell wall that cannot retain crystal violet stain and hence is washed easily when washed with ethanol. Gram-positive bacteria appear purple or blue after staining, whereas Gram-negative bacteria appear pink when observed under a microscope [1].

2. Introducing bacterial biofilms

Bacteria exist in abundance in almost all corners of the Earth, including marine and freshwater, rocks and soil, in man-made/engineered surfaces such as ships, pipelines and living organisms, including humans, animals, birds and plants. Bacteria are either free-living/planktonic or exist in communities embedded in their self-produced extracellular matrix called "Biofilm". It has been projected that on Earth, up to 80% of bacterial cells live the biofilm mode of lifestyle [5]. The biofilm stage is the preferred stage in bacterial lifestyles, principally for species with a pathogenic nature, as the biofilm stage provides resistance against physical, chemical and environmental challenges [6]. Biofilm formation is a complex process with multiple steps starting with the initial adhesion of planktonic bacteria to the surface, aggregation, micro-colony formation and proliferation into the mature biofilm and finally, active disruption of biofilms to release planktonic bacterial cells to progress adhesion at new sites [7]. The biofilm formation process involves various biomolecular pathways; the most prominent one is the cell-to-cell signalling pathway in bacteria and is commonly acknowledged as the Quorum Sensing (QS) system [8]. The QS system in bacteria activates in response to the fluctuations in the bacterial population. As the bacterial cell density increases, bacteria produce chemical signals called "autoinducers" that are recognised by the local population to facilitate communication between their own and different bacterial species [8]. The QS system regulates genes essential for the biosynthesis of various products by bacteria, including biopolymers (polysaccharides, DNA and protein—that are essential for biofilm matrix formation and integrity), virulence factors, biofilm formation and protection against physical (hydrodynamic shear stress), chemical, host immune response and antimicrobial challenges [9–11]. The role of bacteria and its biofilm stage can be beneficial or devastating. Both biofilm applications for beneficial use and biofilm eradication to protect the environment and the health of patients account for a multi-billion-dollar industry annually. Below are the highlights of the pro and cons of bacteria/biofilms.

3. Application of beneficial bacteria in ecosystem and industry

In terms of beneficial bacteria, their applications in the environment and industry are diverse, including maintaining biological balance in natural aquatic and soil ecosystems by remineralisation and restoring nutrients [12]. Rhodococcus spp. of bacterial biofilm has significant application in bioremediation, including cleaning industrial and domestic pollutants in the environment by decaying organic pollutants such as polycyclic aromatic hydrocarbons (e.g. petroleum products) and chlorinated organic compounds from soil and water bodies [13]. Soil bacteria (e.g. *Bacillus subtilis*, *Pseudomonas putida* and *Rhizobium spp.*) that maintain a symbiotic relationship with the plant also promote plant growth by fixing nitrogen in the plant roots, which then converts nitrogen into ammonia essential for plant fitness and development [14]. The use of bacterial secreted by-products in the food and pharmaceuticals industry for commercial use has existed for many decades, such as lipase (e.g. phospholipase) enzyme in making bread (baking) and winemaking brewing) industries, vegetable oil refinement, in the dairy industry to hydrolysis milk fat for cheese production and biodegradation of petroleum products [15]. Microbial amylase is another industrial application enzyme mainly used in the hydrolysing of complex carbohydrates (e.g. starch saccharification) into smaller sugar (glucose and fructose) units in the manufacture of corn syrups [16].

Bacterial biofilms have a more extensive application in biomining, such as the recovery of copper metal and the generation of biogas/coal gas. Some bacterial species, *Leptospirillum ferriphilum*, *Sulfobacillus thermosulfidooxidans* and *Acidithiobacillus,* are used to recover copper from chalcopyrite (CuFeS2); these bacteria catalyse the transformation of solid metal sulfide dissolution to soluble metal sulfates [17]. Methanobacteria is used to produce biogas (methane, carbon dioxide and hydrogen) from organic waste, including cattle and human waste. Biogas's predominant application is used for cooking and water heating in rural India and is also used in the production of electricity [18, 19].

4. The catastrophic impact of biofilms on the health care sector and the environment

Bacterial biofilms cause catastrophic impacts in terms of infection, antimicrobial resistance and associated morbidity and mortality. Statistics show that more than 80% of chronic infections are associated with biofilm-forming microbes [20]. Some of the common infections associated with bacterial biofilms include urinary tract infection, wound infection, infection in diabetic leg ulcers, medical implant-associated infections including surgical site infection and catheter-associated infections, microbial keratitis mainly in people wearing contact lenses, chronic sinusitis, bacterial pneumonia in chronic obstructive pulmonary patients, cystic fibrosis, HIV patients, COVID-19 patients, infective endocarditis, stomach ulcers, tooth decay and periodontitis infection etc. The burden of biofilm-associated infections is responsible for a global economic loss of hundreds and thousands of billions annually [21]. Some of the common bacterial species that are responsible for the above-mentioned health-associated infections include *P. aeruginosa*, *Acinetobacter baumannii*, *Klebsiella pneumoniae*, *Protease mirabilis*, and *Escherichia coli*, *Helicobacter pylori*, *Porphyromonas gingivalis*, *Staphylococcus aureus*, *Streptococcus pyogenes*, *S. pneumoniae* and others. In addition, to being directly detrimental to human health, biofilms are also accountable for an economic loss in agriculture, dairy, livestock and the meat industry. Statistical analysis reveals that biofilm infections in plants (fruits and vegetables) account for up to 10% of the world's food supply loss and are unswervingly responsible for food-borne illnesses [22]. Similarly, a bacterial (e.g., *Streptococcus agalactiae*) infection in cows (bovine mastitis—inflammation of the mammary glands) contributes to an 11% decrease in US total milk production alongside a two billion dollars monetary loss to the US dairy industry [22].

Biofilm-associated corrosion is an enormous problem in multiple sectors, including the marine and shipping industry (damages to the ships) and chemical processing and water treatment industries (water pipelines, heat exchangers and stainless steel tanks). These bacteria can withstand a wide range of pH 4–9 and temperatures 10–50°C [23]. For example, sulphate-reducing bacteria (grow in anoxic conditions) are a prime culprit in the marine industry corrosion; these bacteria influence changes in the physicochemical parameters such as pH of the local environment and redox potential of the metal [24]. It reduces sulfate to metal sulfide, and the production of hydrogen sulfide gas triggers metal corrosion [25]. Microbial-induced corrosion attributes to a negative impact on the man-made infrastructure and loss of billions of dollars annually [25].

This book "Bacterial Biofilms" collected the chapters written by prominent and expert scientists from their respective areas of research and highlighted the pros and cons of bacterial biofilms in different sectors. The content in this book will educate people from different backgrounds, including but not limited to scientists, doctors, infectious diseases specialities, high school and university students and the public.

Author details

Theerthankar Das[1*] and Brandon C. Young[2]

1 Infection Immunity and Inflammation, Sydney Institute for Infectious Diseases, Charles Perkins Centre, School of Medical Science, The University of Sydney, Sydney, Australia

2 Infection Immunity and Inflammation, Charles Perkins Centre, School of Medical Science, The University of Sydney, Sydney, Australia

*Address all correspondence to: das.ashishkumar@sydney.edu.au

References

[1] Salton RJM, Kim K-S. Chapter 2 Structure. In: Baron S, editor. Medical Microbiology, Medical Microbiology—NCBI Bookshelf (nih.gov). 4th ed. Galveston (TX): University of Texas Medical Branch at Galveston; 1996

[2] [Internet]. Available from: prokaryote / procariote | Learn Science at Scitable (nature.com)

[3] [Internet]. Available from: https://study.com/learn/lesson/cellular-respiration-bacteria-process-energy.html#:~:text=Bacteria%20perform%20two%20types%20of,occurs%20in%20the%20plasma%20membrane

[4] [Internet]. Available from: https://www.diffen.com/difference/Aerobic_Respiration_vs_Anaerobic_Respiration

[5] Flemming H-C, Wuertz S. Bacteria and archaea on Earth and their abundance in biofilms. Nature Reviews. Microbiology. 2019;**17**(4):247-260

[6] Santos A, Galdino A, Mello T, Ramos L, Branquinha M, Bolognese A, et al. What are the advantages of living in a community? A microbial biofilm perspective! Memórias do Instituto Oswaldo Cruz. 26 Jul 2018;**113**(9):e180212. DOI: 10.1590/0074-02760180212

[7] Landini P, Antoniani D, Burgess J, Nijland R. Molecular mechanisms of compounds affecting bacterial biofilm formation and dispersal. Applied Microbiology and Biotechnology. 2010;**86**(3):813-823

[8] Federle M, Bassler B. Interspecies communication in bacteria. Journal of Clinical Investigation. 2003;**112**(9):1291-1299

[9] Ueda A, Wood T. Connecting quorum sensing, c-di-GMP, Pel polysaccharide, and biofilm formation in *Pseudomonas aeruginosa* through tyrosine phosphatase TpbA (PA3885). PLoS Pathogens. 2009;**5**(6):e1000483

[10] Ibáñez de Aldecoa A, Zafra O, González-Pastor J. Mechanisms and regulation of extracellular DNA release and its biological roles in microbial communities. Frontiers in Microbiology. 2017;**8**:1390. DOI: 10.3389/fmicb.2017.01390

[11] Eberl L, Riedel K. Mining quorum sensing regulated proteins - role of bacterial cell-to-cell communication in global gene regulation as assessed by proteomics. Proteomics. 2011;**11**(15):3070-3085

[12] Joshi P. Microbial diversity of aquatic ecosystem and its industrial potential. Journal of Bacteriology and Mycology: Open Access. 2016;**3**(1):177-179. DOI: 10.15406/jbmoa.2016.03.00048

[13] Kuyukina MS, Ivshina IB. Application of rhodococcus in bioremediation of contaminated environments. In: Alvarez H editor. Biology of Rhodococcus. Vol 16. Microbiology Monographs. Berlin, Heidelberg: Springer; 2010. DOI: 10.1007/978-3-642-12937-7_9

[14] Antonella Di Benedetto N, Rosaria Corbo M, Campaniello D, Pia Cataldi M, Bevilacqua A, Sinigaglia M, et al. The role of plant growth promoting bacteria in improving nitrogen use efficiency for sustainable crop production: A focus on wheat. AIMS. Microbiology. 2017;**3**(3):413-434

[15] Chandra P, Enespa SR, Arora P. Microbial lipases and their industrial

applications: A comprehensive review. Microbial Cell Factories. 2020;**19**:169. DOI: 10.1186/s12934-020-01428-8

[16] Mehta D, Satyanarayana T. Bacterial and archaeal α-amylases: Diversity and amelioration of the desirable characteristics for industrial applications. Frontiers in Microbiology. 2016;**7**

[17] Buetti-Dinh A, Herold M, Christel S, Hajjami M, Bellenberg S, Ilie O, et al. Systems biology of acidophile biofilms for efficient metal extraction. Scientific Data. 2020;**7**:215. DOI: 10.1038/s41597-020-0519-2

[18] [Internet]. Available from: https://byjus.com/biology/microbes-in-production-of-biogas/

[19] [Internet]. Available from: https://biogas.ifas.ufl.edu/uses.asp#:~:text=The%20biogas%20may%20be%20applied,fuel%20cells%20may%20produce%20electricity

[20] Jamal M, Ahmad W, Andleeb S, Jalil F, Imran M, Nawaz MA, et al. Bacterial biofilm and associated infections. Journal of the Chinese Medical Association. 2018;**81**(1):7-11

[21] Cámara M, Green W, MacPhee C, Rakowska P, Raval R, Richardson M, et al. Economic significance of biofilms: A multidisciplinary and cross-sectoral challenge. npj Biofilms Microbiomes. 2022;**8**:42. DOI: 10.1038/s41522-022-00306-y

[22] [Internet]. Available from: Rise to the biofilm challenge: harmful microbes present opportunities for innovators | PA Consulting

[23] [Internet]. Available from: https://www.corrosionclinic.com/types_of_corrosion/microbiologically_influenced_biological_microbial_corrosion.htm

[24] Videla H. An overview of mechanisms by which sulphate-reducing bacteria influence corrosion of steel in marine environments. Biofouling. 2000;**15**(1-3):37-47

[25] Enning D, Garrelfs J. Corrosion of iron by sulfate-reducing bacteria: New views of an old problem. Applied and Environmental Microbiology. 2014;**80**(4):1226-1236

Section 2

Factors Impacting Bacterial Biofilms Formation

Chapter 2

Bacterial Biofilm and the Medical Impact

Norzawani Jaffar

Abstract

Most pathogenic bacteria species form biofilm as their protective mode of growth, which helps them survive from the bactericidal effect of the antimicrobials or the killing activity of the host immune cells. The bacteria cells' survivability via biofilm formation creates challenges in the medical field in terms of the device and also disease-related to biofilm. The impact of the bacterial biofilm issue is worsening over time, and the association to the high tolerance to the antimicrobial agents leads to increased morbidity and mortality worldwide. This review will highlight the main characteristics of the biofilm, the issue of biofilm in clinical practice, which also covered the pertinence of the biofilm in clinical practice, device-related biofilm disease, oral disease, and the significant bacterial species involved in the biofilm-related infections. Knowledge about the vital role of bacterial biofilm in related disorders will give new insight into the best approaches and alternative treatments for biofilm-related disease.

Keywords: antibiotic resistance, medical device, chronic infections, oral disease

1. Introduction

Microbial biofilm is a microscopic entity that significantly affects human health. It is composed of bacterial colonies within a matrix of extracellular polymeric substances, which protect them from environmental stress, shear stress, detergents, antimicrobial agents, and the host's immune cells. According to the National Institute of Health (NIH), 65% of microbial diseases and 80% of chronic infection is related to biofilm formation [1]. Antibiotics cannot treat several conditions related to biofilm formation due to the high level of biofilm resistance activity. An antibiotic concentration killing effect toward a biofilm might require 1000 times greater than those required to kill the planktonic bacteria cells [2]. In addition, bacterial biofilm causes several diseases in response to both device-related and non-device-related infections. This situation creates challenges for the medical team to provide the best solution or treatment.

Broad heterogeneity of phenotypes developed within a biofilm contributes to the recalcitrance of the sessile bacteria. This condition evolves the bacteria cells inside the biofilm to coordinate and differentiate through the communication system and the releasing of quorum sensing small signaling molecules called autoinducers. Interbacterial communication allows the decision of their density and regulation of the virulence gene expression. This is also the indicator of antibiotic susceptibility profiles of a biofilm. Due to biofilm-cell physiological states, biofilm usually shows high resistance toward most antibiotics. Antibiotics might be effective against the active cells located at the top of the biofilm, in contrast to nutrient-depleted zones at the middle and bottom of the biofilm in which the cell is in the state of dormancy and lack of metabolic activity [3].

The emergence of antibiotic resistance toward biofilm leads to various chronic diseases and is very difficult to treat with efficacy. Most of the recently available antibiotics are not able to resolve the infection. In addition, higher values of minimum bactericidal concentration (MBC) and minimum inhibitory concentration (MIC) used to treat biofilm may result in in-vivo toxicity and other complications. Thus, biofilm formation issues significantly impact human health and the health care industry.

2. Biofilm in clinical practice

A meta-analysis study by Malone et al. (20017) reported the prevalence of biofilms in chronic wounds was 78.2% [4]. This finding supports the clinical assumptions that biofilms appear and are significant in human chronic non-healing wounds. Besides, one of the most prominent clinical-level species is *Staphylococcus aureus* affecting both hospital-acquired and community-acquired infection. The biofilm production of *S. aureus* cells isolated from clinical samples shows the association of the biofilm with methicillin and inducible clindamycin resistance [4]. In addition, MRSA strains showed a higher biofilm production than MSSA strains. Suggesting strong biofilm formation increases the possibility of antibiotic resistance and leads to treatment failures in MRSA infections [4].

Other than that, *Escherichia coli* is reported to lead the urinary tract infection (UTI), contributing to 80 to 90% of all community-acquired and 30 to 50% of all hospital-acquired cases of UTIs [5]. The study of the uropathogenic *E. coli* revealed a high prevalence of biofilm-forming strains of this group of bacteria that are also highly associated with the multi-drug resistant (MDR) phenotype. Out of 200 *E. coli* clinical isolates, 62.5% can produce biofilm, with 93% of the isolates showing varied resistance with amoxicillin and co-trimoxazole, followed by gentamycin (87%), cefuroxime (84%), Nalidixic acid (79%), Amoxicillin clavulanic acid (62.5%), Ciprofloxacin (62%), ceftriaxone (55%), Ceftazidime (54%), chloramphenicol (28%), Nitrofurantoin (25.5%) and Imipenem (0.5%) [5].

This finding represents the burden of the biofilm formation issues, which are highly associated with increased antibiotic resistance. In addition, another meta-analysis study concludes that biofilm formation production by microbial species impacts the blood system infection leads to resistance, persistence, and mortality. Staphylococci biofilm producer shows significantly higher prevalence in the resistant strain, whereas *Candida species* biofilm production highly impacted mortality [6]. High cell density within the biofilm facilitates high rates of horizontal gene transfer between microorganisms through the conjugation process, more frequent within the community inside biofilm than the planktonic bacteria [7].

3. The main characteristic of bacterial biofilm and their resistance to antimicrobial agents

In general, bacterial biofilm shows resistance against antibiotics and human immune systems. The process of biofilm formation initiates with the attachment of the planktonic bacterial cells on the living or non-living surfaces. The attachment will lead to the construction of the micro-colony of the bacteria cells and rise to a three-dimensional structure, followed by biofilm maturation and detachment. The process of biofilm formation until a detachment of the cells is regulated by the cell-to-cell communication known as the quorum-sensing system. Extracellular polymeric substance (EPS) is one of the main components in a biofilm, strengthening the interaction of the microorganism in the biofilm [8]. Typically 65% of the biofilm volume is constituted by the extracellular matrix, partially or mainly composed of polysaccharides, proteins, and nucleic acid [9]. The EPS protects bacteria from environmental stress such as salinity, UV exposure, dehydration, antimicrobial, and phagocytes [10]. Besides, some channels separate the microcolonies inside the biofilm structure to be attached to new niches [1].

There are studies on the resistant mechanism of the bacterial biofilm toward antibiotics. Most of the studies suggest that the production of glycocalyx or EPS matrix and other functions play a prominent role that prevents the penetration of the antimicrobial agents inside the biofilm. Common disinfectant such as chlorine is only 20% or less of the total concentration in the bulk liquid measured inside the biofilm of *P. aeruginosa* and *Klebsiella pneumoniae*. Interestingly, a complete equilibrium with the bulk liquid did not reach even after 1 to 2 hour incubation time [11]. Another study also showed the same finding when the biofilm production of *P. aeruginosa* on a dialysis membrane showed retarded piperacillin diffusion [12].

In contrast, evaluation on *Staphylococcus epidermidis* biofilm that were grown in the same manner show diffusion of rifampicin and vancomycin across the membrane [13]. Thus, this finding might suggest that inhibition of antibiotic absorption cannot be explained by antimicrobial resistance. Other pathways and mechanisms might be occurring inside the biofilm.

In addition, the difference between thin and thick biofilm formation toward antibiotic resistance has been explored. Penetration of the hydrogen peroxide in the thin biofilm of *P. aeruginosa* was observed compared to a viscous biofilm, which shows no penetration of that chemical compound inside the biofilm [14, 15]. Interestingly, the penetration of the hydrogen peroxide in the thick biofilm was observed in the mutant strains of *P. aeruginosa* without *katA* gene, which is the calatase gene that functions to neutralize the hydrogen peroxide [14].

Furthermore, depletion of the nutrient level inside the biofilm will influence the interaction of the bacteria cells against antimicrobials. Generally, during bacterial growth, the transition from exponential to stationary or no growth leads the bacteria to resistance to antibiotics [3]. Due to low nutrient level and high cell density, the planktonic cell of the bacteria starts to aggregate and initiate attachment and biofilm formation. In the biofilm community, bacteria begin to change their mode to slow-growing. These physiological changes might play a role in the insensitivity of the bacterial cells inside the biofilm toward antibiotics.

Biofilm disease includes device-related infection, chronic infection with the absence of a foreign body, and malfunction of medical devices. Biofilm-related disease or infection is complicated to treat and detect at early stages by microbiological analyses. Thus, characterization of the chemical composition of the EPS might expedite the development of new therapies against biofilm related-infection.

4. Biofilm and device-related infection (DRI)

The emergence of device-related infections is highly associated with biofilm-producing bacteria among critical patients in the intensive-care units. DRI is defined as an infection that occurs in a patient with any device (for example, endotracheal tube, intravascular catheter, or indwelling urinary catheter) for at least 48 hours in use before the onset of infection [16]. Most of the DRI reported in the developed country is led by catheter-related bloodstream infections (CRBSI), followed by catheter-associated urinary tract infections (CAUTI) and ventilator-associated pneumonia (VAP) [17]. In addition, another study of the biofilm formation on or in the medical devices that were examined upon removal from the patients or were tested in animal or laboratory systems. Several medical devices may involve biofilm formation, such as central venous catheters, central venous catheter needleless connectors, contact lenses, mechanical heart valves, pacemakers, peritoneal dialysis catheters, prosthetic joints, tympanostomy tubes, and voice prostheses (**Table 1**) [18].

Biofilm formation on medical devices is related to the substratum and cell surface properties. For instance, the characters of glass and various metals that are highly charged hydrophilic materials, water pipes, and environmental surfaces are pretty rough or textured. Some materials might be coated with antimicrobial, such as antibiotic-impregnated catheters [24]. The characteristic of the substratum might have a significant effect on the rate of bacterial adhesion and biofilm formation. The rougher and more hydrophobic materials will develop rapid biofilm formation.

Hydrophobicity of both bacteria and material surfaces may influence the adherence capacity of bacterial cells. Hydrophilic material surfaces are usually more resistant to bacterial attachment than hydrophobic materials [25]. Fletcher and Loeb's (1978) study reported that many marine *Pseudomonas sp*. are attached to hydrophobic plastics with little or surface charge-free like Teflon, polyethylene, polystyrene, poly (ethylene terephthalate). At the same time, very few are attached to hydrophilic and negatively charged substrata like glass, mica, and oxidized plastics [26]. However,

Devices	Causative microorganisms	Burden of illness	References
Contact lenses	*P. aeruginosa, Staphylococcus aureus, E. coli, Staphylococcus epidermidis, species of Candida spp., Serratia spp., Proteus spp.*	Keratitis	Jamal et al. 2018 [1]
Central venous catheter	*Pseudomonas aeruginosa, S. aureus, coagulase-negative staphylococci, Klebsiella pneumoniae, E. coli, Acinetobacter baumanii*	Bloodstream infections (BSI)	Gahlot et al. 2014 [19]
Urinary catheters	*Escherichia coli, Enterococci spp, coagulase negative Staphylococcus, P. aeruginosa, Candida spp., Proteus mirabilis, K. pneumoniae, Morganella morganii.*	Urinary tract infection	Nicolle et al. 2015 [20]
Mechanical heart valves	*Streptococcus spp., S. aureus, S. epidermidis*, gram-negative *Bacillus, Enterococcus, Candida spp. Haemophilus parainfluenzae, Propionibacterium acnes.*	Prosthetic valve endocarditis	Jamal et al. 2018, Gomes et al. 2018 [1, 21]
Implantable prosthetic device	*S. aureus, S. epidermidis, Streptococcus agalactiae, P. aeruginosa, E. coli, P. acnes, Enterococcus faecalis*	Prosthetic joints infection	Benito et al. 2016 [22]
Endotracheal tube	*P. aeruginosa, S. aureus, Candida albicans, Streptococcus spp.*	Ventilator-associated pneumonia	Fernandez-barat et al. 2016 [23]

Table 1.
Common devices related diseases and the microbial etiology.

dental plaque formation in the human oral cavity is reported as far less on hydrophobic compared to hydrophilic surfaces, even after nine days without oral hygiene [27]. In addition, another study by Everaert et al. (1997) showed less biofilm formation on hydrophobic silicone rubber voice prosthesis of laryngectomized patients compared to the hydrophilic surfaces after six weeks in the human body [28]. Thus, the role of hydrophobic material surfaces toward rapid biofilm formation is still unclear.

5. Biofilm in chronic infections

Chronic infections are a significant burden to patients and the healthcare systems. Besides, the economy is also impacted and varies depending on chronic infection due to several treatments failure. It is expected that there will be an increase in chronic infection cases in the future due to an aging population concurrent with the rise in lifestyle diseases such as diabetes which is a significant cause of chronic wounds [29]. Bacterial biofilm has been recognized as responsible for most chronic infections, including otitis, diabetic foot ulcer, rhinosinusitis, chronic pneumonia in cystic fibrosis patients, osteomyelitis, and infective endocarditis [30]. These infections affect millions of people each year, with high mortality and morbidity rate as a consequence. The worse issue of biofilm involvement in infection is due to undetectable species responsible as swabs and scrapes of biofilm samples often show culture-negative. This might be due to the strong association of bacteria within the biofilm or their uncultivability. The same problems occur for implant and catheter-related infections; identifying the bacteria has been almost impossible. Up to this date, bacteria species from a biofilm were considered unculturable. In addition, some pathogenic bacteria that cannot grow the culture media are believed to be activated when present in the host system or environment, and later they can initiate infection [31]. The biofilm infection often finalizes as untreatable, leading to the chronic state of bacterial infections. However, chronic infection will lead to an adaptive inflammatory response, characterized by a high level of mononuclear leucocytes and IgG antibodies [32]. In some cases, such as the cystic fibrosis patient suffering chronic lung infection, the inflammatory response shows the chronic response with continued recruitment of

Diseases	Pathogenesis
Cystic Fibrosis (CF)	*P. aeruginosa* biofilm induces the infiltration PMNs, subsequent tissue damage, and loss of lung function [33].
Infective endocarditis	Bacterial biofilm diminishes the heart valve function and triggers persistent infection to the circulatory system. Detachment of the biofilm might spread to the other systemic system contributes to kidney, brain, and extremities, particularly risk to emboli [34].
Diabetic foot ulcer	Hyperglycemic conditions cause deleterious effects on the innate immune system associated with altered PMNs, impaired phagocytosis, and bactericidal activity against the infections. Thus, bacterial biofilm in the diabetic foot ulcer implicates the failure of the healing process [35].
Chronic rhinosinusitis	Biofilms contribute to the destruction of the epithelial layer and the absence of cilia and continuous local inflammatory response [36].
Osteomyelitis	Biofilm formation and proliferations lead to an inflammatory bone disorder characterized by increased local cytokines and osteoclastogenesis [37].

Table 2.
Examples of biofilm-related chronic infections and suggestive pathogenesis.

polymorphonuclear neutrophils (PMNs) [32]. PMN are the leukocytes critical to the innate immune response against invading pathogens (**Table 2**).

6. Oral diseases

An oral disease associated with bacterial biofilm is periodontal disease. Periodontal disease has been reported by Global Burden of Disease (GBD) 2010 as a global prevalence of 35% for all ages combined and the sixth-most prevalent condition in the world [38]. Initiating biofilm formation at the periodontal area by various pathogenic species of oral bacteria may lead to severe inflammatory disorders that reduce the gum line, bleeding of the gum, and tooth loss. The issue of periodontal disease is not limited to the antibiotic resistance properties of the biofilm but also the aggressive pro-inflammatory response toward the virulence activities of the pathogenic species that reside in the biofilm. In addition, there are associations between periodontal disease and other systemic diseases such as respiratory tract infection, cardiovascular disease, Alzheimer's disease, gastrointestinal and colon-rectal cancer, diabetes and insulin resistance, and adverse pregnancy outcomes [39]. The association of periodontal disease with systemic disease is possible when the progressive inflammatory activity releases toxins or leakage of microbial products enter the bloodstream thru the blood vessel in the pulp chamber of an infected tooth. This agrees with a meta-analysis of 5 prospective cohort studies (86,092 patients) that indicates that individuals with periodontal disease had 1.14 times higher risk of developing coronary heart disease [40]. Whereas for the case of respiratory tract infection and pneumonia, the lung infection might occur due to the accumulation of the pathogens from saliva or oral cavity at the lower airways. Genetically identical respiratory pathogens isolated from dental plaque and bronchoalveolar lavage fluid from the same patient in the ICU indicate that respiratory pathogens' significant reservoir might be associated with dental plaque [41].

7. Significant bacterial species related to a biofilm infection

Biofilm-producing bacteria play a significant role in biofilm-related diseases. The biofilm's high resistance against antimicrobial agents and the host immune system contribute to considerable treatment challenges. Generally, the ability of a microorganism to form biofilms on the human tissue or related medical devices will lead to the association of chronic infection. The most common bacterial species related to biofilm formation in hospital settings are *Enterococcus faecalis, S. aureus, S. epidermidis, Streptococcus viridans, E. coli, K. pneumoniae, Proteus mirabilis, Acinetobacter baumannii,* and *Pseudomonas aeruginosa* [42]. These species may originate from the skin of healthcare workers or patients or might be from the surrounding as simple as tap water to which entry ports are exposed or other sources in the environment. For instance, Staphylococcus species mainly colonize humans' skin and mucous membrane. *S. aureus* and *S. epidermidis* are the prominent aetiologic agents for nosocomial infection, surgical site, and bloodstream infection [43, 44]. The persistence of *S. aureus* biofilm formation is related to antibiotic pressure. This species own the ability to stay in the viable state but is not culturable [45]. Recently, daptomycin has been used as the last resort for treating Gram-positive bacterial infections, including MRSA and Vancomycin-resistant Enterococcus. This is due to its bactericidal activity against these bacteria [46, 47]. Enterococci cause a wide variety of infections in humans, including infection of the endocardium, urinary tract, bloodstream, biliary tract, abdomen, burn wounds, and medical devices [48]. However, the most prevalent is *E. faecalis* due to its biofilm formation ability and several virulence factors related to the persistence of biofilm formation and heterogeneity in antimicrobial resistance acquiring activity [49].

On the other hand, a study of attributable mortality dan morbidity caused by carbapenem-resistant *K. pneumonia* showed that 50% of the 391 patients ended with mortality, with 12.2% of the case being bloodstream infections [50]. In addition, *K. pneumonia* is responsible for many cases of nosocomial infection related to a pyogenic liver abscess or endophthalmitis [51]. Besides that, *P. aeruginosa* and *E. coli* are most prevalent for medical device-associated pathogens. *P. aeruginosa* contributes to 10 to 20% of all nosocomial infections, whereas *E.coli* contributes to 50% of the infections associated with urinary catheters [52, 53]. At the same time, *A. baumannii* emerges with significant pathogenicity due to its multi-drug resistant capacity and the ability to form biofilm on several biotic and abiotic surfaces [54]. This species is rapidly spread in the health care facilities and can stay months on the dry surface on insensate objects [55].

8. Conclusion

Biofilm formation is a natural process employed by several bacteria species. This is part of the adaptation process and survival mechanism in response to their environment. Unfortunately, bacterial biofilm formation develops to impact human health and industries. Evolution to adapt toward the surroundings triggered by an antimicrobial substance during a treatment intervention leads the bacteria cell to manage their survival by acquiring the resistant genes thru several pathways and mechanisms. Applying antibiotics to treat bacteria's biofilm-related infection will lead to another level of resistance activity in the biofilm community as well as toxic effects to the host system. A comprehensive understanding of the biofilm structure organization and the prominent chemical involved might help the researcher elucidate a potent compound or chemical that can degrade or interact with the bacterial biofilm. Alternative methods or therapies other than antibiotics application must be explored to reduce the impact of the bacterial biofilm on human health and the health care industry.

Author details

Norzawani Jaffar
Faculty of Health Sciences, University Sultan Zainal Abidin, Terengganu, Malaysia

*Address all correspondence to: zawanijaffar@unisza.edu.my

References

[1] Jamal M et al. Bacterial biofilm and associated infections. Journal of the Chinese Medical Association. 2018;**81**(1):7-11. DOI: 10.1016/j.jcma.2017.07.012

[2] Haney EF, Trimble MJ, Cheng JT, Vallé Q, Hancock REW. Critical assessment of methods to quantify biofilm growth and evaluate antibiofilm activity of host defence peptides. Biomolecules. 2018;**8**(2):1-22. DOI: 10.3390/biom8020029

[3] Anderl JN, Zahller J, Roe F, Stewart PS. Role of nutrient limitation and stationary-phase existence in Klebsiella pneumoniae biofilm resistance to ampicillin and ciprofloxacin. Antimicrobial Agents and Chemotherapy. 2003;**47**(4):1251-1256. DOI: 10.1128/AAC.47.4.1251-1256.2003

[4] Malone M et al. The prevalence of biofilms in chronic wounds: A systematic review and meta-analysis of published data. Journal of Wound Care. 2017;**26**(1):20-25. DOI: 10.12968/jowc.2017.26.1.20

[5] Katongole P, Nalubega F, Florence NC, Asiimwe B, Andia I. Biofilm formation, antimicrobial susceptibility and virulence genes of Uropathogenic Escherichia coli isolated from clinical isolates in Uganda. BMC Infectious Diseases. 2020;**20**(1):1-6. DOI: 10.1186/s12879-020-05186-1

[6] Pinto H, Simões M, Borges A. Prevalence and impact of biofilms on bloodstream and urinary tract infections: A systematic review and meta-analysis. Antibiotics. 2021;**10**(7):1-24. DOI: 10.3390/antibiotics10070825

[7] Sørensen SJ, Bailey M, Hansen LH, Kroer N, Wuertz S. Studying plasmid horizontal transfer in situ: A critical review. Nature Reviews. Microbiology. 2005;**3**(9):700-710. DOI: 10.1038/nrmicro1232

[8] Branda SS, Vik Å, Friedman L, Kolter R. Biofilms: The matrix revisited. Trends in Microbiology. 2005;**13**(1):20-26. DOI: 10.1016/j.tim.2004.11.006

[9] Del Pozo JL. Biofilm-related disease. Expert Review of Anti-Infective Therapy. 2018;**16**(1):51-65. DOI: 10.1080/14787210.2018.1417036

[10] Hall-Stoodley L, Costerton JW, Stoodley P. Bacterial biofilms: From the natural environment to infectious diseases. Nature Reviews. Microbiology. 2004;**2**(2):95-108. DOI: 10.1038/nrmicro821

[11] De Beer D, Srinivasan R, Stewart PS. Direct measurement of chlorine penetration into biofilms during disinfection. Applied and Environmental Microbiology. 1994;**60**(12):4339-4344. DOI: 10.1128/aem.60.12.4339-4344.1994

[12] Hoyle BD, Alcantara J, Costerton JW. Pseudomonas aeruginosa biofilm as a diffusion barrier to piperacillin. Antimicrobial Agents and Chemotherapy. 1992;**36**(9):2054-2056

[13] Dunne WM, Mason E, Kaplan SL. Diffusion of Rifampin and vancomycin through a Staphylococcus epidermidis biofilm. Antimicrobial Agents and Chemotherapy. 1993;**37**(12):2522-2526

[14] Stewart PS et al. Effect of catalase on hydrogen peroxide penetration into Pseudomonas aeruginosa biofilms. Applied and Environmental Microbiology. 2000;**66**(2):836-838

[15] Cochran WL, Mcfeters GA, and Stewart PS. "Reduced susceptibility of thin Pseudomonas aeruginosa biofilms to hydrogen peroxide and monochloramine". Journal of Applied Microbiology. 2000;**88**:22-30

[16] Singhai M, Malik A, Goyal R. A study on device-related infections with special reference to biofilm production and antibiotic resistance. Journal of Global Infectious Diseases. 2012;**4**(4):193-198. DOI: 10.4103/0974-777X.103896

[17] Horan TC, Andrus M, Dudeck MA. CDC/NHSN surveillance definition of health care-associated infection and criteria for specific types of infections in the acute care setting. American Journal of Infection Control. 2008;**36**(5):309-332. DOI: 10.1016/j.ajic.2008.03.002

[18] Donlan RM. Biofilm formation: A clinically relevant microbiological process. Clinical Infectious Diseases. 2001;**33**(8):1387-1392. DOI: 10.1086/322972

[19] Gahlot R, Nigam C, Kumar V, Yadav G, Anupurba S. Catheter-related bloodstream infections. International Journal of Critical Illness and Injury Science. 2014;**4**(2):162-167. DOI: 10.1097/01.nhh.0000346317.38831.38

[20] Nicolle LE. Infections associated with urinary catheters. In: Schlossberg D, editor. Clinical Infectious Disease. Second ed. Cambridge University Press; 2015. pp. 722-727. DOI: 10.1017/CBO9781139855952.122

[21] Gomes A et al. Sonication of heart valves detects more bacteria in infective endocarditis. Scientific Reports. 2018;**8**(1):1-9. DOI: 10.1038/s41598-018-31029-w

[22] Benito N et al. Time trends in the aetiology of prosthetic joint infections: A multicentre cohort study. Clinical Microbiology and Infection. 2016;**22**(8):732.e1-732.e8. DOI: 10.1016/j.cmi.2016.05.004

[23] Fernández-barat L, Torres A. Biofilms in ventilator-associated pneumonia. Future Microbiology. 2016;**11**(12):1599-1610

[24] Floyd KA, Eberly AR, Hadjifrangiskou M. *Adhesion of bacteria to Surfaces and Biofilm Formation on Medical Devices*. Cambridge, MA, United States: Woodhead Publishing, Elsevier Ltd; 2017

[25] An YH, Friedman RJ. Concise review of mechanisms of bacterial adhesion. Journal of Biomedical Materials Research. 1998;**43**(3):338-348

[26] Fletcher M, Loeb GI. Influence of substratum characteristics on the attachment of a marine pseudomonad to solid surfaces. Applied and Environmental Microbiology. 1979;**37**(1):67-72. DOI: 10.1128/aem.37.1.67-72.1979

[27] Quirynen M et al. The influence of surface free-energy on Planimetric plaque growth in man. Journal of Dental Research. 1989;**68**(5):796-799. DOI: 10.1177/00220345890680050801

[28] Everaert EPJM, Mahieu HF, Chung RPW, Verkerke GJ, Van Der Mei HC, Busscher HJ. A new method for in vivo evaluation of biofilms on surface-modified silicone rubber voice prostheses. European Archives of Oto-Rhino-Laryngology. 1997;**254**(6):261-263. DOI: 10.1007/BF02905983

[29] Narayan KMV, Boyle JP, Geiss LS, Saaddine JB, Thompson TJ. Impact of recent increase in incidence on future diabetes burden: U.S., 2005-2050. Diabetes Care. 2006;**29**(9):2114-2116. DOI: 10.2337/dc06-1136

[30] Burmølle M et al. Biofilms in chronic infections - a matter of opportunity - Monospecies biofilms in multispecies infections. FEMS Immunology and Medical Microbiology. 2010;**59**(3):324-336. DOI: 10.1111/j.1574-695X.2010.00714.x

[31] Brown MRW, Barker J. Unexplored reservoirs of pathogenic bacteria: Protozoa and biofilms. Trends in Microbiology. 1999;**7**(1):46-50. DOI: 10.1016/S0966-842X(98)01425-5

[32] Bjarnsholt T. The role of bacterial biofilms in chronic infections. APMIS. 2013;**121**(Suppl. 136):1-51. DOI: 10.1111/apm.12099

[33] Worlitzsch D et al. Effects of reduced mucus oxygen concentration in airway Pseudomonas infections of cystic fibrosis patients. The Journal of Clinical Investigation. 2002;**109**(3):317-325. DOI: 10.1172/JCI0213870

[34] Vestby LK, Grønseth T, Simm R, Nesse LL. Bacterial biofilm and its role in the pathogenesis of disease. Antibiotics. 2020;**9**(2):1-29. DOI: 10.3390/antibiotics9020059

[35] Pouget C, Dunyach-Remy C, Pantel A, Schuldiner S, Sotto A, Lavigne JP. Biofilms in diabetic foot ulcers: Significance and clinical relevance. Microorganisms. 2020;**8**(10):1-15. DOI: 10.3390/microorganisms8101580

[36] Karunasagar A, Garag SS, Appannavar SB, Kulkarni RD, Naik AS. Bacterial biofilms in chronic Rhinosinusitis and their implications for clinical management. Indian Journal of Otolaryngology and Head & Neck Surgery. 2018;**70**(1):43-48. DOI: 10.1007/s12070-017-1208-0

[37] Beck-Broichsitter BE, Smeets R, Heiland M. Current concepts in pathogenesis of acute and chronic osteomyelitis. Current Opinion in Infectious Diseases. 2015;**28**(3):240-245. DOI: 10.1097/QCO.0000000000000155

[38] Marcenes W et al. Global burden of oral conditions in 1990-2010: A systematic analysis. Journal of Dental Research. 2013;**92**(7):592-597. DOI: 10.1177/0022034513490168

[39] Bui FQ et al. Association between periodontal pathogens and systemic disease. Biomedical Journal. 2019;**42**(1):27-35. DOI: 10.1016/j.bj.2018.12.001

[40] Bahekar AA, Singh S, Saha S, Molnar J, Arora R. The prevalence and incidence of coronary heart disease is significantly increased in periodontitis: A meta-analysis. American Heart Journal. 2007;**154**(5):830-837. DOI: 10.1016/j.ahj.2007.06.037

[41] Heo SM, Sung RS, Scannapieco FA, Haase EM. Genetic relationships between Candida albicans strains isolated from dental plaque, trachea, and bronchoalveolar lavage fluid from mechanically ventilated intensive care unit patients. Journal of Oral Microbiology. 2011;**3**(2011):1-11. DOI: 10.3402/jom.v3i0.6362

[42] Donlan RM. Biofilms and device-associated infections. Emerging Infectious Diseases. 2001;**7**(2):277-281. DOI: 10.3201/eid0702.010226

[43] Zaborowska M, Tillander J, Brånemark R, Hagberg L, Thomsen P, Trobos M. Biofilm formation and antimicrobial susceptibility of staphylococci and enterococci from osteomyelitis associated with percutaneous orthopaedic implants. Journal of Biomedical Materials Research Part B: Applied Biomaterials. 2017;**105**(8):2630-2640. DOI: 10.1002/jbm.b.33803

[44] Cerca N, Pier GB, Vilanova M, Oliveira R, Azeredo J. Quantitative analysis of adhesion and biofilm formation on hydrophilic and hydrophobic surfaces of clinical isolates of Staphylococcus epidermidis. Research in Microbiology. 2005;**156**(4):506-514. DOI: 10.1016/j.resmic.2005.01.007

[45] Pasquaroli S et al. Role of daptomycin in the induction and persistence of the viable but non-culturable state of Staphylococcus aureus biofilms. Pathogens. 2014;**3**(3):759-768. DOI: 10.3390/pathogens3030759

[46] Len O et al. Daptomycin is safe and effective for the treatment of gram-positive cocci infections in solid organ transplantation. Transplant Infectious Disease. 2014;**16**(4):532-538. DOI: 10.1111/tid.12232

[47] Raad I et al. Comparative activities of daptomycin, linezolid, and tigecycline against catheter-related methicillin-resistant Staphylococcus bacteremic isolates embedded in biofilm. Antimicrobial Agents and Chemotherapy. 2007;**51**(5):1656-1660. DOI: 10.1128/AAC.00350-06

[48] Jett BD, Huycke MM, Gilmore MS. Virulence of enterococci. Clinical Microbiology Reviews. 1994;**7**(4):462-478. DOI: 10.1128/CMR.7.4.462

[49] Gold HS. Vancomycin-resistant enterococci: Mechanisms and clinical observations. Clinical Infectious Diseases. 2001;**33**(2):210-219. DOI: 10.1086/321815

[50] Borer A et al. Attributable mortality rate for Carbapenem-resistant Klebsiella pneumoniae Bacteremia. Infection Control and Hospital Epidemiology. 2009;**30**(10):972-976. DOI: 10.1086/605922

[51] Donelli G. *Biofilm-Based Nosocomial Infections*. Basel, Beijing, Wuhan: Shu-Kun Lin (MPDI); 2014. DOI: 10.3390/books978-3-03842-136-8

[52] Ramos GP, Rocha JL, Tuon FF. Seasonal humidity may influence Pseudomonas aeruginosa hospital-acquired infection rates. International Journal of Infectious Diseases. 2013;**17**(9):e757-e761. DOI: 10.1016/j.ijid.2013.03.002

[53] Jacobsen SM, Stickler DJ, Mobley HLT, Shirtliff ME. Complicated catheter-associated urinary tract infections due to Escherichia coli and Proteus mirabilis. Clinical Microbiology Reviews. 2008;**21**(1):26-59. DOI: 10.1128/CMR.00019-07

[54] Longo F, Vuotto C, Donelli G. Biofilm formation in Acinetobacter baumannii. The New Microbiologica. 2014;**37**:119-127

[55] Jawad A, Seifert H, Snelling AM, Heritage J, Hawkey PM. Survival of Acinetobacter baumannii on dry surfaces: Comparison of outbreak and sporadic isolates. Journal of Clinical Microbiology. 1998;**36**(7):1938-1941. DOI: 10.1128/jcm.36.7.1938-1941.1998

Chapter 3

Biofilm Formation by Pathogenic Bacteria: The Role of Quorum Sensing and Physical - Chemical Interactions

Theerthankar Das and Brandon C. Young

Abstract

Pathogenic bacteria cause infectious diseases, mainly when the host (humans, animals, and plants) are colonised by bacteria, especially in its biofilm stage, where it is known to cause chronic infections. Biofilms are associated with resistance to antimicrobial agents, including antibiotics, antiseptics, detergents, and other therapeutic approaches. *Antimicrobial resistance (AMR)* is one of the biggest public health challenges of our time and is termed a 'silent pandemic' by the United Nations. Biofilm formation, pathogenicity and the associated AMR are regulated through a bacterial cell-to-cell communication system termed "Quorum Sensing (QS)". As the bacterial cells sense the fluctuations in their population, they biosynthesise and secrete the signalling molecules called autoinducers (AI). In gram-negative, the signalling molecules are primarily homoserine lactones (AHL) whereas in gram-positive the signalling molecules are autoinducing peptides. The AI binds to receptor and regulator proteins in the bacterial cells to activate the complete QS system, which controls the regulations of various genes that are essential for the biosynthesis of virulence factors, extracellular biopolymers (EPS) production, biofilm formation and bacterial fitness.

Keywords: bacterial biofilms, antibiotic resistance, quorum sensing, *Pseudomonas aeruginosa*, *Staphylococcus aureus*, pyocyanin

1. Introduction

Infectious diseases of humans, animals and plants are caused by the spread of microorganisms, including bacteria, fungi, viruses, protozoa and parasites. Microorganisms that cause disease are called pathogens. Our body (gastrointestinal tract, skin, mucosa of mouth, nose and vagina) is inhabited by numerous bacterial species that form part of the host commensal microflora [1]. However, under certain circumstances, when the host immune system is compromised due to diseases such as HIV, cancer, COVID-19, cystic fibrosis or when the individual has burn injuries, blunt trauma or penetrating trauma (such as through surgery), bacteria can breach the host barriers and colonise to cause infection. Such bacteria are called opportunistic pathogens. Pathogenic bacteria cause infectious diseases, often when they colonise and form biofilms. Biofilms significantly impact human health; it is estimated that 65% of all microbial infections and more than 80% of chronic infections involve biofilm-associated microorganisms [2]. In this chapter, we have discussed a few of the clinically important biofilm-associated infections.

Urinary tract infections (UTIs) are infections involving any part of the urinary tract. They are one of the most common infections, resulting in an estimated 7 million office visits, 1 million emergency department visits and over 100,000 hospitalisations annually in the United States [3]. UTIs are caused by both gram-negative and gram-positive bacteria, with the most common causative agent for both complicated and uncomplicated UTIs being uropathogenic *Escherichia coli* (UPEC), causing approximately 75% and 65% of these cases, respectively, with other notable contributors including *Staphylococcus saprophyticus*, *Enterococcus faecalis*, Group B *Streptococcus* (GBS), *Proteus mirabilis* and *P. aeruginosa* [4]. UPEC, as well as many of the other common uropathogens, establish biofilms on the bladder wall and surfaces of indwelling urinary catheters as a strategy to protect the encased bacteria from the host immune response and intervention with antimicrobial therapy [5, 6].

Microbial keratitis is an infection of the cornea; when mismanaged, this infection can result in scarring of the cornea, permanent loss of vision and even total loss of the eye [7]. In the United States alone, there are nearly 1 million clinical visits for keratitis annually at an estimated cost of US$175 million in direct health care expenditures [8]. Biofilms play an essential role in bacterial keratitis as their presence on contact lenses as well as their storage cases can allow bacteria to survive and eventually spread to corneal epithelium [9]. Biofilm populations have increased resistance to antibiotics and host immune response [10]. Bacterial keratitis is significantly more prevalent than fungal keratitis in the United States and other developed countries and is commonly caused by *S. aureus* and *P. aeruginosa*.

Chronic obstructive pulmonary disease (COPD) is a chronic inflammatory lung disease characterised by poorly reversible airway obstruction and is currently the third leading cause of death worldwide [11]. The lower respiratory tract of COPD patients is often colonised by bacteria, such as *P. aeruginosa*, *Haemophilus influenzae* and *Streptococcus pneumoniae* [12, 13]. Chronic bacterial colonisation is a major factor driving chronic inflammation in COPD patients [14]. Exacerbations are one of the most important manifestations of COPD and are defined as an increase in the inflammation present above the stable state of COPD, and COPD patients are estimated to suffer 1–4 exacerbations annually [15]. Exacerbations are thought to worsen the decline in lung function with increasing exacerbation frequency, are responsible for much of the morbidity and mortality of COPD [16], account for 50%–75% of the total economic burden due to COPD [17] and estimated to be US$32 billion annually in the United States alone [18]. Respiratory infections are the most common cause of severe exacerbations in COPD, with *P. aeruginosa* being one of the most frequently isolated causative microorganisms in severe COPD patients [19, 20].

Seasonal respiratory viruses such as influenza virus and respiratory syncytial virus (RSV) as well as respiratory viruses that have spread in major outbreaks such as SARS-CoV, H_1N_1 Influenza, MERS-CoV and SARS-CoV-2 are a significant cause of morbidity and mortality worldwide. Following the primary viral infection, disruption of the airway epithelium barrier and dysregulation of immune responses promote the colonisation of various bacteria to establish secondary bacterial infections, also known as superinfections, which can have significantly worse clinical outcomes when compared to the initial primary infection [21, 22]. Among COVID-19 patients, secondary bacterial infections can arise due to subsequent colonisation by *Klebsiella pneumoniae*, *Acinetobacter baumannii*, *P. aeruginosa* and other bacteria [23], and it has been observed that patients with these superinfections are seen to have mortality rates twice as high as those without secondary bacterial infections [24].

2. Multiple stages in biofilm formation

Biofilm formation is the most complex stage in the bacterial lifestyle [25]. Compared to the planktonic stage or free-living bacterial cells, bacterial cells encased within biofilms are highly resistant to antimicrobial agents, detergents, host immune responses and environmental and physical stress [26, 27]. Researchers in many publications have widely described the mechanism of biofilm formation [28]. **Figure 1**, in brief, represents schematically bacterial biofilm formation in a hierarchical process.

i. To begin with, motile planktonic bacterial cells travel towards the substratum surface (e.g., mucosal, skin, biomaterials and other non-biotic surfaces) and reversibly adhere. In this step, the motility and adhesion are facilitated by flagella, fimbriae, pili, outer membrane proteins (OMPs) and lipopolysaccharides (LPS). These cell appendages and biomolecules drive non-specific physical-chemical forces (e.g., Lifshitz-van der Waals and electrostatic interactions) [29].

ii. In the second step, bacterial irreversible/strong adhesion to the surface is also driven by bacterial cell appendages, OMPs and LPS. Again, the physicochemical forces drive these interactions (van der Waals, electrostatic interactions, acid-base interactions and hydrophobic forces). These interaction forces promote the transition from initial reversible bacterial adhesion to the irreversible phase, over several minutes by progressive removal of interface water between the bacterial cell surface and substratum or another bacterial cell surface. In addition, bacterial cell surface biopolymers such as proteins and eDNA undergo conformation changes that suit bacterial attachment to the surfaces [29].

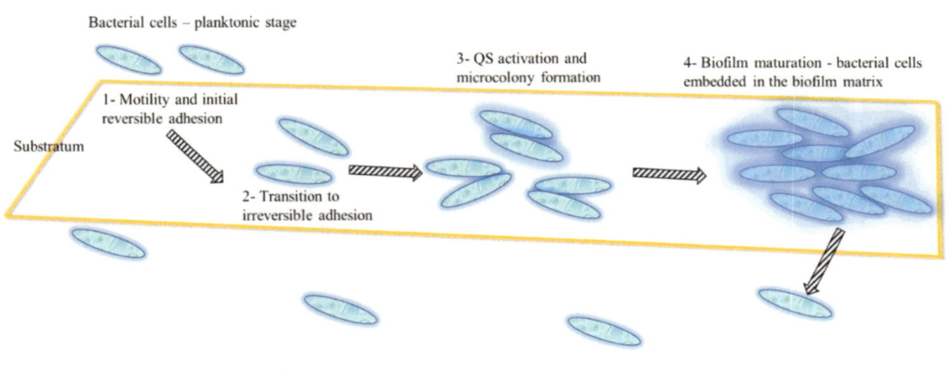

Figure 1.
Schematic showing the five major steps involved in the biofilm formation cycle. The cycle begins with mobility and initial adhesion to the substratum and eventually results in a mature biofilm in which bacteria can disperse as planktonic cells to colonise new sites and repeat the cycle.

iii. In the next step, bacterial cells secrete signalling molecules with increasing bacterial population (e.g., Homoserine lactone, auto-inducing peptides and competence stimulating peptides). These signalling molecules bind with the bacterial cell membrane-bound receptors or/and transcriptional regulatory proteins to initiate the quorum sensing (QS) system in bacteria [29]. QS is essential to trigger bacterial aggregation and microcolony formation.

iv. In the fourth stage, the QS-mediated biosynthesis and secretion of virulence factors and other extracellular compounds, including polysaccharides, eDNA, proteins and metabolites, occur and dictates robust biofilm matrix and maturation of biofilms. The robust biofilm matrix hinders antibiotic penetration into biofilms and can provide resistance against antibiotics for the encased bacteria up to 1000-fold [30]. The biofilm matrix is termed a "house of biofilms' [31].

In the final stage, biofilm ageing and dispersion of mature biofilm as planktonic bacterial cells occur, allowing for bacterial attachment and biofilm formation at new sites through a repeat of the biofilm cycle. The dispersion stage is essential for expanding bacterial colonisation and survival and is triggered through active and passive mechanisms. In the active mechanism, bacteria produce various enzymes/proteins (e.g., DNase I, Alginate lyase, Dispersin B, Exopolysaccharide lyase, protease, surface-protein-releasing enzyme, etc.). These enzymes cleave the biofilm matrix and trigger the release of bacterial cells. The passive dispersal mechanism is mainly the external environment, including nutrient deficiency, QS signals, phagocytosis and antimicrobial agents [32].

3. Physical: Chemical forces influence bacterial adhesion and program biofilm formation

Many studies have acknowledged that the fundamental physical-chemical interaction forces observed throughout the biofilm formation cycle are essential for mature biofilm formation. The physical-chemical interaction forces mediated by bacterial cells or substratum surfaces are purely dependent on the presence of chemical functional groups and the charge of molecules on surfaces. For instance, Das et al. 2012 showed that removing eDNA from *Streptococcus mutans* cell surface via DNase I treatment significantly decreases short-range acid-base interaction forces between bacteria and surface and consequently impaired *S. mutans* adhesion to the glass substratum surface [33, 34]. In another study, Swartjes et al. 2015 showed similar inhibition of *P. aeruginosa* and *S. aureus* adhesion and biofilm formation on DNase I immobilised surfaces [35].

Thermodynamics and extended Derjaguin–Landau–Verwey–Overbeek (DLVO)-analyses theoretically revealed that long-distance van der Waals interaction forces are always favourable or attractive due to the induced dipole interactions. These forces are weak and can range up to hundreds of nanometres and are essential to initially bringing bacteria closer to the substratum [29].

Electrostatic interactions are purely dependent upon the surface charge of bacteria and substratum. Bacterial cell surfaces are generally negatively charged due to the presence of negatively charged biopolymers and cell appendages. Electrostatic interactions would predict repulsion between bacteria and surfaces if the substratum surface also exhibits a negative charge [29, 34], whereas bacteria should rapidly attach to positively charged substratum surface. It is to be noted that many antibiotics (e.g., Gentamicin, tobramycin, etc.) or antimicrobial peptides (bacitracin, colistin/polymyxin E and B) are naturally or engineered to be cationic charged to enhance their interactions with bacterial cells [36]. Also, antimicrobial surfaces are made by immobilising cationic antimicrobial polymers to attract bacterial adhesion and kill without inducing biofilm formation [37]. Electrostatic forces are also influenced by the presence of nutrients such as divalent cations (Ca^{2+} and Mg^{2+}), which promote bacterial interactions, aggregation and biofilm matrix stability by interacting between negatively charged biopolymers within the matrix [38, 39].

Short, ranged acid-base interactions come into action when bacteria are at very close range to the substratum (below 5 nanometres). These forces are influenced by the presence of polar moieties in the molecules; polar groups promote electron-accepting or electron-donating parameters that are essential for bond-strengthening and transition from reversible bacterial adhesion to irreversible adhesion stage. An atomic force microscopic study performed by Das et al. 2011 revealed that bacterial cell surfaces containing eDNA had more vital adhesion forces, multiple minor peaks (due to bond breakage) and a more significant separation distance than DNase I treated bacterial cells [34]. This means eDNA favours bond-strengthening mediated by close-range acid-base interactions (triggers by electron donation and accepting moieties in the eDNA) [29, 34].

Hydrophobic forces are also one significant factor determining bacterial adhesion to the surface and biofilm formation. Studies have shown that hydrophobicity of surfaces (bacteria or substratum) promotes bacterial adhesion and biofilm formation [34, 40, 41]. Hydrophobic forces are strong interactive solid forces compared to van der Waals and hydrogen forces. Garcia-Fernandez et al. 2021 showed that EPS-producing strains of *Streptococcus thermophilus* and *Lactococcus lactis* spp. have a higher water contact angle (hydrophobicity) than EPS-negative mutants [42]. EPS production by these strains is directly related to its robust biofilm formation ability [42]. Contact angle analysis has also revealed a significant change in bacterial cell surface hydrophobicity when subjected to DNase I treatment: *P. aeruginosa* PAO1 strain water contact angle is 65° when exposed to exogenous DNA whereas, when not exposed to exogenous DNA the water contact angle is 44° [34]. Hydrophobic and van der Waals interactions are essential for maintaining biofilm stability by interacting with different biopolymers within the matrix, e.g., carbohydrates and proteins [43]. A study revealed that in Burkholderia *multivorans,* EPS component polysaccharide (EpolC1576) holds many non-polar rhamnoses (6-deoxy sugar) units in its primary structure; these non-polar units influence rhamnose binding with many hydrophobic molecules and are essential for the architecture of three-dimensional biofilm matrix [44].

Mirani et al. 2016 have shown that bacteria can change their cell surface phenotype i.e., hydrophilic to hydrophobic and vice versa when exposed to antibiotics [45]. Their study showed that when *S. aureus* is exposed to a sub-inhibitory concentration of oxacillin, *S. aureus* changes to biofilm mode and its cell surface hydrophobicity increases in contrast to its planktonic phase characterised by more hydrophilic character [45]. Another interesting finding is that in *S. aureus* and *P. aeruginosa* biofilms, the small colony variants (SCVs), which are metabolically inactive (but viable and non-culturable bacterial cells), exhibited hydrophobic properties [46]. These SCVs play a critical role in the persistence of infection and pathogenicity [47, 48].

4. QS mechanism in bacteria

Through the decades of research, it has been well acknowledged that the QS system is an essential phenomenon for the bacterial biofilm lifestyle. The principal purpose of bacterial QS is to control the regulation of gene expression related to bacterial biosynthesis of numerous endo and exogenous molecules critical for necessary bacterial fitness, survival, virulence production, biofilm formation, infection of the host, evading host immune response and antimicrobial agents. QS is a step-by-step mechanism that begins with bacterial population density fluctuations triggering the release of signalling chemical molecules called "autoinducers'. Studies suggest that autoinducers influence bacterial communication (i.e., 'calling distance') at ranges between 5 and 200 μm [49, 50]. Autoinducers could be of different types and classes [51]. For example, most gram-negative bacteria (e.g., *P. aeruginosa, E. coli, A. baumannii, Vibrio Cholera*, etc.) produces homoserine lactone molecules of different molecular weight and carbon length. At the same time, gram-positive bacteria (e.g., Staphylococcus sp. and Streptococcus sp.) produce autoinducing peptides and competence stimulating peptides as their signalling molecules. Once secreted, autoinducers get recognised by bacterial cell membrane-associated or intracellular receptor proteins. In addition to population-based naturally secreting autoinducers/signalling molecules, many other environmental factors, including oxidative stress, antibiotics or antimicrobial chemicals or nutrients, trigger QS in bacteria.

The typical gram-negative and gram-positive QS mechanisms have been illustrated in **Figure 2**.

4.1 *P. aeruginosa* is a classic example of a hierarchical QS system

In most gram-negative bacterial species, *luxI-luxR* genes or homologous genes regulate the QS system. In *P. aeruginosa,* there are four principal QS systems.

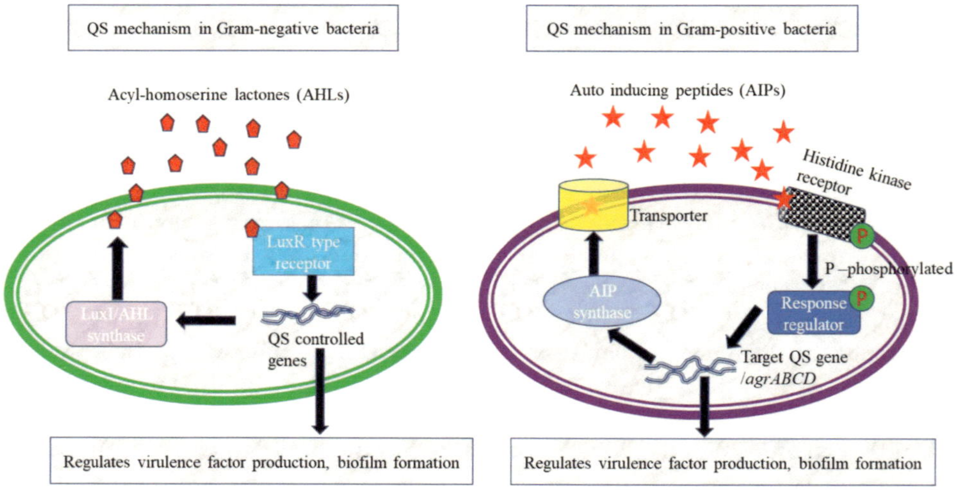

Figure 2.
Schematic showing the quorum sensing (QS) mechanism in gram-negative and gram-positive bacteria. In gram-negative bacteria, the signalling molecule is primarily AHLs, whereas in Gram-positive bacterial species, signalling molecule is primarily by AIPs, followed by bindings of signalling molecules to the receptors in a bacterial cell and triggering activation of QS-controlled genes. Regulation of Qs genes influences virulence factor production and biofilm formation.

First, *lasI/lasR* genes are homologous to the *lux* system and are responsible for the biosynthesis of the chief lactone-based signalling molecule/autoinducer *N*-(3-oxo-dodecanoyl)-L-homoserine lactone (3OC_{12}-HL). The gene *lasI* encodes the autoinducer enzyme LasI, which acts to catalyse the synthesis of the lactone autoinducer (also called AI-1) from substrates 3-oxo-C_{12}-acyl-carrier protein (acyl-ACP) and S-adenosyl-L-methionine [52, 53]. The homoserine lactone molecules are generally lipophilic and freely diffuse through the lipopolysaccharides in the *P. aeruginosa* cell membrane out to the immediate external microenvironment. The AI-1 then binds with the intracellular transcriptional LasR protein (in this case, LasR functions as both AI binding protein and regulatory protein) to activate various virulence factors genes, including *exoprotease (lasA), elastase (lasB), alkaline protease (aprA) and endotoxin A (toxA),* Phospholipase C, heat-labile hemolysin *(plC)*, and *lasI* (for positive autoregulation) [54, 55].

Next in the QS hierarchy is the RhlI-RhlR system. The RhlI (encoded by rhlI) autoinducer synthase enzyme synthesises N-butyryl homoserine lactone (C_4-HSL) binds with transcriptional regulatory protein RhlR. RhlR- C_4-HSL interactions lead to the activation of several other virulence genes, including *rhlAB* (rhamnolipids) and *lasB* (elastase B) in Pseudomonas species [53–55].

The PQS-PqsR QS system is a late QS system responsible for producing a phenazine-based cytotoxic metabolite 1-hydroxy-N-methylphenazine (pyocyanin) [54]. Operons *pqsABCDEHR* and *phnAB* and genes outside these operons are responsible for synthesising the pseudomonas quinolone signal (PQS) autoinducer in a complex multistep process [56]. The receptor for PQS is the PqsR protein (*pqsR*), which is regulated through the AHL-LasR QS system [54, 57–59]. The binding of the PQS autoinducer to the PqsR receptor/regulator protein activates the expression of virulence factors, including *phz* (pyocyanin), which are critical for causing infection. PQS signalling molecules also act as siderophores in chelating ferric ion (Fe^{3+}) and activate the production of siderophore genes *pvd* (pyoverdine) and *pch* (pyochelin) [57–62].

A newly identified class of autoinducer, termed IQS (2-(2-hydroxyphenyl)-thiazole-4-carbaldehyde), has been recognised in *P. aeruginosa* and categorised into a fourth QS system known as the AmbBCDE/IqsR system [63, 64]. This system can integrate environmental stress cues such as phosphate depletion into QS signalling to activate PQS-PqsR signalling in the absence of LasI-LasR activity [65].

QS-mediated toxin biosynthesis induces a severely detrimental effect on the host body. For instance, endotoxin A constrains protein synthesis in the host by impeding protein elongation factor 2 [66]. Exoenzyme S quests on low molecular weight proteins in the host, consequently hindering DNA synthesis and cell morphology [67]. Elastase from *P. aeruginosa* cleaves human leukocyte elastase, human neutrophil elastase and collagens, destroying host tissue elastic properties and impairing wound healing [68, 69]. Production of hemolytic phospholipase C (PlcHR) by *P. aeruginosa* directly interferes with the host protein kinase C signalling pathway (PKC), thus restraining neutrophil burst activity and superoxide ($O_2^{\cdot -}$) production [70]. Neutrophil assembly and production of superoxides at the infection site are essential to fight against *P. aeruginosa* pathogenicity. Thus, PlcHR promotes *P. aeruginosa* survival in host tissue by evading host inflammatory response by restraining neutrophil burst activity [70].

Pyocyanin, a hallmark metabolite of *P. aeruginosa*, gives a unique greenish-blue colour when grown in the lab and is also visible at the infection site. For instance, Green Nail Syndrome (GNS) is a nail infection caused by *P. aeruginosa,* and the presence of pyocyanin (also siderophore pyoverdine) causes the greenish colourisation of nails (chloronychia) [71]. Pyocyanin diffuses into host cells and reduces intracellular thiol antioxidant (glutathione) levels in mammalian cells [72]. *In vitro* study showed pyocyanin induces oxidative stress in cells, hinders human nasal ciliary beat frequency, declines intracellular cyclic AMP and damages epithelium [73]. Pyocyanin has been found in burn wound exudates; from burn wound patients and is known to impair wound healing by triggering cell-cycle arrest and premature senescence (ageing of cells) [74, 75]. Pyocyanin is essential for biofilm matrix stability via intercalation with eDNA [76]. Pyocyanin-DNA binding is necessary to prevent the loss of pyocyanin to the external environment and supports *P. aeruginosa* cells in inner biofilm layers that lack oxygen [77].

4.2 Highlighting QS regulation in gram-positive bacteria

In gram-positive bacteria, the peptide-based QS system is critical in virulence factor production and biofilm formation. For instance, in *Streptococcus* species (*Streptococcus pneumoniae* and *S. mutans*), competence stimulating peptide (CSP) is the primary autoinducer whose synthesis is regulated by *comE* [78]. The CSP gets released extracellularly via the transporter protein ComAB. In the extracellular microenvironment, CSP autoinducers bind with bacterial membrane-bound receptor ComD (transmembrane histidine kinase), causing the phosphorylation (i.e., transfer of phosphate group) of the regulatory protein ComE [78]. ComE undergoes structural modulation and binds with the promoter region of DNA to promote QS regulation genes and virulence factors [79]. CSP-Com mediated QS induces bacterial cell lysis proteins, including murein hydrolases autolysin A and C (LytA and LytC) and Choline-Binding Protein D (CbpD) [80]. These proteins trigger fratricide in the pneumococcal population and trigger virulence factors pneumolysin and *Streptococcus* cell wall constituent lipoteichoic acid (LTA) into the host cell to trigger an immune response [80]. CSP is essential for *Streptococcus-mediated* DNA binding, uptake and transformation from the microenvironment [81] and eDNA-mediated biofilm formation [81]. Other receptors and transcriptional regulatory proteins have also been identified that bind signalling peptides or activate through external environmental factors (oxygen, acid, oxidative stress) and coordinate QS systems in the *Streptococcus* species, including BlpABCSRH, *CiaRH,* HK11/RR11, VicK/VicR and LytST [82, 83]. This QS system is essential for other virulence factor synthesis such as capsular polysaccharides to evade the host immune response (phagocytosis) in *S. pneumoniae*, antibiotic resistance, acid and oxidative stress tolerance and biofilm integrity [84–87].

In *S. aureus*, multiple QS systems have been reported. The primary QS system is coordinated by the global regulatory QS system called accessory gene regulator (agr). Through agr QS system this bacterium deploys a wide collection of virulence factors to establish biofilms and infections [88]. One of the crucial roles of the agr QS system is to encode a signalling circuit that biosynthesis and sense the autoinducers (AI and AIP) and the intracellular effector RNAIII [89]. The autoinducing peptides and agrABCD proteins coordinate the QS system and are essential for expressing exotoxin hemolysin (*hla* and *hlb*), toxic shock syndrome toxins (*tsst*) and controlling biofilm formation and dispersion [90–93]. Other autoinducer binding proteins in *S. aureus* include KdpD/E, KdpD being a receptor protein that binds with autoinducer-2, whereas KdpE is a regulatory protein triggered via phosphorylation [94].

Autoinducer-KdpD/E system regulates capsular polysaccharide biosynthesis in *S. aureus*. VraSR is another two-component signalling system that gets activated via environmental factors, i.e., by sensing the presence of bacterial cell wall inhibitor compounds such as antibiotics [95]. This system's primary role is to regulate cell wall biosynthesis, impair antibiotic effects and develop resistance [95, 96].

5. Anti-QS strategy to encounter bacterial biofilms and their pathogenicity

The introduction of antibiotics (e.g., discovery of penicillin in 1928) into clinical medicine has drastically improved human health, allowing for effective treatment of life-threatening infectious diseases and the ability to perform medical procedures previously avoided due to the high risk of postoperative infections [97, 98]. However, with the immense rise of AMR, existing antibiotics show less effectiveness in treating microbial infections. Developing novel antimicrobial agents and new strategies are critical to overcome biofilms and associated AMR in the medical arena. Antibiotic resistance is rapidly spreading and a major concern, with estimates that by the mid-21st century, antimicrobial resistance could contribute to 10 million deaths each year and cost the global economy US$100 trillion [98].

Widespread antibiotic resistance is driving an intense search for novel therapeutic approaches. Interfering with QS, termed quorum quenching (QQ), has been an area of interest in this space with the aim of inhibiting bacterial virulence and biofilm formation [99]. QS inhibitors can reduce bacterial virulence and alleviate symptoms

Type of Quorum Sensing Inhibitors	Bacteria target	Mechanism of action	References
Halogenated furanone from marine alga *Delisea pulchra.* Ascorbic Acid (Vitamin C) Synthetic furanone (C30 and C56)	*P. aeruginosa, P. mirabilis* and *E.coli*	Competitive antagonist of LasR receptor	[102–105]
Quercetin	*P. aeruginosa, C. violaceum*	Competitive antagonist of LasR receptor	[106, 107]
Curcumin	*C. violaceum, Salmonella enterica, S. marcescens* and *P. aeruginosa*	Competitive antagonist of LuxR type receptors	[108–111]
Dominant-negative competence-stimulating peptide (dnCSP) analog	*S. pneumoniae*	dnCSP competes with CSP for ComD binding	[81, 112]
Lactonase (SsoPox-W263I)	*P. aeruginosa*	Enzymatic degradation of AHL molecules	[113]
QQ antibodies generated with AI-carrier protein immunisation	*P. aeruginosa* and *S. aureus*	Antibodies bind AHL and autoinducing peptides to block their binding to cognate receptors	[114, 115]

Table 1.
Highlighting the anti-QS molecules and their mechanism of action against various bacterial pathogens.

of microbial infections in a non-bactericidal or bacteriostatic manner, hence relaxing selection pressure for resistance to these molecules while also not affecting beneficial bacteria [100, 101]. **Table 1** summarises a few examples of QS inhibiting molecules and their mechanism of action against different pathogenic bacteria.

One historic discovery in QS inhibition was halogenated furanones derived from red alga *Delisea pulchra* [116] and early work demonstrating their impact on QS behaviours such as inducing irregular non-coordinated swarming in *P. mirabilis* [102]. Many furanones are now known to act as competitive inhibitors of LuxR-type receptors in gram-negative bacteria by competing with AHL for binding to reduce QS signalling [103]. Following the discovery of halogenated furanones impact on QS, much research was carried out to test synthetic furanones as a potential treatment for microbial infections and it has shown success within mouse models to reduce *P. aeruginosa* pathogenicity and enhance bacterial clearance within lungs [104].

Ascorbic acid (vitamin C) is a natural furanone relevant to human health. Ascorbic acid has long been known as an important molecule for normal physiological functions, playing important roles as an antioxidant to protect the body from free radicals and improving immune system function by increasing lymphocyte proliferation, natural killer activity and aiding in chemotaxis [117]. Ascorbic acid is now known to be a potent inhibitor of QS within *P. aeruginosa*. It has been shown to inhibit pyocyanin production and attenuate biofilm formation [105].

Flavonoids are a class of polyphenolic secondary metabolites found in plants. Quercetin is a flavonol ubiquitous in vegetables, fruits and plant-derived drinks such as tea and wine [118]. Flavonoids such as quercetin have been extensively studied for their cardioprotective, anticarcinogenic, antioxidant and anti-inflammatory effects [119–121]. Additionally, quercetin is an effective QS inhibitor in *P. aeruginosa,* with research showing it can inhibit biofilm formation and initial bacterial adherence and reduce virulence factor expression [106]. Evidence suggests that quercetin acts as a competitive inhibitor of the LasR receptor, competing with AHL for binding to reduce QS signalling in *P. aeruginosa* [107].

Curcumin is another polyphenol and is the distinctive yellow pigment and a major constituent of turmeric derived from the *Curcuma longa* plant. Curcumin has a rich history in traditional medicine for its use in anti-inflammatory and antimicrobial roles. Recent research has proven curcumin anti-QS in numerous pathogens. In *Chromobacterium violaceum,* curcumin inhibits violacein pigment production controlled by QS [108]. In *Salmonella* serovar Montevideo, curcumin is seen to inhibit biofilm formation, and in *Serratia marcescens*, it can completely inhibit swarming motility [109]. In *P. aeruginosa,* curcumin attenuates biofilm formation and down-regulates virulence factors such as pyocyanin and elastase [110]. Silico analysis suggests that curcumin also acts as a competitive antagonist of LuxR-type receptors [111].

Gram-positive bacteria such as *S. pneumoniae* participate in QS through secreting oligopeptides as autoinducers. The competence regulon is a QS circuit present within *S. pneumoniae* and is centred on the competence stimulating peptide (CSP), the AI oligopeptide [122]. Two main CSP variants exist, CSP1 and CSP2, which bind to their corresponding histidine kinase receptors ComD1 and ComD2 to drive virulence factor production and biofilm formation [123, 124]. Synthetic peptide analogues have been explored to inhibit QS in peptide-based QS systems. Dominant-negative competence-stimulating peptides (dnCSPs) are one such example. They can reduce virulence factor expression *in vitro* and attenuate pneumococcus infections in mice by competing with CSP for ComD binding [81, 112].

QS inhibition can also be achieved by enzymatic degradation of AIs. This mechanism has been a major focus within QS inhibition research for gram-negative bacteria, and many QQ enzymes from prokaryotic and eukaryotic origins have been discovered [125]. QQ enzymes targeting AHL in gram-negative principally involve four types of enzymes, AHL-lactonases and decarboxylases hydrolyse the lactone ring, whilst AHL-acylase and deaminase cleave the acyl side chain, ultimately leading to reduced AHL-Lux receptor binding and decay of the QS signalling [125]. Many research examples of QQ enzymes show success in QS inhibition within many different bacteria; in one example, an engineered lactonase originally isolated from *Sulfolobus solfataricus* was seen to reduce virulence in clinical isolates of *P. aeruginosa* with pyocyanin production, protease secretion and biofilm formation all inhibited [113].

QQ antibodies are a novel approach to QS inhibition. AHLs and autoinducing peptides have low molecular weights; consequently, they are poorly immunogenic and not expected to elicit an antibody-based immune response [125]. However, hapten–carrier strategies can overcome this lack of immunogenicity by attaching AHL molecules to carrier proteins before immunisation. Miyairi et al. synthesised a carrier protein-conjugated 3-oxo-C12-HSL (*P. aeruginosa* HSL) and immunised mice prior to intranasal challenge with *P. aeruginosa* [114]. Immunisation generated high titres of specific antibodies to 3-oxo-C12-HSL, which was strongly associated with a survival benefit in mice [114]. Bacterial numbers in the lungs did not differ between control and immunised groups, and the increased survival of immunised mice was suggested to be through blocking an excessive pro-inflammatory host response through suppression of virulence factors under QS control [114]. In a similar approach, antibodies targeting Staphylococcal autoinducing peptides (AIPs) show potent QQ abilities and increasing protection of mice challenged with *S. aureus* [115].

6. Concluding remarks

Biofilm formation by opportunistic pathogens and its associated AMR has a catastrophic effect on society. Despite extensive research on bacterial biofilms carried out over the past century and AMR in the past few decades, we are yet to fully understand bacterial biofilms and the bacterial strategy to evade host immune responses and antibiotic therapy. The discovery of the QS mechanism in bacterial lifestyle is ground-breaking research that has revealed various behaviours and processes under its control, including adaption to physical and chemical stress, expression of genes that regulate extracellular polymeric substances, metabolite production, the integrity of biofilm matrix, efflux pumps to reduce intracellular antibiotic concentration and various antibiotic cleaving enzymes such as beta-lactamase and macrolide esterases, etc. The discovery and use of natural QS inhibiting molecules such as plant-based curcumin, vitamin C, polyphenols (flavonoids) from green tea and furanone from red algae, as well as the subsequent development of synthetic molecules have provided an innovative strategy to tackle bacterial infection and AMR and may play a critical role in the future to address to the continual spread of AMR in many clinically important bacteria and their increasing burden on human health.

There is a multitude of factors that influence the rise of bacterial-associated infections, AMR and consequently mortality. In developing countries, the burden is disproportionately high due to various factors, including high population density, inadequate and unaffordable healthcare, poor education leading to inappropriate use of antibiotics (e.g., prescribing antibiotics against common cold and seasonal viral infections), political factors including poor governance that does not provide the necessary infrastructure and policies related to healthcare, sanitation, hygiene, etc. Tangible measures are essential for governments and corporate sectors to ensure the availability of basic facilities to circumvent the increase in bacterial-associated infections, AMR and its associated mortality and morbidity. Developing innovative ideas, new drugs or improving existing drugs through increased financial support to research institutes, universities and the pharmaceutical industry is critical to addressing AMR and ultimately improving global health.

Author details

Theerthankar Das[1*] and Brandon C. Young[2]

1 Infection, Immunity and Inflammation Theme, Sydney Institute for Infectious Diseases, Charles Perkins Centre, School of Medical Science, The University of Sydney, Sydney, Australia

2 Infection, Immunity, and Inflammation Theme, Charles Perkins Centre, School of Medical Science, The University of Sydney, Sydney, Australia

*Address all correspondence to: das.ashishkumar@sydney.edu.au

References

[1] Belizario JE, Napolitano M. Human microbiomes and their roles in dysbiosis, common diseases, and novel therapeutic approaches. Frontiers in Microbiology. 2015;**6**:1050

[2] Jamal M, Ahmad W, Andleeb S, Jalil F, Imran M, Nawaz MA, et al. Bacterial biofilm and associated infections. Journal of the Chinese Medical Association. 2018;**81**(1):7-11

[3] Simmering JE, Tang F, Cavanaugh JE, Polgreen LA, Polgreen PM. The increase in hospitalizations for urinary tract infections and the associated costs in the United States, 1998-2011. Open Forum Infectious Diseases. Feb 24, 2017;**4**(1):ofw281. DOI: 10.1093/ofid/ofw281. PMID: 28480273; PMCID: PMC5414046

[4] Flores-Mireles AL, Walker JN, Caparon M, Hultgren SJ. Urinary tract infections: Epidemiology, mechanisms of infection and treatment options. Nature Reviews Microbiology. 2015;**13**(5):269-284

[5] Anderson GG, Palermo JJ, Schilling JD, Roth R, Heuser J, Hultgren SJ. Intracellular bacterial biofilm-like pods in urinary tract infections. Science. 2003;**301**(5629):105-107

[6] Rosen DA, Hooton TM, Stamm WE, Humphrey PA, Hultgren SJ. Detection of intracellular bacterial communities in human urinary tract infection. PLoS Medicine. 2007;**4**(12):e329

[7] Nguyen V, Lee GA. Management of microbial keratitis in general practice. Australian Journal of General Practice. 2019;**48**(8):516-519

[8] Collier SA, Gronostaj MP, MacGurn AK, Cope JR, Awsumb KL, Yoder JS, et al. Estimated burden of keratitis--United States, 2010. Morbidity and Mortality Weekly Report. 2014;**63**(45):1027-1030

[9] Wu YT, Zhu H, Willcox M, Stapleton F. Removal of biofilm from contact lens storage cases. Investigative Ophthalmology & Visual Science. 2010;**51**(12):6329-6333

[10] Urwin L, Okurowska K, Crowther G, Roy S, Garg P, Karunakaran E, et al. Corneal infection models: Tools to investigate the role of biofilms in bacterial keratitis. Cell. 2020;**9**(11):2450

[11] Quaderi SA, Hurst JR. The unmet global burden of COPD. Global Health, Epidemiology and Genomics. 2018;**3**:e4

[12] Garcia-Vidal C, Almagro P, Romani V, Rodriguez-Carballeira M, Cuchi E, Canales L, et al. Pseudomonas aeruginosa in patients hospitalised for COPD exacerbation: A prospective study. European Respiratory Journal. 2009;**34**(5):1072-1078

[13] Taylor AE, Finney-Hayward TK, Quint JK, Thomas CM, Tudhope SJ, Wedzicha JA, et al. Defective macrophage phagocytosis of bacteria in COPD. European Respiratory Journal. 2009;**35**(5):1039-1047

[14] Singh R, Mackay AJ, Patel AR, Garcha DS, Kowlessar BS, Brill SE, et al. Inflammatory thresholds and the species-specific effects of colonising bacteria in stable chronic obstructive pulmonary disease. Respiratory Research. 2014;**15**(1):114

[15] Miravitlles M, Mayordomo C, Artés M, Sánchez-Agudo L, Nicolau F, Segú J. Treatment of chronic obstructive

pulmonary disease and its exacerbations in general practice. Respiratory Medicine. 1999;**93**(3):173-179

[16] Halpin DM, Decramer M, Celli B, Kesten S, Liu D, Tashkin DP. Exacerbation frequency and course of COPD. International Journal of Chronic Obstructive Pulmonary Disease. 2012;**7**:653-661

[17] Dhamane A, Moretz C, Zhou Y, Burslem K, Saverno K, Jain G, et al. COPD exacerbation frequency and its association with health care resource utilization and costs. International Journal of Chronic Obstructive Pulmonary Disease. 2015;**10**:2609-2618

[18] Guarascio AJ, Ray SM, Finch CK, Self TH. The clinical and economic burden of chronic obstructive pulmonary disease in the USA. ClinicoEconomics and Outcomes Research. 2013;**5**:235-245

[19] Papi A, Bellettato CM, Braccioni F, Romagnoli M, Casolari P, Caramori G, et al. Infections and airway inflammation in chronic obstructive pulmonary disease severe exacerbations. American Journal of Respiratory and Critical Care Medicine. 2006;**173**(10):1114-1121

[20] Domenech A, Puig C, Martí S, Santos S, Fernández A, Calatayud L, et al. Infectious etiology of acute exacerbations in severe COPD patients. Journal of Infection. 2013;**67**(6):516-523

[21] McCullers JA. The co-pathogenesis of influenza viruses with bacteria in the lung. Nature Reviews Microbiology. 2014;**12**(4):252-262

[22] Manohar P, Loh B, Nachimuthu R, Hua X, Welburn SC, Leptihn S. Secondary bacterial infections in patients with viral pneumonia. Frontiers in Medicine. 2020;**7**:420

[23] Sharifipour E, Shams S, Esmkhani M, Khodadadi J, Fotouhi-Ardakani R, Koohpaei A, et al. Evaluation of bacterial co-infections of the respiratory tract in COVID-19 patients admitted to ICU. BMC Infectious Diseases. 2020;**20**(1):646

[24] Pourajam S, Kalantari E, Talebzadeh H, Mellali H, Sami R, Soltaninejad F, et al. Secondary bacterial infection and clinical characteristics in patients with COVID-19 admitted to two intensive care units of an academic Hospital in Iran during the first wave of the pandemic. Frontiers in Cellular and Infection Microbiology. 2022;**12**:784130

[25] Donlan RM. Biofilms: Microbial life on surfaces. Emerging Infectious Diseases. 2002;**8**(9):881-890

[26] Roy R, Tiwari M, Donelli G, Tiwari V. Strategies for combating bacterial biofilms: A focus on anti-biofilm agents and their mechanisms of action. Virulence. 2018;**9**(1):522-554

[27] Saxena P, Joshi Y, Rawat K, Bisht R. Biofilms: Architecture, resistance, quorum sensing and control mechanisms. Indian Journal of Microbiology. 2018;**59**(1):3-12

[28] Thi M, Wibowo D, Rehm B. Pseudomonas aeruginosa Biofilms. International Journal of Molecular Sciences. 2020;**21**(22):8671

[29] Carniello V, Peterson BW, van der Mei HC, Busscher HJ. Physico-chemistry from initial bacterial adhesion to surface-programmed biofilm growth. Advances in Colloid and Interface Science. 2018;**261**:1-14

[30] Mah TF. Biofilm-specific antibiotic resistance. Future Microbiology. 2012;**7**(9):1061-1072

[31] Flemming HC, Wingender J. The biofilm matrix. Nature Reviews Microbiology. 2010;**8**(9):623-633

[32] Kaplan JB. Biofilm dispersal: Mechanisms, clinical implications, and potential therapeutic uses. Journal of Dental Research. 2010;**89**(3):205-218

[33] Das T, Sharma PK, Busscher HJ, van der Mei HC, Krom BP. Role of extracellular DNA in initial bacterial adhesion and surface aggregation. Applied and Environmental Microbiology. 2010;**76**(10):3405-3408

[34] Das T, Krom BP, van der Mei HC, Busscher HJ, Sharma PK. DNA-mediated bacterial aggregation is dictated by acid–base interactions. Soft Matter. 2011;**7**(6):2927

[35] Swartjes JJ, Das T, Sharifi S, Subbiahdoss G, Sharma PK, Krom BP, et al. A functional DNase I coating to prevent adhesion of bacteria and the formation of biofilm. Advanced Functional Materials. 2013;**23**(22):2843-2849

[36] Internet - https://www.msdmanuals.com/en-au/professional/infectious-diseases/bacteria-and-antibacterial-drugs/polypeptide-antibiotics-bacitracin,-colistin,-polymyxin-b

[37] Gottenbos B, Grijpma DW, van der Mei HC, Feijen J, Busscher HJ. Antimicrobial effects of positively charged surfaces on adhering Gram-positive and Gram-negative bacteria. Journal of Antimicrobial Chemotherapy. 2001;**48**(1):7-13

[38] Nishikawa M, Kobayashi K. Calcium prevents biofilm dispersion in Bacillus subtilis. Journal of Bacteriology. 2021;**203**(14):e0011421

[39] Das T, Sehar S, Koop L, Wong YK, Ahmed S, Siddiqui KS, et al. Influence of calcium in extracellular DNA mediated bacterial aggregation and biofilm formation. PLoS One. 2014;**9**(3):e91935

[40] De-la-Pinta I, Cobos M, Ibarretxe J, Montoya E, Eraso E, Guraya T, et al. Effect of biomaterials hydrophobicity and roughness on biofilm development. Journal of Materials Science: Materials in Medicine. 2019;**30**(7):77

[41] Das T, Sharma PK, Krom BP, van der Mei HC, Busscher HJ. Role of eDNA on the adhesion forces between Streptococcus mutans and substratum surfaces: Influence of ionic strength and substratum hydrophobicity. Langmuir. 2011;**27**(16):10113-10118

[42] Garcia-Fernandez N, Hassan A, Anand S. Effect of exopolysaccharides produced by dairy starter cultures on biofilms formed on reverse osmosis membranes. JDS Communications. 2021;**2**(3):104-109

[43] Quiocho F. Molecular features and basic understanding of protein-carbohydrate interactions: The arabinose-binding protein-sugar complex. Current Topics in Microbiology and Immunology. 1988;**139**:135-148. DOI: 10.1007/978-3-642-46641-0_5

[44] Bellich B, Distefano M, Syrgiannis Z, Bosi S, Guida F, Rizzo R, et al. The polysaccharide extracted from the biofilm of Burkholderia multivorans strain C1576 binds hydrophobic species and exhibits a compact 3D-structure. International Journal of Biological Macromolecules. 2019;**136**:944-950

[45] Mirani ZA, Naz S, Khan F, Aziz M, Asadullah KMN, Khan SI. Antibacterial fatty acids destabilize hydrophobic and multicellular aggregates of biofilm in S. aureus. The Journal of Antibiotics. 2016;**70**(2):115-121

[46] Mirani ZA, Fatima A, Urooj S, Aziz M, Khan MN, Abbas T. Relationship of cell surface hydrophobicity with biofilm formation and growth rate: A study on Pseudomonas aeruginosa, Staphylococcus aureus, and Escherichia coli. Iranian Journal of Basic Medical Sciences. 2018;**21**(7):760-769

[47] Kahl BC, Becker K, Löffler B. Clinical significance and pathogenesis of staphylococcal small colony variants in persistent infections. Clinical Microbiology Reviews. 2016;**2**:401-427

[48] Mirani ZA, Aziz M, Khan SI. Small colony variants have a significant role in the stability and persistence of Staphylococcus aureus biofilms. The Journal of Antibiotics. 2015;**68**:98-105

[49] Gantner S, Schmid M, Dürr C, Schuhegger R, Steidle A, Hutzler P, et al. In situ quantitation of the spatial scale of calling distances and population density-independent N-acylhomoserine lactone-mediated communication by rhizobacteria colonized on plant roots. FEMS Microbiology Ecology. 2006;**56**(2):188-194

[50] Darch SE, Simoska O, Fitzpatrick M, Barraza JP, Stevenson KJ, Bonnecaze RT, et al. Spatial determinants of quorum signaling in a Pseudomonas aeruginosa infection model. Proceedings of the National Academy of Sciences. 2018;**115**(18):4779-4784

[51] Hense B, Schuster M. Core principles of bacterial autoinducer systems. Microbiology and Molecular Biology Reviews. 2015;**79**(1):153-169

[52] Gould TA, Schweizer HP, Churchill ME. Structure of the Pseudomonas aeruginosa acyl-homoserinelactone synthase LasI. Molecular Microbiology. 2004;**53**(4):1135-1146

[53] Pearson JP, Pesci EC, Iglewski BH. Roles of Pseudomonas aeruginosa las and rhl quorum-sensing systems in control of elastase and rhamnolipid biosynthesis genes. Journal of Bacteriology. 1997;**179**(18):5756-5767

[54] Turkina MV, Vikström E. Bacteria-host crosstalk: Sensing of the quorum in the context of Pseudomonas aeruginosa infections. Journal of Innate Immunity. 2018;**11**(3):263-279

[55] Ochsner UA, Reiser J. Autoinducer-mediated regulation of rhamnolipid biosurfactant synthesis in Pseudomonas aeruginosa. Proceedings of the National Academy of Sciences. 1995;**92**(14):6424-6428

[56] Gellatly S, Hancock R. Pseudomonas aeruginosa: New insights into pathogenesis and host defences. Pathogens and Disease. 2013;**67**(3):159-173

[57] Bredenbruch F, Nimtz M, Wray V, Morr M, Müller R, Häussler S. Biosynthetic pathway of Pseudomonas aeruginosa 4-Hydroxy-2-Alkylquinolines. Journal of Bacteriology. 2005;**187**(11):3630-3635

[58] Gallagher LA, McKnight SL, Kuznetsova MS, Pesci EC, Manoil C. Functions required for extracellular quinolone signaling by Pseudomonas aeruginosa. Journal of Bacteriology. 2002;**184**(23):6472-6480

[59] Déziel E, Lépine F, Milot S, He J, Mindrinos MN, Tompkins RG, et al. Analysis of Pseudomonas aeruginosa 4-hydroxy-2-alkylquinolines (HAQs) reveals a role for 4-hydroxy-2-heptylquinoline in cell-to-cell communication. Proceedings of the National Academy of Sciences. 2004;**101**(5):1339-1344

[60] Mossialos D, Meyer JM, Budzikiewicz H, Wolff U, Koedam N, Baysse C, et al. Quinolobactin, a new Siderophore of Pseudomonas fluorescens ATCC 17400, the production of which is repressed by the cognate Pyoverdine. Applied and Environmental Microbiology. 2000;**66**(2):487-492

[61] Bredenbruch F, Geffers R, Nimtz M, Buer J, Häussler S. The Pseudomonas aeruginosa quinolone signal (PQS) has an iron-chelating activity. Environmental Microbiology. 2006;**8**(8):1318-1329

[62] Diggle SP, Matthijs S, Wright VJ, Fletcher MP, Chhabra SR, Lamont IL, et al. The Pseudomonas aeruginosa 4-quinolone signal molecules HHQ and PQS play multifunctional roles in quorum sensing and iron entrapment. Chemistry & Biology. 2007;**14**(1):87-96

[63] Lee J, Wu J, Deng Y, Wang J, Wang C, Wang J, et al. A cell-cell communication signal integrates quorum sensing and stress response. Nature Chemical Biology. 2013;**9**(5):339-343

[64] Lee J, Zhang L. The hierarchy quorum sensing network in Pseudomonas aeruginosa. Protein & Cell. 2014;**6**(1):26-41

[65] Pérez-Pérez M, Jorge P, Pérez Rodríguez G, Pereira M, Lourenço A. Quorum sensing inhibition in Pseudomonas aeruginosa biofilms: New insights through network mining. Biofouling. 2017;**33**(2):128-142

[66] Iglewski BH, Kabat D. NAD-dependent inhibition of protein synthesis by Pseudomonas aeruginosa toxin. Proceedings of the National Academy of Sciences. 1975;**72**(6):2284-2288

[67] Iglewski BH, Sadoff J, Bjorn MJ, Maxwell ES. Pseudomonas aeruginosa exoenzyme S: An adenosine diphosphate ribosyltransferase distinct from toxin a. Proceedings of the National Academy of Sciences. 1978;**75**(7):3211-3215

[68] Peters JE, Park SJ, Darzins A, Freck LC, Saulnier JM, Wallach JM, et al. Further studies on Pseudomonas aeruginosa LasA: Analysis of specificity. Molecular Microbiology. 1992;**6**(9):1155-1162

[69] Schmidtchen A, Holst E, Tapper H, Björck L. Elastase-producing Pseudomonas aeruginosa degrade plasma proteins and extracellular products of human skin and fibroblasts, and inhibit fibroblast growth. Microbial Pathogenesis. 2003;**34**(1):47-55

[70] Terada LS, Johansen KA, Nowbar S, Vasil AI, Vasil ML. Pseudomonas aeruginosa hemolytic phospholipase C suppresses neutrophil respiratory burst activity. Infection and Immunity. 1999;**67**(5):2371-2376

[71] Internet - https://www.aocd.org/page/GreenNailSyndrome

[72] O'Malley YQ, Reszka KJ, Spitz DR, Denning GM, Britigan BE. Pseudomonas aeruginosa pyocyanin directly oxidizes glutathione and decreases its levels in airway epithelial cells. American Journal of Physiology-Lung Cellular and Molecular Physiology. 2004;**287**(1):L94-L103

[73] Kanthakumar K, Taylor G, Tsang KW, Cundell DR, Rutman A, Smith S, et al. Mechanisms of action of Pseudomonas aeruginosa pyocyanin on human ciliary beat in vitro. Infection and Immunity. 1993;**61**(7):2848-2853

[74] Muller M, Li Z, Maitz PK. Pseudomonas pyocyanin inhibits wound repair by inducing premature cellular senescence: Role for p38

mitogen-activated protein kinase. Burns. 2009;**35**(4):500-508

[75] Gonzalez MR, Fleuchot B, Lauciello L, Jafari P, Applegate LA, Raffoul W, et al. Effect of human burn wound exudate on Pseudomonas aeruginosa virulence. mSphere. 2016;**1**(2):e00111-e00115

[76] Das T, Kutty SK, Tavallaie R, Ibugo AI, Panchompoo J, Sehar S, et al. Phenazine virulence factor binding to extracellular DNA is important for Pseudomonas aeruginosa biofilm formation. Scientific Reports. 2015;**5**:8398

[77] Saunders SH, Tse EC, Yates MD, Otero FJ, Trammell SA, Stemp ED, et al. Extracellular DNA promotes efficient extracellular electron transfer by Pyocyanin in Pseudomonas aeruginosa biofilms. Cell. 2020;**182**(4):919-932.e19

[78] Steinmoen H, Knutsen E, Håvarstein LS. Induction of natural competence in Streptococcus pneumoniae triggers lysis and DNA release from a subfraction of the cell population. Proceedings of the National Academy of Sciences. 2002;**99**(11):7681-7686

[79] Ween O, Gaustad P, Havarstein LS. Identification of DNA binding sites for ComE, a key regulator of natural competence in Streptococcus pneumoniae. Molecular Microbiology. 1999;**33**(4):817-827

[80] Eldholm V, Johnsborg O, Haugen K, Ohnstad HS, Havarstein LS. Fratricide in Streptococcus pneumoniae: Contributions and role of the cell wall hydrolases CbpD, LytA and LytC. Microbiology (Reading). 2009;**155**(Pt 7):2223-2234

[81] Zhu L, Lau GW. Inhibition of competence development, horizontal gene transfer and virulence in Streptococcus pneumoniae by a modified competence stimulating peptide. PLoS Pathogens. 2011;**7**(9):e1002241

[82] Valente C, Dawid S, Pinto FR, Hinds J, Simões AS, Gould KA, et al. The blp locus of Streptococcus pneumoniae plays a limited role in the selection of strains that can Cocolonize the human nasopharynx. Applied and Environmental Microbiology. 2016;**82**(17):5206-5215

[83] Mascher T, Zähner D, Merai M, Balmelle N, de Saizieu AB, Hakenbeck R. The Streptococcus pneumoniae cia regulon: CiaR target sites and transcription profile analysis. Journal of Bacteriology. 2003;**185**(1):60-70

[84] Giammarinaro P, Sicard M, Gasc AM. Genetic and physiological studies of the CiaH-CiaR two-component signal-transducing system involved in cefotaxime resistance and competence of Streptococcus pneumoniae. Microbiology. 1999;**145**(8):1859-1869

[85] Haas W, Kaushal D, Sublett J, Obert C, Tuomanen EI. Vancomycin stress response in a sensitive and a tolerant strain of Streptococcus pneumoniae. Journal of Bacteriology. 2005;**187**(23):8205-8210

[86] Mascher T, Heintz M, Zähner D, Merai M, Hakenbeck R. The CiaRH system of Streptococcus pneumoniae prevents lysis during stress induced by treatment with Cell Wall inhibitors and by mutations in pbp2x involved in β-lactam resistance. Journal of Bacteriology. 2006;**188**(5):1959-1968

[87] Ibrahim YM, Kerr AR, McCluskey J, Mitchell TJ. Control of virulence by the two-component system CiaR/H is mediated via HtrA, a major virulence factor of Streptococcus

pneumoniae. Journal of Bacteriology. 2004;**186**(16):5258-5266

[88] Bibalan MH, Shakeri F, Javid N, Ghaemi A, Ghaemi EA. Accessory gene regulator types of Staphylococcus aureus isolated in Gorgan, north of Iran. Journal of Clinical and Diagnostic Research. 2014;**8**(4):DC07-DC09

[89] Wang B, Muir TW. Regulation of virulence in Staphylococcus aureus: Molecular mechanisms and remaining puzzles. Cell Chemical Biology. 2016;**23**(2):214-224

[90] Lyon GJ, Novick RP. Peptide signaling in Staphylococcus aureus and other Gram-positive bacteria. Peptides. 2004;**25**(9):1389-1403

[91] Tan L, Li SR, Jiang B, Hu XM, Li S. Therapeutic targeting of the Staphylococcus aureus accessory gene regulator (agr) system. Frontiers in Microbiology. 2018;**9**:55

[92] Queck SY, Jameson-Lee M, Villaruz AE, Bach TH, Khan BA, Sturdevant DE, et al. RNAIII-independent target gene control by the agr quorum-sensing system: Insight into the evolution of virulence regulation in Staphylococcus aureus. Molecular Cell. 2008;**32**(1):150-158

[93] Periasamy S, Joo HS, Duong AC, Bach TH, Tan VY, Chatterjee SS, et al. How Staphylococcus aureus biofilms develop their characteristic structure. Proceedings of the National Academy of Sciences. 2012;**109**(4):1281-1286

[94] Zhao L, Xue T, Shang F, Sun H, Sun B. Staphylococcus aureus AI-2 quorum sensing associates with the KdpDE two-component system to regulate capsular polysaccharide synthesis and virulence. Infection and Immunity. 2010;**78**(8):3506-3515

[95] Kato Y, Suzuki T, Ida T, Maebashi K. Genetic changes associated with glycopeptide resistance in Staphylococcus aureus: Predominance of amino acid substitutions in YvqF/VraSR. Journal of Antimicrobial Chemotherapy. 2010;**65**(1):37-45

[96] Gardete S, Wu SW, Gill S, Tomasz A. Role of VraSR in antibiotic resistance and antibiotic-induced stress response in Staphylococcus aureus. Antimicrobial Agents and Chemotherapy. 2006;**50**(10):3424-3434

[97] Hutchings MI, Truman AW, Wilkinson B. Antibiotics: Past, present and future. Current Opinion in Microbiology. 2019;**51**:72-80

[98] O'Neill J. Tackling Drug-Resistant Infections Globally: Final Report and Recommendations. Government of the United Kingdom; 2016. Available from: https://apo.org.au/node/63983. Internet - https://amr-review.org/sites/default/files/160518_Final%20paper_with%20cover.pdf

[99] Santhakumari S, Ravi AV. Targeting quorum sensing mechanism: An alternative anti-virulent strategy for the treatment of bacterial infections. South African Journal of Botany. 2019;**120**:81-86

[100] Haque S, Ahmad F, Dar SA, Jawed A, Mandal RK, Wahid M, et al. Developments in strategies for quorum sensing virulence factor inhibition to combat bacterial drug resistance. Microbial Pathogenesis. 2018;**121**:293-302

[101] Rasmussen TB, Bjarnsholt T, Skindersoe ME, Hentzer M, Kristoffersen P, Köte M, et al. Screening for quorum-sensing inhibitors (QSI) by use of a novel genetic system, the QSI selector. Journal of Bacteriology. 2005;**187**(5):1799-1814

[102] Gram L, de Nys R, Maximilien R, Givskov M, Steinberg P, Kjelleberg S. Inhibitory effects of secondary metabolites from the red alga Delisea pulchra on swarming motility of Proteus mirabilis. Applied and Environmental Microbiology. 1996;**62**(11):4284-4287

[103] Ren D, Bedzyk LA, Ye RW, Thomas SM, Wood TK. Differential gene expression shows natural brominated furanones interfere with the autoinducer-2 bacterial signaling system of Escherichia coli. Biotechnology and Bioengineering. 2004;**88**(5):630-642

[104] Wu H, Song Z, Hentzer M, Andersen JB, Molin S, Givskov M, et al. Synthetic furanones inhibit quorum-sensing and enhance bacterial clearance in Pseudomonas aeruginosa lung infection in mice. Journal of Antimicrobial Chemotherapy. 2004;**53**(6):1054-1061

[105] El-Mowafy SA, Shaaban MI, Abd El Galil KH. Sodium ascorbate as a quorum sensing inhibitor of Pseudomonas aeruginosa. Journal of Applied Microbiology. 2014;**117**(5):1388-1399

[106] Ouyang J, Sun F, Feng W, Sun Y, Qiu X, Xiong L, et al. Quercetin is an effective inhibitor of quorum sensing, biofilm formation and virulence factors in Pseudomonas aeruginosa. Journal of Applied Microbiology. 2016;**120**(4):966-974

[107] Gopu V, Meena CK, Shetty PH. Quercetin influences quorum sensing in food borne bacteria: In-vitro and In-silico evidence. PLoS One. 2015;**10**(8):e0134684

[108] Bali EB, Erkan Türkmen KE, Erdönmez D, Sağlam N. Comparative study of inhibitory potential of dietary phytochemicals against quorum sensing activity of and biofilm formation by Chromobacterium violaceum 12472, and swimming and swarming behaviour of Pseudomonas aeruginosa PAO1. Food Technology and Biotechnology. 2019;**57**(2):212-221

[109] Santos CA, Lima EM, Franco BD, Pinto UM. Exploring phenolic compounds as quorum sensing inhibitors in foodborne bacteria. Frontiers in Microbiology. 2021;**12**:735931

[110] Rudrappa T, Bais HP. Curcumin, a known phenolic from Curcuma longa, attenuates the virulence of Pseudomonas aeruginosa PAO1 in whole plant and animal pathogenicity models. Journal of Agricultural and Food Chemistry. 2008;**56**(6):1955-1962

[111] Deryabin D, Galadzhieva A, Kosyan D, Duskaev G. Plant-derived inhibitors of AHL-mediated quorum sensing in bacteria: Modes of action. International Journal of Molecular Sciences. 2019;**20**(22):5588

[112] Koirala B, Lin J, Lau GW, Tal-Gan Y. Development of a dominant negative competence-stimulating peptide (dnCSP) that attenuates Streptococcus pneumoniae infectivity in a mouse model of acute pneumonia. Chembiochem. 2018;**19**(22):2380-2386

[113] Guendouze A, Plener L, Bzdrenga J, Jacquet P, Rémy B, Elias M, et al. Effect of quorum quenching lactonase in clinical isolates of Pseudomonas aeruginosa and comparison with quorum sensing inhibitors. Frontiers in Microbiology. 2017;**8**:227

[114] Miyairi S, Tateda K, Fuse ET, Ueda C, Saito H, Takabatake T, et al. Immunization with 3-oxododecanoyl-l-homoserine lactone–protein conjugate protects mice from lethal Pseudomonas aeruginosa lung infection.

Journal of Medical Microbiology. 2006;**55**(10):1381-1387

[115] Park J, Jagasia R, Kaufmann GF, Mathison JC, Ruiz DI, Moss JA, et al. Infection control by antibody disruption of bacterial quorum sensing signaling. Chemistry and Biology. 2007;**14**(10):1119-1127

[116] De Nys R, Wright AD, König GM, Sticher O. New halogenated furanones from the marine alga delisea pulchra (cf. fimbriata). Tetrahedron. 1993;**49**(48):11213-11220

[117] Chambial S, Dwivedi S, Shukla KK, John PJ, Sharma P. Vitamin C in disease prevention and cure: An overview. Indian Journal of Clinical Biochemistry. 2013;**28**(4):314-328

[118] Boots A, Haenen G, Bast A. Health effects of quercetin: From antioxidant to nutraceutical. European Journal of Pharmacology. 2008;**585**(2-3):325-337

[119] Chen YW, Chou HC, Lin ST, Chen YH, Chang YJ, Chen L, et al. Cardioprotective effects of quercetin in cardiomyocyte under ischemia/reperfusion injury. Evidence-Based Complementary and Alternative Medicine. 2013;**2013**:1-16

[120] Jung M, Bu SY, Tak KH, Park JE, Kim E. Anticarcinogenic effect of quercetin by inhibition of insulin-like growth factor (IGF)-1 signaling in mouse skin cancer. Nutrition Research and Practice. 2013;**7**(6):439-445

[121] Ravichandran R, Rajendran M, Devapiriam D. Antioxidant study of quercetin and their metal complex and determination of stability constant by spectrophotometry method. Food Chemistry. 2014;**146**:472-478

[122] Håvarstein LS, Coomaraswamy G, Morrison DA. An unmodified heptadecapeptide pheromone induces competence for genetic transformation in Streptococcus pneumoniae. Proceedings of the National Academy of Sciences. 1995;**92**(24):11140-11144

[123] Hava DL, Camilli A. Large-scale identification of serotype 4 Streptococcus pneumoniae virulence factors. Molecular Microbiology. 2002;**45**(5):1389-1406

[124] Oggioni MR, Trappetti C, Kadioglu A, Cassone M, Iannelli F, Ricci S, et al. Switch from planktonic to sessile life: A major event in pneumococcal pathogenesis. Molecular Microbiology. 2006;**61**(5):1196-1210

[125] Kalia VC. Quorum sensing inhibitors: An overview. Biotechnology Advances. 2013;**31**(2):224-245

Chapter 4

Biofilm and Quorum Sensing in *Helicobacter pylori*

Tarik Aanniz, Wissal Bakri, Safae El Mazouri, Hajar Wakrim, Ilham Kandoussi, Lahcen Belyamani, Mouna Ouadghiri and Azeddine Ibrahimi

Abstract

Helicobacter pylori (*H. pylori*) is a gram-negative bacterium living in the human gastrointestinal tract considered as the most common cause of gastritis. *H. pylori* was listed as the main risk factor for gastric cancer. Triple therapy consisting of a proton pump inhibitor and combinations of antibiotics is the main treatment used. However, this line of therapy has proven less effective mainly due to biofilm formation. Bacteria can regulate and synchronize the expression of multiple genes involved in virulence, toxin production, motility, chemotaxis, and biofilm formation by quorum sensing (QS), thus contributing to antimicrobial resistance. Henceforth, the inhibition of QS called quorum quenching (QQ) is a promising target and alternative to fight *H. pylori* resistance to antimicrobials. Many phytochemicals as well as synthetic compounds acting as quorum quenchers in *H. pylori* were described *in vitro* and *in vivo.* Otherwise, many other compounds known as quorum quenchers in other species and inhibitors of biofilm formation in *H. pylori* could act as quorum quenchers in *H. pylori.* Here, we summarize and discuss the latest findings on *H. pylori*'s biofilm formation, QS sensing, and QQ mechanisms.

Keywords: biofilm, *Helicobacter pylori*, quorum sensing, bacterial resistance, chemoreceptor, quorum quenching

1. Introduction

Helicobacter pylori (*H. pylori*) is a microaerophilic, spiral-shaped, gram-negative bacterium that belongs to Epsilonproteobacteria [1]. *H. pylori* establishes about 50% life-long infections. While it is asymptomatic in 85% of cases, individuals with chronic gastritis linked to *H. pylori* have a 10–20% chance to develop peptic ulcers and 1% chance to develop gastric carcinoma [2]. Barry Marshall and Robin Warren were the first to successfully isolate and culture *H. pylori* from the human stomach in 1983 [3]. The pair later conducted self-ingestion experiments that confirmed *H. pylori*'s colonization of the human stomach, thereby inducing inflammation of the gastric mucosa. Marshall first reported the development of persistent gastritis after ingestion, which was treated with doxycycline and bismuth subsalicylate [4].

These findings promoted more research, which ended up showing that high amount of H. pylori in the stomach promotes multiple gastrointestinal troubles, including chronic gastritis, peptic ulcer disease, gastric mucosa-associated lymphoid tissue (MALT) lymphoma, and gastric cancer [3].

In the early 1980s, Robin Warren and Barry Marshall showed for the first time that a bacterium named H. pylori could be associated with cancer development. In 2005, the Nobel Prize in Physiology or Medicine was awarded to R. Warren and B. Marshall for the "*discovery of the bacterium H. pylori and its role in gastritis and peptic ulcer disease*."

Furthermore, the International Agency for Research on Cancer classified *H. pylori* in group 1 of carcinogens [5]. It has been shown that *H. pylori* infection may as well be correlated with insulin resistance, the increase of total and low-density lipoprotein cholesterol, and the decrease of high-density lipoprotein [6]. Due to differences in socioeconomic and hygienic conditions, *H. pylori* prevalence varies between and within countries. In general, it is estimated to range from 85–95% in developing countries and between 30% and 50% in developed countries [7]. The prevalence of the infection cannot be summarized in a single figure due to unreliable diagnostic methods in some regions, poor representation of some countries, and differences in data quality [8].

Currently, the first line therapy used to treat *H. pylori* infection is a combination of proton pump inhibitors (PPIs) with amoxicillin or metronidazole and clarithromycin. This triple therapy fails in about 20–30% of cases, requiring the use of a quadruple therapy consisting of a PPI, bismuth, tetracycline, and metronidazole [9, 10]. Nevertheless, an alarming increase in multidrug-resistant strains of *H. pylori* to ampicillin, penicillin, co-amoxiclav, amoxicillin, clarithromycin, metronidazole, tetracycline, doxycycline, erythromycin, and doxycycline has been reported [11–13]. This is ascribed to antibiotic abuse, therapeutic failures, and phenotypical mechanisms promoting resistance and/or tolerance to antimicrobials, notably, biofilm formation [14, 15]. Biofilm formation is a process in which organisms firmly adhere to abiotic, and/or biotic surfaces then grow together to form a complex community that often forms a special structure through four stages: (i) reversible bacterial adhesion; (ii) irreversible adhesion; (iii) formation and maturation of matrix; and (iv) dispersal of cells [16]. Biofilms mainly consist of extracellular polymeric substances composed of polysaccharides, proteins, nucleic acids, and lipids forming a protective barrier against adverse conditions and decreasing the penetration of antibiotics [17]. In *H. pylori,* flagella play a major role in biofilm formation in the gastrointestinal tract [18].

Most bacteria use quorum sensing (QS) as a communication system, relying on the secretion and perception of small molecules called auto-inducers (AIs) [19, 20]. The QS system can activate and/or regulate gene expression of many phenotypes that can be problematic for humans, i.e., biofilm formation, so that bacteria as a group can jointly cope with changes in the surrounding environment, resulting in adverse consequences such as drug resistance and virulence [21, 22]. A new tactic for outsmarting bacteria called quorum quenching (QQ) is currently explored to reduce their virulence without interfering with their growth, causing less Darwinian selection pressure for bacterial resistance [23]. This paradigm shift has become a promising antibacterial strategy, which not only prevents the development of antimicrobial resistance but also the disturbance of human gastrointestinal microflora, as well as the prevention of adverse side effects commonly associated with the available treatment [24]. Since the main steps of QS are the production and detection of signal molecules, QQ can interfere with this system in different ways, either intracellularly or extracellularly by application of inhibitors of AI biosynthesis and perception [25], application of AI antagonists (mimicking AIs), chemical inactivation of AI, sequestering antibodies [26] or macromolecules such as cyclodextrins [27], and degrading enzymes [28]. This strategy showed promising effect *in vitro* and *in vivo*, as well as synergistic effects with antibiotics by increasing bacterial susceptibility to antibiotics [29].

Here, we summarize the biofilm formation regulated by the QS system involved in the antimicrobial resistance in *H. pylori*. Meanwhile, we also provide the latest development of QS inhibitors (QSIs) or QQ enzymes (QQEs) as a potential strategy for the design of new antimicrobial agents to manage *H. pylori* infections.

2. Biofilm formation in *H. pylori*

Biofilms have been recognized as a microbial sessile community, irreversibly attached to either animate and inanimate objects [30]. Biofilms are contained in a self-produced extracellular polysaccharide (EPS) layer. This matrix is commonly rich in proteins including enzymes, polysaccharides (1–2%), nucleic acids (<1%), and water (up to 97%) [31]. Temperature, pH, osmolarity, UV radiation, desiccation, oxygen tension, and nutrient availability are all environmental stressors that directly affect the phenotype of biofilms [16, 32]. *In vitro* analyses have further confirmed that *H. pylori* biofilms reduce drug permeability and decrease the susceptibility to antibiotics. In fact, cells in the bacterial biofilm are 10–100 times more resistant toward antimicrobial agents than cells in a planktonic state [33, 34]. *H. pylori* colonizing the stomach has developed three patterns of drug resistance, including single drug resistance (SDR), heteroresistance (HR), and multidrug resistance (MDR), which probably overlap and are linked in their molecular mechanisms and their clinical implications [35–42].

Factors	References
Flagella and pili	[18]
Outer membrane vesicles (OMV)	[43]
Extracellular DNA (e-ADN)	[43]
Adhesin (outer membrane proteins namely Hop & Hom)	[51]
Lipopolysaccharides (LPS)	[52]
Flagellar proteins	[52]
Efflux pumps	[53]
Enzymes regulating pH (urease and arginase)	[54]
luxS gene	[54]
Chemoreceptors	[54]
Toxin-antitoxin system proteins	[55]
H. pylori neutrophil-activating protein (HP-NAP)	[55, 56]
Mannose-related proteoglycans (proteomannans)	[57]

Table 1.
Factors involved in the formation of biofilms in Helicobacter pylori.

In the human stomach, *H. pylori* biofilms are found on the surface of gastric mucosa. Once introduced into the stomach, *H. pylori* appears in a spiral form, which is very mobile and associated with the colonization of new niches [43–46]. Subsequently, it comes into contact with the mucin layer that covers the epithelial cells, resulting in tension-dependent adhesion between the mucin and *H. pylori* [47]. After an efficient adhesion and multiplication, a morphological transformation occurs, which is accompanied by the creation of multiple shapes (spiral, rod, curved, coccoid, and filamentous forms) to establish a biofilm [48]. However, in the case of prolonged colonization, all biofilm cells eventually transform into a coccoid form involved in survival and greater tolerance to adverse environmental factors [49, 50]. Biofilm formation in *H. pylori* involves many factors shown in **Table 1**.

3. Biofilm formation and QS in *H. pylori*

The discovery of QS in *Vibrio fischeri* and *Vibrio harveyi*, two species that achieve bioluminescence using QS signaling molecules, sparked research into this complex signaling system [58]. The regulation of gene expression under QS control was investigated in multiple gram-negative bacteria species, including *H. pylori* [52, 59, 60]. For *H. pylori*, QS is involved in motility, biofilm development, and antibiotic resistance [32, 47, 59, 61]. Once biofilm formation is elicited from planktonic cells, the aggregated cells surrounded with extracellular polymeric substances (EPS) modify their phenotype, exchange genetic material, produce AI, and provide physical protection [33]. Owing to the formation of biofilms, *H. pylori* infections became typically persistent and rarely resolved by traditional antimicrobial therapies [34].

Overall, the QS system includes the following steps: (i) AI production; (ii) excretion of AI to the surrounding environment; (iii) sensing and binding of the AI to receptors at high cell density; (iv) retrieval of the receptor-signal complex from the cell and its binding to the promoter region; and (v) activation of genes expression [62, 63]. There are four different signals involved in QS. The most common are N-acyl homoserine lactones (AHLs), also known as autoinducer-1 (AI-1), which are fatty acid derivatives produced and used by gram-negative bacteria [64], while gram-positive bacteria use peptides or modified peptides. Furanosyl borate diesters or autoinducer-2 (AI-2) are derived from the recycling of S-adenosyl-homocysteine and used by both gram-positive and gram-negative bacteria [64]. There is also the autoinducer-3 (AI-3), which allows the cross-talking with mammalian epinephrine host cell signaling systems [65].

H. pylori, when located in the gastric mucosa, responds to several specific chemical signals. The chemotactic response is mediated by chemoreceptors called chemotaxis proteins [59]. *H. pylori* genome encodes four chemoreceptors: TlpA (effector; arginine, bicarbonate), TlpB (effector; AI2, urea, hydroxyurea, formamide acetamide.), TlpC (effector; unknown), and TlpD (effector; hydrogen peroxide) [66]. The *H. pylori* QS network involves the chemoreceptor TlpB responding to the AI-2 signaling molecule, a class of furanosyl borate diesters synthesized by the LuxS protein [59, 66] (**Figure 1**). The 4,5-dihydroxy-2,3-pentanedione (DPD), which is the precursor of AI-2 in H. pylori, is produced by LuxS protein [67]. First, LuxS produces the homocysteine through the cleavage of S-ribosylhomocysteine (SRH), which is a part of the S-adenosylmethionine (SAM) pathway. The process involves two main enzymes, i.e., 5′-methylthioadenosine/adenosylhomocysteine nucleosidase (MTAN) and metalloenzyme [68]. The DPD generated is rearranged into an assortment of

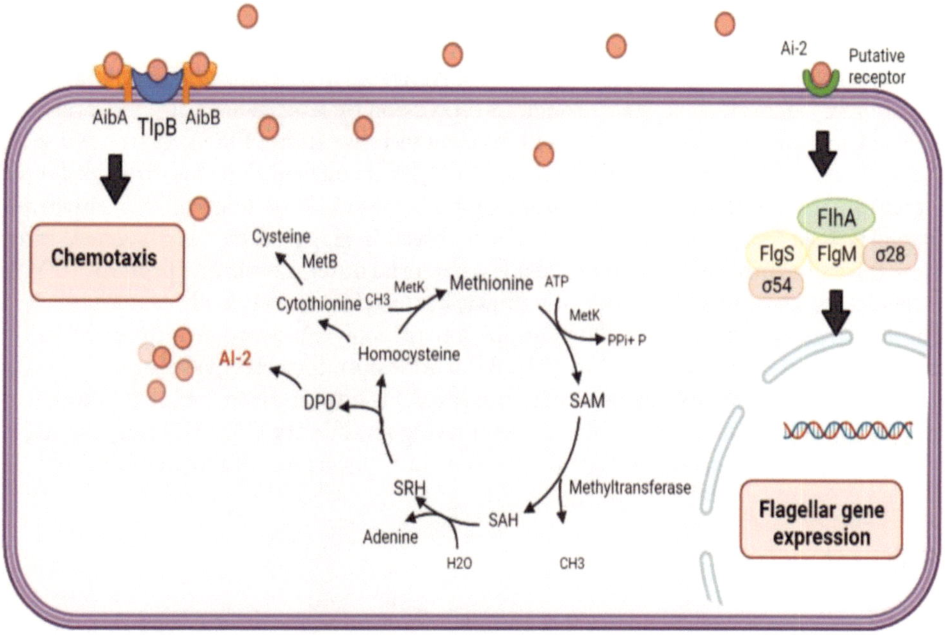

Figure 1.
QS in Helicobacter pylori: LuxS produces AI-2 from the methyl cycle. At high cell density, high concentration of AI-2 in the environment bind to TlpB to active chemotaxis. The binding to the periplasmic proteins AibA and AibB active chemorepulsion. Moreover, AI-2 signals upstream of FlhA manages the branching pathways of gene expression under control of FlgS, FlgM, and σ28 proteins.

chemically related molecules known as AI-2 through a process of dehydration and cyclization [69]. Usually, there are two types of chemoreceptor binding to their AIs, either through direct binding with AI or through interactions with AI binding proteins that transduce signals to the chemoreceptor [70]. In *H. pylori*, TlpB does not bind to AI-2 *in vitro* with high affinity and requires two periplasmic binding proteins, AibA and AibB, which bind to AI-2 independently. AibA and AibB are conserved at greater than 95% identity at the amino acid sequence level in all species of *H. pylori* [32]. The structures of AibA and AibB are not yet elucidated. However, protein sequence homology identifies AibA as homologous to dipeptide binding proteins (39% identity to *E. coli* dipeptide binding protein (PDB ID: 1DPP) and AibB as homologous to proteins of *E. coli* molybdate binding (36% identity to the periplasmic molybdate binding protein of *Azotobacter vinelandii* (PDB ID: 1ATG) [32].

The QS system regulates several mechanisms to assure *H. pylori* colonization in the harsh conditions of the stomach. These include flagellar motility, chemotaxis, and the cag pathogenicity island (Cag PAI) expression, which are all involved in biofilm formation [18, 32, 60]. This indicates that the QS system regulates the various stages of biofilm development from the initial adhesion to the final detachment of the cells [46, 69]. The deletion of luxS gene altered the expression of flagellar genes, i.e., flaA, flaE, flhA, and fliI [69]. Otherwise, the addition of AI-2 or DPD restored the altered phenotype and transcription of these genes. This evidenced that AI-2 is involved in flagellar morphology in H. pylori as it influences the first steps of the flagellar gene expression (**Figure 1**) [69]. The presence of flagella provides motility that enhances the recruitment of planktonic cells to the biofilm, a crucial step in biofilm formation [18].

CagA protein, encoded by cag PAI, has been identified to be induced in *H. pylori* biofilms [54]. A significant decrease in biofilm biomass was observed following mutations in cagA and cag PAI, confirming its important role in biofilm formation [52]. The QS system regulates the cag PAI through its repression by AI-2, which, in turn, attenuates inflammatory response [60]. The type IV secretion system (T4SS), also encoded by cag PAI, is essential in direct cell–cell contact [71]. It is believed that this direct cell–cell contact can also control the biofilm behavior in *H. pylori* [33]. While cag PAI is involved in bacteria-host interaction, it could also be involved in H. pylori bacteria-bacteria interaction, as well as biofilm formation. Besides, bacterial outer membrane proteins (OMPs) are crucial for ion transport, osmotic stability, bacterial virulence, and adherence. Adhesion to gastric cell mediated by Omp18, a peptidoglycan-associated lipoprotein precursor, was reported in H. pylori [72]. After adhesion, the cell envelope gene (lpxD) is upregulated [73]. *H. pylori* urease enzyme (ureA) is important for pH regulation; it prevents the acidification of the biofilm, increasing its stability [74, 75]. Thus, omp18, lpxD, and ureA genes could be directly involved in *H. pylori* biofilm formation [76].

4. QQ in *H. pylori*

In *H. pylori*, AI-2 has been involved in the regulation of motility, type IV secretion, and, most importantly, biofilm formation [32]. The QS plays a critical role in multidrug resistance of *H. pylori* by upregulating both biofilm-associated matrix and efflux pump genes to improve bacterial resistance [77]. Cells in the bacterial biofilm are 100–1000 times more resistant toward antimicrobials than cells in a planktonic state [34]. The inhibition of QS results in a decrease in biofilm formation, making bacteria more susceptible [78].

Since the main component of QS is the production and detection of signal molecules, QQ can interfere with this system in different ways, either intracellularly or extracellularly. It includes: (i) the inhibition of signal synthesis; (ii) the inhibition of signal transmission; (iii) the enzymatic degradation of AI; and (iv) the inhibition of signal detection [25, 28] (**Figure 2**). These strategies showed promising effect *in vitro* and *in vivo*, as well as synergistic effects with traditional antibacterial treatments by increasing bacterial susceptibility to antibiotics [79].

To date, few *H. pylori* QSIs were described, whether synthetic or produced by living organisms, such as plants, animals, and bacteria [80–82]. Flavonoids, i.e., naringenin, quercetin, myricetin, baicalein, catechin, flavone, and turmeric, exhibited promising antibiofilm and antiadhesive properties against *H. pylori* [83–88] (**Table 2**). Notably, a study conducted to assess the effect of *Acorus calamus* on *H. pylori* cultures demonstrated strong antibiofilm and antiadhesive properties [89]. Molecular interaction studies were later performed by the same group of researchers through molecular docking of β-sitosterol, a phytobioactive component of *A. calamus*, toward QS proteins ToxB, DnaA, PhnB, and Sip. Exceptionally high binding affinity and molecular interaction were exhibited, linking the antibiofilm properties of *A. calamus* to the inhibition of QS proteins by β-sitosterol [89]. The most direct and effective way to inhibit the QS system is the enzymatic degradation of the QS molecules, which stops signal transduction [93]. In gram-negative bacteria, two types of hydrolases were described, namely, AHL-lactonase and AHL-acylase. Today, few studies investigated the enzymatic lysis of QS signals in H. pylori. By degrading AHL produced by H. pylori, N-acylhomoserine lactonase produced by Bacillus licheniformis inhibited the biofilm formation and attenuate virulence [90].

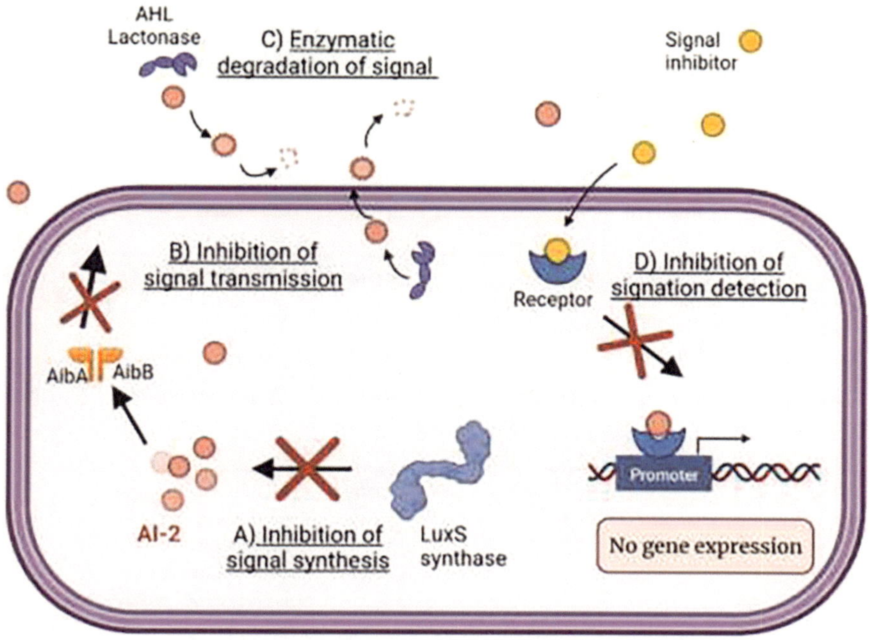

Figure 2.
Different ways to inhibit QS in Helicobacter pylori.

Quencher	Effect on *H. pylori*	Test	Mechanism of QQ	Reference
β-sitosterol (*Acorus calamus)*	Antibiofilm, Antibacterial	*in silico & in vitro*	AI-2 antagonist	[89]
N-acylhomoserine lactonase (*Bacillus licheniformis)*	Antibiofilm & antibacterial	*in vitro*	Degradation of AHL (Ais)	[90]
Methylthio-DADMe-immucillin-A	MTAN inhibitor	*in silico*	Binding to the MTAN target	[91]
Parachlorophenylthio-DADMe-immucillin-A	MTAN inhibitor	*in silico*	Binding to the MTAN target	[91]
-SH Furanosyl Borate Diester	Antibiofilm, Antibacterial[2]	*in silico*	AI-2 antagonist	[92]

Table 2.
QSIs and QQEs in Helicobacter pylori.

Another effective way to inhibit QS is the blockage of signaling cascade through the inactivation of downstream response regulators. The precursor SRH of AI-2 results from the action of MTAN on SAH. The inhibition of MTAN induces an accumulation of 5-methylthioadenosine (MTA) and SAH, which, in turn, inhibits AI-2 production [91, 94]. *In silico* testing of DADMe-ImmA derivatives further confirmed this as a viable QQ technique, since it displayed MTAN inhibition by tight binding to the receptor [95]. More *in silico* studies investigated the possibility of designing furanosyl borate diester derivatives from its pharmacophore modeling by substituting the –OH groups of AI-2 and DPD by -SH making it a potent competitive inhibitor to AI-2 [92].

Based on previous studies, various phytochemicals from medicinal plants with known antibiofilm activity could act *via* inhibition of QS in *H. pylori* (**Table 3**). Baicalin

Molecule	Effect on *H. pylori*	Test	Possible mechanism	Reference
Baicalin	Antibiofilm Adhesion inhibition Bactericidal Virulence reduction Urease inhibition	*in vivo*	Reduction of binding and colonization Suppression urease and blockade of sulfhydryl group.	[83, 88]
Quercetin (*V. rotundifolia*)	Antibiofilm Growth inhibition	*in vitro*	QSI in *P. aeruginosa*	[84]
Catechin (*Chamomilla recutita*)	Antibiofilm Growth inhibition Urease inhibition Membrane disruption	*in vivo*	QSI in *P. fluorescens*	[86]
Naringenin (*H. rosa sinensis*)	Antibiofilm Bactericidal	*in vitro*	QSI in *P. aeruginosa*	[96]
Turmeric (*C. longa*)	Antibiofilm Antiadhesive Immunostimulant (igG toward *H. pylori*)	*in vitro*	Inhibition of AHL production in *A. sobria* Interaction with LuxI Down-regulation of LuxI-type & LuxR	[97, 98]
Proantho-cyanidins (*Vaccinium oxycoccus*)	Antibiofilm, Bacteriostatic, Inhibits siallylactose-specific (S-fimbriae)	*in vitro & in vivo*	Inhibition of AHL production Anti-QS regulators in *P. aeruginosa*	[98]
Emodin (*A. vera*)	Antibiofilm Antiadhesion Affects n-acetyl transferase	*in vitro*	Inhibition of the HefA gene	[99]
Niclosamide	Antibiofilm Bacteriostatic, Decreasing the secretion of IL-8, Disruption of *H. pylori* proton motive force.	*in vitro & in vivo*	QSI in *P. aeruginosa* Affects transcription of QS genes in *P. aeruginosa*	[100]

Table 3.
Inhibitors of biofilm formation potentially via inhibition of QS in Helicobacter pylori.

from medicinal plants exhibited, *in vivo*, bactericidal and antiadhesive activities as well as limited urease production and reduced vacA gene expression, leading to virulence reduction. Baicalin limited the bacterial adhesion and colonization and enhanced bacterial sensitivity *via* suppression of urease and blockage of the sulfhydryl group. This makes Baicalin a potential quorum quencher in H. pylori [83, 88]. Quercetin from *Vitis rotundifolia* inhibited the growth of *H. pylori* [84], while in *P. aeruginosa,* quercetin inhibited AHL production suggesting its action through QQ against *H. pylori.* In parallel, catechin was described as a quorum quencher in *P. fluorescens* suggesting its potential inhibition of QS in *H. pylori.* Catechin from *Chamomilla recutita* inhibited the growth of *H. pylori* and urease production in *H. pylori* (which increases bacterial sensitivity) as well as caused membrane disruption [86]. Naringenin produced by *Hibiscus rosa sinensis* showed a potent bactericidal effect to MDR bacteria and also the inhibition of growth and biofilm formation in *H. pylori* [96]. Moreover, naringenin exhibited a potent competition with AHL for binding in *P. aeruginosa*. Taken together, it seems that naringenin inhibits biofilm formation in *H. pylori* by acting as quorum quencher. Turmeric (*Curcuma longa*) exhibited a good antibiofilm effect toward *H. pylori* [97, 101]. Besides, turmeric decreased AHL production in Aeromonas sobria and limited interaction with LuxI-type synthases and downregulated LuxI-type and LuxR-type genes in various bacterial species. This makes turmeric a potential quencher toward H. pylori. Vaccinium oxycoccus produces proanthocyanidins with antibiofilm and bacteriostatic activities against H. pylori [98]. Proanthocyanidins also limited the siallylactose-specific (S-fimbriae) adhesion of H. pylori to human mucus, erythrocytes, and gastric epithelial cells. In P. aeruginosa, proanthocyanidins was shown to inhibit AI production and to limit the activation of QS transcriptional regulators. Taken together, proanthocyanidins could be considered as a potent quorum quencher in H. pylori.

5. Conclusion

Despite the advancements in the medical field, the treatment of *H. pylori* infections has lost its efficacy. *H. pylori* QS-mediated behavior is the main contributor to bacterial survival and pathogenicity. The significance of bacterial communication in the expression of pathogenic factors makes QS a great target to treat *H. pylori* infection or increase antibiotic efficacy by synergy. In the past two decades, researchers have discovered plenty of QSI agents that can prevent biofilm formation and decrease virulence. The development of new QSI/QQE that can be combined with antibiotics has been a hot topic in the antibacterial research field. More studies are required to demonstrate their mechanisms of action and the optimal doses of the QS inhibitory compounds that are safe and effective.

Conflict of interest

The authors declare no conflict of interest.

Author details

Tarik Aanniz[1*], Wissal Bakri[1], Safae El Mazouri[1], Hajar Wakrim[1], Ilham Kandoussi[1], Lahcen Belyamani[2], Mouna Ouadghiri[1] and Azeddine Ibrahimi[1]

1 Medical Biotechnology Laboratory (MedBiotech), Bioinova Research Center, Rabat Medical and Pharmacy School, Mohammed V[th] University, Rabat, Morocco

2 Emergency Department, Military Hospital Mohammed Vth, Rabat Medical and Pharmacy School, Mohammed V[th] University, Rabat, Morocco

*Address all correspondence to: tarik.aanniz@gmail.com

References

[1] Pohl D, Peter MK, Valentine B, Karoline W. Review of current diagnostic methods and advances in *Helicobacter pylori* diagnostics in the era of next generation sequencing. World Journal of Gastroenterology. 2019;**25**(32):4629-4660. DOI: 10.3748/wjg.v25.i32.4629

[2] Peek RM Jr, Jean EC. *Helicobacter* infection and gastric neoplasia. The Journal of Pathology. 2006;**208**(2): 233-248. DOI: 0.1002/path.1868

[3] Pincock S. Nobel prize winners robin warren and Barry Marshall. The Lancet. 2005;**366**(9495):1429. DOI: 10.1016/S0140-6736(05)67587-3

[4] Marshall BJ. One hundred years of discovery and rediscovery of *Helicobacter pylori* and its association with peptic ulcer disease. In: Mobley HLT, Mendz GL, Hazell SL, editors. *Helicobacter pylori*: Physiology and Genetics. 1st ed. Washington, DC: ASM Press; 2001. pp. 19-26. DOI: 10.1128/9781555818005

[5] "Schistosomes, Liver Flukes and *Helicobacter pylori*. IARC Working Group on the Evaluation of Carcinogenic Risks to Humans. Lyon, 7-14 June 1994." IARC Monographs on the Evaluation of Carcinogenic Risks to Humans. 1994;**61**:1-241

[6] Buzás GM. Metabolic consequences of *Helicobacter pylori* infection and eradication. World Journal of Gastroenterology. 2014;**20**(18):5226-5234. DOI: 10.3748/wjg.v20.i18.5226

[7] Burucoa C, Axon A. Epidemiology of *Helicobacter pylori* infection. Helicobacter. 2017;**22**:e12403. DOI: 10.1111/hel.12403

[8] Hooi JKY, Lai WY, Ng WK, Suen MMY, Underwood FE, Tanyingoh D, et al. Global prevalence of *Helicobacter pylori* infection: Systematic review and meta-analysis. Gastroenterology. 2017;**153**(2):420-429. DOI: 10.1053/j.gastro.2017.04.022

[9] Jaka H, Andreas M, Christa K, Stephen EM. Predictors of triple therapy treatment failure among *H. pylori* infected patients attending at a tertiary Hospital in Northwest Tanzania: A prospective study. BMC Infectious Diseases. 2019;**19**(1):447. DOI: 10.1186/s12879-019-4085-1

[10] Malfertheiner P, Megraud F, O'Morain CA, Atherton J, Axon ATR, Bazzoli F, et al. Management of *Helicobacter pylori* infection--the Maastricht IV/Florence consensus report. Gut. 2012;**61**(5):646-564. DOI: 10.1136/gutjnl-2012-302084

[11] Almeida NJMR, Donato MM, Luxo C, Cardoso O, Cipriano MA, Marinho C, et al. *Helicobacter pylori* antimicrobial resistance rates in the central region of Portugal. Clinical Microbiology and Infection: The Official Publication of the European Society of Clinical Microbiology and Infectious Diseases. 2014;**20**(11):1127-1133. DOI: 10.1111/1469-0691.12701

[12] De Palma GZ, Mendiondo N, Wonaga A, Viola L, Ibarra D, Campitelli E, et al. Occurrence of mutations in the antimicrobial target genes related to levofloxacin, clarithromycin, and amoxicillin resistance in *Helicobacter pylori* Isolates from Buenos Aires City. Microbial Drug Resistance. 2017;**23**(3):351-358. DOI: 10.1089/mdr.2015.0361

[13] Mabeku LBK, Bille BE, Zemnou CT, Nguefack LDT, Leundji H. Broad

Spectrum resistance in *Helicobacter pylori* isolated from gastric biopsies of patients with dyspepsia in Cameroon and efflux-mediated multiresistance detection in MDR isolates. BMC Infectious Diseases. 2019;**19**(1):880. DOI: 10.1186/s12879-019-4536-8

[14] Apolinaria G, Salas-jara MJ, Herrera C, González C. Biofilm and *Helicobacter pylori*: From environment to human host. World Journal of Gastroenterology. 2014;**20**(19):5632-5638. DOI: 10.3748/wjg.v20.i19.5632

[15] Cammarota G, Branca G, Ardito F, Sanguinetti M, Ianiro G, Cianci R, et al. Biofilm demolition and antibiotic treatment to eradicate resistant *Helicobacter pylori*: A clinical trial. Clinical Gastroenterology and Hepatology. 2010;**8**(9):817-820. DOI: 10.1016/j.cgh.2010.05.006

[16] Wen Y, Wang Y, Liu L, He J. Biofilms: The microbial 'protective clothing' in extreme environments. International Journal of Molecular Sciences. 2019;**20**(14):3423. DOI: 10.3390/ijms20143423

[17] Percival SL, Suleman L. Biofilms and *Helicobacter pylori*: Dissemination and persistence within the environment and host. World Journal of Gastrointestinal Pathophysiology. 2014;**5**(3):122. DOI: 10.4291/wjgp.v5.i3.122

[18] Hathroubi S, Zerebinski J, Ottemann KM. *Helicobacter pylori* biofilm involves a multigene stress-biased response, including a structural role for flagella. MBio. 2018;**9**(5):e01973-18. DOI: 10.1128/mBio.01973-18

[19] Diggle SP, Griffin AS, Campbell GS, West SA. Cooperation and conflict in quorum-sensing bacterial populations. Nature. 2007;**450**(7168):411-414. DOI: 10.1038/nature06279

[20] Li YH, Tian X. Quorum sensing and bacterial social interactions in biofilms. Sensors. 2012;**12**(3):2519-2538. DOI: 10.3390/s120302519

[21] Dong YH, Wang LH, Xu JL, Zhang HB, Zhang XF, Zhang LH. Quenching quorum-sensing-dependent bacterial infection by an N-acyl homoserine lactonase. Nature. 2001;**411**(6839):813-817. DOI: 10.1038/35081101

[22] Miller BL, Bassler B. Quorum sensing in bacteria. Annual Review of Microbiology. 2001;**55**:165-199. DOI: 10.1146/annurev.micro.55.1.165

[23] Maeda T, García-Contreras R, Pu M, et al. Quorum quenching quandary: Resistance to antivirulence compounds. The ISME Journal. 2012;**6**:493-501. DOI: 10.1038/ismej.2011.122

[24] Huedo P, Coves X, Daura X, Gibert I, Yero D. Quorum sensing Signaling and quenching in the multidrug-resistant pathogen *Stenotrophomonas Maltophilia*. Frontiers in Cellular and Infection Microbiology. 2018;**8**(122):1-8. DOI: 10.3389/fcimb.2018.00122

[25] Tang K, Zhang XH. Quorum quenching agents: Resources for Antivirulence therapy. Marine Drugs. 2014;**12**(6):3245-3282. DOI: 10.3390/md12063245

[26] Park J, Jagasia R, Kaufmann GF, Mathison JC, Ruiz DI, Moss JA, et al. Infection control by antibody disruption of bacterial quorum sensing Signaling. Chemistry & Biology. 2007;**14**(10):1119-1127. DOI: 10.1016/j.chembiol.2007.08.013

[27] Kato N, Morohoshi T, Nozawa T, Matsumoto H, Ikeda T. Control of gram-negative bacterial quorum sensing with Cyclodextrin immobilized cellulose ether

gel. Journal of Inclusion Phenomena and Macrocyclic Chemistry. 2006;**56**(1):55-59. DOI: 10.1007/s10847-006-9060-y

[28] Fetzner S. Quorum Quenching Enzymes. Journal of Biotechnology. 2015;**201**:2-14. DOI: 10.1016/j.jbiotec.2014.09.001

[29] Ilk S, Sağlam N, Özgen M, Korkusuz F. Chitosan nanoparticles enhances the anti-quorum sensing activity of kaempferol. International Journal of Biological Macromolecules. 2017;**94**:653-662. DOI: 10.1016/j.ijbiomac.2016.10.068

[30] Irie Y, Parsek MR. Quorum sensing and microbial biofilms. Current Topics in Microbiology and Immunology. 2008;**322**:67-84. DOI: 10.1007/978-3-540-75418-3_4

[31] Cammarota G, Sanguinetti M, Gallo A, Posteraro B. Biofilm formation by *Helicobacter pylori* as a target for eradication of resistant infection. Alimentary Pharmacology & Therapeutics. 2012;**36**(3):222-230. DOI: 10.1111/j.1365-2036.2012.05165.x

[32] Yonezawa H, Osaki T, Kamiya S. Biofilm formation by *Helicobacter pylori* and its involvement for antibiotic resistance. BioMed Research International. 2015;**2015**:914791. DOI: 10.1155/2015/914791

[33] Binkowska A, Biernat M, Dus I, Gosciniak G. The role of biofilm formation in pathogenesis of *Helicobacter pylori* infections. Przeglad Gastroenterologiczny. 2013;**8**(1):27-30. DOI: 10.5114/pg.2013.34179

[34] Mah TF, O'Toole GA. Mechanisms of biofilm resistance to antimicrobial agents. Trends in Microbiology. 2001;**9**(1):34-39. DOI: 10.1016/s0966-842x(00)01913-2

[35] Savarino V, Zentilin P, Bisso G, Pivari M, Bilardi C, Biagini R, et al. Optimal duration of therapy combining ranitidine bismuth citrate with clarithromycin and metronidazole in the eradication of *Helicobacter pylori* infection. Alimentary Pharmacology & Therapeutics. 1999;**13**(1):43-47. DOI: 10.1046/j.1365-2036.1999.00436.x

[36] Torres J, Camorlinga-Ponce M, Pérez-Pérez G, Madrazo-De la Garza A, Dehesa M, González-Valencia G, et al. Increasing multidrug resistance in *Helicobacter pylori* strains isolated from children and adults in Mexico. Journal of Clinical Microbiology. 2001;**39**(7):2677-2680. DOI: 10.1128/JCM.39.7.2677-2680.2001

[37] Bardhan KD, Morton D, Perry MJ, Sanders DS, Morris P, Rowland A, et al. Ranitidine bismuth citrate with clarithromycin alone or with metronidazole for the eradication of *Helicobacter pylori*. Alimentary Pharmacology & Therapeutics. 2001;**15**(8):1199-1204. DOI: 10.1046/j.1365-2036.2001.01040.x

[38] Zullo A, De Francesco V, Hassan C, Morini S, Vaira D. The sequential therapy regimen for *Helicobacter pylori* eradication: A pooled-data analysis. Gut. 2007;**56**(10):1353-1357. DOI: 10.1136/gut.2007.125658

[39] Tanih NF, Ndip LM, Clarke AM, Ndip RN. An overview of pathogenesis and epidemiology of *Helicobacter pylori* infection. African Journal of Microbiology Research. 2010;**4**(6):426-436

[40] Boyanova L, Davidkov L, Gergova G, Kandilarov N, Evstatiev I, Panteleeva E, et al. *Helicobacter pylori* susceptibility to fosfomycin, rifampin, and 5 usual antibiotics for *H. pylori* eradication. Diagnostic Microbiology and Infectious

Disease. 2014;**79**(3):358-361. DOI: 10.1016/j.diagmicrobio.2014.03.028

[41] Wueppenhorst N, Stueger HP, Kist M, Glocker EO. High secondary resistance to quinolones in German *Helicobacter pylori* clinical isolates. The Journal of Antimicrobial Chemotherapy. 2013;**68**(7):1562-1566. DOI: 10.1093/jac/dkt061

[42] Bolor-Erdene M, Namdag B, Yamaoka Y, Jav S. Antibiotic resistance of *Helicobacter pylori* in Mongolia. The Journal of Infection in Developing Countries. 2017;**11**(11):887-894. DOI: 10.3855/jidc.8619

[43] Azevedo NF, Pinto AR, Reis NM, Vieira MJ, Keevil CW. Shear stress, temperature, and inoculation concentration influence the adhesion of water-stressed *Helicobacter pylori* to stainless steel 304 and polypropylene. Applied and Environmental Microbiology. 2006;**72**(4):2936-2941. DOI: 10.1128/AEM.72.4.2936-2941.2006

[44] Yang JC, Wang HL, Chern HD, Shun CT, Lin BR, Lin CJ, et al. Role of omeprazole dosage and cytochrome P450 2C19 genotype in patients receiving omeprazole-amoxicillin dual therapy for *Helicobacter pylori* eradication. Pharmacotherapy: The Journal of Human Pharmacology and Drug Therapy. 2011;**31**(3):227-238. DOI: 10.1592/phco.31.3.227

[45] Rhee KH, Park JS, Cho MJ. *Helicobacter pylori*: Bacterial strategy for incipient stage and persistent colonization in human gastric niches. Yonsei Medical Journal. 2014;**55**(6):1453-1466. DOI: 10.3349/ymj.2014.55.6.1453

[46] Krzyzek P, Gosciniak G. A proposed role for diffusible signal factors in the biofilm formation and morphological transformation of *Helicobacter pylori*. The Turkish Journal of Gastroenterology. 2018;**29**(1):7-13. DOI: 10.5152/tjg.2017.17349

[47] Anderson JK, Huang JY, Wreden C, Goers Sweeney E, Goers J, Remington SJ, et al. Chemorepulsion from the quorum signal Autoinducer-2 promotes *Helicobacter pylori* biofilm dispersal. MBio. 2015;**6**(4):e00379-15. DOI: 10.1128/mBio.00379-15

[48] Bavington C, Page C. Stopping bacterial adhesion: A novel approach to treating infections. Respiration. 2005;**72**(4):335-344. DOI: 10.1159/000086243

[49] Krzyżek P, Paluch E, Gościniak G. Synergistic therapies as a promising option for the treatment of antibiotic-resistant *Helicobacter pylori*. Antibiotics. 2020;**9**(10):658. DOI: 10.3390/antibiotics9100658

[50] Krzyżek P, Biernat MM, Gościniak G. Intensive formation of coccoid forms as a feature strongly associated with highly pathogenic *Helicobacter pylori* strains. Folia Microbiologica. 2019;**64**(3):273-281. DOI: 10.1007/s12223-018-0665-5

[51] Servetas SL, Doster RS, Kim A, Windham IH, Cha JH, Gaddy JA, et al. ArsRS-dependent regulation of homB contributes to *Helicobacter pylori* biofilm formation. Frontiers in Microbiology. 2018;**9**:1497. DOI: DOI. 10.3389/fmicb.2018.01497

[52] Wong EHJ, Ng CG, Chua EG, Tay ACY, Peters F, Marshall BJ, et al. Comparative genomics revealed multiple *Helicobacter pylori* genes associated with biofilm formation *in vitro*. PLoS One. 2016;**11**(11):1-16. DOI: 10.1371/journal.pone.0166835

[53] Ge X, Cai Y, Chen Z, Gao S, Geng X, Li Y, et al. Bifunctional enzyme SpoT

is involved in biofilm formation of *Helicobacter pylori* with multidrug resistance by upregulating efflux pump Hp1174 (gluP). Antimicrobial Agents and Chemotherapy. 2018;**62**(11):e00957-18. DOI: 10.1128/AAC.00957-18

[54] Shao C, Sun Y, Wang N, Yu H, Zhou Y, Chen C, et al. Changes of proteome components of *Helicobacter pylori* biofilms induced by serum starvation. Molecular Medicine Reports. 2013;**8**(6):1761-1766. DOI: 10.3892/mmr.2013.1712

[55] Cárdenas-Mondragón MG, Ares MA, Panunzi LG, Pacheco S, Camorlinga-Ponce M, Girón JA. De la Cruz MA. Transcriptional profiling of type II toxin–antitoxin genes of *Helicobacter pylori* under different environmental conditions: Identification of HP0967–HP0968 system. Frontiers in Microbiology. 2016;**7**:1872. DOI: 10.3389/fmicb.2016.01872

[56] Satin B, Del Giudice G, Della Bianca V, Dusi S, Laudanna C, Tonello F, et al. The neutrophil-activating protein (HP-NAP) of *Helicobacter pylori* is a protective antigen and a major virulence factor. The Journal of Experimental Medicine. 2000;**191**(9):1467-1476. DOI: 10.1084/jem.191.9.1467

[57] Bessa LJ, Grande R, Iorio DD, Giulio MD, Campli ED, Cellini L. *Helicobacter pylori* free-living and biofilm modes of growth: Behavior in response to different culture media. APMIS. 2013;**121**(6):549-560. DOI: 10.1111/apm.12020

[58] Nealson KH, Hastings JW. Bacterial bioluminescence: Its control and ecological significance. Microbiological Reviews. 1979;**43**(4):496-518. DOI: 10.1128/MMBR.43.4.496-518.1979

[59] Rader BA, Wreden C, Hicks KG, Sweeney EG, Ottemann KM, Guillemin K. *Helicobacter pylori* perceives the quorum-sensing molecule AI-2 as a Chemorepellent *via* the chemoreceptor TlpB. Microbiology. 2011;**157**(9):2445-2455. DOI: 10.1099/mic.0.049353-0

[60] Wen Y, Huang H, Tang T, Yang H, Wang X, Huang X, et al. AI-2 represses CagA expression and bacterial adhesion, attenuating the *Helicobacter pylori*-induced inflammatory response of gastric epithelial cells. Helicobacter. 2021;**26**(2):e12778. DOI: 10.1111/hel.12778

[61] Challa S, Neelapu NRR. Quorum sensing in *Helicobacter pylori*: Role of biofilm and its implications for antibiotic resistance and immune evasion. In: Implication of Quorum Sensing System in Biofilm Formation and Virulence. 1st ed. Singapore: Springer; 2019. pp. 277-286. DOI: 10.1007/978-981-13-2429-1_18

[62] Deng Y, Wu J, Tao F, Zhang LH. Listening to a new language: DSF-based quorum sensing in gram-negative bacteria. Chemical Reviews. 2011;**111**:160-173. DOI: 10.1021/cr100354f M.B

[63] Duran N, Justo GZ, Duran M, Brocchi M, Cordi L, Tasic L, et al. Advances in *Chromobacterium violaceum* and properties of violacein-its main secondary metabolite: A review. Biotechnology Advances. 2016;**34**:1030-1045. DOI: 10.1016/j.biotechadv.2016.06.003

[64] Hemmati F, Salehi R, Ghotaslou R, Kafil HS, Hasani A, Gholizadeh P, et al. Quorum quenching: A potential target for antipseudomonal therapy. Infection and Drug Resistance. 2020;**13**:2989-3005. DOI: 10.2147/IDR.S263196

[65] Asfour HZ. Anti - quorum sensing natural compounds. Journal of Microscopy and Ultrastructure. 2018;**6**(1):1-10. DOI: 10.4103/JMAU.JMAU

[66] Ortega Á, Zhulin IB, Krell T. Sensory repertoire of bacterial chemoreceptors. Microbiology and Molecular Biology Reviews. 2017;**81**(4):e00033-e00017. DOI: 10.1128/MMBR.00033-17

[67] Bassler BL, Greenberg EP, Stevens AM. Cross-species induction of luminescence in the quorum-sensing bacterium *Vibrio harveyi*. Journal of Bacteriology. 1997;**179**(12):4043-4045. DOI: 10.1128/jb.179.12.4043-4045.1997

[68] Forsyth MH, Cover TL. Intercellular communication in *Helicobacter pylo*ri: LuxS is essential for the production of an extracellular Signaling molecule. Infection and Immunity. 2000;**68**(6):3193-3199. DOI: 10.1128/IAI.68.6.3193-3199.2000

[69] Rader BA, Campagna SR, Semmelhack MF, Bassler BL, Guillemin K. The quorum-sensing molecule autoinducer 2 regulates motility and Flagellar morphogenesis in *Helicobacter pylori*. Journal of Bacteriology. 2007;**189**(17): 6109-6117. DOI: 10.1128/JB.00246-07

[70] Miller ST, Xavier KB, Campagna SR, Taga ME, Semmelhack MF, Bassler BL, et al. *Salmonella typhimurium* recognizes a chemically distinct form of the bacterial quorum-sensing signal AI-2. Molecular Cell. 2004;**15**:677-687. DOI: 10.1016/j.molcel.2004.07.020

[71] Mattar R, Marques SB, Monteiro MDS, Dos Santos AF, Iriya K, Carrilho FJ. *Helicobacter pylori* cag Pathogenicity Island genes: Clinical relevance for peptic ulcer disease development in Brazil. Journal of Medical Microbiology. 2007;**56**(1):9-14. DOI: 10.1099/jmm.0.46824-0

[72] Voland P, Hafsi N, Zeitner M, Laforsch S, Wagner H, Prinz C. Antigenic properties of HpaA and Omp18, two outer membrane proteins of *Helicobacter pylori*. Infection and Immunity. 2003;**71**(7):3837-3843. DOI: 10.1128/IAI.71.7.3837-3843.2003

[73] Kim N, Marcus EA, Wen Y, Weeks DL, Scott DR, Jung HC, et al. Genes of *Helicobacter pylori* regulated by attachment to AGS cells. Infection and Immunity. 2004;**72**(4):2358-2368. DOI: 10.1128/IAI.72.4.2358-2368.2004

[74] Dunne C, Dolan B, Clyne M. Factors that mediate colonization of the human stomach by *Helicobacter pylori*. World Journal of Gastroenterology. 2014;**20**(19):5610-5624. DOI: 10.3748/wjg.v20.i19.5610

[75] Coticchia JM, Sugawa C, Tran VR, Gurrola J, Kowalski E, Carron MA. Presence and density of *Helicobacter pylori* biofilms in human gastric mucosa in patients with peptic ulcer disease. Journal of Gastrointestinal Surgery. 2006;**10**(6):883-889. DOI: 10.1016/j.gassur.2005.12.009

[76] Lucio-Sauceda DG, Urrutia-Baca VH, Gomez-Flores R, La Garza-Ramos D, Angélica M, Tamez-Guerra P, et al. Antimicrobial and anti-biofilm effect of an electrolyzed Superoxidized solution at neutral-PH against *Helicobacter pylori*. BioMed Research International. 2019;**2019**:6154867. DOI: 10.1155/2019/6154867

[77] De Francesco V, Zullo A, Hassan C, Giorgio F, Rosania R, Ierardi E. Mechanisms of *Helicobacter pylori* antibiotic resistance: An updated appraisal. World Journal of Gastrointestinal Pathophysiology. 2011;**2**(3):35-41. DOI: /10.4291/wjgp.v2.i3.35

[78] Cushnie TT, Lamb AJ. Recent advances in understanding the antibacterial properties of flavonoids. International Journal of Antimicrobial

Agents. 2011;**38**(2):99-107. DOI: 10.1016/j.ijantimicag.2011.02.014

[79] Ning Q, Wang D, You J. Joint effects of antibiotics and quorum sensing inhibitors on resistance development in bacteria. Environmental Science: Processes & Impacts. 2021;**23**:995-1005. DOI: 10.1039/D1EM00047K

[80] Ahumedo M, Velásquez M, Reyes RV. Virtual screening: Identification of compounds with possible quorum sensing agonistic activity in *Pseudomonas aeruginosa*. Vitae. 2017;**24**(2):89-101. DOI: 10.17533/udea.vitae.v24n2a02

[81] Miró-Canturri A, Ayerbe-Algaba R, Smani Y. Drug repurposing for the treatment of bacterial and fungal infections. Frontiers in Microbiology. 2019;**10**:41-52. DOI: 10.3389/fmicb.2019.00041

[82] O'Mahony R, Al-Khtheeri H, Weerasekera D, Fernando N, Vaira D, Holton J, et al. Bactericidal and anti-adhesive properties of culinary and medicinal plants against *Helicobacter pylori*. World Journal of Gastroenterology. 2005;**11**(47):7499-7507. DOI: 10.3748/wjg.v11.i47.7499

[83] Chen ME, Su CH, Yang JS, Lu CC, Hou YC, Wu JB, et al. Baicalin, Baicalein, and *lactobacillus Rhamnosus* JB3 alleviated *Helicobacter pylori* infections *in vitro* and *in vivo*. Journal of Food Science. 2018;**83**(12):3118-3125. DOI: 10.1111/1750-3841.14372

[84] Brown JC, Wang J, Kasman L, Jiang X, Haley-Zitlin V. Activities of Muscadine grape skin and quercetin against *Helicobacter pylori* infection in mice. Journal of Applied Microbiology. 2011;**110**(1):139-146. DOI: 10.1111/j.1365-2672.2010.04870.x

[85] Krzyżek P, Junka A, Słupski W, Dołowacka-Jóźwiak A, Płachno BJ, Sobiecka A, et al. Antibiofilm and antimicrobial-enhancing activity of *Chelidonium majus* and *Corydalis cheilanthifolia* extracts against multidrug-resistant *Helicobacter pylori*. Pathogens. 2021;**10**(8):1033. DOI: 10.3390/pathogens10081033

[86] Shin JE, Kim JM, Bae EA, Hyun YJ, Kim DH. *In vitro* inhibitory effect of flavonoids on growth, infection and Vacuolation of *Helicobacter pylori*. Planta Medica. 2005;**71**(3):197-201. DOI: 10.1055/s-2005-837816

[87] Jiang X, Jiang C, Huang C, Chen G, Jiang K, Huang B, et al. Berberine combined with triple therapy versus triple therapy for *Helicobacter pylori* eradication: A Meta-analysis of randomized controlled trials. Evidence-based Complementary and Alternative Medicine. 2018;**2018**:8716910. DOI: 10.1155/2018/8716910

[88] Yu XD, Zheng RB, Xie JH, Su JY, Huang XQ, Wang YH, et al. Biological evaluation and molecular docking of Baicalin and Scutellarin as *Helicobacter pylori* urease inhibitors. Journal of Ethnopharmacology. 2015;**162**:69-78. DOI: 10.1016/j.jep.2014.12.041

[89] Prasad A, Devi AT, Prasad MN, Zameer F, Shruthi G, Shivamallu C. Phyto anti-biofilm elicitors as potential inhibitors of *Helicobacter pylori*. 3 Biotech. 2019;**9**(2):53. DOI: 10.1007/s13205-019-1582-2

[90] Gopalakrishnan V, Masanam E, Ramkumar VS, Baskaraligam V, Selvaraj G. Influence of N-Acylhomoserine lactonase silver nanoparticles on the quorum sensing system of *Helicobacter pylori*: A potential strategy to combat biofilm formation. Journal of Basic Microbiology. 2020;**60**(3):207-215. DOI: 10.1002/jobm.201900537

[91] Harijan RK, Hoff O, Ducati RG, Firestone RS, Hirsch BM, Evans GB, et al. Selective inhibitors of *Helicobacter pylori* Methylthioadenosine Nucleosidase and human Methylthioadenosine phosphorylase. Journal of Medicinal Chemistry. 2019;**62**(7):3286-3296. DOI: 10.1021/acs.jmedchem.8b01642

[92] Kumar M, Naaz A, Sharma A, Teotia D, Nandi S. Pharmacophore Modeling of Furanosyl borate Diester: An attempt to combat antibiotic resistance induced *Helicobacter pylori*. Quorum sensing mediated peptic and duodenal ulcer. Journal of Bioanalysis & Biomedicine. 2018;**10**(6):114-117. DOI: 10.4172/1948-593x.1000218

[93] Zhou L, Zhang Y, Ge Y, Zhu X, Pan J. Regulatory mechanisms and promising applications of quorum sensing-inhibiting agents in control of bacterial biofilm formation. Frontiers in Microbiology. 2020;**11**:1-11. DOI: 10.3389/fmicb.2020.589640

[94] Ronning DR, Iacopelli NM, Mishra V. Enzyme-ligand interactions that drive active site rearrangements in the *Helicobacter pylori* 5′-Methylthioadenosine/S-Adenosylhomocysteine Nucleosidase. Protein Science. 2010;**19**(12):2498-2510. DOI: 10.1002/pro.524

[95] Wang S, Haapalainen AM, Yan F, Du Q, Tyler PC, Evans GB, et al. A picomolar transition state analogue inhibitor of MTAN as a specific antibiotic for *Helicobacter pylori*. Biochemistry. 2012;**51**(35):6892-6894. DOI: 10.1021/bi3009664

[96] Tran TH, Huynh TTH, Thuy NTL, Minh NVH, Nguyen TMN, Thi LMN. Growth-inhibiting, bactericidal, Antibiofilm, and urease inhibitory activities of *Hibiscus rosa sinensis* L. flower constituents toward antibiotic sensitive-and resistant-strains of *Helicobacter pylori*. ACS Omega. 2020;**5**(32):20080-20089. DOI: 10.1021/acsomega.0c01640

[97] Vetvicka V, Vetvickova J, Fernandez-Botran R. Effects of curcumin on *Helicobacter pylori* infection. Annals of Translational Medicine. 2016;**4**(24):1-7. DOI: 10.21037/atm.2016.12.52

[98] Shmuely H, Ofek I, Weiss EI, Rones Z, Houri-Haddad Y. Cranberry components for the therapy of infectious disease. Current Opinion in Biotechnology. 2012;**23**(2):148-152. DOI: 10.1016/j.copbio.2011.10.009

[99] Cellini L, Di Bartolomeo S, Di Campli E, Genovese S, Locatelli M, Di Giulio M. *In vitro* activity of aloe Vera inner gel against *Helicobacter pylori* strains. Letters in Applied Microbiology. 2014;**59**(1):43-48. DOI: 10.1111/lam.12241

[100] Tharmalingam N, Port J, Castillo D, Mylonakis E. Repurposing the anthelmintic drug Niclosamide to combat *Helicobacter pylori*. Scientific Reports. 2018;**8**(1):1-12. DOI: 10.1038/s41598-018-22037-x

[101] Pattiyathanee P, Vilaichone RK, Chaichanawongsaroj N. Effect of curcumin on *Helicobacter pylori* biofilm formation. African Journal of Biotechnology. 2009;**8**(19):5106-5115. DOI: 10.4314/ajb.v8i19.65231

Chapter 5

Mechanism Involved in Biofilm Formation of *Enterococcus faecalis*

Ajay Kumar Oli, Palaksha K. Javaregowda, Apoorva Jain and Chandrakanth R. Kelmani

Abstract

Enterococci are commensal bacteria in the gastrointestinal flora of animals and humans. These are an important global cause of nosocomial infections. A Biofilm formation constitutes an alternative lifestyle in which microorganisms adopt a multi-cellular behavior that facilitates and prolongs survival in diverse environmental niches. The species of enterococcus forms the biofilm on biotic and abiotic surfaces both in the environment and in the healthcare settings. The ability to form biofilms is among the prominent virulence properties of enterococcus. The present chapter highlights the mechanisms underlying in the biofilm formation by enterococcus species, which influences in causing development of the diseases.

Keywords: biofilm, *Enterococcus faecalis*, pathogenesis, microcolony, quorum sensing

1. Introduction

Gram Positive bacterium has been renowned as a pathogen of hospitals acquired infectious. One among these bacteria is *Enterococcu*s species. *Enterococcus* species are ubiquitous, commensally inhabitants of the gastrointestinal tract of humans and animals. These can be frequently isolated from the environmental sources such as soil, surface water, raw plant and animal products. Even these can screen from female genital tract, oropharynx and skin. *Enterococcus sps* belongs to the gram positive, facultative anaerobic cocci with an optimum growth temperature of 35°C [1]. There are around 36 species of enterococci have been reported; conversely 26 species are associated with human infection. The most predominant human pathogen is *Enterococcus faecalis*, even *Enterococcus faecium* is one of the important pathogen which is prevalent increasing as hospital acquired infections. The other remaining enterococci species only accounts 5% of infections [2–4]. Some few examples of enterococcus species which are associated with human infections, *E. avium, E. cecorum, E. cassseliflavus, E. durans, E. gallinarum, E. raffinosus* [5, 6].

E. faecalis has now become the most common nosocomial pathogen and its virulence is increasing in clinical isolates. The presence and function of different suggested characteristics related virulence have been reported [7, 8]. The factor which influences the virulence is mediated through gelatinase production, enterococcus surface protein (ESP), aggregation substance (AS), and biofilm formation [9].

It cause the following infections such as pelvic and abdominal infections, infections in the mouth especially after root canal surgery, infections in open wounds, a lesser known form of meningitis called enterococcal meningitis, infections in the blood called bacteremia and urinary tract infections.

Biofilms are surface attached, organized microbial communities made up of sessile cells (bacteria and /or fungi) embedded in an extracellular matrix composed of polysaccharides, DNA and other components.

2. Chronological background on biofilm

Generally bacterial cell grow in two modes; biofilm formation through aggregate and planktonic cell. It associated with microorganism in which cells stick to each other on a surface encased within matrix of extracellular polymeric substance produced by bacteria itself [10]. Antoni van Leeuwenhoek, the Dutch research, who discovered the simple microscope and observed 'animalcule' on surfaces of tooth and this event is known as discovery of biofilm. Characklis, in the year 1973 phrase that biofilms are not only tenacious but even resist to disinfectants (e.g. chlorine). In 1978, Costerton, defined the term biofilm and explained the importance of biofilm. Biofilms can be found in nature in all places like waste water, labs, and hospital settings. It forms as floating mat on the surface of liquid on both living and non-living surfaces [11].

3. Components of biofilm

Biofilm are produced from different group of organisms, the microbes cells produces the extracellular polymeric substances (EPS) such as DNA <1%, Polysaccharides 1–2%, proteins(includes enzymes) with <1–2%, RNA <1% and water with 97% are the major part of biofilm which is responsible for the flow of nutrients inside biofilm matrix [12]. The main two components of the biofilm that is water channel for nutrients transport and a region of densely packed cells having no prominent pores in it [12]. Another way microbial cells in which biofilms are arranged with significant different physiology and physical properties. They will access of antibiotics and human immune system. The organism that produces biofilm has capability to bear and neutralize antimicrobial agents and result in prolonged treatment. The bacteria which produces the biofilm, switch on the genes that can activate the expression of stress genes which in turn switch to resistant phenotypes due to certain changes examples are as follows cell density, nutritional, temperature, pH and osmolarity. When the biofilm water channels are compared with system of circulations showed that biofilms are considered primitive multi-cellular organism [13, 14]. The compositions of biofilms like DNA, proteins, polysaccharides and water will signify the biofilm integrity and making it resistant against different environmental factors [15].

4. Epidemiology of biofilm formation by *Enterococcus faecalis*

In the worldwide, the prevalence of production of biofilm varies to different part. The study reported in Rome, Italy, 80% of *E. faecalis* isolates have ability to form biofilms in the infected patients [16]. In India, a study has showed that 52% of *E. faecalis* isolated screened from clinical samples has showed the biofilm formation [17]. In China, Shenzhen Nanshan Hospital, the prevalence of *E. faecalis* biofilm formation has showed 50.4% (57/113) in urinary tract infection isolates [18]. The biofilm formation in case of food isolates were less with 60% non-biofilm producers. The major ability in formation of biofilm was endodontic isolates with 73.7% was observed in the Department of Operative Dentistry and Periodontology, University of Freiburg Medical Center, Germany [19].

A study carried out Ahvaz teaching hospital, Iran demonstrated that high frequency 63% of biofilm formation in clinical isolates [20]. The *E. faecalis* bacterial isolated from patient with complicated UTI from department of Urology, Okayama University, Japan has showed the biofilm formation 64 (18.2%) and 156 (44.3%) exhibited strong and medium respectively [21]. A study reported at Malaysia, the *E. faecalis* isolates has showed the biofilm formation of 49% [22]. In the United Kingdom, 100% *E. faecalis* isolates produced biofilms, these isolates were from intravascular catheter-related bloodstream infections (CRBI) found to produce more biofilm than enterococcal isolates that cause non-CRBI [23]. A 93% of *E. faecalis* strains isolated from clinical samples especially fecal isolates have showed more biofilm formation in the United States [24]. In Spain, 57% of *E. faecalis* clinical isolates represent the biofilm production [25]. Tertiary care hospital in India showed 26% isolates of *E. faecalis* having capability in forming biofilm [26].

5. Pathogenesis of biofilm in causing disease

Generally infectious is connected with biofilm primarily confine to particular location and though time detachment may occur. Further, the detached biofilms may result in bloodstream or urinary tract infections or in the production of blockage of blood flow [26]. In another side cells in biofilms are mostly resistant to antimicrobial agents and the host immune system. *E. faecalis* isolates which produces biofilms is 1000 times more resistant to antibodies, antimicrobial agents and phagocytosis process than non-biofilm producers. Consequently, infections caused from *E. faecalis* associated with biofilm aggravated in this case [27, 28].

In endocarditis infection a complex biofilm formed by *E. faecalis* and host components will be formed on cardiac valve. These biofilms causes disease is through three basic mechanisms. Firstly, the biofilms physically disrupts valve function and may cause leakage. Second, detachment of biofilm can be carried to a terminal point in the circulation and formation of emboli (blockage of the blood vessel). Finally, the biofilm provides continuous infection of the bloodstream even during antibiotic treatment. These can cause recurrent fever, chronic systemic inflammation and lead to other infection also [27, 29].

6. Mechanism steps involved in *E. faecalis* biofilm formation

It comprises of four stages; initial attachment, microcolony formation, biofilm maturation (which is in part governed by quorum sensing) and dispersal.

7. Initial attachment

A surface adhesion is the first step in establishing a biofilm, and a number of surface adhesions, proteases, and lipids are involved. The endocarditis and biofilm-associated pilus (Ebp), which is composed of subunits A, B, and C, mediates the adherence of biofilms on surface *in-vitro* and *in-vivo* [30–35]. The deletion of ebpABC attenuates binding to platelets, fibrinogen and collagen, reduces initial attachment, and thus impairs biofilm formation *in-vitro* [30, 32, 33].

In addition, Ebp contributed to early biofilm formation in *in-vivo* models of urinary tract infection (UTI), catheter associated UTI (CAUTI), and infectious endocarditis, in which bacteria with deletions of pilus components were substantially attenuated [30, 32, 33, 36]. Additionally, the absence of surface adhesions, such as aggregation substance (Agg), enterococcol surface protein (ESP), and adhesion to collagen from *E. faecalis* (Ace), reduced adhesion to cultured human cells and prevented biofilm formation *in-vivo* [37–41]. Bacteria deficient for Esp showed reduced initial attachment and decreased bladder colonization in a UTI ascending model, which is not unexpected since Esp binds fibrinogen and collagen, and these ligands are present in the bladder because Esp binds fibrinogen and collagen, and these ligands are present in the bladder [41, 42].

Ace is also involved in interacting with collagen, laminin, and dentin and deletion of Ace resulted in reduced colonization in rat endocarditis and UTI models [43–47]. As a result, Ace deletion in the peritonitis model did not reduce bacterial burden suggesting Ace-mediated biofilm formation is not relevant to peritoneal infection. By disparity, deletion of Agg reduced adherence to renal epithelial cells [38, 39], binding to lipoteichoic acid (LTA) of other *E. faecalis* cells (and therefore inter-bacterial clumping) and bacterial titers recovered from endocarditis vegetation on aortic heart valves. Agg cannot colonize the urinary tract, suggesting that Agg-mediated biofilms aren't necessary for ascending UTI's [48, 49].

In-vitro, biofilm associated glycolipid synthesis A (BgsA) contributes to initial adhesion and biofilm development, but its role *in-vivo* is unknown [50]. The extracellular secreted protein encoded by salB (Saga-Like Protein B) increased fibronectin and collagen binding but decreased biofilm formation paradoxically, which has hypothesized to be owing to the salB mutant cells decreased hydrophobicity. These investigations suggest that a variety of variables play a role in the initial attachment of bacteria, and that their contribution is likely to vary depending on the surface to which the bacteria adhere. As a result, focusing on a single component as anti-adherence or anti-biofilm strategy is unlikely to totally prevent enterococcal biofilm formation [37].

8. Microcolony formation

Bacteria proliferate and produce modest amounts of biofilm matrix to form aggregates known as microcolonies after first adhesion [51]. However, the enterococcal mechanisms that drive the establishment of microcolonies are unknown, and no transcriptome data from early-stage biofilms or microcolonies is available. The importance of microcolonies for gut colonization has been demonstrated. *E. faecalis* colonization of the stomach of germ free mice resulted in discrete microcolonies covered in a fibrous sweater-like matrix within a week, rather than the largely 2D biofilm sheets (2–3 cells high) that are normally observed in biofilm models *in-vitro* [52].

Despite the fact that microcolonies are commonly assumed to be a temporary stage of early biofilm production, these data imply that microcolonies may represent a mature biofilm stage in this niche that is particularly crucial for gut colonization. In addition, *in-vitro* enterococcal microcolonies emerge in response to antibiotic therapy [53, 54]. Biofilms treated with sub-inhibitory levels of daptomycin began to restructure extensively into microcolonies as early as 8 hours after drug exposure, in contrast to typical biofilm sheets. Even in the absence of antibiotics, deletion mutants of eapOX, which encodes a glycosyl-transferase involved in the formation of cell wall associated rhamnopolysaccharide (Epa), developed microcolonies *in-vitro.* In contrast to the monolayer biofilms, these epaOX microcolonies had lower structural integrity, as shown by their facile separation following washing.

9. Biofilm growth and maturation

Active growth and synthesis of extracellular matrix components such as extracellular DNA (eDNA), polysaccharides, LTA, and extracellular proteases are required for biofilm development. eDNA is the best studied matrix component of enterococcal biofilms:eDNA can be found at the bacterial septum, as part of intercellular filamentous structures, and as part of the larger biofilm matrix, and its release from cells is controlled by autolysin Atla [55–57].

eDNA-associated cells showed no significant cell lysis and had a membrane potential [55], implying that eDNA is liberated from metabolically active cells. As a result, DNase treatment decreased biofilm stability and increased detachment [58, 59], whereas atlA deletion decreased eDNA release and biofilm formation [56]. Despite the lack of evidence that eDNA influences the spatial organization of enterococcal biofilms (as has been postulated for other bacterial species), eDNA remains a potential therapeutic target.

Biofilm production is also aided by non-proteinaceous cell surface components such as glycoproteins, polysaccharides, and modified lipids. The dltABCD operons are involved in the production of D-alanine esters of LTA, which are an important component of Gram-positive bacteria's cell wall, and deletion of this operons decreased biofilm formation *in-vitro*, decreased adherence to epithelial cells, and increased susceptibility to antimicrobial peptides [60]. Biofilm on plastic D (BopD), a potential sugar-binding transcriptional regulator, also promotes to biofilm development *in-vitro* [61].

The deletion of bopABC, which is located upstream of bopD, boosted biofilm growth in glucose but decreased biofilm growth and colonization levels in the murine gut, implying that the ability to utilize maltose is required for biofilm growth in the gut. MprF2, a paralogue of multiple peptide resistance factor (MprF), was likewise found to promote eDNA release and biofilm formation [61–63]. MprF2 reduces the net positive charge of the membrane via aminoacylating phosphatidylglyceroal to mediate electrostatic repulsion of cationic antimicrobial peptides.

While deletion of MprF2 had no effect on biofilm persistence in a mouse bacteremia model, deletion of both MprF1 and MprF2 reduced biofilm persistence in a wound infection model, suggesting that cell membrane charge may play a role in biofilm formation and pathogenicity *in-vivo* [63, 64]. These findings back up the theory that cell surface glycoproteins, membrane phosphatidylglycerol, and polysaccharides all play a role in biofilm development.

The quorum sensing response regulator FsrA regulates matrix remodeling by upregulating the expression of gelE, SprE, and altA [57, 58, 65–67]. The proteases gelE and sprE were found to diminish biofilm formation *in-vitro* and bacterial load in numerous *in-vivo* models [68–71]. However, in a rabbit endocarditis model, loss of gelE alone increased fibrinous matrix formation in aotic vegetation, leading to endocarditis as shown in the **Table 1** [70].

Name of the Gene	Gene code	Role
D-alanine- d-alanine ligase	*ddl*	It involved in metabolism process (d-ala) especially for bacterial peptidoglycan biosynthesis. Its role in cell wall integrity and biofilm formation.
Cytolysin	*cyl*	It a secreted toxin expressed in response to pheromones, contributes to the pathogenicity of *E. faecalis* by causing blood hemolysis.
Gelatinase	*gelE*	It hydrolyzes the gelatin and ability to damage host tissues plays a vital role in spreading of enterococci in their host. It promotes the aggregation of the cells in microcolonies which constitutes the initial step of biofilm formation.
Serine protease	*sprE*	It hydrolyzes the casein, quorum sensing and autolysis (release of eDNA)
Fecal streptococci regulator locus genes	*fsrA, fsrB, fsrC*	It the major quorum sensing in *E. faecalis*, the*fsr* regulator locus, is encoded by*fsrA, fsrB* and*fsrC* genes which regulate the expression of both gelatinase and serine protease. It controls biofilm development through regulating the production of gelatinase.
Biofilm associated pili	*ebp*	It is the protein organelles, anchored to the surface of the bacterium, that interact with the external environment. It role in biofilm formation, initial attachment and IE.
Adhesion to collagen of *E. faecalis*	*ace*	A surface protein that facilitates the bacterial adherence to collagen is the adhesion to collagen of *E. faecalis*. It play key role in adherence and colonization process.
Aggregation substance	*agg*	A surface protein expressed in response to pheromone induction that mediates the adherence of *E. faecalis* to renal epithelial cells. It plays important role in adherence to and colonization of host tissues.
Enterococcal fibronectin-binding protein A	*efbA*	It is an adhesin, localized on the outer surface of *E. faecalis* that confers adhesion to immobilized fibronectin.
Enterococcal surface protein	*esp*	It promotes primary attachment and biofilm formation.
LuxS/autoincuder −2 (AI-2) quorum sensing system	*luxS*	It plays role in interspecies communication and involved in bacterial virulence, persistence infections and biofilms

Table 1.
Different quorum sensing genes signaling molecules involved in Enterococcus quorum sensing system and virulence factors production.

In-vitro, sprE deletion increased autolysis and eDNA release and accelerated biofilm development, but gelE deletion inhibited eDNA releaseand elevated ace expression, which may increase surface attachment but make the biofilm detachable [71, 72].

10. Quorum sensing

Population density-dependent signaling influences biofilm formation [73, 74]. Despite the fact that quorum sensing and peptide pheromone signaling are known to coordinate gene expression and direct enterococcus biofilm growth, there have been few research on these tiny signaling molecules and secondary messengers in enterococci. The cCF10 peptide pheromone, which facilitates the transfer of the conjugative plasmid pCF10, is an exception. This plasmid has the ability to transfer antibiotic resistance genes as well as virulence determinants like Agg across cells [75–79]. The buildup of cCF10, which stimulates conjugation proteins, is required for pCF10 transfer. The mechanism underpinning peptide pheromone-mediated gene regulation and plasmid transfer has been well documented, and it was recently demonstrated in mice to promote pCF10 transmission between *E. faecalis* cells in the gut [79, 80]. The immature peptide pheromones cAD1 and cCF10 are processed by the membrane protease Eep. Eep also facilities the proteolytic processing of RsiV, the anti-sigma factor for sigV, resulting in improved stress resistance. A sigV mutant showed similar symptoms, indicating that Eep is involved in the regulation of sigV production [81–83].

In-vitro, Eep, together with AhrC and the ArgR family transcriptional regulators, leads to biofilm formation, and deletion of the genes encoding either protein lowered bacterial burden in UTI and endocarditis models [84–86]. Furthermore, eep deletion mutants develop tiny aggregates unlike wild-type biofilms. FsrABC is another quorum-sensing system. FsrC is a membrane sensor kinase that detects density-dependent accumulation of the FsrB peptide and triggers a signal to the FsrA response regulator [87]. Because this system controls multiple biofilm-related genes and operons (such as bopABCD, ebpABC, GelE, and SprE), knocking down fsrABC entirely eliminates biofilm formation [88]. FsrD, a precursor for the cyclic peptide gelatinase biosynthesis activating pheromone (GBAP), is also controlled by the Fsr quorum sensing system as shown in the **Table 1** [89]. Finally, autoinducer 2 (Al-2) is involved in *E. faecalis* biofilm formation and is produced by S-ribosylhomocysteinelyase (LuxS). *In-vitro* biofilm development of *E. faecalis* is increased by Al-2 supplementation, while luxS deletion causes aberrant biofilm production with aggregation a dense structure, in contrast to the confluent monolayers of wild type *in-vitro* biofilms [90, 91].

11. Factors influencing for the formation of biofilms in *E. faecalis*

11.1 Dlt gene

A Lipoteichoic Acid, component of *E. faecalis*, the most common organism in root canals, develops colonies on the dentin surface (LTA). LTA is a biofilm-forming component of *E. faecalis* that functions as a receptor molecule on receptor cells during the aggregation process. *E. faecalis* antigen recognizes immune cells via pattern recognition receptors (PRRs) and induces the release of proinflammatory cytokines like TNF alpha (TNFα), interleukin 1 beta (IL-1β), IL-6, and IL-8 [92]. LTA causes cells to produce cytokines, which is followed by the activation of Nuclear Factors kβ (NF-kβ), which promotes cytokines release as shown in the **Table 2** [93].

The release of these cytokines causes the dlt gene in LTA to fabricate D-alanine instantly, causing other bacteria to assist in the formation of biofilms [94, 95]. The D-Ala-LTA gene is triggered by the surface protein of Gram-Positive bacteria. Cationic homeostasis and autolytic activity are controlled by this gene. Additionally, it is involved in the assimilation of metal cations as well as the electromechanical repair of bacterial cell walls [94]. These capabilities will enhance bacterial cell system transfer while even increasing autolytic activity. The host's defense system will be weakened by the modified tick.

Factors	Function
dlt gene	It as acts biofilm forming component during aggregation process. It causes cells to produce cytokines. It controls cationic homeostasis and autolytic activity
Cytolysin lytic enzymes	It is the virulence factors, play role in lysing erythrocytes and collagen fragmentation. The *cylLL* and *cylLS* genes on cytolysin promoted for longer survive of *E. faecalis.*
Hyaluronidase	It acts as toxin protein for the progression of host tissue increase damage and inflammation. It beneficial protein for the development of *E. faecalis.*
Dentine Matrix	It increases the enhancement of biofilm formation through dentin. It also resists the antimicrobial treatment by delay penetration of the drug through the biofilm matrix by altering/changing the physiological shaper of biofilm growth in dentin.
Nutrients	Glucose is the major determinate in the formation of *E. faecalis.* It utilizes as the carbon source and hydrolyzes the substrate for its survival.
Environmental	Physicochemical properties of the surface may exert a strong influence on the rate and extent of attachment. Temperature, cations, and presence of antimicrobial agents influence the attachment. The optimum temperature 37°C, pH -8.5 increase the production biofilm formation.

Table 2.
Factors influencing for the formation of biofilms in E. faecalis.

11.2 Cytolisin lytic enzymes

A lytic enzyme operated on by cytolysin is the one of *E. faecalis* bacteria's virulence factors. Apart from lysing erythrocytes, collagen fragmentation caused by this enzyme can cause tissue injury at the site of inflammation. The cylLL and cylLs genes on cytolysin promote this role, allowing *E. faecalis* to survive longer. *E. f*aecalis is the most common microbe found in root canals [92, 96]. Other bacteria will be inhibited by *E. faecalis* cytolysin. The cylLL and cylLS genes in *E. faecalis* cytolysin encode structural cytolysin subunits. They create cytolysin in anaerobic circumstances and respond to oxygen depletion in root canals by producing cytolysin as shown in the **Table 2**.

11.3 Hyaluronidase

Hyaluronidase is a protein to be found in *E. faecalis* that helps the bacteria and toxins progress to the host tissue. Other bacteria will continue to migrate from the root canal to the periapical lesions as a result of hyaluronidase. Furthermore, hyaluronidase stimulates the production of toxins by other bacteria, which increases damage and inflammation. This stipulation is very beneficial for the development of *E. faecalis* [97, 98].

11.4 Dentine matrix structurization

E. faecalis will increase resistance to antimicrobial treatments by increasing the biofilm structural characteristics at the primary site of *E. faecalis* invasion, notably dentin. As a result, *E. faecalis* is known to delay antimicrobial agent penetration through the biofilm matrix by altering the growth rate of other microbes in biofilm development and encouraging changes in the physiological shape of biofilm growth in dentin.

When *E. faecalis* is cultivated in nutrient-poor media, it forms thicker biofilms than when cultured in nutrient-rich media [99]. Under stress inducing mechanism in other bacteria that can cause a more resilient *E. faecalis* biofilm. Besides *E. faecalis* biofilms profitably renew themselves. Furthermore, *E. faecalis* will receive vital carbon by hydrolyzing the substrate required for survival [23].

E. faecalis will continue to grow and develop in environments with or without oxygen with extreme alkaline pH by penetrating cell membrane ions and increasing the cytoplasmic's buffer capacity [100]. The pH balance of the biofilm is always maintained by bacteria by assimilation of protons into the cell, resulting in a lower internal cell pH. As a result, the dentin buffer capacity is unable to keep the pH in the dentinal tubule constant, and *E. faecalis* survives [101].

Other investigations found in *E. faecalis* that the ability to promote apatite re-deposition in the forming biofilm is responsible for its persistence after root canal therapy. Besides this, the dentin matrix is composed of chlorapatite $Ca_5(PO_4)_3$ [102]. Different varieties of apatite have different dissolving tolerances. Till date, chlorapatite has been considered as a weaker apatite than hydroxyapatite and fluorapatite in terms of nanostructure [102, 103]. Although it is known that calcium hydroxide can stimulate the formation of hard tissue by raising the Ca^{2+} ion to increase defense through dentin mineralization, the type of apatite that makes up the host dentin will influence the results [104, 105].

However, no further research into the drug resistance of this inorganic dentin material's nanostructures has been done. Furthermore, dentin deterioration is not solely dependent on inorganic elements. Collagen makes up 20% of the organic dentin, which accounts for 85% of the total [103]. Gelatinase, an *E. faecalis* virulence component, is required for hydrolyzing host collagen, High gelatinase levels have been linked to dentin organic matrix degradation [106, 107].

11.5 Tolerance for antimicrobial therapy

Antimicrobial therapy is known to be limited to eliminating free microbes but not to remove cells bound to the biofilm so that re-infection can occur [100]. As a root canal medication, calcium hydroxide is currently the most popular option among dentists. *E. faecalis* is known to be resistant to calcium hydroxide. This is a serious clinical problem. Every root canal treatment failure, which is documented widely, has linked to *E. faecalis* [101]. Calcium hydroxide is known to prevent the acid reaction that happens as a result of the inflammatory response. This lactic acid generated by osteoclasts to absorb hard tissue will be neutralized by the alkaline pH [102, 103].

12. Conclusion

Enterococcus faecalis is one of the most predominant organism in nosocomial infection and also developed the drug resistance. The intrinsic virulence factors *E. faecalis* are associated in biofilm formation and other environmental factor and signals are alarming the biofilm formation. A genome wide study is required to know the role of genetic and environmental factors in development of biofilm and mounting the superior strategies for biofilm control in *E. faecalis* isolates.

Author details

Ajay Kumar Oli[1*], Palaksha K. Javaregowda[1], Apoorva Jain[1]
and Chandrakanth R. Kelmani[2]

1 Department of Biomedical Science, SDM Research Institute for Biomedical Sciences, Shri Dharmasthala Manjunatheshwara University, Dharwad, Karnataka, India

2 Department of Biotechnology, Gulbarga University, Jnana Ganga campus, Kalaburagi, Karnataka, India

*Address all correspondence to: ajay.moli@hotmail.com

References

[1] Murray BE. The life and times of the Enterococcus. Clinical Microbiology Reviews. 1990;**3**:46-45

[2] Gordon S, Swenson JM, Hill BC, Pigott NE, Facklam RR, Cooksey RC, et al. Antimicrobial susceptibility patterns of common and unusual species of enterococci causing infections in the United States. Journal of Clinical Microbiology. 1992;**30**:2373-2378

[3] Ruoff KL, de La Maza L, Mj M, Spargo JD, Ferraro MJ. Species identities of enterococci isolated from clinical specimens. Journal of Clinical Microbiology. 1990;**28**:435-437

[4] Facklam RR, Collins MD. Identification of Enterococcus species isolated from human infections by a conventional test scheme. Journal of Clinical Microbology. 1989;**27**:731-734

[5] Kohler W. The present state of species within the genera Streptococcus and Enterococcus. International Journal of Medical Microbiology. 2007;**297**(3):133-150

[6] Transupawat S, Sukontasing S, Lee JS. *Enterococcus thailandicus* Spnov, isolated from fermented sausage ("mum") in Thailand. International Journal of Systematic and Evolutionary Microbiology. 2008;**58**(7):1630-1364

[7] Klare I, Konstable C, Muller BS, Werner G, Strommenger B, Kettlitz C, et al. Spread of ampicillin/vancomycin-resistant *Enterococcus faecium* of the epidemic virulent clonal complex-17 carrying the gene esp and hy1 in German hospitals. European Journal of Clinical Microbiology & Infectious Diseases. 2005;**24**(12):815-825

[8] Mundy LM, Sahm DF, Gilmore M. Relationships between enterococcal virulence and antimicrobial resistance. Clinical Microbiology Reviews. 2000;**13**:513-522

[9] Carolina BC, Cr MC, Juliana C, Pedro AA. Presence of virulence factors in *Enterococcus faecalis* and *Enterococcus faecium* susceptible and resistant to vancomycin. Memórias do Instituto Oswaldo Cruz. 2013;**108**(5): 590-595

[10] Hall-Stoodley L et al. Bacterial biofilms: From the natural environment to infectious diseases. Nature Reviews. Microbiology. 2004;**2**:95-108

[11] Costern J et al. Bacterial biofilms: A common cause of persistent infections. Science. 1999;**284**:1318-1322

[12] Available from: http://www.horizonpress.com/biofilms

[13] Fux C et al. Survival strategies of infectious biofilms. Trends in Microbiology. 2005;**13**:34-40

[14] Gilbert P et al. Biofilm susceptibility to antimicrobials. Advances in Dental Research. 1997;**11**:160-167

[15] Vinodkumar CS et al. Utility of lytic bacteriophage in the treatment of multidrug-resistant *Pseudomonas aeruginosa* septicaemia in mice. Indian Journal of Pathology & Microbiology. 2008;**51**:360-366

[16] Baldassarri L, Betuccini L, Ammedolia MG, Cocconcelli P, Arciola CR, Montanaro L, et al. Receptor-mediated endocytosis of biofilm- forming *Enterococcus faecalis* by rat peritoneal macrophage. The Indian Journal of

Medical Research. 2004;**199**(Suppl): 131-135

[17] Oli AK, Raju S, Rajeshwari S, Nagaveni S, Kelmani Chandrakanth R. Biofilm formation by multidrug resistant *Enterococcus faecalis* (MDEF) originated from clinical samples. Journal of Microbiology and Biotechnology Research. 2012;**2**(2):284-288

[18] Jin XZ, Bing B, Zhi-wei L, Zhang-ya P, Wei-ming Y, Zhong C, et al. Characterization of biofilm formation by *Enterococcus faecalis* isolates derived from urinary tract infections in China. Journal of Medical Microbiology. 2018;**67**(1):60-67

[19] Annette CA, Jonas D, Ingrid H, Lamprini K, Johan W, Elmar H, et al. *Enterococcus faecalis* from food, clinical specimens, and Oral sites: Prevalence of virulence factors in association with biofilm formation. Frontiers in Microbiology. 2015;**6**(1534)

[20] Fatemeh S, Hajar H, Saeed K, Golshan M, Aram AD, Ahamd FS. Virulence determinants and biofilm formation in clinical isolates of Enterococcus: A cross-sectional study. Journal of Acute Disease. 2020;**9**(1):27-32

[21] Yuko S, Reiko K, Ritsuko M, Koichi M, Hiromi K. Clinical implications of biofilm formation by *Enterococcus faecalis* in the urinary tract. Acta Medica Okayama. 2005;**59**(3):79-87

[22] Poh LW, Ramliza R, Rukman AH. Antibiotic susceptibility patterns, biofilm formation and esp gene among clinical enterococci: Is there any association. International Journal of Environmental Research and Public Health. 2019;**16**:3439

[23] Sandoe JA, Witherden IR, Cove JH, Hertiage J, Wilcox MH. Correlation between enterococcal biofilm formation in vitro and medical-device related infection potential in vivo. Journal of Medical Microbiology. 2003;**52**:547-550

[24] Mohamed JA, Huang W, Nallapareddy SR, Teng F, Murray BE. Influence of origin of isolates, especially endocarditis isolates, and various genes on biofilm formation by *Enterococcus faecalis*. Infection and Immunity. 2004;**72**:3658-3663

[25] Toledo AA, Valle J, Solano C, Arrizubieta MJ, Cucarella C, Lamata M, et al. The enterococcal surface protein, Esp, is involved in *Enterococcus faecalis* biofilm formation. Applied and Environmental Microbiology. 2001;**67**:4538-4545

[26] Prakash VP. Clinical prevalence, identification and molecular characterization of enterococci, PhD thesis, Pondicherry University India. 2005;1-150

[27] Parsek M, Singh P. Bacterial biofilms: An emerging link to disease pathogenesis. Annual Review of Microbiology. 2003;**57**:677-701

[28] Oliveria M, Santos V, Fernandes A, Bernardo F, Vilela C. Antimicrobial resistance and in vitro biofilm-forming ability of enterococci from intensive and extensive farming broilers. Poultry Science. 2010;**89**:1065-1069

[29] Donlan R. Biofilms: Microbial life on surfaces. Emerging infectious Diseases. 2002;**9**(8):881-890

[30] Mohamed J, Huang W, Nallapareddy S, Teng F, Murray B. Influence of origin of isolates especially endocarditis isolates, and various genes on biofilm formation by *E. faecalis*. Infection and immunity. 2004;**72**(6):3653-3663

[31] Nallapareddy SR et al. Endocarditis and biofilm associated pili of *Enterococcus faecalis*. The Journal of Clinical Investigation. 2006;**116**:2799-2807

[32] Bourgogne A, Thomson LC, Murray BE. Bicarbonate enhances expression of the endocarditisand biofilm associated pilus locus, ebpR-ebpABC, in *Enterococcus faecalis*. BMC Microbiology. 2010;**10**:17

[33] Nallapareddy SR et al. Conservation of Ebp- type pilus genes among enterococci and demonstration oftheir role in adherence of *Enterococcus faecalis* to human platelets. Infection and Immunity. 2011;**79**:2911-2920

[34] Nallapareddy SR, Singh KV, Sillanpaa J, Zhao M, Murray BE. Relative contributions of Ebp pili and the collagen adhesin ace to hostextracellular matrix protein adherence andexperimental urinary tract infection by *Enterococcus faecalis* OG1RF. Infection and Immunity. 2011;**79**:2901-2910

[35] Singh KV, Nallapareddy SR, Murray BE. Importance of the ebp (endocarditis and biofilm associated pilus) locus in the pathogenesis of *Enterococcus faecalis* ascending urinary tract infection. The Journal of Infectious Diseases. 2007;**195**:1671-1677

[36] Nielsen HV et al. The metal ion dependent adhesionsite motif of the *Enterococcus faecalis* EbpA pilin mediates pilus function in catheter-associated urinary tract infection. MBio. 2012;**3**:e00177-e00112

[37] Nielsen HV et al. Pilin and sortase residues critical for endocarditis and biofilm- associated pilusbiogenesis in *Enterococcus faecalis*. Journal of Bacteriology. 2013;**195**:4484-4495

[38] Mohamed JA, Teng F, Nallapareddy SR, Murray BE. Pleiotrophic effects of 2 *Enterococcus faecalis* sagA- like genes, salA and salB, which encode proteins that are antigenic during human infection,on biofilm formation and binding to collagen type I and fibronectin. The Journal of Infectious Diseases. 2006;**193**:231-240

[39] Rozdzinski E, Marre R, Susa M, Wirth R, Muscholl SA. Aggregation substance mediated adherence of *Enterococcus faecalis* to immobilized extracellular matrix proteins. Microbial Pathogenesis. 2001;**30**:211-220

[40] Sussmuth SD et al. Aggregation substancepromotes adherence, phagocytosis, and intracellularsurvival of *Enterococcus faecalis* within human macrophages and suppresses respiratory burst. Infection and Immunity. 2000;**68**:4900-4906

[41] Sillanpaa J et al. Characterization of the ebp (fm) pilus encoding operon of *Enterococcus faecium* and its role in biofilm formation and virulence in a murine model of urinary tract infection. Virulence. 2010;**1**:236-246

[42] Toledo AA et al. The enterococcal surface protein, Esp, is involved in *Enterococcus faecalis*biofilm formation. Applied and Environmental Microbiology. 2001;**67**:4538-4545

[43] Shankar N et al. Role of *Enterococcus faecalis* surface protein Esp in the pathogenesis of ascending urinary tract infection. Infection and Immunity. 2001;**69**:4366-4372

[44] Nallapareddy SR, Qin X, Weinstock GM, Hook M, Murray BE. *Enterococcus faecalis*adhesin, ace, mediates attachment to extracellular matrix proteins collagen type IV and laminin as well as collagen type I. Infection and Immunity. 2000;**68**:5218-5224

[45] Nallapareddy SR, Singh KV, Duh RW, Weinstock GM, Murray BE. Diversity of ace,a gene encoding a microbial surface component recognizing adhesive matrix molecules, from different strains of *Enterococcus faecalis* and evidence for production of ace during human infections. Infection and Immunity. 2000;**68**:5210-5217

[46] Kowalski WJ et al. *Enterococcus faecalis* adhesin, Ace, mediates attachment to particulate dentin. Journal of Endodontia. 2006;**32**:634-637

[47] Singh KV, Nallapareddy SR, Sillanpaa J, Murray BE. Importance of the collagen adhesin acein pathogenesis and protection against *Enterococcus faecalis* experimental endocarditis. PLoS Pathogens. 2010;**6**:e1000716

[48] Lebreton F et al. Ace, which encodes an adhesin in*Enterococcus faecalis*, is regulated by Ers and is involved in virulence. Infection and Immunity. 2009;**77**:2832-2839

[49] Waters CM et al. An amino-terminal domain of*Enterococcus faecalis* aggregation substance isrequired for aggregation, bacterial internalizationby epithelial cells and binding to lipoteichoic acid. Molecular Microbiology. 2004;**52**:1159-1171

[50] JohnsonJR CC, Hirt H, Waters C, Dunny G. Enterococcal aggregation substance andbinding substance are not major contributors tourinary tract colonization by *Enterococcus faecalis* in a mouse model of ascending unobstructed urinary tract infection. Infection and Immunity. 2004;**72**:2445-2448

[51] Theilacker C et al. Glycolipids are involved inbiofilm accumulation and prolonged bacteraemia in *Enterococcus faecalis*. Molecular Microbiology. 2009;**71**:1055-1069

[52] Monds RD, O'Toole GA. The developmental model of microbial biofilms: Ten years of aparadigm up for review. Trends in Microbiology. 2009;**17**:73-87

[53] Barnes AMT et al. *Enterococcus faecalis* readily colonizes the entire gastrointestinal tract and formsbiofilms in a germ- free mouse model. Virulence. 2017;**8**:282-296

[54] Dale JL, Nilson JL, Barnes AMT, Dunny GM. Restructuring of *Enterococcus faecalis* biofilmarchitecture in response to antibiotic induced stress. NPJ Biofilms Microbiomes. 2017;**3**:15

[55] Dale JL, Cagnazzo J, Phan CQ, Barnes AM, Dunny GM. Multiple roles for *Enterococcus faecalis* glycosyltransferases in biofilm associated antibiotic resistance, cell envelope integrity, and conjugative transfer. Antimicrobial Agents and Chemotherapy. 2015;**59**:4094-4105

[56] Barnes AM, Ballering KS, Leibman RS, Wells CL, Dunny GM. *Enterococcus faecalis* produces abundant extracellular structures containing DNA in the absence of cell lysis during early biofilm formation. MBio. 2012;**3**:e00193-e00112

[57] Guiton PS et al. Contribution of autolysin and sortase a during *Enterococcus faecalis* DNAdependentbiofilm development. Infection and Immunity. 2009;**77**:3626-3638

[58] Thomas VC et al. A fratricidal mechanism isresponsible for eDNA release and contributes to biofilm development of *Enterococcus faecalis*. Molecular Microbiology. 2009;**72**:1022-1036

[59] Dunny GM, Hancock LE, Shankar N. In: Gilmore M, editor. *Enterococci: From*

Commensals to Leading Causesof Drug Resistant Infection. Boston: Massachusetts Eye and Ear Infirmary; 2014

[60] Vorkapic D, Pressler K, Schild S. Multifacetedroles of extracellular DNA in bacterial physiology. Current Genetics. 2016;**62**:71-79

[61] Fabretti F et al. Alanine esters of enterococcallipoteichoic acid play a role in biofilm formation andresistance to antimicrobial peptides. Infection and Immunity. 2006;**74**:4164-4171

[62] Hufnagel M, Koch S, Creti R, Baldassarri L, Huebner J. A putative sugar binding transcriptional regulator in a novel gene locus in *Enterococcus faecalis* contributes to production of biofilm and prolonged bacteremia in mice. The Journal of Infectious Diseases. 2004;**189**:420-430

[63] Creti R, Koch S, Fabretti F, Baldassarri L, Huebner J. Enterococcal colonization of the gastro-intestinal tract: Role of biofilm and environmental oligosaccharides. BMC Microbiology. 2006;**6**:60

[64] Bao Y et al. Role of mprF1 and mprF2 in the pathogenicity of *Enterococcus faecalis*. PLoS One. 2012;**7**:e38458

[65] Chong KKL et al. *Enterococcus faecalis* modulates immune activation and slows healing during wound infection. The Journal of Infectious Diseases. 2017;**216**:1644-1654

[66] Waters CM, Antiporta MH, Murray BE, Dunny GM. Role of the *Enterococcus faecalis* GelE protease in determination of cellular chain length, supernatant pheromone levels, and degradation offibrin and misfolded surface proteins. Journal of Bacteriology. 2003;**185**:3613-3623

[67] Qin X, Singh KV, Xu Y, Weinstock GM, Murray BE. Effect of disruption of a gene encoding an autolysin of *Enterococcus faecalis* OG1RF. Antimicrobial Agents and Chemotherapy. 1998;**42**:2883-2888

[68] Kristich CJ et al. Development and use of anefficient system for random mariner transposon mutagenesis to identify novel genetic determinants of biofilm formation in the core *Enterococcus faecalis* genome. Applied and Environmental Microbiology. 2008;**74**:3377-3386

[69] Hancock LE, Perego M. Systematic inactivation and phenotypic characterization of two component signal transduction systems of *Enterococcus faecalis* V583. Journal of Bacteriology. 2004;**186**:7951-7958

[70] Kristich CJ, Li YH, Cvitkovitch DG, Dunny GM. Esp independent biofilm formation by *Enterococcus faecalis*. Journal of Bacteriology. 2004;**186**:154-163

[71] Thurlow LR et al. Gelatinase contributes to the pathogenesis of endocarditis caused by *Enterococcus faecalis*. Infection and Immunity. 2010;**78**:4936-4943

[72] Thomas VC, Thurlow LR, Boyle D, Hancock LE. Regulation of autolysis dependent extracellular DNA release by *Enterococcus faecalis* extracellular proteases influences biofilm development. Journal of Bacteriology. 2008;**190**:5690-5698

[73] Pinkston KL et al. The Fsr quorum-sensing system of *Enterococcus faecalis* modulates surface display of the collagen- binding MSCRAMM ace through regulation of gelE. Journal of Bacteriology. 2011;**193**:4317-4325

[74] Krasteva PV, Giglio KM, Sondermann H. Sensing the messenger:

The diverse ways that bacteria signal through c- di-GMP. Protein Science. 2012;**21**:929-948

[75] Camilli A, Bassler BL. Bacterial small moleculesignaling pathways. Science. 2006;**311**:1113-1116

[76] Cook LC, Federle MJ. Peptide pheromone signaling in *Streptococcus* and *Enterococcus*. FEMS Microbiology Reviews. 2014;**38**:473-492

[77] Li YH, Tian X. Quorum sensing and bacterial socialinteractions in biofilms. Sensors. 2012;**12**:2519-2538

[78] Cook L et al. Biofilm growth alters regulation of conjugation by a bacterial pheromone. Molecular Microbiology. 2011;**81**:1499-1510

[79] Antiporta MH, Dunny GM. ccfA, the genetic determinant for the cCF10 peptide pheromone in *Enterococcus faecalis* OG1RF. Journal of Bacteriology. 2002;**184**:1155-1162

[80] Dunny GM. The peptide pheromone inducible conjugation system of *Enterococcus faecalis* plasmid pCF10: Cell- cell signaling, gene transfer, complexityand evolution. Philosophical Transactions of the Royal Society B. 2007;**362**:1185-1193

[81] Hirt H et al. *Enterococcus faecalis* sex pheromonec CF10 enhances conjugative plasmid transfer in vivo. MBio. 2018;**9**:e00037-e00018

[82] An FY, Sulavik MC, Clewell DB. Identificationand characterization of a determinant (eep) on the *Enterococcus faecalis* chromosome that is involvedin production of the peptide sex pheromone cAD1. Journal of Bacteriology. 1999;**181**:5915-5921

[83] Chandler JR, Dunny GM. Characterization of the sequence specificity determinants required forprocessing and control of sex pheromone by the intramembrane protease Eep and the plasmid encoded protein PrgY. Journal of Bacteriology. 2008;**190**:1172-1183

[84] Varahan S, Iyer VS, Moore WT, Hancock LE. Eep confers lysozyme resistance to *Enterococcusfaecalis* via the activation of the extracytoplasmic function sigma factor SigV. Journal of Bacteriology. 2013;**195**:3125-3134

[85] Frank KL et al. AhrC and Eep are biofilm infectionassociatedvirulence factors in *Enterococcus faecalis*. Infection and Immunity. 2013;**81**:1696-1708

[86] Frank KL et al. Evaluation of the *Enterococcus faecalis* biofilm associated virulence factors AhrC and Eep in rat foreign body osteomyelitis and in vitro biofilm associated antimicrobial resistance. PLoS One. 2015;**10**:e0130187

[87] Frank KL et al. Use of recombinase based in vivoexpression technology to characterize *Enterococcus faecalis* gene expression during infection identifiesin vivo- expressed antisense RNAs and implicates the protease Eep in pathogenesis. Infection and Immunity. 2012;**80**:539-549

[88] Ali L et al. Molecular mechanism of quorum- sensingin *Enterococcus faecalis*: Its role in virulence and therapeutic approaches. International Journal of Molecular Sciences. 2017;**18**:960

[89] Hancock LE, Perego M. The *Enterococcus faecalis* fsr two-component system controls biofilm developmentthrough production of gelatinase. Journal of Bacteriology. 2004;**186**:5629-5639

[90] Nakayama J et al. Revised model for *Enterococcus faecalis* fsr quorum sensing

system: The small open reading frame fsrD encodes the gelatinase biosynthesis activating pheromone pro-peptide corresponding to staphylococcal agrd. Journal of Bacteriology. 2006;**188**:8321-8326

[91] Shao C et al. LuxS dependent AI-2 regulates versatile functions in *Enterococcus faecalis* V583. Journal of Proteome Research. 2012;**11**: 4465-4475

[92] He Z et al. Effect of the quorum-sensing luxS gene on biofilm formation by *Enterococcus faecalis*. European Journal of Oral Sciences. 2016;**124**:234-240

[93] Kayaoglu G, Orstavik D. Virulence factors of *Enterococcus faecalis*: Relationship of endodontic disease. Critical Reviews in Oral Biology and Medicine. 2004;**15**(5):308-320

[94] Albiger B, Dahlberg S, Henriques-Normark B, Normark S. Role of the innate immune system in host defence against bacterial infections: Focus on the toll-like receptors. Journal of Internal Medicine. 2007;**261**:511-528

[95] Neuhaus FC, Baddiley J. A continuum of anionic charge: Structures and functions of D-alanyl-teichoic acids in gram-positive bacteria. Microbiology and Molecular Biology Reviews. 2003;**67**:686-723

[96] Fabretti F, Theilacker C, Baldassarri L, Kaczynski Z, Kropec A, Holst O, et al. Alanine esters of enterococcal lipoteichoic acid play a role in biofilm formation and resistance to antimicrobial peptides. Infection and Immunity. 2006;**74**:4164-4171

[97] Distel JW, Hatton JF, Gillespie MJ. Biofilm formation in medicated root canals. Journal of Endodontia. 2002;**28**:689-693

[98] Abou-Rass M, Bogen G. Microorganisms in closed periapical lesions. International Endodontic Journal. 1998;**31**:39-47

[99] Sunde PT, Olsen I, Debelian GJ, Tronstad L. Microbiota of periapical lesions refractory to endodontic therapy. Journal of Endodontia. 2002;**28**:304-310

[100] Shin SJ, Lee JI, Baek SH, Lim SS. Tissue levels of matrix metalloproteinases in pulps and periapical lesions. Journal of Endodontia. 2002;**28**:313-315

[101] Athanassiadis B, Abbott PV, Walsh LJ. The use of calcium hydroxide, antibiotics and biocides as antimicrobial medicaments in endodontics. Australian Dental Journal. 2007;**52**(1 Suppl):S64-S82

[102] Evan M, Davies JK, Sundqvist G, Fidgor D. Mechanisms involved in the resistance of the *Enteococcus faecalis* to calcium hydroxide. International Endodontic Journal. 2002;**35**:221-228

[103] Nasution AI, Soraya C, Nati S, Alibasyah ZM. Effect of ethylene diamine tetra acetic acid and rcprep to microstrain of human root dentin. Journal of International Oral Health. 2016;**8**(1):32-33

[104] Avery JK, Chiego DJ. Essential of Oral histology and embryology. In: A Clinical Approach. 3rd ed. Missouri: Mosby Elsevier; 2006. pp. 108-113

[105] Cwikla S, Bellanger M, Giguere S, Fox A, Verticci F. Dentinal tubulus disinfection using three calcium hydroxide formulation. Journal of Endodontia. 2000;**31**:50-52

[106] Estrella C, Pimenta FC, Ito IY, Bamman LL. Antimicrobial evaluation of calcium hydroxide in infected dentinal tubules. Journal of Endodontia. 1999;**25**:416-418

[107] Tjaderhane L, Palosaari H, Wahlgren J, Larmas M, Sorsa T, Salo T. Human odontoblast culture method: The expression of collagen and matrix metalloproteinases (MMPs). Advances in Dental Research. 2001;**15**:55-58

Chapter 6

Biofilm Development in Gram-Positive and Gram-Negative Bacteria

Deepak Dwivedi and Trishla Sehgal

Abstract

Biofilms are the communities of microorganisms, especially bacteria attached to a biotic or abiotic surface. These biofilms live in a self-sustained matrix and produce different substances called extracellular polymeric substances (EPS) which are responsible for the pathogenicity of a number of bacteria such as *Pseudomonas aeruginosa, Staphylococcus aureus, Vibrio cholerae, Klebsiella pneumoniae, Escherichia coli,* etc. These EPS substance makes it difficult to eradicate the biofilm present on the surface. Biofilm formation is a five-step process. Biofilms can be monospecies or multispecies. In biofilms, cells communicate via Quorum Sensing (QS). QS is the regulation of gene expression in bacteria with respect to changes in cell population density. In QS, bacteria produce various signaling molecules called Auto-inducers (AI). AI concentration increases as the bacterial population increases. Bacteria respond to these AIs results in an alteration of gene expression, which results in the release of various virulence factors. QS involves a two-component signaling process which is different for both Gram-positive and Gram-negative bacteria. QS and EPS make the bacteria resistant to various antibiotics, which make the eradication difficult and hence requires more effective treatment. This article discusses the biofilm structure, phenomenon of biofilm formation, signaling, and pathogenicity to highlight the understanding of processes involved in biofilm formation.

Keywords: biofilm, exopolysaccharides, quorum sensing, *Staphylococcus aureus*, *Pseudomonas aeruginosa*, pathogenicity

1. Introduction

Microorganisms exist in nature primarily attached to biotic and abiotic surfaces. This is possible due to the development of biofilm. Biofilms are the group of microorganisms living within a self-produced matrix of polymeric substances which get attached to several surfaces [1]. Biofilms are different from the planktonic form of bacteria. Planktonic forms are the free-living forms of bacteria. Bacteria try to switch this planktonic form to biofilm due to a number of advantages which includes protection against environmental stresses such as extreme pH, oxygen, osmotic shock, heat, freezing, UV radiation, predators, etc [2]. Biofilm contains a group of microorganisms irreversibly attached to and grow on a surface. The substances produced by these microbes are known as extracellular polymeric substances (EPS) result in the alteration in the phenotype of the organism with respect to growth rate and gene transcription [3].

Biofilms are found to be present on liquid surfaces as floating mat and in a submerged state as well [4]. Biofilms appear either beneficial or detrimental. Biofilms are considered beneficial as these degrade hazardous substances which are present in the soil, but are detrimental to food and slaughterhouse equipment and are also found responsible for the pathogenesis of a number of diseases [5]. Biofilm has been used for the remediation of heavy metals for a long time. EPS as being poly-anionic in nature, forms complexes with positively charged metals (cations) result in metal immobilization within the exopolymeric network. Extracellular enzymatic activities in EPS assist the detoxification of heavy metals by transforming and subsequently participating in exopolymeric mass [6]. Microorganisms in biofilm help in the production and degradation of organic matter, remediation of environmental pollutants, nitrogen cycle, sulfur, and many metals. Some of the literature revealed that microbial biofilms are involved in sewage purification also [7].

Biofilms can grow on surfaces of many medical implants such as sutures, catheters, dental implants, etc [8]. Biofilm formation is an important virulence mechanism in the pathogenesis of many medically important organisms such as *Pseudomonas aeruginosa*, *Staphylococcus aureus*, *Escherichia coli*, etc [9–11] infections including biofilm formation such as vaginitis, colitis, gingivitis, otitis, urethritis, etc [12–14]

Biofilms are communities of bacteria embedded in the EPS matrix. EPS is composed mainly of a complex mixture of proteins, lipids, nucleic acids i.e. extracellular DNA (e-DNA) and polysaccharides [15]. EPS helps the biofilm to withstand mechanical stress. Biofilms are viscoelastic in nature and EPS provides physical support against mechanical and chemical stresses [16].

Depending on the interaction between surface and constituent cells, biofilms can be categorized as monolayer or multilayer [17]. Flagellum and pilus present on the surface of cells increase the attachment of bacteria to the surface which accelerates the formation of biofilm monolayer. In another type, the microbial adhesion is synthesized with the simultaneous transition to the permanent attachment [17]. When microorganisms are able to adhere to a surface and also to each other, they often develop multilayer biofilm. It has been noted in many cases that the bacterial surface characteristics lead to repulsion [17].

2. Biofilm structure

The structure of biofilm consists of matrix of EPS which comprises e-DNA, polysaccharides, and proteins [18]. Channels in this biofilm allow water, air, and nutrients transport to all parts of the biofilm [19].

Exopolysaccharides: These are the high molecular-weight sugar polymers that are secreted outside the matrix act as a scaffold for proteins, nucleic acids, carbohydrates, and lipids to adhere to the surface [20]. Mannose, galactose, and glucose are the most abundant carbohydrates in EPS. Most of the exopolysaccharides are not biofilm specific but their production increases as an environmental stress response.

Extracellular Proteins: This is another major class of EPS. These are found attached to the surface and polysaccharides to help with biofilm formation and stabilization. E.g. Amyloids play a supportive role in biofilm formation. Fap amyloids in *P. aeruginosa* lead to cell aggregation and increased biofilm formation [21]. The dispersal and

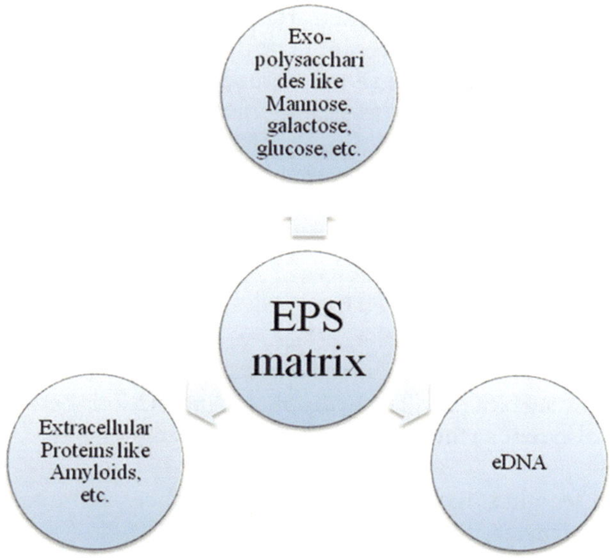

Figure 1.
Components of EPS matrix.

detachment of biofilm also require some enzymes which release biofilm cells and initiate a new biofilm lifecycle. For E.g. Dsp B protein is responsible for the detachment of *Actinobacillus pleuropneumoniae* biofilms [22].

e-DNA: It comes from both lyzed cells and also actively secreted [23]. It plays an important role in biofilm formation critical for attachment. It interacts with receptors present on the substratum surface to facilitate adhesion [24]. It also coordinates with the cell movement in twitching motility mediated *P. aeruginosa* biofilm expansion [25]. It also inhibits the transportation of antibiotics within biofilm thus protects the bacteria within the biofilm. E.g. In *Staphylococcus epidermis*, e-DNA inhibits the transportation of vancomycin and thus protect the biofilm [26]. Vancomycin is a glycopeptide antibiotic that penetrates the biofilm and kills the growing biofilm including gram-positive bacteria. **Figure 1** shows components of the EPS matrix.

3. Steps of biofilm formation

Biofilms are three-dimensional communities of microorganisms that adhere to a surface and form a matrix of EPS. Both gram-positive and gram-negative bacteria develop biofilm but the most common species are *E. faecalis*, *S. aureus*, *S. epidermidis*, *S. viridans*, *E. coli*, *K. pneumoniae*, *P. mirabilis,* and *P. aeruginosa* [27]. Biofilm formation takes place over five main stages including: 1. Initial reversible attachment; 2. Irreversible attachment; 3. Maturation Stage I; 4. Maturation Stage II and 5. Dispersion [28, 29].

1. *Initial reversible attachment*: Bacteria generally adhere to a surface that is rich in organic molecules (e.g. nutrients, salivary proteins, large macromolecules). These molecules promote the adherence of bacteria to the surface. Initial attachment is mediated through weak van der Waals force which later turns to stronger dipole-dipole interaction, hydrogen, ionic or hydrophobic interactions. There is

a stronger adhesin-receptor mediated attachment. It is an attachment between adhesins, adhesive structures present on the surface of microorganisms and receptors, complementary adhesive structures present on the surface of host cells [6]. These interactions are mediated through the surface structures present on the bacterial cell such as fimbriae, flagella, lipopolysaccharides (LPS), outer membrane proteins (OMPs), and exopolysaccharides [30].

2. *Irreversible attachment*: Initial reversible attachment further changes to the irreversible attachment. In this stage, the forces of attraction are greater than the forces of repulsion. Initially immobilized bacterial cells attach to the surface irreversibly [31]. The structures present on the surface overcome the physical repulsive forces of the electrical double layer of the cell and consolidate the interaction between bacteria and the surface [32]. The hydrophobic interactions between the surface and bacteria also reduce the repulsive forces between them [4].

In the first and second stages, bacteria reversibly adhere to the surface which is further replaced by irreversible interaction.

3. *Maturation Stage I*: The bacterial cells start communicating in this stage by the production of AI signals which results in the expression of biofilm-specific genes [33]. The bacteria start producing EPS which stabilizes the biofilm. In this stage, the thickness of biofilm increases up to 10 μm.

4. *Maturation Stage II*: In this stage, the thickness of biofilm further increases to 100μm. Multispecies microconsortia develops on the surface which results in

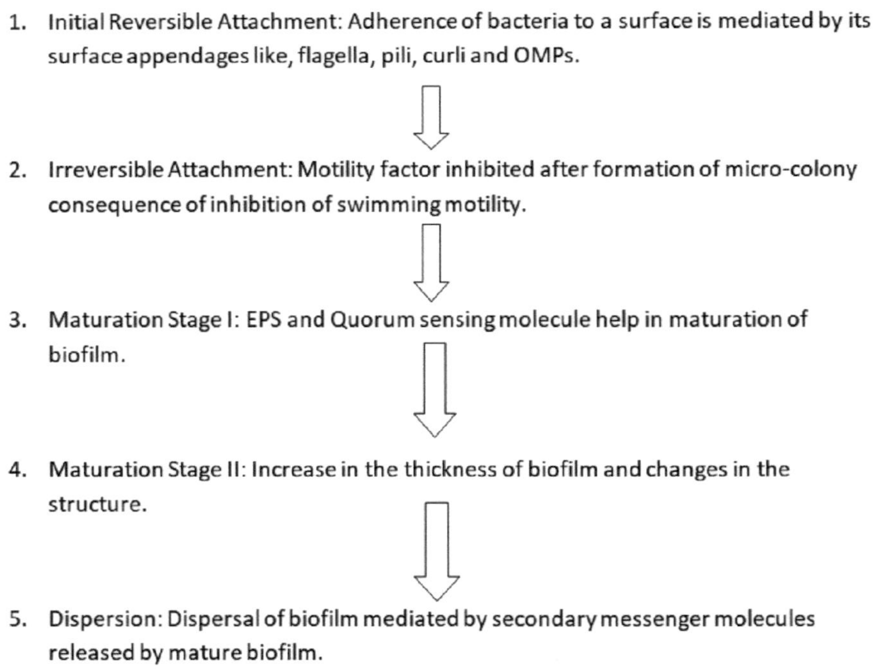

Figure 2.
Stages of biofilm formation.

increase in substrate exchange between bacteria, distribution of metabolic products, and removal of toxic end-products produced by the bacteria [34]. Syntrophic association develops between distinct bacteria in which these utilize certain substrates as energy sources [34]. In this stage, biofilm adapts with the external conditions by manipulating its structure, physiology, and metabolism.

5. *Dispersion*: In this stage, dispersion of bacteria takes place and bacteria return to motile form [35]. In this stage, the microbial community produces different saccharolytic enzymes which break the biofilm stabilizing polysaccharides that releases the bacteria present on the top of the biofilm and colonize to the new surface. The microorganism upregulates the expression of flagella proteins and bacteria return to motile form to translocate to the new site. **Figure 2** shows the process of biofilm formation.

4. Quorum sensing

QS in bacteria is the regulation of gene expression with respect to the fluctuations in the cell-population density. In QS, bacteria produce chemical signal molecules called AI which increase in concentration as a function of cell density [36]. Bacterial populations coordinate their gene expression by producing and responding to a variety of intra and inter-cellular signals called AIs [37]. Microorganisms communicate by producing and responding to small diffusible molecules AIs that acts as signals. When a single bacterium releases AIs into the environment, the concentration is too low to be detected but when mass bacteria releases AIs, the concentration reaches a threshold level which allows the bacteria to sense a critical cell mass, and in response to this it activates or represses target genes. Many classes of AIs have been described to date and N-acyl homoserine lactones (AHLs) are most studied AIs of gram-negative bacteria. A class of AIs termed AI-2 with unknown structure in most cases and the peptides of gram-positive bacteria are most studied [38].

5. Quorum sensing in gram-negative bacteria

In gram-negative bacteria, the QS circuit involves at least two regulatory proteins called LuxR and LuxI. These proteins bind with the protein receptor bound to the bacterial cell membrane/wall. The signaling molecules bind with the receptor proteins then enter the cell. The LuxI protein is responsible for the biosynthesis of AHL, which is utilized as signaling molecules. The AHL concentration increases with the increase in cell population density. The LuxR protein is responsible for binding to cognate AHL AIs that have achieved a threshold concentration; these complexes also activate target gene transcription. The following **Figure 3** shows protein involved in QS and signaling pathway in gram-negative bacteria.

6. Quorum sensing in *Pseudomonas aeruginosa*

P. aeruginosa can be best understood in terms of the virulence factors regulated and the role of QS plays in pathogenicity. *P. aeruginosa* is found to be an opportunistic pathogen as it primarily infects individuals who are immune-compromised, such as

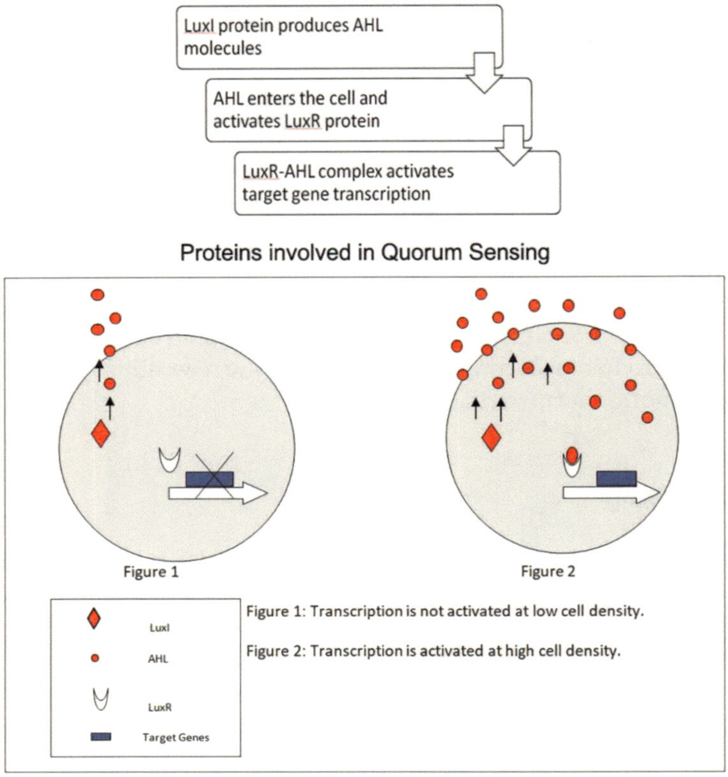

Figure 3.
Proteins & two-component signaling pathway in gram-negative bacteria.

patients with cancer or AIDS or those having breaches in normal barriers caused by burns, indwelling medical devices, or prolonged use of broad-spectrum antibiotics [39]. *P. aeruginosa* is an impressive armament of both cell-associated and extracellular virulence factors. *P. aeruginosa* involves two intertwined QS systems in virulence, biofilm development, and many other processes. Iglewski and colleagues discovered the first system (Las) consists of LasI encoded acyl-HSL synthase and the LasR encoded transcriptional activator. LasI is homologous to LuxI. A number of investigators found the second system (Rhl) consists of an rhlI-encoded acyl-HSL synthase and an rhlR-encoded transcriptional activator. In the respective QS systems, each produces and responds to a specific acyl-HSL; LasI directs the synthesis of 3-oxo-dodecamoyl-HSL (3-oxo-C12-HSL) and RhlI directs the synthesis of butyryl-HSL (C4-HSL) [40].

Using *P. aeruginosa,* lasI, and rhlI double mutant recently, Whiteley et al identified nearly 40 QSc genes that showed a fivefold or greater response to exogenously added acyl-HSL signals. On the basis of the pattern of the responses to cells grown in presence of Las signal, 3-oxo-C12-HSL and/or the Rhl signal, CH-HSL, the QSc genes were classified. A number of early QSc genes were found that responded immediately to exogenously added signals suggesting that these genes behave like the Lux genes of *V. fischeri* and the carbapenem biosynthesis genes of Ervinia. By seminal observations, a number of proteins have been found that support this hypothesis including the stationary phase sigma factors RpoS, RsmA, a third LuxR homolog (QScR), and stringent response proteins RelA, all of them are involved in modulating the expression of genes. QScR gene was found to be the negative regulator of both rhlI and

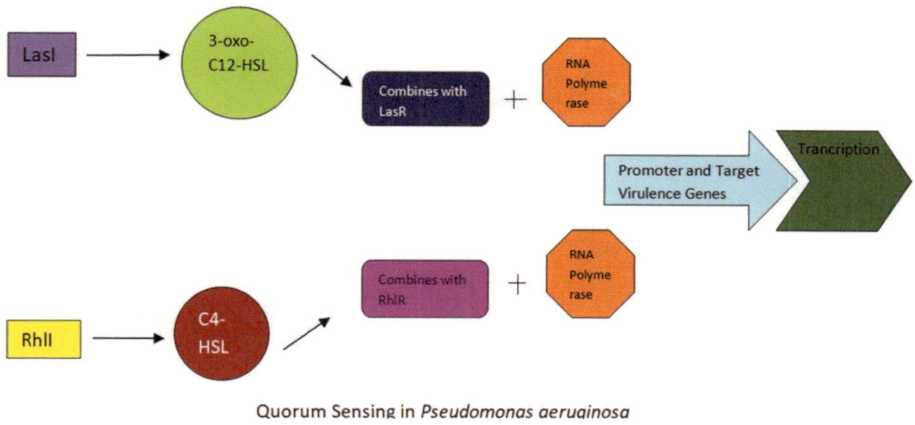

Figure 4.
Quorum sensing in P. aeruginosa.

lasI genes. In *P. aeruginosa,* early activation of QSc genes and premature synthesis of signals like C4-HSL and 3-oxo-C12-HSL were found in QScR mutant varieties. Overexpression of rsmA gene product resulted in decreased production of QSc virulence factors and acyl-HSLs whereas rsmA deletion led to early activation of LasI and thus the early synthesis of 3-oxo-C12-HSL [41].

Expression of a number of virulence factors is regulated by QS in *P. aeruginosa* and QS plays an important role in the pathogenicity of this organism. This presumption has been confirmed by using a number of different animal models. A lasR deficient strain of *P. aeruginosa* was found to have decreased virulence compared to that of the parent in a neonatal mouse model of pneumonia. Analysis of the *P. aeruginosa* mutant varieties such as lasI mutant, rhlI mutant, and a lasI, rhlI double mutant in the same model revealed markedly decreased virulence and the most remarkable reduction was found in the double I mutant variety [42]. **Figure 4** shows the QS in *P. aeruginosa.*

7. Quorum sensing in gram-positive bacteria

QS systems are found to be involved in the pathogenicity and biofilm formation of a number of gram-positive bacteria and these systems use different signal molecules from those of gram-negative bacteria which produce AHLs as AIs. In gram-positive bacteria, no AHL production has been observed in biofilm. Small post-translationally processed peptide signal molecules are used by the gram-positive bacteria QS system. These peptide signals interact with the sensor element of a histidine kinase two-component signal transduction system. Development of bacterial competence in *B. subtilis* and *S. pneumoniae*, conjugation in *E. faecalis,* and virulence in *S. aureus* is regulated by using QS system. A wide variety of disease states caused by *S. aureus* ranges from mild skin infections to life-threatening endocarditis. The virulence of this organism is dependent on the temporal expression of a diverse array of virulence factors which include cell-associated products, such as collagen and fibronectin-binding protein A, and secreted products including lipases, proteases, alpha-toxin, toxin-1, beta-hemolysin, and enterotoxin [43]. **Figure 5** shows the signaling pathway in gram-positive bacteria.

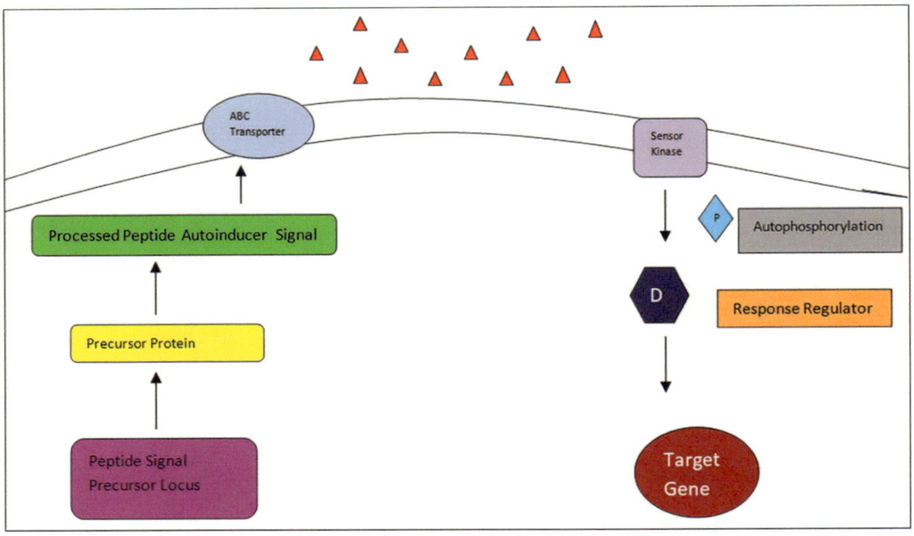

Figure 5.
Signaling pathway in gram-positive bacteria.

8. Quorum sensing in *Staphylococcus aureus*

Surface proteins involved in attachment during the early stages of *S. aureus* infection (collagen and fibronectin-binding protein) and defense protein (protein A) predominate. Expression of *S. aureus* surface proteins is decreased and secreted

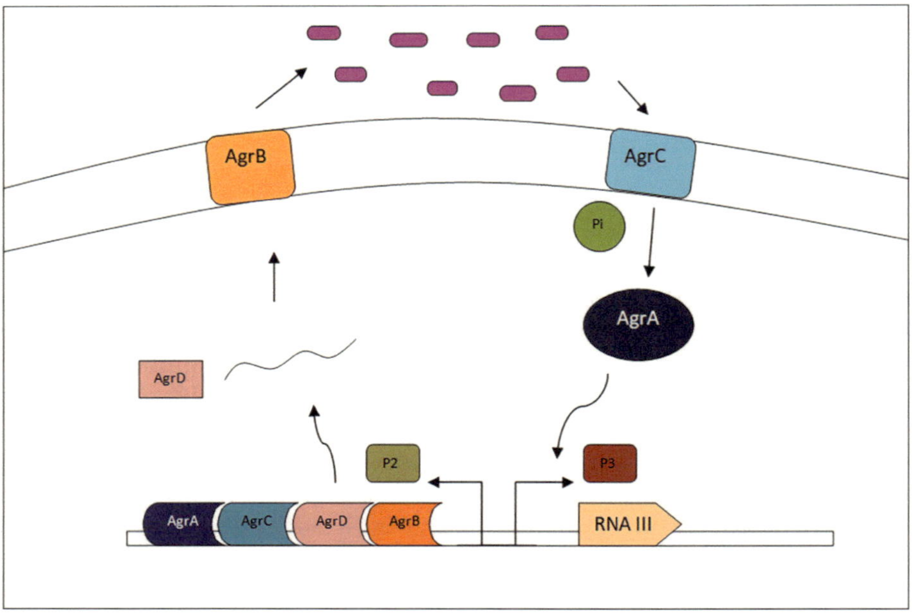

Figure 6.
Quorum sensing in S. aureus.

proteins are preferentially expressed when once a high cell density is achieved at the infection site. Two pleiotropic regulatory gene loci called agr (accessory gene regulator) and sar (staphylococcus accessory gene regulator) determine the genetic basis for this temporal gene expression [44].

The agr locus of *S. aureus* consists of two promoters P2 and P3 with two divergent operons, RNAII and RNAIII. The RNAII operon contains the agr BDCA genes which encode the response regulator (AgrA) and signal transducer (AgrC), and AgrB and AgrD which are involved in generating the QS signal molecule. δ-hemolysin is encoded by the RNAIII and is itself a regulatory RNA that plays a key role in agr response. In response to the octapeptide signal molecule, the AgrC signal transducer is autophosphorylated during *S. aureus* QS, which in turn leads to the phosphorylation of the AgrA response regulator. The transcription of RNAIII is stimulated by phosphorylated AgrA and in turn RNAIII upregulates the expression of numerous *S. aureus* exoproteins as well as the agr BDCA locus. The latter leads to a rapid increase in the synthesis and the export of the octapeptide signal molecules. The AgrA gene product (AgrA) functions as a regulatory DNA-binding protein to induce the expression of both RNAII and RNAIII operons of the agr locus at the second regulatory locus [45, 46]. **Figure 6**: Showing the QS in *S. aureus.*

9. Role of biofilm in pathogenesis

Biofilms play a major role in the pathogenesis of many diseases [47]. A large number of nosocomial infections result due to the colonization of bacteria on the surface. Almost 95% of urinary tract infections are associated with urinary catheters which include *S. aureus* infections. *S. aureus* and *P. aeruginosa* are responsible for frequent biofilm infections.

10. *Pseudomonas aeruginosa* pathogenicity

P. aeruginosa is a gram-negative bacterium that is found to be responsible for a number of infections. It is an opportunistic human pathogen capable of causing both acute and chronic infections [48]. The lungs are one of the common niches for its colonization. It is found to be associated with respiratory infections like cystic fibrosis, lung infections [49]. Its greater adaptability and opportunistic sense enable its association with other infections also like wounds, burns, etc. [50]. Multidrug-resistant *P. aeruginosa* is emerging nowadays which makes the treatment more difficult. *P. aeruginosa* shows resistance to a number of antibiotics like β-lactams, aminoglycosides, quinolones, etc due to mechanisms such as low outer membrane permeability, efflux system, inactivating enzymes like β-lactamases [51]. It can also acquire resistance genes from other micro-organisms by horizontal gene transfer such as in the case of biofilm [52].

P. aeruginosa shows adaptation which is related to complex mechanisms. A number of factors are found to be responsible for the pathogenic potential of bacteria which play a key role in biofilm formation and dispersion. These include flagella, pili, enzymes like proteases, siderophores like pyoverdine, surfactants like rhamnolipids and toxins like exotoxin A and pyocyanin, etc. [53].

11. *Staphylococcus aureus* pathogenicity

Both gram-positive and gram-negative bacteria are found to be pathogenic in nature. *S. aureus* is a gram-positive bacteria frequently found on the mucosal surface of the nose and respiratory tract and skin [51]. It is easily transmitted by direct contact. It is also found to be methicillin-resistant which makes it difficult to treat. Methicillin is a narrow-spectrum β-lactam antibiotic of the penicillin family. *S. aureus* is very often found to be associated with nosocomial infections. Multidrug-resistant *S. aureus* (MRSA) has the ability to evolve and adapt easily which is being considered as a threat according to W.H.O [54]. In addition to this, MRSA is also developing resistance to other antibiotics via mutations and horizontal gene transfer [55]. It has been reported that the presence of *S. aureus* in heterogeneous biofilms increases the rate of plasmid horizontal transfer which increases the resistance of antibiotics in biofilm [56]. *S. aureus* shows the ability to survive host-defense mechanisms through different factors such as cell wall-anchored proteins like clumping factors, fibronectin-binding protein A, collagen adhesion which enables tissue attachment, evasion, and biofilm formation [57]. Extracellular toxins (including hemolysin, leukotoxin, entero-toxin) and enzymes (including coagulase, proteases, staphylokinase) help in tissue penetration and host invasion [58]. Surface-associated factors are down-regulated and surfactants are also expressed in the later stages which lead to biofilm dispersion and the spread of infection [59].

12. Conclusion

Biofilms are made up of bacteria that consist of monospecies or multispecies. Bacterial biofilms are found to be present on a number of surfaces and for this purpose, bacteria secrete and produce EPS matrix which makes adherence easier. Biofilm formation has become a ubiquitous phenomenon found on both living and non-living surfaces. In this biofilm, bacteria interact by producing various toxins, virulence factors that are pathogenic in nature. Both gram-positive and gram-negative bacteria show different QS systems. QS leads the bacteria to evade the immune response and increase cell density. QS is found to be responsible for the virulence shown by the bacteria. Many bacteria show virulence characteristics such as *S. aureus*, *P. aeruginosa*, *E. faecalis, V. cholerae*, *S. pneumoniae*, etc. *S. aureus* produces alpha-hemolysin, toxins, various proteases whereas *P. aeruginosa* is found to produce exoenzymes, cell-cell spacing and sis also resistant to chloramphenicol. *S. aureus* and *P. aeruginosa* are two of the most common bacteria which show biofilm formation. These bacterial biofilms are difficult to eradicate from the surface due to strong adhesive forces and resistance against a number of antibiotics. Current therapeutic approaches are not effective to prevent biofilm formation and thus there is a requirement for new strategies and drugs for the treatment of biofilm infection.

Abbreviations

EPS	Extracellular polymeric substances
AI	Auto-inducers
AHL	N-acyl homoserine lactones
MRSA	Multidrug resistant *S.aureus*
QS	Quorum sensing

Author details

Deepak Dwivedi[1]* and Trishla Sehgal[2]

1 Minor Forest Produce Processing and Research Center, Bhopal, Madhya Pradesh, India

2 Vedanta Testing and Research Laboratory, Bhopal, Madhya Pradesh, India

*Address all correspondence to: microbio.deep@gmail.com

References

[1] Hurlow J, Couch K, Laforet K, Bolton L, Metcalf D, Bowler P. Clinical biofilms: A challenging frontier in wound care. Advances in Wound Care. 2015;**4**(5):295-301. DOI: 10.1089/wound.2014.0567

[2] Annous BA, Fratamico PM, Smith JL. Quorum sensing in biofilms: Why bacteria behave the way they do. Journal of Food Sciences. 2009;**74**(1):R24-R37. DOI: 10.1111/j.1750-3841.2008.01022.x

[3] Donlan RM. Biofilm formation: A clinically relevant microbiological process. Clinical Infectious Diseases. 2001;**33**(8):1387-1392. DOI: 10.1086/322972

[4] Vasudevan R. Biofilms: Microbial cities of scientific significance. Journal of Microbiology & Experimentation. 2014;**1**(3):00014

[5] Peterson BW, He Y, Ren Y, Zerdoum A, Libera MR, Sharma PK, et al. Viscoelasticity of biofilms and their recalcitrance to mechanical and chemical challenges. FEMS Microbiology Reviews. 2015;**39**(2):234-245. DOI: 10.1093/femsre/fuu008

[6] Rodney M, Donlan. Biofilms: Microbial life on surfaces. Emerging Infectious Diseases. 2002;**8**(9):881-890. DOI: 10.3201/eid0809.020063

[7] Cortes ME, Consuegra J, Sinisterra RD. Biofilm formation, control and novel strategies for eradication. Science Against Microbial Pathogens: Communicating Current Research and Technological Advances. 2011;**2**:896-905

[8] Khatoon Z, McTiernan CD, Suuronen EJ, Mah TF, Alarcon EI. Bacterial biofilm formation on implantable devices and approaches to its treatment and prevention. Heliyon. 2018;**4**(12):e01067. DOI: 10.1016/j.heliyon.2018.e01067

[9] Gellatly SL, Hancock RE. *Pseudomonas aeruginosa*: New insights into pathogenesis and host defenses. Pathogens and Disease. 2013;**67**(3):159-173. DOI: 10.1111/2049-632X.12033

[10] Gordon RJ, Lowy FD. Pathogenesis of methicillin-resistant Staphylococcus aureus infection. Clinical Infectious Diseases: An Official Publication of the Infectious Diseases Society of America. 2008;**46**(5):S350-S359. DOI: 10.1086/533591

[11] Beloin C, Roux A, Ghigo JM. Escherichia coli biofilms. Current Topics in Microbiology and Immunology. 2008;**322**:249-289. DOI: 10.1007/978-3-540-75418-3_12

[12] Machado D, Castro J, Palmeira-de-Oliveira A, Martinez-de-Oliveira J, Cerca N. Bacterial vaginosis biofilms: Challenges to current therapies and emerging solutions. Frontiers in Microbiology. 2016;**6**:1528

[13] Vieira Colombo AP, Magalhães CB, Hartenbach FA, Martins do Souto R, Maciel da Silva-Boghossian C. Periodontal-disease-associated biofilm: A reservoir for pathogens of medical importance. Microbial Pathogenesis. 2016;**94**:27-34. DOI: 10.1016/j.micpath.2015.09.009

[14] Delcaru C, Alexandru I, Podgoreanu P, et al. Microbial biofilms in urinary tract infections and prostatitis: Etiology, pathogenicity, and combating strategies. Pathogens. 2016;**5**(4):65. DOI: 10.3390/pathogens5040065

[15] Skandamis PN, Nychas GJ. Quorum sensing in the context of food microbiology. Applied and Environmental Microbiology. 2012;**78**(16):5473-5482. DOI: 10.1128/AEM.00468-12

[16] Flemming HC, Wingender J. The biofilm matrix. Nature Reviews Microbiology. 2010;**8**:623-633. DOI: 10.1038/nrmicro2415

[17] Karatan E, Watnick P. Signals, regulatory networks, and materials that build and break bacterial biofilms. Microbiology and Molecular Biology Reviews. 2009;**73**(2):310-347. DOI: 10.1128/MMBR.00041-08

[18] Flemming HC, Neu TR, Wozniak DJ. The EPS matrix: The "house of biofilm cells". Journal of Bacteriology. 2007;**189**(22):7945-7947. DOI: 10.1128/JB.00858-07

[19] Zhang X, Bishop PL, Kupferle MJ. Measurement of polysaccharides and proteins in biofilm extracellular polymers. Water Science and Technology. 1998;**37**(45):345-348. DOI: 10.1016/S0273-1223(98)00127-9

[20] Rabin N, Zheng Y, Opoku-Temeng C, Du Y, Bonsu E, Sintim HO. Biofilm formation mechanisms and targets for developing antibiofilm agents. Future Medicinal Chemistry. 2015;**7**(4):493-512. DOI: 10.4155/fmc.15.6

[21] Dueholm MS, Søndergaard MT, Nilsson M, Christiansen G, Stensballe A, Overgaard MT, et al. Expression of Fap amyloids in *Pseudomonas aeruginosa*, *P. fluorescens*, and *P. putida* results in aggregation and increased biofilm formation. Microbiologyopen. 2013;**2**(3):365-382. DOI: 10.1002/mbo3.81

[22] Kaplan JB, Velliyagounder K, Ragunath C, Rohde H, Mack D, Knobloch JK, et al. Genes involved in the synthesis and degradation of matrix polysaccharide in Actinobacillus actinomycetemcomitans and Actinobacillus pleuropneumoniae biofilms. Journal of Bacteriology. 2004;**186**(24):8213-8220. DOI: 10.1128/JB.186.24.8213-8220.2004

[23] Hamilton HL, Domínguez NM, Schwartz KJ, Hackett KT, Dillard JP. Neisseria gonorrhoeae secretes chromosomal DNA via a novel type IV secretion system. Molecular Microbiology. 2005;**55**(6):1704-1721. DOI: 10.1111/j.1365-2958.2005.04521.x

[24] Das T, Sharma PK, Busscher HJ, van der Mei HC, Krom BP. Role of extracellular DNA in initial bacterial adhesion and surface aggregation. Applied and Environmental Microbiology. 2010;**76**(10):3405-3408. DOI: 10.1128/AEM.03119-09

[25] Gloag ES, Turnbull L, Huang A, Vallotton P, Wang H, Nolan LM, et al. Self-organization of bacterial biofilms is facilitated by extracellular DNA. Proceedings of the National Academy of Sciences of the United States of America. 2013;**110**(28):11541-11546. DOI: 10.1073/pnas.1218898110

[26] Otto M. Staphylococcus epidermidis pathogenesis. Methods in Molecular Biology. 2014;**1106**:17-31. DOI: 10.1007/978-1-62703-736-5_2

[27] Sauer K, Camper AK, Ehrlich GD, Costerton JW, Davies DG. *Pseudomonas aeruginosa* displays multiple phenotypes during development as a biofilm. Journal of Bacteriology. 2002;**184**(4):1140-1154. DOI: 10.1128/jb.184.4.1140-1154.2002

[28] Stoodley P, Cargo R, Rupp CJ, Wilson S, Klapper I. Biofilm material properties as related to shear-induced deformation and detachment

phenomena. Journal of Industrial Microbiology & Biotechnology. 2002;**29**(6):361-367. DOI: 10.1038/sj.jim.7000282

[29] Beachey EH. Bacterial adherence: Adhesin-receptor interactions mediating the attachment of bacteria to mucosal surface. The Journal of Infectious Diseases. 1981;**143**(3):325-345. DOI: 10.1093/infdis/143.3.325

[30] Garrett TR, Bhakoo M, Zhang Z. Bacterial adhesion and biofilms on surfaces. Progress in Natural Science. 2008;**18**(9):1049-1056. DOI: 10.1016/j.pnsc.2008.04.001

[31] Kumar CG, Anand SK. Significance of microbial biofilms in food industry: A review. International Journal of Food Microbiology. 1998;**42**(1-2):9-27. DOI: 10.1016/s0168-1605(98)00060-9

[32] Tribedi P, Sil AK. Cell surface hydrophobicity: A key component in the degradation of polyethylene succinate by Pseudomonas sp. AKS2. Journal of Applied Microbiology. 2014;**116**(2):295-303. DOI: 10.1111/jam.12375

[33] Davey ME, O'toole GA. Microbial biofilms: From ecology to molecular genetics. Microbiology and Molecular Biology Reviews: MMBR. 2000;**64**(4):847-867. DOI: 10.1128/MMBR.64.4.847-867.2000

[34] Hall-Stoodley L, Costerton JW, Stoodley P. Bacterial biofilms: From the natural environment to infectious diseases. Nature Reviews: Microbiology. 2004;**2**(2):95-108. DOI: 10.1038/nrmicro821

[35] Camilli A, Bassler BL. Bacterial small-molecule signaling pathways. Science (New York, N.Y.). 2006;**311**(5764):1113-1116

[36] Miller MB, Bassler BL. Quorum sensing in bacteria. Annual Review of Microbiology. 2001;**55**:165-199. DOI: 10.1146/annurev.micro.55.1.165

[37] Diggle SP, Gardner A, West SA, Griffin AS. Evolutionary theory of bacterial quorum sensing: When is a signal not a signal? Philosophical transactions of the Royal Society of London Series B, Biological Sciences. 2007;**362**(1483):1241-1249. DOI: 10.1098/rstb.2007.2049

[38] Costerton JW, Stewart PS, Greenberg EP. Bacterial biofilms: A common cause of persistent infections. Science (New York, N.Y.). 1999; **284**(5418):1318-1322

[39] Jaime E. Fergie, et al. *Pseudomonas Aeruginosa* Bacteremia in Immunocompromised children: Analysis of factors associated with a poor outcome. Clinical Infectious Diseases. 1994;**18**(3):390-394

[40] Pearson JP, Gray KM, Passador L, Tucker KD, Eberhard A, Iglewski BH, et al. Structure of the autoinducer required for expression of Pseudomonas aeruginosa virulence genes. Proceedings of the National Academy of Sciences of the United States of America. 1994;**91**(1):197-201. DOI: 10.1073/pnas.91.1.197

[41] Pessi G, Williams F, Hindle Z, Heurlier K, Holden MT, Cámara M, et al. The global posttranscriptional regulator RsmA modulates production of virulence determinants and N-acylhomoserine lactones in Pseudomonas aeruginosa. Journal of Bacteriology. 2001;**183**(22):6676-6683. DOI: 10.1128/JB.183.22.6676-6683.2001

[42] Pearson JP, Feldman M, Iglewski BH, Prince A. *Pseudomonas aeruginosa* cell-to-cell signaling is required for

virulence in a model of acute pulmonary infection. Infection and Immununity. 2000;**68**(7):4331-4334. DOI: 10.1128/IAI.68.7.4331-4334.2000

[43] Dunny GM, Leonard BA. Cell-cell communication in gram-positive bacteria. Annual Review of Microbiology. 1997;**51**:527-564. DOI: 10.1146/annurev.micro.51.1.527

[44] Cheung AL, Koomey JM, Butler CA, Projan SJ, Fischetti VA. Regulation of exoprotein expression in *Staphylococcus aureus* by a locus (sar) distinct from agr. Proceedings of the National Academy of Sciences of the United States of America. 1992;**89**:6462-6466. DOI: 10.1073/pnas.89.14.6462

[45] Lina G, Jarraud S, Ji G, Greenland T, Pedraza A, Etienne J, et al. Transmembrane topology and histidine protein kinase activity of AgrC, the agr signal receptor in *Staphylococcus aureus*. Molecular Microbiology. 1998;**28**(3):655-662. DOI: 10.1046/j.1365-2958.1998.00830.x

[46] Rechtin TM, Gillaspy AF, SchumacherMA,BrennanRG,SmeltzerMS, Hurlburt BK. Characterization of the SarA virulence gene regulator of *Staphylococcus aureus*. Molecular Microbiology. 1999;**33**(2):307-316. DOI: 10.1046/j.1365-2958.1999.01474.x

[47] Sharma G, Rao S, Bansal A, Dang S, Gupta S, Gabrani R. Pseudomonas aeruginosa biofilm: Potential therapeutic targets. Biologicals. 2014;**42**(1):1-7. DOI: 10.1016/j.biologicals.2013.11.001

[48] Mulcahy LR, Isabella VM, Lewis K. *Pseudomonas aeruginosa* biofilms in disease. Microbial Ecology. 2014;**68**(1):1-12. DOI: 10.1007/s00248-013-0297-x

[49] Silby MW, Winstanley C, Godfrey SA, Levy SB, Jackson RW. Pseudomonas genomes: Diverse and adaptable. FEMS Microbiology Reviews. 2011;**35**(4):652-680. DOI: 10.1111/j.1574-6976.2011.00269.x

[50] Pang Z, Raudonis R, Glick BR, Lin TJ, Cheng Z. Antibiotic resistance in *Pseudomonas aeruginosa*: Mechanisms and alternative therapeutic strategies. Biotechnology Advances. 2019;**37**(1):177-192. DOI: 10.1016/j.biotechadv.2018.11.013

[51] Jorge P, Magalhães AP, Grainha T, Alves D, Sousa AM, Lopes SP, et al. Antimicrobial resistance three ways: Healthcare crisis, major concepts and the relevance of biofilms. FEMS Microbiology Ecology. 2019;**95**(8):fiz115

[52] Kazmierczak BI, Schniederberend M, Jain R. Cross-regulation of Pseudomonas motility systems: The intimate relationship between flagella, pili and virulence. Current Opinion in Microbiology. 2015;**28**:78-82. DOI: 10.1016/j.mib.2015.07.017

[53] Wendlandt S, Schwarz S, Silley P. Methicillin-resistant *Staphylococcus aureus*: A food-borne pathogen? Annual Review of Food Science and Technology. 2013;**4**:117-139. DOI: 10.1146/annurev-food-030212-182653

[54] Watkins RR, Holubar M, David MZ. Antimicrobial resistance in methicillin-resistant *Staphylococcus aureus* to newer antimicrobial agents. Antimicrobial Agents and Chemotherapy. 2019;**63**(12):e01216-e01219. DOI: 10.1128/AAC.01216-19

[55] Venkatesan N, Perumal G, Doble M. Bacterial resistance in biofilm-associated bacteria. Future Microbiology. 2015;**10**(11):1743-1750. DOI: 10.2217/fmb.15.69

[56] Foster TJ, Geoghegan JA, Ganesh VK, Höök M. Adhesion, invasion and evasion: The many functions of the surface proteins of *Staphylococcus aureus*. Nature Reviews: Microbiology. 2014;**12**(1):49-62. DOI: 10.1038/nrmicro3161

[57] Kong C, Neoh HM, Nathan S. Targeting *Staphylococcus aureus* toxins: A potential form of anti-virulence therapy. Toxins (Basel). 2016;**8**(3):72. DOI: 10.3390/toxins8030072

[58] Lister JL, Horswill AR. *Staphylococcus aureus* biofilms: Recent developments in biofilm dispersal. Frontiers in Cellular and Infection Microbiology. 2014;**4**:178. DOI: 10.3389/fcimb.2014.00178

[59] Hartmann A, Rothballer M, Hense BA, Schröder P. Bacterial quorum sensing compounds are important modulators of microbe-plant interactions. Frontiers in Plant Science. 2014;**5**:131. DOI: 10.3389/fpls.2014.00131

Chapter 7

Molecular Pathogenesis and Clinical Impact of Biofilms in Surgery

Roger Bayston

Abstract

Biofilms are responsible for chronic persistent infections and are a major problem in implant surgery. The microbial pathogenesis, treatment and prevention of biofilm infections is reviewed.

Keywords: biofilm infections, biofilm phenotype, small colony variants, prevention of biofilm infections

1. Introduction

Though the "discovery" of biofilms is ascribed to Anton van Leeuwenhoek in 1676 using a novel magnifying device, and possibly to Robert Hooke two decades earlier, and biofilms were recognised in a marine setting about a century ago, they were of no medical interest until two studies described them in a medical device and in sputum in 1972 and 1974 respectively. The latter was a description of aggregates of *Pseudomonas aeruginosa* in secretions from the lungs of people with cystic fibrosis [1], and led to a burgeoning of research into *Ps aeruginosa* infection in that field.

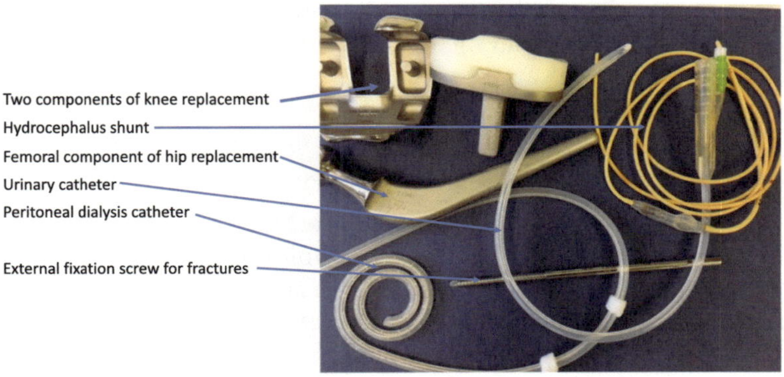

Figure 1.
Examples of implantable devices.

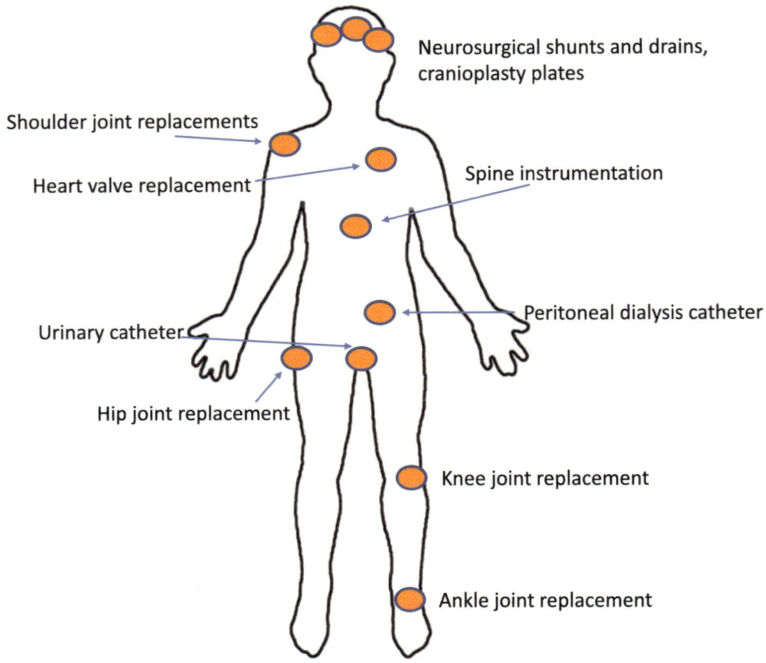

Figure 2.
Anatomical sites of common implantable devices.

Through a meeting with Costerton, Højby studied these aggregates and the term "Biofilm" was made popular by Costerton in 1987 [2], though the term was originally used by Mack et al. [3] to describe "biofilm" on a water filter. However, many biofilm infections occur in association with implanted materials and devices, and their use has become much more common since the middle 1900's. The first biofilm reported in a medical device was found in a shunt to treat hydrocephalus in 1972 [4]. This discovery explained the difficulty in successfully treating these infections non-surgically with antibiotics alone, and the report demonstrated the extracellular matrix of the biofilm in vitro and in vivo and carried out investigations to suggest that it was a glycosaminoglycan. This was later confirmed by important studies in 1996 [5]. Implantable biomaterials and devices are now widely used in modern surgery, and the list is extensive (**Figures 1** and **2**).

2. Biofilm definitions

Many definitions of "biofilm" are found in the literature, and they can be based on either structure or function. Many of the definitions and their accompanying images are derived from in vitro models, and the appearance of mushroom-like structures and water-channels are not seen in biofilms occurring in vivo [6]. A definition based on functional aspects of biofilms is more useful in a medical context. This could be reduced to a population of bacteria or other micro-organisms, often associated with a surface, and enveloped in an extracellular matrix, showing insusceptibility to antimicrobials and to the host immune system, and ability to persist for long periods.

2.1 Biofilm phenotypes

The basis of this functional definition is the paucity of nutrients, including iron, and oxygen in the depths of the biofilm leading to a bacterial stress response caused by a crisis in energy generation and transport [7]. The bacterial stress response is mediated by the intracellular signal sigma-B. The bacterial response to this is to downregulate all synthetic functions not needed in biofilm mode, such as cell wall material, toxin and other non-essential protein synthesis, and DNA replication. These are the targets for common antibiotics, and beta-lactams, glycopeptides, aminoglycosides, macrolides and fluoroquinolones all become significantly less effective against biofilm bacteria. Other factors contribute to the lack of effect of antibiotics, including a slowing of their penetration into the biofilm, though this is rarely a major factor. The bacterial stress response results in significantly reduced cell metabolic activity and loss of some synthetic activities leading to auxotrophy for heme and menadione, and sometimes other substances such as thymidine. This biofilm phenotype is crucial to the clinical impact of biofilm infections; the colonies of biofilm bacteria when grown from clinical samples in the laboratory are typically less than ten times the size of their planktonic counterparts, and are known as small colony variants or SCV. The molecular control and regulation of biofilm phenotype has been described in detail by Proctor et al. [8]. SCV are important in biofilm infections not only because their metabolism leads to antibiotic insusceptibility, but because, though they can be internalised by professional and non-professional phagocytes, they are not killed and survive inside the phagocytic cells. Auxotrophic SCV of *Staphylococcus aureus* for heme and menadione, that do not produce alpha-toxin, are more able to survive intracellularly, and supplementation of intracellular populations of *S aureus* in vitro with menadione resulted in restoration of alpha-toxin production and reduced intracellular survival [9, 10]. SCV are not always auxotrophic and considerable variation occurs, but intracellular survival is a common feature. Many also show reduced susceptibility to aminoglycosides, and exposure to gentamicin can induce SCV formation [11]. Some SCV are the result of mutations in the genes concerned with electron transport, and these do not revert to parent forms whereas other forms of SCV appear to be phenotypic variants that revert to parent forms when the stress factor is withdrawn [8, 12]. SCV of gram negative bacteria have been known for decades, having been produced in the laboratory from exposure to antibacterial chemicals [13, 14]. However, more recently capnophilic (carbon dioxide—dependent) SCV of *Escherichia coli* have been isolated from a patient with a urinary tract infection, though no information on biofilm involvement was given [15]. A report of septic shock in a patient from whose urine capnophilic *Proteus mirabilis* SCV were isolated again did not state that biofilms were involved [16] but the patient had chronic renal stones, known to be associated with biofilms [17]. *P mirabilis* is an important uropathogen as it is highly motile and is capable of enzymatically hydrolysing urea into ammonia, thus being highly inflammatory as well as alkalinising the urine. The rising pH causes crystallisation of calcium and magnesium phosphates [18], and the *P mirabilis* biofilm typically consists of a mesh of bacteria, their extracellular matrix and phosphate crystals. These biofilms are obviously different in composition from those consisting mainly of bacteria and their products, and another example of such complex biofilms is the vegetations found in native valve endocarditis. Here the lesion consists largely of a matrix of platelets and fibrin, with bacteria, usually viridans streptococci, embedded in it. The lesion usually begins as a response to damage to the endocardium, which is then colonised by bacteria from the bloodstream, becoming progressively built up of fibrin and platelets with rafts of bacteria interspersed [19, 20]. A similar situation arises with prosthetic heart valves. In both cases, SCVs have been reported [21, 22] as well as other auxotrophic variants [23].

The biofilm phenotype, and SCV in particular, are important in treatment of biofilm infections. Surviving intracellular bacteria are protected from further immune assault and from most therapeutic antibiotics, which do not accumulate inside host cells sufficiently to kill SCV [24]. These factors mean that the amount of antibiotic required to kill bacteria in biofilm mode is typically 500–1000 times the minimum inhibitory concentration as measured in the clinical laboratory. Such concentrations are not achievable by intravenous or oral therapy, and eradication of biofilm infection usually requires extensive surgery to debride the site and to remove all surgical hardware.

2.2 Biofilm development

Development of biofilms in surgery depends on a sequence of events. Initially, the causative bacteria must be able to gain access to the site of biofilm formation, usually an implantable device. In modern surgery most device pathogens originate on the patient's skin or mucous membranes, consisting mainly of coagulase-negative staphylococci (CoNS), typically *Staphylococcus epidermidis*, and *Cutibacterium acnes*. Conventional pre-operative skin preparation reduces but does not eradicate these bacteria, and the importance of relatively small numbers of bacteria in the operation field has been shown by an experiment in human volunteers, where various "doses" of *S aureus* were inoculated into incisions to determine how many bacteria were necessary to produce an abscess [25]. In one group, "foreign" material in the form of sutures were also introduced into the incision, and the number of bacteria required to form an abscess in those cases was 10,000 times fewer. This study, which is unlikely to be repeated in a modern setting, is extremely important in illustrating the role played by implantable materials and devices in infection in modern surgery.

The sequence of events involved in development of a biofilm infection involving a surgically implanted device are (**Figure 3**):

Access to the device from the source. Though heavy contamination of the air in the operating environment has historically been associated with surgical infection, modern operating room design and ventilation has meant that this source has declined in importance, and most surgical infections are caused by bacteria originating on the patient's skin or mucous membranes. Bacteria reach the incision from the cut edges of

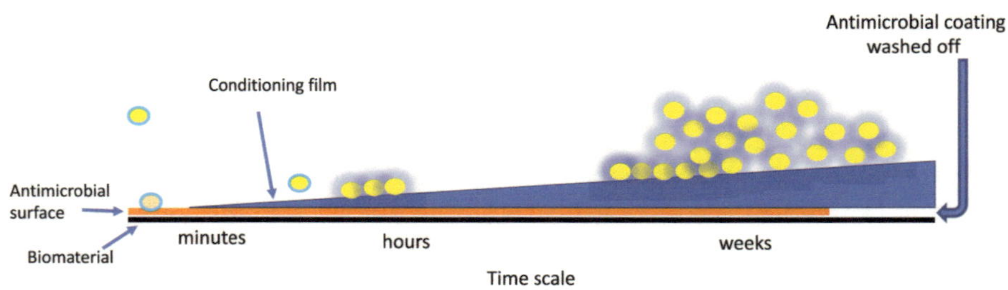

Figure 3.
Sequence of events in development of biofilm infection. Here implant has an antimicrobial coating, but within minutes this is covered by a glycoprotein conditioning film produced by the patient. This usually prevents the activity of the coating and bacteria now adhere to the conditioning film. Within a few hours the attached bacteria begin to produce an extracellular matrix and to multiply. Powerful antibacterial activity is essential now, as after this point, it is almost inevitable that a biofilm will develop, within a few weeks.

the skin, or from contamination from surrounding skin surfaces, during surgery. The causative bacteria are therefore often present when the device is implanted.

Attachment to the device. Many bacteria possess adhesins on their surfaces that allow them to attach to biomaterials (vitronectin—binding protein etc) but more often they employ specific adhesins for the glycoproteins, platelets and other host-derived materials that rapidly coat all implanted materials [26, 27]. *S aureus* possesses specific adhesins for fibrinogen, fibronectin, laminin, thrombospondin, bone sialoprotein and other host-derived components of the conditioning film. These bacterial surface adhesins are known as MSCRAMMs (Microbial Surface Component Recognising Adhesive Matrix Molecules) [28] and they can be found in other organisms such as *S epidermidis* and enterococci [29]. Gram negative bacteria often attach by means of swarming or twitching motility over the new surface [30], some using twitching motility by Type IV pili [31, 32], and this might be particularly important in biofilm formation on urinary catheters. In addition, *Ps aeruginosa* uses a von Willebrand Factor-like surface factor in twitching motility over biomaterial surfaces [33].

Once bacteria have attached to the surface or conditioning film, they begin to proliferate and to develop intercellular adhesins such as polysaccharide intercellular adhesin (PIA) in staphylococci. This substance is integral to further development of biofilm, and is encoded by the *ABDC* operon, and regulated by *icaR*. At this stage, bacterial stress responses are operating in response to limitation of nutrients and oxygen and the biofilm phenotype is appearing [34]. It is important to note that the bacterial stress response, mediated by Sigma B, downregulates *icaR* and increases PIA production, and the stress response can be provoked by external factors such as antibiotics as well as nutrient starvation. Once the biofilm phenotype has developed, the biofilm is stable and is not susceptible to host immune activity or to antimicrobials. There is often a lag phase of about 14–28 days before the biofilm reaches functional maturity, during which it might be more susceptible to antimicrobials [35].

Clear understanding of the sequence of events and periods of risk is essential for effective planning of preventative measures.

3. Prevention of biofilm infections

3.1 Surgical considerations

Since the days of Semmelweis, Lister and others in the mid–to late 1800s, personal hygiene of the surgeon, aseptic technique and antisepsis have become accepted norms. Since the 1950s, when bacteria-laden operating room air was identified as a major factor in surgical infection [36], greatly improved practices and ventilation systems have made this a minor source. Two main forms of ventilation are in use in modern operating rooms: plenum, and laminar flow with high efficiency particulate air (HEPA) filtration. While it is clear that the numbers of airborne bacteria are significantly reduced when laminar flow is used [37] there has never been a clear causative link between either this reduction or the actual bacteria and surgical infection, leading the USA CDC to downgrade their initial recommendation [38]. More recently, reports have appeared of small but significantly increased infection rates when laminar flow is used [39, 40] and this appears to be due to flaws in its design and manner of use [41]. For most types of implant surgery, plenum (conventional) ventilation appears to be satisfactory so long as other precautions are taken (**Figure 4**).

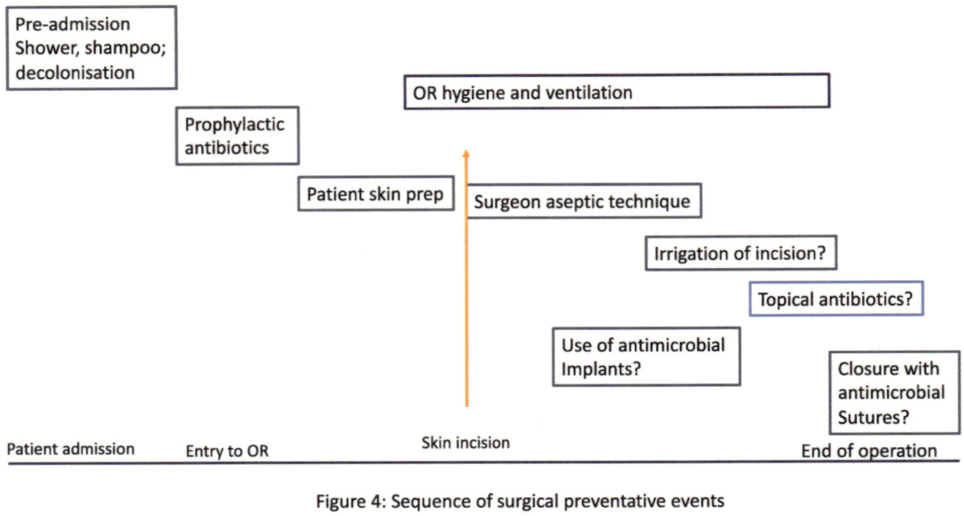

Figure 4.
Sequence of surgical preventative events.

Care bundles have been proposed for infection reduction in various healthcare settings. A bundle is a collection of interventions that are expected to contribute to reduced risk of infection, but which singly might have weak or no evidence base. A measure such as ensuring that only three people are present in the operating room during a procedure is not supported by any clear evidence but it is intuitively likely to be beneficial if only in reinforcing operating room discipline. A bundle must be directed towards behaviour change on the part of relevant staff members, and it works best if they contribute to its content, and formally agree to abide by it. Some bundles insist on contents being evidence-based, but the quality of evidence is usually very weak for individual components. However, when bundles are properly applied, they are often very effective in reducing surgical infection [42, 43] and in any case they and their contents should form part of a well-managed surgical discipline. Usually no single component can be identified to explain their success, but clinical trial evidence has shown that violations of the bundle are associated with re-emergence of infection [43].

As the major source of pathogens is the patient's skin, attention has been directed towards the effectiveness of preoperative skin preparation. Two main antiseptics are in use: chlorhexidine and povidone iodine. Each can be formulated in water or 70% alcohol. A report by the World Health Organisation (WHO) favouring chlorhexidine [44] has been called into question on the basis of quality of evidence [45]. However, sampling is usually by swabbing of the skin surface, and almost none of the many studies on surgical skin preparation explore the effectiveness of any agent on bacteria resident in the dermis, though an early study showed that full thickness skin biopsy was necessary [46]. This has since been confirmed [47, 48]. When skin biopsy is used, neither antiseptic in alcohol is able to eradicate resident skin bacteria, and though reduced, the remaining numbers are often sufficient to cause a biomaterial-associated infection [25]. Two studies on the penetration of both aqueous and alcoholic chlorhexidine into human skin using full thickness biopsy have found it to be minimal [49, 50]. Further measures are therefore necessary. Some researchers have investigated the effect of antiseptic-soaked material to protect the incision from the skin edges during surgery, and while this is commonly used, there have been no quantitative studies to show benefit. Intravenous antibiotics are almost universally used in surgery, ideally as a single dose 30–60 min before incision, but extra doses are commonly used postoperatively though they offer no benefit over that of the single pre-operative dose. Antibiotic prophylaxis is undoubtedly highly effective in reducing infection risk in many types of surgery, including colorectal surgery [51] and orthopaedic surgery [52] but probably less so in neurosurgery due to limited penetration of systemic antibiotics intracranially. However, it is probably inevitable that a small number of bacteria will reach the implant during operation, and further measures have been directed to attempts to eradicate these. As knowledge of attached bacteria and biofilms has shown that very high concentrations of antibiotics are necessary, some surgeons have used either antiseptic or antibiotic irrigation [53, 54], or have simply added antibiotic powder to the incision before closure [55–57] with successful reduction in infection rates and complications. This intervention gives extremely high local antibiotic levels not reachable by systemic administration, yet avoids most of the complications associated with the latter method.

3.2 Antimicrobial biomaterials

Other methods of prevention accept that despite efforts, bacteria will reach the implant, and aim to prevent their attachment or to kill them when attached. Various "anti-fouling" surfaces have been investigated with the aim of allowing host cell and tissue proliferation but preventing bacterial attachment [58, 59] but none of these has yet reached clinical application, largely because of the complex relationship between implant surface, host tissue environment, and bacterial surface adhesins. Biomaterials designed to kill bacteria that do attach to them have generally included coatings of silver, antiseptic or antibiotic and combinations of these, often with a vehicle to bind the antimicrobial to the biomaterial surface. Such coatings have several disadvantages. The normal host reaction to the implant of deposition of plasma proteins [26, 27] also obliterates the antimicrobial coating in many cases, making it ineffective. Silver is susceptible to this due its avidity for proteins [60], and it can also be inactivated by chloride [61] which is abundant in the human body. Silver ions have also been shown to be cytotoxic in certain conditions [62]. Clinical studies on silver-processed devices give very variable results, and there is doubt about their cost-effectiveness in wound dressings [63]. A recent randomised controlled trial of silver-containing catheters intended to reduce ventriculitis in people with hydrocephalus shunts found no difference from plain catheters [64]. Another randomised controlled trial of silver-processed urinary catheters again found no significant difference from plain catheters [65]. In both of these clinical settings, biofilms play a key role, and the goal is to prevent bacterial proliferation and biofilm development on the catheters. Both have fluid containing proteins and chloride flowing through them.

Another approach has been impregnation of catheter material with antimicrobials. Though the impregnation processes differ, two catheter types can be considered: those containing rifampicin and minocycline, and those containing rifampicin and clindamycin. The first type has been used in central venous catheters [66] and external ventricular drains [67]. The second type has been used in hydrocephalus shunts and external ventricular drains. In all cases they have shown effectiveness in reducing device -related infection. The advantage of impregnation over coatings is that they give a long duration of activity: coatings are usually washed away by fluid after a few days, whereas the surface of an impregnated material is continually replenished by

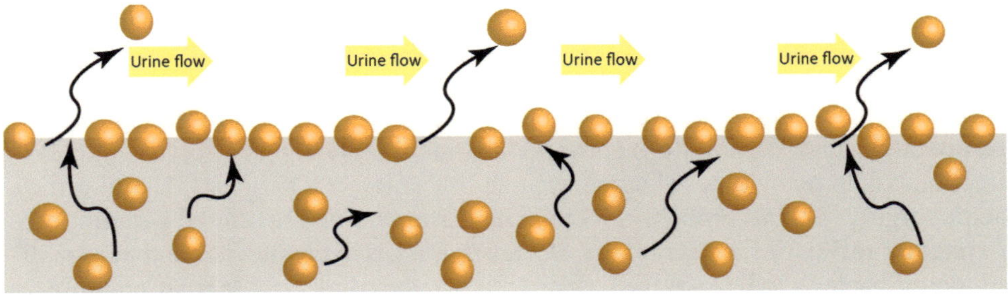

Figure 5.
Principle of impregnated biomaterial. Antimicrobial molecules are motile within the device matrix and can migrate to the surface to replace those removed by fluid flow.

migrating antimicrobials until the depot in the material is depleted, usually several weeks later (**Figure 5**). This is important when the implantable device is at risk of contamination for an extended period.

3.3 Importance of source of infection and period of risk

In order to formulate an effective preventive strategy, knowledge of the source and nature of device pathogens and the period during which the device is at risk is essential (**Table 1**). As many biofilm infections are caused by micro-organisms originating in or on the patient, a knowledge of the distribution of these is useful. The normal bacterial flora of the skin differs according to age and sex, but particularly depending on the anatomical site. The most common bacteria found on the skin are staphylococci, particularly members of the CoNS. These are typified by *S epidermidis* which is broadly distributed over the body surfaces, but other species such as *Staphylococcus*

Implant/device	Duration of use	Main source of pathogens	Period of risk	
			At insertion	During use
Hydrocephalus shunt	indefinite	Patient's skin	++	—
External ventricular drain	Few days-weeks	Patient's skin/ environment	±	++
Joint replacement	Indefinite	Patient's skin	++	±
Urinary catheter 1	<28 days	Patient/environment	±	++
Urinary catheter 2	~90 days	Patient/environment	±	++
Peritoneal dialysis catheter	Indefinite	Patient/environment	±	++
Vascular graft	Indefinite	Patient	++	+
Prosthetic heart valve	Indefinite	Patient	+	++
Spinal instrumentation	Indefinite	Patient	++	±
Venous access device	Days—months	Patient/environment	±	++
Sutures	Days	Patient/healthcare worker	+	±

Table 1.
Periods of risk of infection of common implantable devices.

capitis have preferred sites such as the head and neck. *C acnes* is an important pathogen in the context of implant infections, but it is a good example of the importance of specific topographical distribution in determining the important pathogens in particular implants. *C acnes* is found on the upper body and head (**Figure 6**) [68], and it is therefore not surprising that devices implanted in these areas show a significantly higher incidence of *C acnes* infection. Examples are neurosurgical shunts and drains [69, 70], spine instrumentation [71], breast implants [72] and shoulder arthroplasty [73, 74]. Implants in other sites such as urinary catheters are at risk from a different microbial profile, as the pathogens originate in the large intestine, and *E coli*, *Klebsiella pneumoniae* and *P mirabilis* are the most common.

The time at which the implant is at risk of microbial contamination also varies. While there is always a risk at the time of implantation, in some implants this is the main time, and the risk of subsequent contamination is proportionally small. Examples of this are hydrocephalus shunts and joint replacements. In other implants the risk at insertion is significantly outweighed by that during use. Examples are external ventricular drains (EVD) for raised intracranial pressure, urinary catheters, venous access catheters and peritoneal dialysis catheters, all of which can be contaminated from environmental sources or from the hands of staff or users during use. Other examples are vascular grafts and prosthetic heart valves, which are at risk from hematogenous seeding from bacteria entering the bloodstream at a distant site.

When planning strategies for prevention of biofilm infections involving antimicrobials, it is therefore important to match the antimicrobial to the most likely pathogen(s). If systemic antimicrobial prophylaxis is contemplated, then the adverse effects of this must be taken into consideration if there is a need for prolonged

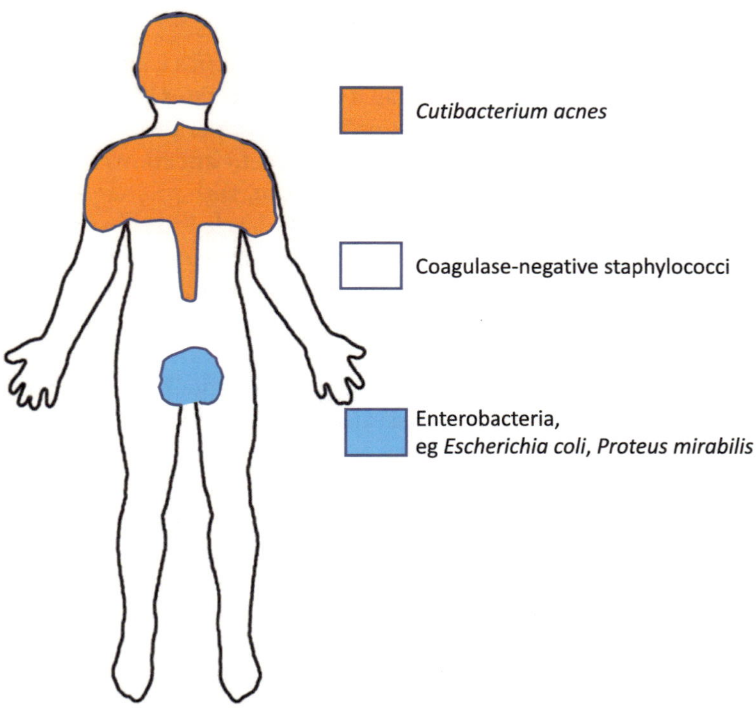

Figure 6.
Topographical distribution of common biofilm pathogens (after Grice et al. [68]).

administration due to extended period of risk. If antimicrobial materials or devices are to be used, these must address not only the likely pathogen(s) but also the duration of protective activity required.

International guidelines indicate that for most surgical procedures, any systemic antimicrobial prophylaxis should be administered as one dose 30–60 min before start of surgery [75, 76]. Extension of this prophylaxis beyond 24 hours does not reduce surgical infection further, but it does increase the incidence of acute kidney injury and *Clostridioides difficile* infection [77], which is a life-threatening colitis associated with over-use of antibiotics. Where the period of risk extends beyond the insertion procedure, such as in EVD, long courses of systemic antibiotics are often given until the drain is removed. This has been shown in some cases to reduce brain infections, but at a cost. A randomised study comparing the use of plain catheters and prolonged systemic antibiotics with antimicrobial-impregnated catheters and one dose of antibiotic at insertion found no difference in the brain infection rate, which was low in each group, but there were three cases of *C difficile* infection in the prolonged antibiotics group, one patient requiring total colectomy [78].

4. Treatment of biofilm infections

The difficulty in treating biofilm infections in surgery emphasises the importance of effective prevention. However, this is not always possible. The nature of the biofilm phenotype and its implications for antibiotic treatment mean that further surgery is almost inevitable, and this usually involves removal of the device. This might be relatively simple, as in the case of a venous access catheter or a urinary catheter, but it can be both surgically complicated and hazardous, as in the case of spinal instrumentation or prosthetic heart valves.

Attempts to eradicate established biofilm with antibiotics usually fail. A comparison of treatment regimens for hydrocephalus shunt infections showed that results with shunt removal and antibiotics were significantly superior to those with antibiotics alone [79]. Successful treatment of joint replacement infections relies on device removal and extensive debridement of infected tissue, with prolonged antibiotic therapy. However, understanding of biofilm biology has led to advances in this area. The biofilm phenotype takes a few weeks to "mature" to the point where full insusceptibility to antibiotics is expressed, and this has been exploited in development of a regimen for treatment of prosthetic joint infection when the diagnosis can be made within 3–4 weeks of insertion [80]. In this regimen, known as Debridement, Antibiotics and Implant Retention (DAIR), surgical treatment of the infected joint prosthesis is carried out on a planned basis after careful investigation to establish the causative micro-organism and its antimicrobial susceptibilities, to allow consultation with specialists including Microbiology/Infectious Diseases, and to determine that the implant is stable (**Figure 7**). Infections due to multi-drug-resistant bacteria, fungi or multiple bacteria are not suitable for this approach. During the operation, the prosthetic components are exposed and the acetabular module is removed, leaving the main metal prosthesis in place. All infected tissue is removed and samples are sent for microbiological examination. Copious irrigation with antiseptic is applied, and biodegradable antibiotic—eluting beads can be inserted to provide high local concentrations. The choice of antibiotic in the beads should be made in consultation with a microbiologist. The joint is then closed and a long postoperative course of suitable antibiotics is then started [81]. The success rate of DAIR compared to conventional

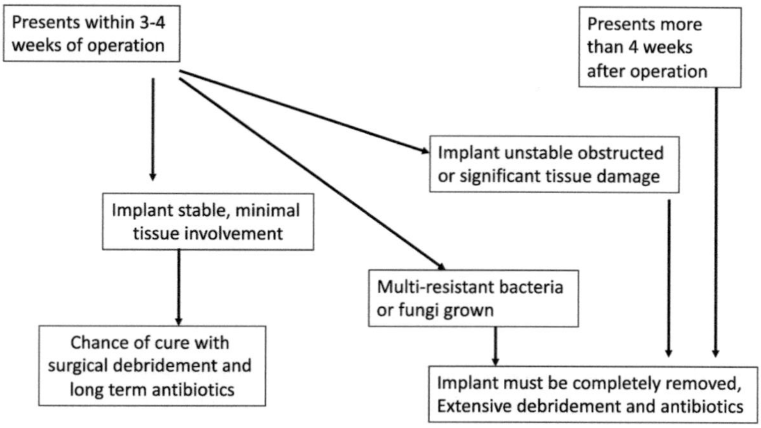

Figure 7.
Possibility of retention of infected implant based on knowledge of biofilm phenotype maturation (based on Zimmerli and Trampuz, 2004) [80].

full implant removal and replacement is slightly lower. Moreover, despite the very thorough surgical debridement and long courses of antibiotics, often for over a year, relapse can occur [82], illustrating the difficulty in eradication of biofilms. DAIR spares the patient the much more extensive surgical removal of the main implant components, and the second surgery to inert fresh implants a few weeks later.

5. Diagnosis of biofilm infections

5.1 Clinical features

Most biofilm infections in surgery are chronic and persistent, sometimes for many years [83]. It is important to distinguish between "late infection," implying an infection contracted long after surgery, such as hematogenously, and "delayed infection," meaning that the infection appears long after surgery even though it was contracted at the operation. Delayed infection in spine instrumentation is usually due to infection with CoNS or *C acnes* [84, 85]. A similar situation is found in shoulder arthroplasty [86]. Generally, more virulent bacteria such as *S aureus* are associated with either early-presenting or with hematogenous infections. The delay of months or years between initial surgical implantation and appearance of symptoms [84] has led to doubt about the surgical origins of some infections but this has now been largely dispelled. However, the need for prolonged follow-up and vigilance must be emphasised.

Acute postoperative biofilm infections usually appear within days or weeks of surgery, with failure of wound healing, drainage of pus or other fluid from the wound, local pain and swelling, fever and general illness. Delayed or chronic infections of joint prostheses present with persistent pain and restricted mobility, local swelling and sometimes a sinus. In the absence of a sinus, diagnosis might be delayed as it is often difficult to distinguish infective from mechanical complications. Aspiration of synovial fluid often gives a diagnosis but sensitivity is low [87, 88]. Delayed infection in spine instrumentation similarly presents with persistent pain, tenderness and possibly a draining sinus. Delayed infections in hydrocephalus shunts are very uncommon now that the preferred route of drainage is to the abdomen (ventriculoperitoneal, VP), but the ventriculo-atrial (VA) route is still used in some cases. In VP shunts infection usually presents within a few months as it leads to obstruction, but this does not happen in VA shunts and symptoms might not appear, or at least become recognisable, for several years. During this time, bacteria are being discharged from the biofilm in the shunt into the bloodstream, and this might give rise to periods of ill-health or sporadic fevers. It also provokes production of antibodies to the bacteria, and eventually the concentrations of circulating antigen and antibody, and therefore immune complexes, become so high that they precipitate on basement membranes of joints, renal glomeruli, alveoli and microvascular system. The presenting clinical picture can therefore be a confusing array of disorders from hematuria, hemorrhagic skin rashes, arthropathy, and chronic cough [89, 90]. Clinical diagnosis can therefore be very difficult, and a high level of suspicion is needed. Aspiration of cerebrospinal fluid from the shunt often gives the diagnosis, but blood cultures can be negative in the later stages.

5.2 Laboratory methods

Depending on the site of the infection and presence of an implant, sometimes blood cultures are positive, indicating systemic spread of the infection, and risk of sepsis. Blood inflammatory markers such as erythrocyte sedimentation rate (ESR) and C-reactive protein (CRP) levels are usually raised. Swab cultures from the wound might yield the infecting pathogen, but they might be misleading due to contamination [91]. Surgical exploration of the incision and deeper layers allows tissue samples to be taken and these are more likely to yield the pathogen(s). Such samples should always be taken during debridement surgery [92], using fresh instruments for each of up to six separate samples [81, 93]. In view of the anaerobic preference of *C acnes* and its slow growth, cultures should be incubated anaerobically for up to 10 days [94]. The way in which tissue samples are processed in the laboratory is important. Simply rubbing them on a culture plate or incubating them in a fluid culture is prone to contamination and gives poor yield, leading to under-diagnosis of infection. Tissue should be homogenised but the method of doing this is also important [95]. When hardware such as joint replacement or spinal instrumentation components are removed, these should be seen as valuable samples. Sonication to remove the biofilm has been shown to significantly increase the culture positivity rate [96, 97]. A further aid to laboratory diagnosis has been PCR [98] especially when applied to tissue homogenates or hardware sonicates. However, if PCR is used in an attempt to certify eradication of infection before re-insertion of a prosthesis, residual DNA from bacteria successfully killed by antibiotic therapy can give false positive results suggesting ongoing active infection. This can be overcome by use of a modified PCR method that detects DNA only from live bacteria [99].

6. Conclusions

The impact of biofilm infections in surgery on healthcare systems, economies and personal lives of patients is immense. The financial cost can only be estimated and published figures do not usually take into account "unseen" costs such as loss of earnings due to disability, increased dependency, and financial burden on carers.

The physical and mental trauma of surgery such as joint replacement, reconstructive breast implant or hydrocephalus treatment can be made unimaginably worse by postoperative biofilm infection.

The significant difficulty in successfully treating biofilm infections with antibiotics, due largely to the biofilm phenotype, is now well recognised, and the importance of commensal bacteria previously thought to be harmless, such as *S epidermidis* and *C acnes*, is becoming more widely known. However, surgical device removal remains the mainstay of treatment, and new approaches that allow implant retention are needed. Prevention of biofilm infections is crucial, and biomaterials that either reduce bacterial attachment, such as those coated with novel synthetic polymers [100] or those designed to kill bacteria on contact [66, 67] are now in clinical use. Many other biomaterials approaches are in development, and considerable strides have been made in this direction but further progress is being slowed by unrealistic commercial and regulatory barriers [101].

Author details

Roger Bayston
School of Medicine, University of Nottingham, Nottingham, United Kingdom

*Address all correspondence to: roger.bayston@nottingham.ac.uk

References

[1] Høiby N. *Pseudomonas aeruginosa* infection in cystic fibrosis. Relationship between mucoid strains of *Pseudomonas aeruginosa* and the humoral immune response. Acta Pathologica et Microbiologica Scandinavica. 1974;**82B**:551-558

[2] Costerton JW, Cheng K-J, Geesey GG, Ladd TI, Nickel JC, Dasgupta M, et al. Bacterial biofilms in nature and disease. Annual Reviews in Microbiology. 1987;**41**:435-464. DOI: 10.1146/annurev.mi.41.100187.002251

[3] Mack WN, Mack JP, Ackerson AO. Microbial film development in a trickling filter. Microbial Ecology. 1975;**2**:215-226. DOI: 10.1007/BF02010441

[4] Bayston R, Penny SR. Excessive production of mucoid substance in Staphylococcus SIIA: A possible factor in colonisation of Holter shunts. Developmental Medicine & Child Neurology. 1972;**14**(Suppl 27):25-28. DOI: 10.1111/j.1469-8749.1972.tb09769.x

[5] Mack D, Fischer W, Krokotsch A, Leopold K, Hartmann R, Egge H, et al. The intercellular adhesin involved in biofilm accumulation of Staphylococcus epidermidis is a linear beta-1,6-linked glycosaminoglycan: Purification and structural analysis. Journal of Bacteriology. 1996;**178**:175-183. DOI: 10.1128/jb.178.1.175-183.1996

[6] Bjarnsholt T, Alhede M, AlhedeM, Eickhardt-SørensenSR, MoserC, Kühl M, et al. The in vivo biofilm. Trends in Microbiology. 2013;**21**:466-474. DOI: 10.1016/j.tim.2013.06.002

[7] Proctor R. Bacterial energetics and antimicrobial resistance. Drug Resistance Updates. 1998;**1**:227-235. DOI: 10.1016/s1368-7646(98)80003-4

[8] Proctor RA, Kriegeskorte A, Kahl BC, Becker K, Löffler B, Peters G. Staphylococcus aureus small colony variants (SCVs): A road map for the metabolic pathways involved in persistent infections. Frontiers in Cellular and Infection Microbiology. 2014;**4**:1-8. DOI: 10.3389/fcimb.2014.00099

[9] Vann JM, Proctor RA. Cytotoxic effects of ingested *Staphylococcus aureus* on bovine endothelial cells: Role of *S aureus* α-hemolysin. Microbial Pathogenesis. 1988;**4**:443-453. DOI: 10.1016/0882-4010(88)90029-0

[10] Balwit JM, van Langevelde P, Vann JM, Proctor RA. Gentamicin-resistant menadione and hemin auxotrophic *Staphylococcus aureus* persist within cultured endothelial cells. The Journal of Infectious Diseases. 1994;**170**:1033-1037. DOI: 10.1093/infdis/170.4.1033

[11] Pelletier LL, Richardson M, Feist M. Virulent gentamicin-induced small colony variants of *Staphylococcus aureus*. The Journal of Laboratory and Clinical Medicine. 1979;**94**:324-334

[12] Vulin C, Leimer N, Huemer M, Ackerman M, Zinkernagel AS. Prolonged bacterial lag time results in small colony variants that represent a sub-population of persisters. Nature Communications. 2018;**9**:4074. DOI: 10.1038/s41467-018-06527-0

[13] Bryan LE, Kwan S. Aminoglycoside-resistant mutants of *Pseudomonas aeruginosa* deficient in cytochrome d, nitrate reductase, and aerobic transport. Antimicrobial Agents and Chemotherapy. 1981;**19**:958-964. DOI: 10.1128/aac.19.6.958

[14] Colwell CA. Small colony variants of *Escherichia coli*. Journal of Bacteriology. 1946;**52**:417-422. DOI: 10.1128/jb.52.4.417-422.1946

[15] Park YJ, Phuong NL, Pinto NA, Kwon JM, D'Souza R, Byun J-H, et al. Urinary tract infection caused by a small colony variant form of capnophilic *Escherichia coli* leading to misidentification and non-reactions in antimicrobial susceptibility tests. Antimicrobial Resistance and Infection Control. 2018;**7**:139. DOI: 10.1186/s13756-018-0438-6

[16] Kikuchi M, Suzuki Y, Okada S, Sato A, Kengo O, Matsumoto T. Septic shock caused by a carbon dioxide-dependent and extended spectrum lactamase-producing Proteus mirabilis small colony variant in a long -term bedridden patient. Journal of Infection and Chemotherapy. 2022;**28**:455-458. DOI: 10.1016/j.jiac.2021.12.013

[17] Marcus RJ, Post JC, Stoodley P, Hall-Stoodley L, RL MG, Sureshkumar KK, et al. Biofilms in nephrology. Expert Opinion on Biological Therapy. 2008;**8**:1159-1166. DOI: 10.1517/14712598.8.8.1159

[18] Stickler DJ. Clinical complications of urinary catheters caused by crystalline biofilms: Something needs to be done. Journal of Internal Medicine. 2014;**276**:120-129. DOI: 10.1111/joim.12220

[19] Durack DT. Experimental bacterial endocarditis. IV. Structure and evolution of very early lesions. The Journal of Pathology. 1975;**115**:81-89. DOI: 10.1002/path.1711150204

[20] Jung C-J, Yeh C-Y, Shun C-T, Hsu R-B, Cheng H-W, Lin C-S, et al. Platelets enhance biofilm formation and resistance of endocarditis-inducing streptococci on the injured heart valve. The Journal of Infectious Diseases. 2012;**205**:1066-1075. DOI: 10.1093/infdis/jis021

[21] Maduka-Ezeh A, Seville MT, Kusne S, Vikram HR, Blair JE, Greenwood-Quaintance K, et al. Thymidine auxotrophic *Staphylococcus aureus* small-colony variant endocarditis and left ventricular assist device infection. Journal of Clinical Microbiology. 2012;**50**:1102-1105. DOI: 10.1128/JCM.01170-11

[22] Benes J, Dzupova O, Setina M, Feuereisl R, Svec P, Pantucek R. Relapsing endocarditis caused by *Enterococcus faecalis* forming small colony variants. Scandinavian Journal of Infectious Diseases. 2013;**45**:800-803. DOI: 10.3109/00365548.2013.800227

[23] Madison G, Golamari R, Bhattacharya P. Endocarditis caused by Abiotrophia and Granulicatella species. In: Firstenberg MS, editor. Advanced Concepts in Endocarditis. London: IntechOpen; 2018. DOI: 10.5772/intechopen.74252

[24] Thwaites GE, Gant V. Are bloodstream leucocytes Trojan horses for the metastasis of *Staphylococcus aureus*? Nature Reviews. Microbiology. 2011;**9**:215-222. DOI: 10.1038/nrmicro2508

[25] Elek SD, Conen PE. The virulence of *Staphylococcus pyogenes* for man. A study of the problems of wound infection. British Journal of Experimental Pathology. 1957;**38**:573-586

[26] Gristina AG. Biomaterial-centered infection: Microbial adhesion versus tissue integration. Science. 1987;**237**:1588-1595. DOI: 10.1126/science.3629258

[27] Vaudaux P, Pittet D, Haeberli A, Lerch PG, Morgenthaler JJ, Proctor RA, et al. Fibronectin is more active than fibrin

of fibrinogen in promoting *Staphylococcus aureus* adherence to inserted intravascular catheters. The Journal of Infectious Diseases. 1993;**167**:633-641. DOI: 10.1093/infdis/167.3.633

[28] Foster TJ. The MSCRAMM family of Cell-wall_-anchored surface proteins of Gram positive cocci. Trends in Microbiology. 2019;**27**:927-941. DOI: 10.1016/j.tim.2019.06.007

[29] Arrecubieta C, Lee M-H, Macey A, Foster TJ, Lowy FD. SdrF, a Staphylococcus epidermidis surface protein, binds type I collagen. The Journal of Biological Chemistry. 2007;**282**:18767-18776. DOI: 10.1074/jbc.M610940200

[30] Harshey RM. Bacterial motility on a surface: Many ways to a common goal. Annual Review of Microbiology. 2003;**57**:249-273. DOI: 10.1146/annurev.micro.57.030502.09101

[31] Merz AJ, So M, Sheetz MP. Pilus retraction powers bacterial twitching motility. Nature. 2000;**407**:98-102

[32] Jarrell KF, Stark M, Nair DB, Chong JP. Flagella and pili are both necessary for efficient attachment of *Methanococcus maripaludis* to surfaces. FEMS Microbiology Letters. 2011;**319**:44-50. DOI: 10.1111/j.1574-6968.2011.02264.x

[33] Kuchma SL, Ballok AE, Merritt JH, Hammond JH, Lu W, Rabinowitz JD, et al. Cyclic-di-GMP-mediated repression of swarming motility by *Pseudomonas aeruginosa*: The *pilY1* gene and its impact on surface-associated behaviors. Journal of Bacteriology. 2010;**192**:2950-2964. DOI: 10.1128/JB.01642-09

[34] Costerton JW, Stewart PS, Greenberg EP. Bacterial biofilms: A common cause of persistent infections. Science. 1999;**284**:1318-1322. DOI: 10.1126/science.284.5418.1318

[35] Widmer AF. New developments in diagnosis and treatment of infection in orthopedic implants. Clinical Infectious Diseases. 2001;**33**(Suppl 2):S94-S106. DOI: 10.1086/321863

[36] Blowers R, Mason GA, Wallace KR, Walton M. Control of wound infection in a thoracic surgery unit. The Lancet. 1955;**266**:786-794. DOI: 10.1016/S0140-6736(55)92385-0

[37] Babb JR, Lynam P, Ayliffe GAJ. Risk of airborne transmission in an operating theatre containing four ultraclean air units. Journal of Hospital Infection. 1995;**31**:159-168. DOI: 10.1016/0195-6701(95)90062-4

[38] Sehulster L, Chinn RYW. Guidelines for environmental infection control in health-care facilities. MMWR Recommendations and Report. 2003;**52**(RR10):1-42

[39] Brandt C, Hott U, Sohr D, Daschner F, Gastmeier P, Rüden H. Operating room ventilation with laminar airflow shows no protective effect on the surgical site infection rate in orthopedic and abdominal surgery. Annals of Surgery. 2008;**248**:695-700. DOI: 10.1097/SLA.0b013e31818b757d

[40] Hooper GJ, Rothwell AG, Frampton C, Wyatt MC. Does the use of laminar flow and space suits reduce early deep infection after total hip and knee replacement? Journal of Bone and Joint Surgery. 2011;**93-B**:85-90. DOI: 10.1302/0301-620X.93B1

[41] Jain S, Reed M. Laminar air flow handling systems in the operating room. Surgical Infections. 2019;**20**:1-8. DOI: 10.1089/sur.2018.258

[42] Choksey MS, Malik IA. Zero tolerance to shunt infections: Can it be achieved? Journal of Neurology,

Neurosurgery, and Psychiatry. 2004;**75**: 87-91

[43] Korinek A-M, Reina M, Boch AL, Rivera AO, de Bels D, Puybasset L. Prevention of external ventricular drain—Related ventriculitis. Acta Neurochirurgica (Wien). 2005;**147**:39-46. DOI: 10.1007/s00701-004-0416-z

[44] Allegranzi B, Bischoff P, de Jonge S, Kubilay NZ, Zayed B, Gomes SM, et al. New WHO recommendations on preoperative measures for surgical site infection prevention: An evidence-based global perspective. The Lancet Infectious Diseases. 2016;**16**:e276-e287. DOI: 10.1016/S1473-3099(16)30398-X

[45] Allegranzi B, Widmer A. WHO's recommendation for surgical skin antisepsis is premature. The Lancet Infectious Diseases. 2017;**17**:1023-1024

[46] Selwyn S, Ellis H. Skin bacteria and skin disinfection reconsidered. British Medical Journal. 1972;**5792**:136-140. DOI: 10.1136/bmj.1.5793.136

[47] Sato S, Sakuragi T, Dan K. Human skin flora as a potential source of epidural abscess. Anesthesiology. 1996;**85**:1276-1282. DOI: 10.1097/00000542-199612000-00008

[48] Lee MJ, Pottinger PS, Butler-Wu S, Bumgarner RE, Russ SM, Matsen FA. Propionibacterium persists in the skin despite standard surgical preparation. Journal of Bone and Joint Surgery. 2014;**96**:1447-1450. DOI: 10.2106/JBJS.M.01474

[49] Karpanen TJ, Worthington T, Conway BR, Hilton AC, Elliott TSJ, Lambert PA. Penetration of chlorhexidine into human skin. Antimicrobial Agents and Chemotherapy. 2008;**52**:3633-3636. DOI: 10.1128/AAC.00637-08

[50] Karpanen TJ, Worthington T, Conway BR, Hilton AC, Elliott TSJ, Lambert PA. Permeation of chlorhexidine from alcoholic and aqueous solutions within excised human skin. Antimicrobial Agents and Chemotherapy. 2009;**53**:1717-1719. DOI: 10.1128/AAC.01289-08

[51] Nelson RL, Gladman E, Barbateskovic M. Antimicrobial prophylaxis for colorectal surgery. Cochrane Database of Systematic Reviews. 2014;**5**. DOI: 10.1002/14651858.CD001181.pub4

[52] Prokuski L. Prophylactic antibiotics in orthopaedic surgery. The Journal of the American Academy of Orthopaedic Surgeons. 2008;**16**:283-293. DOI: 10.5435/00124635-200805000-00007

[53] Norman G, Atkinson RA, Smith TA, Rowlands C, Rithalia AD, Crosbie EJ, et al. Intracavity lavage and wound irrigation for prevention of surgical site infection. Cochrane Database of Systematic Reviews. 2017;**10**:CD012234. DOI: 10.1002/14651858.CD012234.pub2

[54] Papadakis M. Wound irrigation for preventing surgical site infections. World Journal of Methodology. 2021;**11**:222-227. DOI: 10.5662/wjm.v11.i4.222

[55] Abdullah KG, Attiah MA, Olsen AS, Richardson A, Lucas TH. Reducing surgical site infections following craniotomy: Examination of the use of topical vancomycin. Journal of Neurosurgery. 2015;**123**:1600-1604. DOI: 10.3171/2014.12.JNS142092

[56] Beckman JM, Amankwah EK, Tetreault LL, Perlman SA, Tuite GF. Complications associated with bacitracin powder in surgical wounds. Journal of Neurosurgery: Pediatrics. 2015;**16**:719-725. DOI: 10.3171/2015.4.PEDS14699

[57] Haimoto S, Schar RT, Nishimura Y, Hara M, Wakabayashi T, Ginsberg HJ. Reduction in surgical site infection with suprafascial intrawound application of vancomycin powder in instrumented posterior spinal fusion: A retrospective case-control study. Journal of Neurosurgery. Spine. 2018;**29**:193-198. DOI: 10.3171/2017.12.SPINE17997

[58] Francolini I, Vuotto C, Piozzi A, Donelli G. Antifouling and antimicrobial biomaterials: An overview. Acta Pathologica, Microbiologica et Immunologica Scandinavica. 2017;**125**:392-417. DOI: 10.1111/apm.12675

[59] Ozcan E, Mondal A, Douglass M, Hopkins SP, Garren M, Devine R, et al. Bioinspired ultra-low fouling coatings on medical devices to prevent device-associated infections and thrombosis. Journal of Colloid and Interface Science. 2022;**608**:1015-1024. DOI: 10.1016/j.jcis.2021.09.183

[60] Mulley G, Jenkins ATA, Waterfield NR. Inactivation of the antibacterial and cytotoxic properties of silver ions by biologically relevant compounds. PLoS One. 2014;**9**:e94409. DOI: 10.1371/journal.pone.0094409

[61] Li Y, Zhao J, Shang E, Xia X, Niu J, Crittenden J. Effects of chloride ions on dissolution, ROS generation, and toxicity of silver nanoparticles under UN irradiation. Environmental Science & Technology. 2018;**52**:4842-4849. DOI: 10.1021/acs.est.7b04547

[62] Hemmerlein JB, Treretola SO, Kraus MA, Mendonca MS, Desmond LA. In vitro cytotoxicity of silver-impregnated collagen cuffs designed to decrease infection in tunneled catheters. Radiology. 1997;**204**:363-367

[63] Anon. Silver dressings-do they work? Drug and Therapeutics Bulletin. 2010;**48**:38-42

[64] Mallucci CL, Jenkinson MD, Conroy EJ, Hartley JC, Brown M, Dalton J, et al. Antibiotic or silver versus standard ventriculoperitoneal shunts (BASICS: A multicentre, single-blinded, randomised trial and economic evaluation). Lancet. 2019;**394**:1530-1539. DOI: 10.1016/S0140-6736(19)31603-4

[65] Pickard R, Lam T, Maclennan G, Starr K, Kilonzo M, McPherson G, et al. Antimicrobial catheters for reduction of symptomatic urinary tract infection in adults requiring short-term catheterisation in hospital: A multicentre randomised controlled trial. Lancet. 2012;**380**:1927-1935. DOI: 10.1016/S0140-6736(12)61380-4

[66] Darouiche RO, Berger DH, Khardori N, Robertson CS, Wall MJ, Metzler MH, et al. Comparison of antimicrobial impregnation with tunneling of longterm central venous catheters. A randomized controlled trial. Annals of Surgery. 2005;**242**:193-200. DOI: 10.1097/01.sla.0000171874.29934.61

[67] Zabramski JM, Whiting D, Darouiche RO, Horner TG, Olson J, Robertson C, et al. Efficacy of antimicrobial -impregnated external ventricular drain catheters: A prospective randomized, controlled trial. Journal of Neurosurgery. 2003;**98**:725-730. DOI: 10.3171/jns.2003.98.4.0725

[68] Grice EA, Kong HH, Conlan S, Deming CB, Davies J, Young AC, et al. Topographical and temporal diversity of the human skin microbiome. Science. 2009;**324**:1190-1192. DOI: 10.1126/science.1171700

[69] Viravaghavan R, Jantausch B, Campos J. Late-onset central nervous system shunt infections with *Propionibacterium acnes*: Diagnosis and management. La Clinica Pediatrica. 2004;**43**:393-397

[70] Carneiro IM, Pereira AS, Pinto S, Prata F, Faria CC, Marques JC. *Propionibacterium acnes*: Cause of cerebrospinal fluid shunt infection. Pediatric Infectious Disease. 2018;**37**:e168-e169. DOI: 10.1097/INF.0000000000001786

[71] Shiono Y, Nagai S, Kakinuma H, Sasaki A, Funao H, Kuramoto T, et al. Delayed *Propionibacterium acnes* surgical site infections occur only in the presence of an implant. Scientific Reports. 2016;**6**:1-10. DOI: 10.1038/srep32758

[72] Lee M, Ponjara G, McLeod K, Chong S. Breast implant illness: A biofilm hypothesis. Plastic and Reconstructive Surgery. Global Open. 2020;**8**:e2755. DOI: 10.1097/GOX.0000000000002755

[73] Kadler BK, Mehta SS, Funk L. *Propionibacterium acnes* infection after shoulder surgery. International Journal of Shoulder Surgery. 2015;**9**:139-144. DOI: 10.4103/0973-6042.167957

[74] Millet PJ, Yen Y-M, Price CS, Horan MP, van der Meijden OA, Elser F. *Propionibacterium acnes* infection as an occult cause of postoperative shoulder pain. Clinical Orthopaedics and Related Research. 2011;**469**:2824-2830. DOI: 10.1007/s11999-011-1767-4

[75] McDonald M, Grabsch E, Marshall C, Forbes A. Single -versus multiple-dose antimicrobial prophylaxis for major surgery: A systematic review. ANZ Journal of Surgery. 1998;**68**:388-396. DOI: 10.1111/j. 1445-2197.1998.tb04785.x

[76] Bratzler DW, Dellinger EP, Olsen KM, Perl TM, Auwaerter PG, Bolon MK, et al. Clinical Practice Guidelines for antimicrobial prophylaxis in surgery. American Journal of Health-System Pharmacy. 2013;**70**:195-283. DOI: 10.2146/ajhp120568

[77] Branch-Elliman W, O'Brien W, Strymish J, Itani K, Wyatt C, Gupta K. Association of duration and type of surgical prophylaxis with antimicrobial-associated adverse events. JAMA Surgery. 2019;**154**:590-598. DOI: 10.1001/jamasurg.2019. 0569

[78] Wong GKC, Ip M, Poon WS, Mak CWK, Ng RYT. Antibiotics-impregnated ventricular catheter versus systemic antibiotics for prevention of nosocomial CSF and non-CSF infections: A prospective randomised clinical trial. Journal of Neurology, Neurosurgery, and Psychiatry. 2010;**81**:1064-1067. DOI: 10.1136/jnnp.2009.198523

[79] James HE, Walsh JW, Wilson HD, Connor JD, Bean JR, Tibbs PA. Prospective randomized study of therapy in cerebrospinal fluid shunt infection. Neurosurgery. 1980;7:459-463. DOI: 10.1227/00006123-198011000-00006

[80] Zimmerli W, Trampuz A, Ochsner PE. Prosthetic-joint infections. The New England Journal of Medicine. 2004;**351**:1645-1654. DOI: 10.1056/nejmra040181

[81] Vaz K, Taylor A, Kendrick B, Alvand A. A guide to debridement, antibiotics, and implant retention. Annals of Joint. 2022;**7**:1-5. DOI: 10.21037/aoj-20-89

[82] Byren I, Bejon P, Atkins BL, Angus B, Masters S, McLardy-Smith P, et al. One hundred and twelve infected arthroplasties treated with "DAIR" (debridement, antibiotics and implant retention): Antibiotic duration and outcome. The Journal of Antimicrobial Chemotherapy. 2009;**63**:1264-1271. DOI: 10.1093/jac/dkp107

[83] Bjarnsholt T. The role of biofilms in chronic infections. Acta Pathologica, Microbiologica et Immunologica

Scandinavica. 2013;**121**:1-58. DOI: 10.1111/apm.12099

[84] Richards BR, Emara KM. Delayed infections after posterior TSRH spinal instrumentation for idiopathic scoliosis: Revisited. Spine. 2001;**26**:1990-1996. DOI: 10.1097/00007632-200109150-00009

[85] Hahn F, Zbinden R, Min K. Late implant infections caused by *Propionibacterium acnes* in scoliosis surgery. European Spine Journal. 2005;**14**:783-788. DOI: 10.1007/s00586-004-0854-6

[86] Kanafani ZA, Sexton DJ, Pien BC, Varkey J, Basmania C, Kaye KS. Postoperative joint infections due to *Propionibacterium* species: A case control study. Clinical Infectious Diseases. 2009;**49**:1083-1085. DOI: 10.1086/605577

[87] Yee DKH, Chiu KY, Yan CH, Ng FY. Review article: Joint aspiration for diagnosis of periprosthetic infection. Journal of Orthopaedic Surgery. 2013;**21**:236-240. DOI: 10.1177/230949901302100225

[88] Chalmers PN, Sporer SM, Levine BR. Correlation of aspiration results with periprosthetic sepsis in revision total hip arthroplasty. The Journal of Arthroplasty. 2014;**29**:438-442. DOI: 10.1016/j.arth.2013.06.020

[89] Bayston R, Swinden J. The aetiology and prevention of shunt nephritis. Zeitschrift für Kinderchirurgie. 1979;**28**:377-384

[90] Legoupil N, Ronco P, Berenbaum F. Arthritis -related shunt nephritis in an adult. Rheumatology. 2003;**42**:698-699. DOI: 10.1093/rheumatology/keg151

[91] Aggrawal VK. Swab cultures are not as effective as tissue cultures for diagnosis of periprosthetic joint infection. Clinical Orthopaedics and Related Research. 2013;**471**:3196-3203

[92] Parvizi J, Gehrke T, Chen AF. Proceedings of the international consensus on periprosthetic joint infection. Bone and Joint Journal. 2013;**95B**:1450-1452

[93] Atkins BL, Athanasou N, Deeks JJ, Crook DWM, Simpson H, Peto TEA, et al. Prospective evaluation of criteria for microbiological diagnosis of prosthetic-joint infection at revision arthroplasty. Journal of Clinical Microbiology. 1998;**36**:2932-2939

[94] Butler-WuSM,BurnsEM,PottingerPS, Magaret AS, Rakeman JL, Matsen FA, et al. Optimization of periprosthetic culture for diagnosis of *Propionibacterium acnes* prosthetic joint infection. Journal of Clinical Microbiology. 2011;**49**:2490-2495. DOI: 10.1128/JCM.00450-11

[95] Askar M, Ashraf W, Scammell B, Bayston R. Comparison of different human tissue processing methods for maximization of bacterial recovery. European Journal of Clinical Microbiology & Infectious Diseases. 2019;**38**:149-155. DOI: 10.1007/s10096-018-3406-4

[96] Trampuz A, Piper KE, Mj J, Hanssen AD, Unni KK, Osmon DR, et al. Sonication of removed hip and knee prostheses for diagnosis of infection. The New England Journal of Medicine. 2007;**357**:654-663

[97] Yano MH, Klautau GB, da Silva CB, Nigro S, Avanzi O, Mercadente MT, et al. Improved diagnosis of infection associated with osteosynthesis by use of sonication of fracture fixation implants. Journal of Clinical Microbiology. 2014;**52**:4176-4182. DOI: 10.1128/JCM.02140-14

[98] Qu X, Zhai Z, Li H, Liu X, Zhu Z, Wang Y, et al. PCR-based diagnosis of

prosthetic joint infection. Journal of Clinical Microbiology. 2013;**51**:2742-2746. DOI: 10.1128/JCM.00657-13

[99] Askar M, Sajid M, Nassif Y, Ashraf W, Scammell B, Bayston R. Propidium monoazide-polymerase chain reaction for detection of residual periprosthetic joint infection in two-stage revision. Molecular Biology Reports. 2019;**46**:6463-6470. DOI: 10.1007/s11033-019-05092-z

[100] Hook AL, Chang C-Y, Yang J, Lucket J, Cockayne A, Atkinson S, et al. Combinatorial discovery of polymers resistant to bacterial attachment. Nature Biotechnology. 2012;**30**:868-875. DOI: 10.1038/nbt.2316

[101] Busscher HJ, Alt V, van der Mei HC, Fagette PH, Zimmerli W, Moriarty TF, et al. A transatlantic perspective on stagnation in clinical translation of antimicrobial strategies for the control of biomaterial-implant-associated infection. ACS Biomaterials Science & Engineering. 2018;**5**:402-406. DOI: 10.1021/acsbiomaterials.8b01071

Section 3

Antimicrobial Resistance and Anti Biofilm Strategies

Chapter 8

Bacterial Biofilm: Contribution to AMR and Approaches to Tackle

Meenakshi Sharma, Pragati Yadav and Deepika Tripathi

Abstract

The brisk emergence of resistant microbes is occurring worldwide, endangering the efficacy of various antimicrobial agents. The overprescription of antimicrobial drugs results in the emergence of mutant strains of drug-resistant pathogens challenging the existing antimicrobial regime. Moreover, the outbreak of the pandemic has emphasized the necessity to consider the coinfections and antimicrobial resistance crisis as a vital motive of morbidity and mortality. Therefore, the prevention of such infections is much better than the eradication of the same. Thus, herein, we aim at providing a comprehensive list that can be used as an alternative class of antibacterial agents by exploiting the activity of various phytochemicals. The antibiofilm activity of various classes of phytochemicals would be projected for both the eradication and the prevention of biofilm formation in the presence of selected compounds. This chapter visualizes antimicrobial resistance as a matter of grave concern and one of the greatest threats to global health, food security, and development today.

Keywords: biofilm, antimicrobial resistance, phytochemicals, antibacterial resistance

1. Introduction

Antimicrobials can be synthetic or natural molecules that have the efficacy to kill microorganisms effectively. The tolerance toward antimicrobials has emerged as a major challenge for scientists and doctors across healthcare sectors, and it is becoming a serious threat worldwide. Since the late 1960s, the situation is intensified by decline in the search of novel drugs, as testing new drugs and finally its acceptance requires long time periods by the authorities for commercialization [1]. Antimicrobial resistance (AMR) in pathogenic microbes is the threatening global health problem with the biggest threat to human health, and the world is suffering without any significant and effective antibiotics [2]. It occurs when bacteria, viruses, fungi, and parasites change over time and, now, no longer respond to antibiotics.

In other words, microbes become resistant to antibiotics and cause reinfection. Sometimes, it is impossible to treat such infection, and it ultimately increases the risk of disease spread, severe illness, and even becomes fatal day by day. According to recent studies and World Health Organization (WHO)'s reference, the antimicrobial resistant microbes are also referred to as "superbugs" sometimes. According to 2014 World Health Organization (WHO) report, "Antimicrobial Resistance: Global Report on Surveillance," the problem is "so serious that it threatens the achievements of

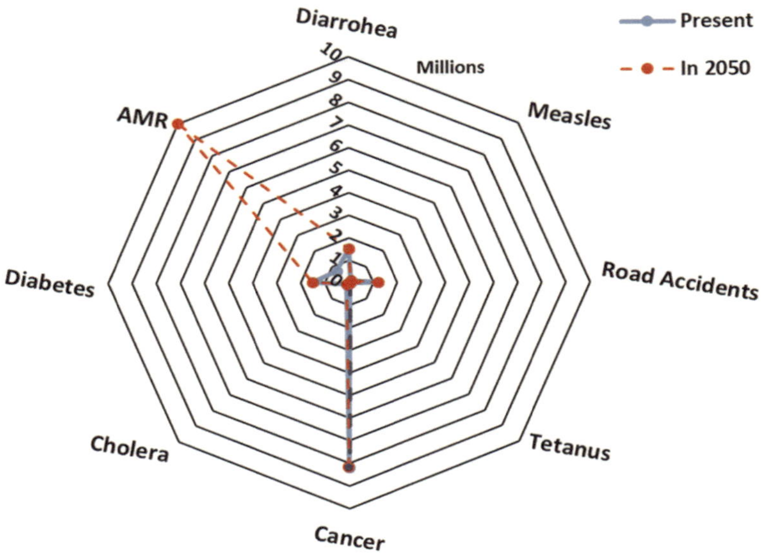

Figure 1.
The number of deaths per year (in millions) as per data provided by the report on the AMR review by Hala Audi in 2014.

modern medicine. A post-antibiotic era—in which common infections and minor injuries can kill—is a very real possibility for the 21st century" [3]. **Figure 1**, based on the report presented by Hala Audi in 2014 [3], shows the number of deaths (in million) versus various causes of death in the present age, and the number of deaths due to AMR is estimated to be increased from 700,000 at present to 10 million deaths per year in 2050 [4].

One of the major reasons contributing to the emergence of AMR is the overuse of antibiotics. At present, most of the antimicrobial compounds target the necessary microbial physiological processes, thereby exerting strong selection pressure on microbes that promote the emergence and spread of drug-resistant strains. Recently, researchers have targeted their research toward finding novel solutions to overcome AMR by targeting the cause of resistance. Phytochemicals, such as alkaloids, flavonoids, quinones, tannins, coumarins, terpenes, lectins, and saponins, have exerted potential antibacterial activities against sensitive as well as resistant pathogens [5, 6]. In this chapter, we have focused on AMR in bacteria, their mechanism of action specifically biofilm formation, and the probable ways to tackle them with emphasis on phytochemicals.

2. Antibacterial resistance

With the discovery of new antibiotics, resistance closely follows and develops continuously. The first antibiotic, penicillin (discovered in 1928), was soon followed by the identification of penicillinase, which led to the discovery of new β-lactams. Similarly, the discovery of sulfonamides, in 1937, led to the resistance in late 1930s. Therefore, each and every new discovery of antibiotics led to the emergence of resistance, leading to decreased treatment options and ultimately rise in morbidity and mortality [7]. The antibacterial resistance is an ever-evolving genetic phenomenon that may be due to genetic mutations or horizontal gene transfer.

The multidrug resistance (MDR) in bacteria is increasing rapidly (**Table 1**), and in 2017, the WHO has categorized and prioritized the drug-resistant bacteria as "critical, high, and medium" for research of new antibiotics. The list includes carbapenem

Bacteria name	Resistant antibiotics	Illnesses caused	References
Gram-negative bacteria			
Acinetobacter baumannii	Carbapenem-resistant	Severe pneumonia, urinary tract infection (UTI), bloodstream infections.	[8, 9]
Pseudomonas aeruginosa	Fluoroquinolone-resistant β-Lactams resistance	Generalized inflammation and sepsis pneumonia, septic shock, skin and soft tissue infections, UTI, gastrointestinal infections.	[10]
Enterobacteriaceae	Carbapenem-resistant Third-generation cephalosporin-resistant	Multiple enteric problems	[11, 12]
Helicobacter pylori	Clarithromycin-resistant	Stomach inflammation and ulcers may lead to stomach cancer.	[13, 14]
Campylobacter	Fluoroquinolone-resistant	Diarrhea, dysentery	[15]
Salmonella spp.	Fluoroquinolone-resistant	Enteritis, osteomyelitis, meningitis, and osteitis.	[16, 17]
Neisseria gonorrhea	Fluoroquinolone-resistant Third generation cephalosporin-resistant	Gonorrhea	[18, 19]
Haemophilus influenzae	Ampicillin-resistant	Pneumonia, bloodstream infection, meningitis, epiglottitis, cellulitis, and infectious arthritis	[20]
Shigella spp.	Fluoroquinolone-resistant	Dysentery	[21]
Klebsiella pneumoniae	Ceftazidime-avibactam	Pneumonia, urinary tract infections, bacteremia, and liver abscesses.	[22]
Gram-positive bacteria			
Enterococcus spp.	Vancomycin-resistant Ampicillin-penicillin and cephalosporin-resistant Fluoroquinolone-resistant Other resistance to aminoglycoside like tobramycin, kanamycin, and gentamicin	UTI, bacterial endocarditis, diverticulitis, and meningitis.	[23–27]
Staphylococcus aureus	Methicillin-resistant Vancomycin intermediate and resistant Other antibiotics resistance like linezolid and daptomycin	Pneumonia, meningitis, osteomyelitis, endocarditis, bacteremia, sepsis, toxic shock syndrome	[28, 29]

Bacteria name	Resistant antibiotics	Illnesses caused	References
Clostridium difficile	Fluoroquinolone-resistant	Severe diarrhea and other intestinal diarrhea.	[30]
Clostridium perfringens	Streptomycin Lincomycin Trimethoprim-sulfamethoxazole	Food poisoning (gastroenteritis) and clostridial myonecrosis.	[31, 32]
Streptococcus pneumoniae	Penicillin non susceptible Macrolide resistance β-Lactams resistance Fluoroquinolone-resistant MDR pneumococcus Resistance to other antibiotics like tetracycline and doxycycline	Pneumonia, meningitis, bacteremia, otitis media, sinusitis	[33–35]
Streptococcus spp.	Penicillin β-Lactams resistance Macrolides Fluoroquinolone-resistant Streptogramins Erythromycin	Bacteremia, sepsis, pneumonia, and meningitis	[36–38]
Bacillus spp.	Penicillin resistance Ampicillin resistance Cephalosporins resistance Trimethoprim resistance	Anthrax, food poisoning syndromes, septicemia, endocarditis, meningitis, and infections of wounds, the ears, eyes, RT, urinary tract, and gastrointestinal tract	[39]
Corynebacterium diphtheria	Chloramphenicol Sulfonamides Tetracyclines resistance	Diphtheria and pharyngitis	[40–42]
Listeria monocytogenes	Tetracyclines resistance Fluoroquinolones resistance	Listeriosis, diarrhea, muscle aches, etc.	[43]

Table 1.
List of bacteria showing antibacterial resistance and the illness caused by them.

resistant (*Acinetobacter baumannii*, *Pseudomonas aeruginosa*, *Enterobacteriaceae*, ESBL-producing) as critical priority; *Enterococcus faecium* (vancomycin-resistant), *Staphylococcus aureus* (methicillin-resistant and vancomycin-intermediate and resistant), *Helicobacter pylori* (clarithromycin-resistant), *Campylobacter* spp. (fluoroquinolone-resistant), *Salmonellae* (fluoroquinolone-resistant), and *Neisseria gonorrhoeae* (cephalosporin-resistant, fluoroquinolone-resistant) as high priority; and *Streptococcus pneumoniae* (penicillin nonsusceptible), *Haemophilus influenzae* (ampicillin-resistant), *Shigella* spp. (fluoroquinolone-resistant) as medium priority drug-resistant bacteria.

The antibiotics have a specific site of action in the bacterial cells, as shown in **Figure 2**. The antibiotic can cause defect in cell wall synthesis, inhibition of DNA

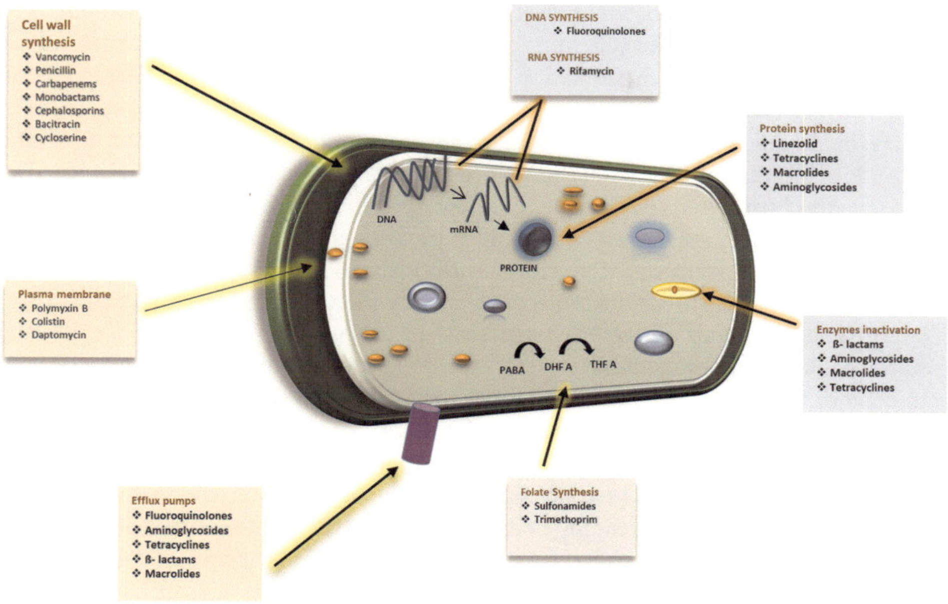

Figure 2.
Different classes of antibiotics and their site of action in the bacterial cell.

gyrase, topoisomerase IV, and translocation inhibition (via 30S ribosome subunit) leading to formation of nonfunctional proteins or protein synthesis inhibition (via 50S ribosome subunit) [44].

3. Mechanism of antibacterial resistance

Antibacterial resistance means that the bacterial cell is capable of escaping the effects of drugs by various mechanisms. These resistant mechanisms can be general like modification in structure, which results in the hindrance of drug attachment to bacterial cells, attainment of aminoglycoside modifying enzyme, neutralizing or pumping the antibodies outside by efflux pumps, mutation of DNA gyrase, decrease in the affinity to antibiotics, methylation and/or mutation of 23S rRNA, alteration of target sites like penicillin-binding proteins (PBPs), and inactivation of antibiotics. The specific mechanisms, such as the production of lactamases for the enzymatic degradation of lactam antibodies and affecting the susceptibility and affinity of the target sites as in gram-positive bacteria [45, 46], are also present. The mechanism can be either intrinsic or extrinsic resistance, which helps bacteria to acquire new resistant genes. Apart from these well-known genetic mechanisms, biofilm-formation- and quorum sensing (QS)-related responses are other important features that help bacteria to gain resistance. In this chapter, we will discuss about the role of biofilm and its formation in detail.

3.1 Biofilm

Biofilms are a complex three-dimensional densely packed architectural network of microbes residing inside the polymeric matter secreted by them on several biotic and abiotic surfaces. The biofilm concept was given in 1971 by Marshall et al. [47], and later,

Fletcher, Characklis, and Costerton described it as follows: "Biofilm is the unique pattern of growth in the life cycle of microbes that provides specific properties, advantages, and a higher level of organization to the free-living bacterial cells during colonization" [48]. According to the National Institutes of Health (NIH), 65% of microbial and 80% of chronic infections are linked to biofilm forming bacteria as compared to planktonic cells. The biofilm formation gives bacteria protection from antibiotics, disinfectants, and host defense system, thus showing resistance to them. For biofilm production, some bacteria adjust their gene expression and some use quorum-sensing systems. In both the gram-negative and gram-positive bacteria, quorum-sensing (QS) mechanisms exist, but the signal molecules used by them to transmit information are different. The QS signals of bacteria participate in various physiological processes such as motility, plasmid conjugation, biofilm formation, and antibiotic resistance to help them cope in the adverse environmental situations. The QS system comprises autoinducing peptides (AIPs), autoinducer-2 (AI-2), and acyl-homoserine lactones (AHLs) [49]. The presence of glycocalyx, outer membrane structure, efflux pumps, heterogeneity in growth rate, genetic adaptation, metabolic state, and metabolism of cells within a biofilm are the leading causes of biofilm that acquire resistance against antimicrobials [50]. As biofilms have extracellular polymeric substances (EPSs) that surround the cells, they provide protection to the microbial cells against harsh growth conditions [51]. EPSs are constituted of lipids, proteins, extracellular DNA, and polysaccharides. The biofilm formation is a multistep process, starting with attachment to the biotic or abiotic surface, forming a microcolony and then finally forming a three-dimensional structure, which, after maturation, starts the detachment of bacterial cells for another cycle of biofilm formation via attachment (**Figure 3**).

3.1.1 Attachment to the surfaces

The first initial step is the attachment step, which is further divided into a two-stage process: initial reversible attachment and irreversible attachment [52]. Biofilm

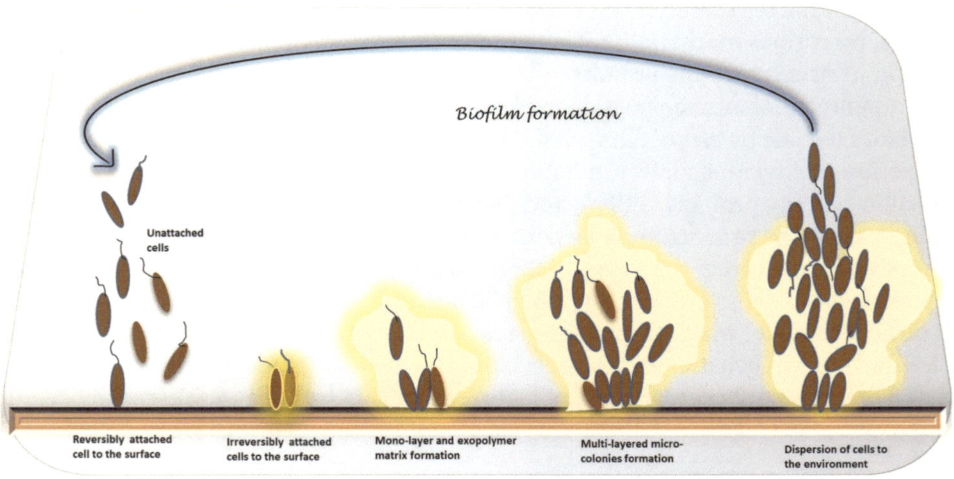

Figure 3.
Stages of biofilm formation: the formation begins with a reversible attachment of the planktonic cells (dark brown ovals) followed by the adhesion to the surface (light brown). The bacteria then form a monolayer and irreversibly attach by producing an extracellular matrix. Next, a microcolony is formed where multilayers appear. During later stages, the biofilm matures, and finally, some cells start to detach and the biofilm (shown in yellow) disperses, releasing planktonic cells for re-attachment.

formation begins by the preliminary reversible attachment of the planktonic microbial cells to the biotic or abiotic surface followed by adhesion. Bacteria will then start to form a monolayer and will produce an extracellular matrix (also known as slime) for protection. In this stage, the formation of microcolonies takes place, which shows significant growth and cell-cell communication for example quorum sensing. Now, the biofilm grows and the attachment is irreversible.

3.1.2 Maturation

This step initiates the cell growth that results in small colonies of microorganisms forming a characteristic "toadstool"-like structure. Bacteria within biofilm communities perform specialized functions after communicating via QS to each other. As the biofilm matures, more DNA, proteins, polysaccharides, etc., also known as biofilm scaffolds, are secreted by the bacteria residing within the biofilm. As the stage progresses, a heterogeneous physicochemical environment—mediated by van der Waals forces and hydrophobic and electrostatic interactions—is developed via the cell-to-cell interaction, which provides the embedded-cell-specialized physiological features. This environment inside the biofilm leads to specialized characters to the residing microbes for differentiation into the mature bacterial community for the final dispersion of the planktonic form [53].

3.1.3 Dispersion

After the biofilm maturation, some cells of mature biofilm start detaching and disperse into the environment as planktonic cells; this planktonic stage is considered as more sensitive to antimicrobials and immune responses. Therefore, dispersion is a very promising path for biofilm control. This mechanism is cyclic as the released microbial planktonic cells have the potential to again start a new biofilm formation cycle.

4. Approaches to tackle

The resistance of pathogenic microbes against the known drug is becoming a global problem. These pathogens also acquire resistance toward various drugs and, thus, termed as multidrug resistance (MDR). These MDR bacteria pose a major threat to community and health care as hospital-acquired secondary infections lead to longer stay in hospitals and complications. The common examples are *S. pneumoniae*, *E. faecium*, and *S. aureus*. Thus, active research for novel antibiotics or novel targets such as dodecyl deoxy glycosides, teixobactin, 2-((3-(3,6-dichloro-9H-carbazol-9-yl)-2-hydroxypropyl) amino)-2 (hydroxymethyl)propane1,3-diol (DCAP), and malacidins to combat such bacterial infections is the need of an hour. Moreover, natural compounds of either plant origin or microbial by-products as antimicrobials, such as cannabinoids, antimicrobial peptides, and odilorhabdins, are promising aspects of this research. The combinatorial strategy giving synergistic effect is also being used to tackle AMR such as probiotics and bacteriophages. Of these various strategies, this chapter will focus on plant products or phytochemicals that are being researched for their use to combat AMR by targeting various resistance mechanisms such as biofilm, quorum sensing, etc. (**Figure 4**).

Many present studies focus on the strategy for screening various phytochemicals, the method in the identification of their bioactive components, their further investigations, and various approaches that could be adopted to prevent the lethal

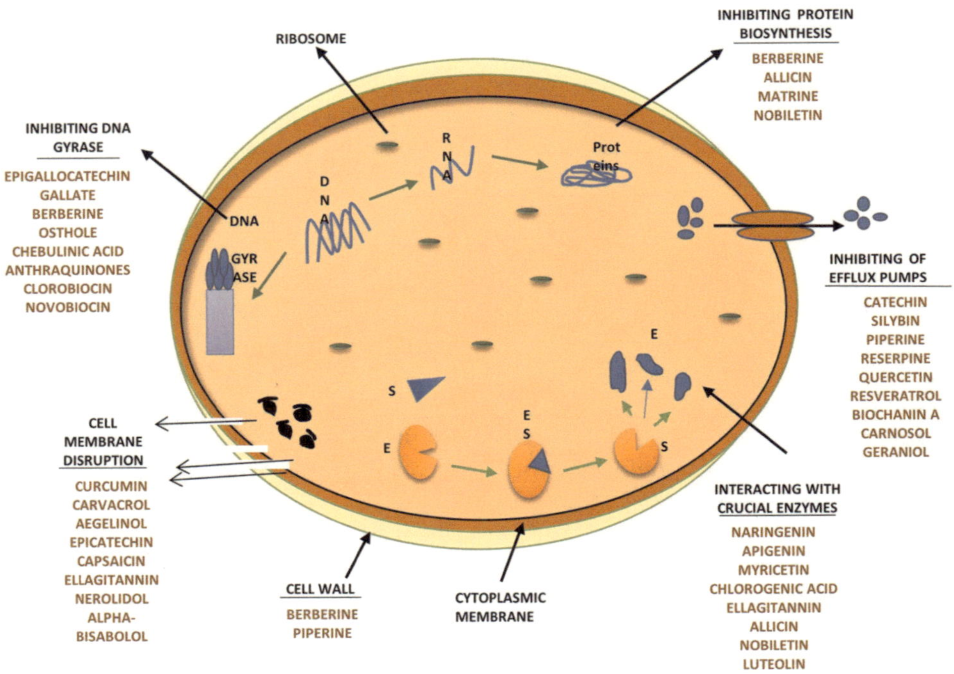

Figure 4.
Different types of phytochemicals and their site of action in the bacterial cell.

consequences of multidrug resistance. Phytochemicals have an immense potential to combat bacterial infections by disrupting the bacterial membrane, inhibition of cell wall or protein synthesis, interference with intermediary metabolism, damage to the synthesis and function of DNA/RNA, and normal cell communication interruption and induction of coagulated cytoplasmic constituents without any pronounced side effect. Major phytochemical classes studied are alkaloids, flavonoids, quinones, tannins, coumarins, terpenes, lectins, and saponins [5, 6]. **Table 2** depicts in detail the structure and common name of phytochemicals with their known mechanism of action.

4.1 Phenolics and polyphenols

These is a diverse group of aromatic secondary metabolites consisting of flavonoids, quinones, tannins, and coumarins involved in plant defense mechanisms. They exhibit antibacterial properties against various bacteria. Among all flavonols, phenolic acids show maximum activities because they can interact with the cytoplasmic membrane, inhibit bacterial virulence factors including enzymes and toxins, suppress biofilm formation, reduce the pH values, reduce the extracellular polysaccharide activity, exert synergistic effects with conventional antibiotics, and finally can act as EP inhibitors [77].

4.1.1 Flavonoids

Flavonoids are the main constituent of common edible part of plant, such as fruits, vegetables, nuts, and seeds. These are known to possess various biological activities, such as anti-inflammatory, antioxidant, and antitumor activity, which is now a new

Source	Structures	Common name	Active compounds	Mechanism of action	Activity K/n against	References
Polyphenols						
Rosa rugosa	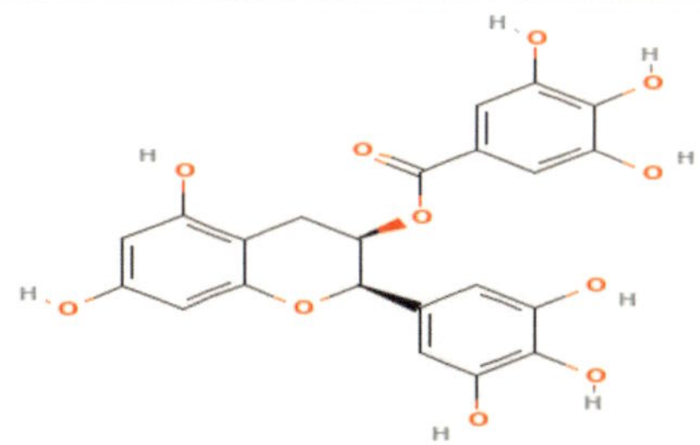 $C_{22}H_{18}O_{11}$	*Beach rose*	Epigallocatechin gallate	Inhibition of swarming motility and biofilm formation	*E. coli*, *P. aeruginosa*	[54]
	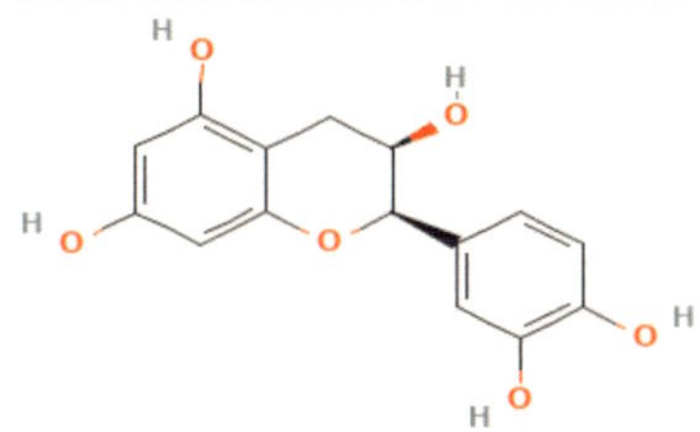 $C_{15}H_{14}O_6$		Epicatechin			
Allium sativum *Arachis hypogaea*	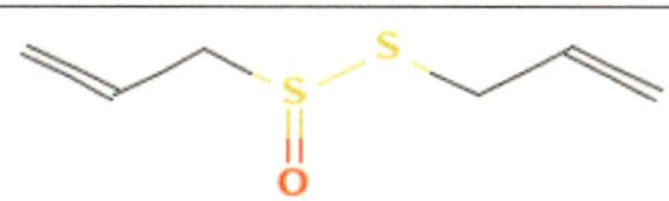 $C_6H_{10}OS_2$	Garlic	Allicin	Inhibition of biofilms formation, inhibition of expression of bacterial virulence factor and antagonized the activity of LuxR, AhyR, and TraR receptor.	*S. aureus*, *E. coli*, *P. aeruginosa*, *K. pneumonia*	[55–57]

Source	Structures	Common name	Active compounds	Mechanism of action	Activity K/n against	References
	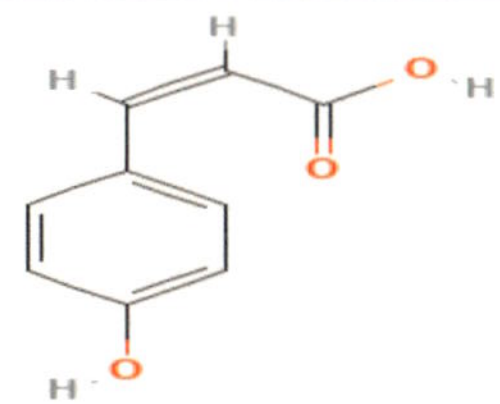 $C_9H_8O_3$	Peanuts	cis-p-coumaric acid			
Buchanania lanzan Spreng	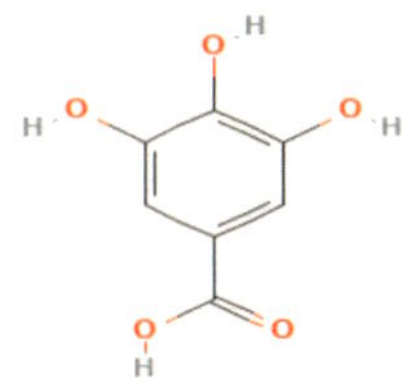 $C_7H_6O_5$ or $C_6H_2(OH)_3COOH$	Char, chironji	Gallic acid	Reduction of biofilms formation	*E. coli*, *P. aeruginosa*	[58]
Curcuma longa	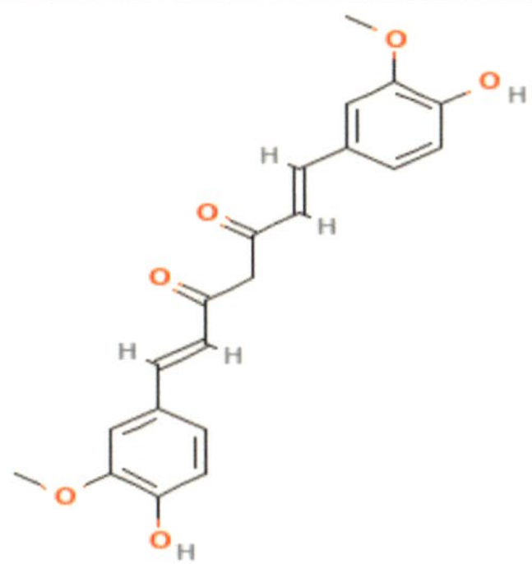 $C_{21}H_{20}O_6$	Turmeric/ haldi	Curcumin	Antimicrobial activity, Biofilm inhibition, Anti-MDR	*S. aureus*, *S. mutans*, *H. pylori*, *A. baumannii*	[59–62]

DOI: http://dx.doi.org/10.5772/intechopen.103951

Source	Structures	Common name	Active compounds	Mechanism of action	Activity K/n against	References
Terminalia chebula	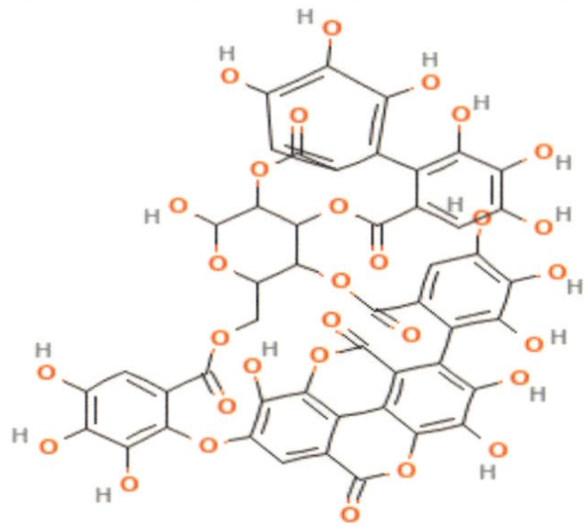 $C_{48}H_{28}O_{30}$	Chebulic myrobalan	Terchebulin	Antimicrobial activity or anti-MDR	*A. baumannii*	[63]
	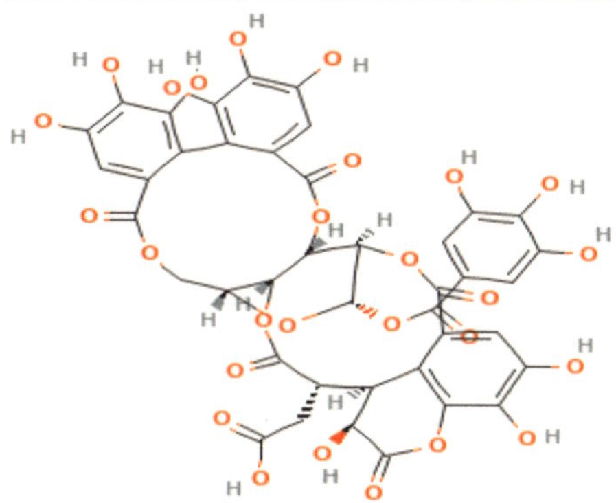 $C_{41}H_{30}O_{27}$		Chebulagic acid			

Source	Structures	Common name	Active compounds	Mechanism of action	Activity K/n against	References
	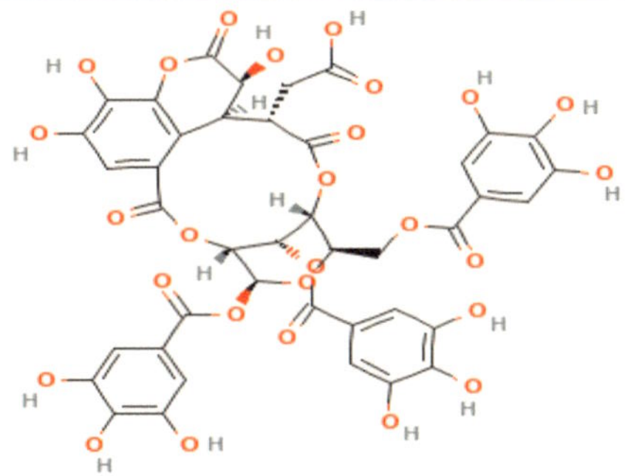 $C_{41}H_{32}O_{27}$		Chebulinic acid			
Alkaloids						
Piper nigrum L. *Piper longum L.*	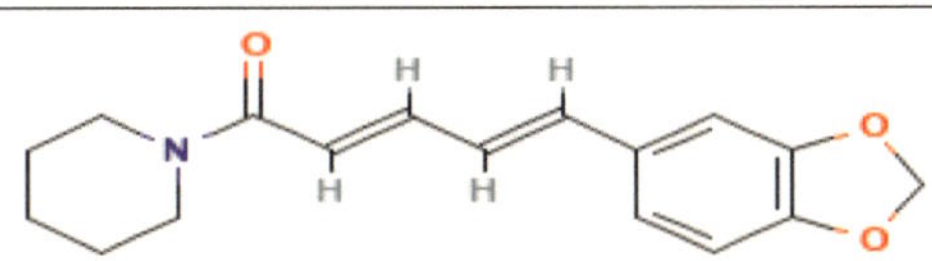 $C_{17}H_{19}NO_3$	Black pepper, long pepper	Piperine	EP inhibitor, delay biofilm formation	MRSA, *C. albicans*	[64, 65]
Sanguinaria canadensis	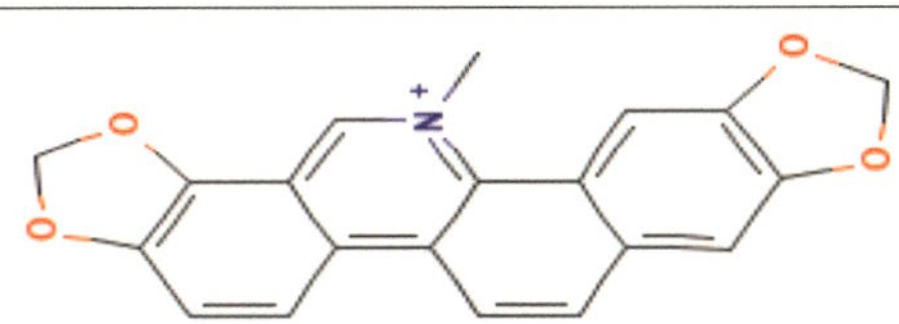 $C_{20}H_{14}NO_4^+$	Bloodroots	Sanguinarine	Antimicrobial activity	MRSA	[66]

DOI: http://dx.doi.org/10.5772/intechopen.103951

Source	Structures	Common name	Active compounds	Mechanism of action	Activity K/n against	References
Coffea arabica, *Theobroma cacao*, *Cola acuminata*, etc.	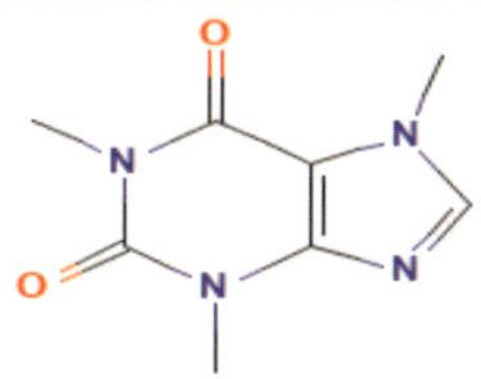 $C_8H_{10}N_4O_2$	Coffee beans, cocoa beans, cola nut	Caffeine	Interaction with the quorum sensing proteins and inhibiting biofilm formation	*P. aeruginosa*	[67, 68]
Berberis vulgaris, *Berberis aquifolium*, *Hydrastis canadensis*, etc.	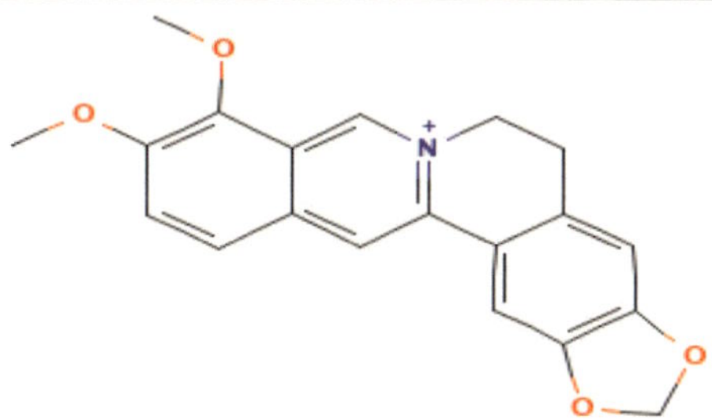 $C_{20}H_{18}NO_4^+$	Barberry, Oregon grape, Goldenseal	Berberine	Protein and DNA synthesis inhibitor	*E. coli*, *C. albicans*	[69]
Coumarins						
Citrus paradisi	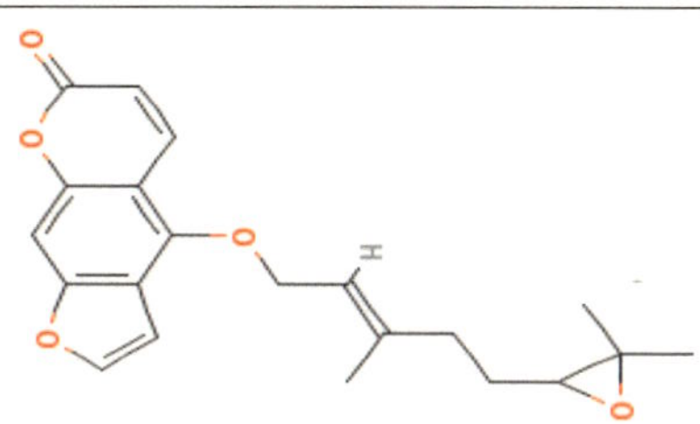 $C_{21}H_{22}O_5$	Grapefruits	Epoxybergamottin	EB inhibitor	MRSA	[70]

Source	Structures	Common name	Active compounds	Mechanism of action	Activity K/n against	References
Cnidium monnieri, *Angelica pubescens*	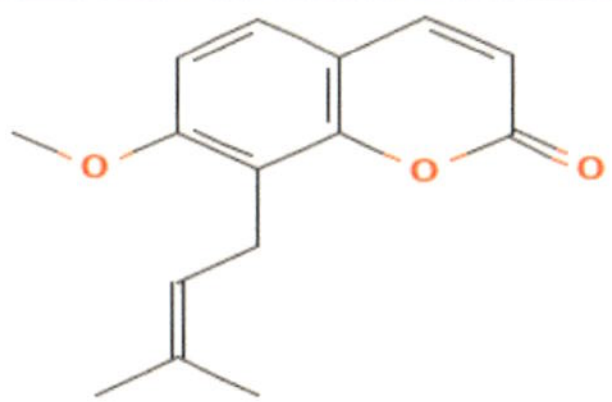 $C_{15}H_{16}O_3$	Monnier's snow parsley	Osthol	DNA gyrase inhibitor	*B. subtilis*, *S. aureus*, *K. pneumoniae*, MRSA	[71, 72]
Aegle marmelos	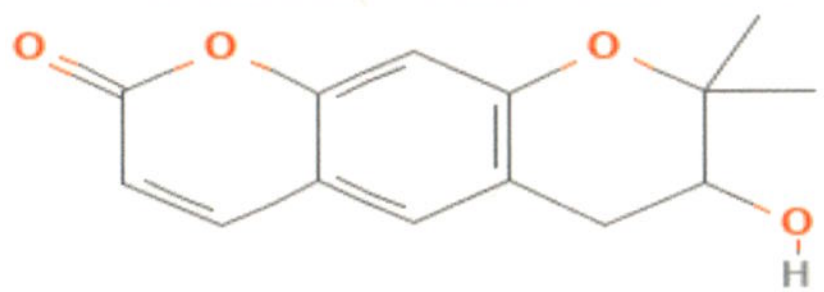 $C_{14}H_{14}O_4$	Bael	Aegelinol	DNA gyrase inhibitor	*S. typhi*, *E. aerogenes*, *E. cloacae*, *S. aureus*	[73]
Terpenes						
Thymus vulgaris	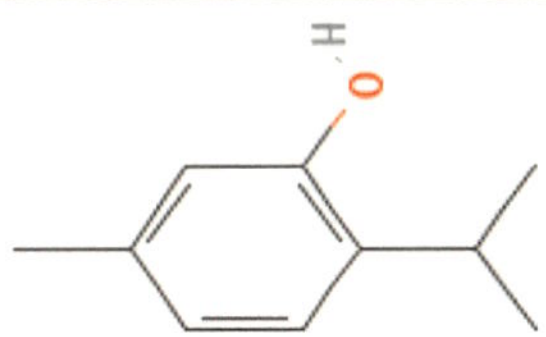 $C_{10}H_{14}O$	Garden thyme	Thymol	Anti-fungal activity	*Candida* sp. (*C. albicans*, *C. glabrata*, *C. krusei*)	[74]
Cymbopogon nardus, *Cymbopogon citratus*, etc.	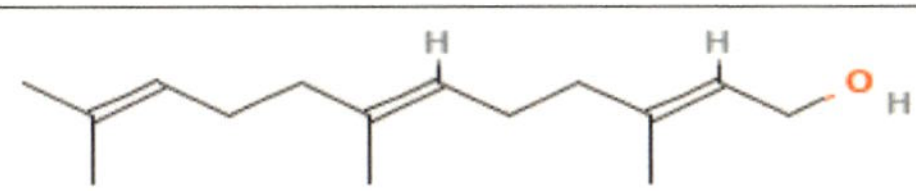 $C_{15}H_{26}O$	Citronella grass, Lemongrass	Farnesol	Cell membrane disturbance	*S. aureus*	[75]

DOI: http://dx.doi.org/10.5772/intechopen.103951

Source	Structures	Common name	Active compounds	Mechanism of action	Activity K/n against	References
Origanum vulgare, *Thymus vulgaris*, *Lepidium flavum* etc.	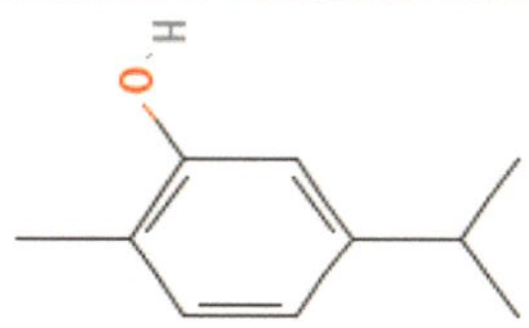 $C_{10}H_{14}O$	Oregano, Thyme, pepperwort	Carvacrol	Anti-microbial activity, Biofilm inhibition, Cell membrane disturbance, EP inhibitor	*B. subtilis*, *B. cereus*, *E. coli*, *S. aureus*, *S. enterica*, *S. enteritidis*, *C. perfringens*, *L. monocytogenes P. aeruginosa*	[76]

Table 2.
List of phytochemicals, their source, mode of action, and the sensitive microbe.

therapeutic interest. Flavonoids are the pigments that are responsible for colors in fruits, leaves, and flowers and belong to the polyphenol family. Flavonoids show interesting properties in controlling plant growth and development by interacting in a complex manner with the various plant growth hormones [78].

Flavonoid can be classified on the basis of biosynthesis such as chalcones, flavanones, flavan-3-ols, and flavan-3,4-diols, which are both intermediates in biosynthesis and end products that can accumulate in plant tissues. Other classes are only known as end products of biosynthesis such as anthocyanidins, proanthocyanins, flavones, and flavanols. Two additional classes of flavonoids are those in which the 2-phenyl side chain of flavanone isomerizes to the third position, giving rise to isoflavones and related isoflavonoids. Flavonoids have many medicinal activities; therefore, they have been reported to have many useful properties including anti-inflammatory activity, enzyme inhibition, and antimicrobial activity [79, 80].

4.1.2 Quinones

Quinones are aromatic ring compounds with two ketone substitutions. The major targets of quinones in the microbial cells are cell wall polypeptides, surface-exposed adhesin proteins, and membrane-bound enzymes. Naphthoquinones is one of the largest groups of plant secondary metabolites that exhibit many biological activities.

4.1.3 Tannins

Tannins are found in almost all plant parts, and they possess different antibacterial and antifungal activities. The possible mechanism of antimicrobial efficiency is due to the inactivation of cell envelope transport proteins and microbial adhesins [5].

4.2 Alkaloids

Alkaloids contain variable chemical structures and generally are heterocyclic nitrogen compounds. They tend to exhibit different biological activities, including analgesic effects and antibacterial properties. Therefore, they play a significant role in treating many infectious diseases. The most critical alkaloid groups are aporphines, isoquinolines, quinolones, and phenanthrenes exhibiting suitable antibacterial activities [81]. Their mode of action might be due to the inhibition of repair mechanisms and DNA synthesis, the enzymatic alterations affecting physiological processes, the inhibition of the bacterial nucleic acid and protein synthesis, the modification of the bacterial cell membrane permeability, the damage of the cell membrane and cell wall, the inhibition of bacterial metabolism, and the inhibition of efflux pumps [82–84]. The alkaloids, such as harmane and berberine, results in impaired cell division and ultimately cell death as they possess the ability to intercalate with DNA [85].

4.3 Coumarins

Coumarins are produced naturally by many plants as well as microorganisms, and chemically, they are aromatic benzopyrones, benzene fused with alpha pyrone rings. Some recent studies also have suggested that coumarins are capable of suppressing quorum-sensing meshwork of bacterial pathogens and affect their ability to form biofilm and virulence factor formations.

4.4 Terpenes

Terpenes are naturally occurring hydrocarbons of either cyclic or open-chain structure, such as sesquiterpenes and monoterpenes. Their oils and compounds have several pharmacological activities, such as antitumor, antiviral, antibacterial, antifungal, anti-inflammatory, antiparasitic, and antioxidant properties [86]. Essential oils (EOs) from medicinal plants have shown anti-QS effects, and EOs produced by aromatic plants have been observed to be effective against biofilms. Preferentially, monoterpenes could impact the membrane structures via increasing the permeability and fluidity, thereby changing the topology of proteins leading to the disturbances in the respiratory chain [87].

5. Conclusion

AMR is becoming a primary cause of morbidity and mortality worldwide, and the resistant microbes are mounting and phenomenal according to the geographic area and the extent of resistance [88]. The infectious agents and diseases that were thought to be controlled by drugs are again emerging with more force against these treatments. The recurrence of resistant microbes, importantly in developing countries, is due to the accessibility of drugs without valid prescription. The golden example is the re-emergence of tuberculosis (in 1980s), which has emerged as multidrug resistant and escalated by HIV infection [89, 90]. The trouble and seriousness in treating MDR strains requires the utilization of a few, some of the time six to seven distinct, drugs. Few mechanisms leading to resistance are the modification of drug targets, the limiting uptake of drug, the active efflux of drug, or the inactivation of drug. Another major well-known resistance mechanism is the biofilm formation.

The protective layers build in the biofilm are a major setback in the treatment of biofilm-related infections, which leads to the ineffectiveness of the existing antibiotics. These layers limit the antibiotic penetration, and thus, the community of sedentary cells survives even in the presence of antibiotics effective against their motile counterparts [53]. Many pieces of evidence suggest that the medicinal plants hold great promise in search of novel antimicrobial agents, and the phytochemicals obtained are very effective in the treatment of infections. Moreover, the plants are cheap, readily available, and almost have minimum side effects. These properties of medicinal plants have gained attention in recent years, for the herbal-based medicines as therapeutics. However, studies are still needed to ensure the safety of antimicrobial phytochemicals and its mechanism of action. Till date, the mechanism of action and the activity related to the structure of phytochemicals have been largely elusive and need further attention [91].

To overcome AMR effectively, all combating new strategies should be practically delivered at all levels, such as community, national, and global levels. Active research to investigate the AMR, its mechanism, strategies to overcome resistance, and leading the novel antimicrobial candidates to clinical practice should be continued. It is important to understand that the distribution, driving force, and the solutions for AMR are different in different countries. Therefore, different approaches are required in high-income countries as compared with low-income countries.

Acknowledgements

Meenakshi Sharma acknowledges the support of grant received from IOE, University of Delhi, Delhi [No./IoE/2021/12/FRP].

Conflict of interest

The authors declare no conflict of interest.

Author details

Meenakshi Sharma*, Pragati Yadav and Deepika Tripathi
Genetics of Aging, Dr. B.R. Ambedkar Center for Biomedical Research (ACBR), University of Delhi, Delhi, India

*Address all correspondence to: meenakshisharma@acbr.du.ac.in

References

[1] Spellberg B, Powers JH, Brass EP, Miller LG, Edwards JE. Trends in Antimicrobial Drug Development: Implications for the Future [Internet]. 2004. Available from: https://academic.oup.com/cid/article/38/9/1279/317267

[2] Boucher HW, Talbot GH, Bradley JS, Edwards JE, Gilbert D, Rice LB, et al. Bad bugs, no drugs: No ESKAPE! An Update from the Infectious Diseases Society of America. Clinical Infectious Diseases. 2009;**48**(1):1-12

[3] World Health Organization. Antimicrobial Resistance: Global Report on Surveillance. World Health Organization. 2014. https://apps.who.int/iris/handle/10665/112642

[4] Bauer KA, Kullar R, Gilchrist M, File TM. Antibiotics and adverse events. Current Opinion in Infectious Diseases. 2019;**32**(6):553-558

[5] Gupta PD, Birdi TJ. Development of botanicals to combat antibiotic resistance. Journal of Ayurveda and Integrative Medicine. 2017;**8**(4):266-275

[6] Radulovic NS, Blagojevic PD, Stojanovic-Radic ZZ, Stojanovic NM. Antimicrobial plant metabolites: Structural diversity and mechanism of action. Current Medicinal Chemistry. 2013;**20**(7):932-952

[7] Davies J, Davies D. Origins and evolution of antibiotic resistance. Microbiology and Molecular Biology Reviews. 2010;**74**(3):417-433

[8] Alcántar-Curiel MD, Rosales-Reyes R, Jarillo-Quijada MD, Gayosso-Vázquez C, Fernández-Vázquez JL, Toledano-Tableros JE, et al. Carbapenem-resistant acinetobacter baumannii in three tertiary care hospitals in Mexico: Virulence profiles, innate immune response and clonal dissemination. Frontiers in Microbiology. 2019;**20**:10

[9] Lin M-F. Antimicrobial resistance in *Acinetobacter baumannii*: From bench to bedside. World Journal of Clinical Cases. 2014;**2**(12):787

[10] Pachori P, Gothalwal R, Gandhi P. Emergence of antibiotic resistance *Pseudomonas aeruginosa* in intensive care unit; a critical review. Genes & Diseases. 2019;**6**(2):109-119

[11] Lutgring JD. Carbapenem-resistant Enterobacteriaceae: An emerging bacterial threat. Seminars in Diagnostic Pathology. 2019;**36**(3):182-186

[12] Paterson DL. Resistance in gram-negative bacteria: Enterobacteriaceae. The American Journal of Medicine. 2006;**119**(6):S20-S28

[13] Flores-Treviño S, Mendoza-Olazarán S, Bocanegra-Ibarias P, Maldonado-Garza HJ, Garza-González E. *Helicobacter pylori* drug resistance: Therapy changes and challenges. Expert Review of Gastroenterology & Hepatology. 2018;**12**(8):819-827

[14] Savoldi A, Carrara E, Graham DY, Conti M, Tacconelli E. Prevalence of antibiotic resistance in *Helicobacter pylori*: A systematic review and meta-analysis in World Health Organization Regions. Gastroenterology. 2018;**155**(5):1372-1382

[15] Smith JL, Fratamico PM. Fluoroquinolone resistance in campylobacter. Journal of Food Protection. 2010;**73**(6):1141-1152

[16] Cuypers WL, Jacobs J, Wong V, Klemm EJ, Deborggraeve S, van

Puyvelde S. Fluoroquinolone resistance in Salmonella: Insights by whole-genome sequencing. Microbial Genomics. 2018;**4**(7):e000195

[17] Crump JA, Sjölund-Karlsson M, Gordon MA, Parry CM. Epidemiology, clinical presentation, laboratory diagnosis, antimicrobial resistance, and antimicrobial management of invasive Salmonella infections. Clinical Microbiology Reviews. 2015;**28**(4): 901-937

[18] Ghanem KG, Giles JA, Zenilman JM. Fluoroquinolone-resistant Neisseria gonorrhoeae: The Inevitable Epidemic. Infectious Disease Clinics of North America. 2005;**19**(2):351-365

[19] Bala M, Sood S. Cephalosporin resistance in Neisseria gonorrhoeae. Journal of Global Infectious Diseases. 2010;**2**(3):284

[20] Tristram S, Jacobs MR, Appelbaum PC. Antimicrobial Resistance in *Haemophilus influenzae*. Clinical Microbiology Reviews. 2007; **20**(2):368-389

[21] Qin T, Qian H, Fan W, Ma P, Zhou L, Dong C, et al. Newest data on fluoroquinolone resistance mechanism of Shigella flexneri isolates in Jiangsu Province of China. Antimicrobial Resistance and Infection Control. 2017; **6**(1):97

[22] Zhang P, Shi Q, Hu H, Hong B, Wu X, Du X, et al. Emergence of ceftazidime/avibactam resistance in carbapenem-resistant Klebsiella pneumoniae in China. Clinical Microbiology and Infection. 2020; **26**(1):124

[23] Jubeh B, Breijyeh Z, Karaman R. Resistance of gram-positive bacteria to current antibacterial agents and overcoming approaches. Molecules. 2020;**25**(12):2888

[24] Miller WR, Munita JM, Arias CA. Mechanisms of antibiotic resistance in enterococci. Expert Review of Anti-Infective Therapy. 2014;**12**(10): 1221-1236

[25] Lee T, Pang S, Abraham S, Coombs GW. Antimicrobial-resistant CC17 Enterococcus faecium: The past, the present and the future. Journal of Global Antimicrobial Resistance. 2019; **16**:36-47

[26] Fontana R, Aldegheri M, Ligozzi M, Lopez H, Sucari A, Satta G. Overproduction of a low-affinity penicillin-binding protein and high-level ampicillin resistance in Enterococcus faecium. Antimicrobial Agents and Chemotherapy. 1994;**38**(9):1980-1983

[27] Rubinstein E, Keynan Y. Vancomycin-resistant enterococci. Critical Care Clinics. 2013;**29**(4):841-852

[28] Dweba CC, Zishiri O, el Zowalaty M. Methicillin-resistant *Staphylococcus aureus*: Livestock-associated, antimicrobial, and heavy metal resistance. Infection and Drug Resistance. 2018;**11**:2497-2509

[29] Gardete S, Tomasz A. Mechanisms of vancomycin resistance in *Staphylococcus aureus*. The Journal of Clinical Investigation. 2014;**124**(7): 2836-2840

[30] Spigaglia P. Recent advances in the understanding of antibiotic resistance in *Clostridium difficile* infection. Therapeutic Advances in Infectious Disease. 2016;**3**(1):23-42

[31] Chon J-W, Seo K-H, Bae D, Park J-H, Khan S, Sung K. Prevalence, toxin gene profile, antibiotic resistance, and

molecular characterization of *Clostridium perfringens* from diarrheic and non-diarrheic dogs in Korea. Journal of Veterinary Science. 2018;**19**(3):368

[32] Hamza D, Dorgham S, Hakim A. Toxinotyping and antimicrobial resistance of *Clostridium perfringens* isolated from processed chicken meat products. Journal of Veterinary Research. 2017;**61**(1):53-58

[33] Mamishi S, Moradkhani S, Mahmoudi S, Hosseinpour-Sadeghi R, Pourakbari B. Penicillin-Resistant trend of Streptococcus pneumoniae in Asia: A systematic review. Iranian Journal of Microbiology. 2014;**6**(4):198-210

[34] Zivich PN, Grabenstein JD, Becker-Dreps SI, Weber DJ. Streptococcus pneumoniae outbreaks and implications for transmission and control: A systematic review. Pneumonia. 2018;**10**(1):11

[35] Jacobs MR. Drug-resistant Streptococcus pneumoniae: Rational antibiotic choices. The American Journal of Medicine. 1999;**106**(5):19-25

[36] Haenni M, Lupo A, Madec JY. Antimicrobial resistance in *Streptococcus* spp. Microbiology Spectrum. 2018;**6**(2). DOI: 10.1128/microbiolspec.ARBA-0008-2017. PMID: 29600772

[37] Bolukaoto JY, Monyama CM, Chukwu MO, Lekala SM, Nchabeleng M, Maloba MRB, et al. Antibiotic resistance of Streptococcus agalactiae isolated from pregnant women in Garankuwa, South Africa. BMC Research Notes. 2015;**8**(1): 364

[38] Nakajima T, Nakanishi S, Mason C, Montgomery J, Leggett P, Matsuda M, et al. Population structure and characterization of viridans group streptococci (VGS) isolated from the upper respiratory tract of patients in the community. The Ulster Medical Journal. 2013;**82**(3):164-168

[39] Helgason E, Økstad OA, Caugant DA, Johansen HA, Fouet A, Mock M, et al. *Bacillus anthracis*, *Bacillus cereus*, and *Bacillus thuringiensis*—One Species on the Basis of Genetic Evidence. Applied and Environmental Microbiology. 2000;**66**(6):2627-2630

[40] Rockhill RC, Hadiputranto H, Siregar SP, Muslihun B. Tetracycline resistance of Corynebacterium diphtheriae isolated from diphtheria patients in Jakarta, Indonesia. Antimicrobial Agents and Chemotherapy. 1982;**21**(5):842-843

[41] Alibi S, Ferjani A, Boukadida J, Cano ME, Fernández-Martínez M, Martínez-Martínez L, et al. Occurrence of Corynebacterium striatum as an emerging antibiotic-resistant nosocomial pathogen in a Tunisian hospital. Scientific Reports. 2017;7(1):9704

[42] Trost E, Blom J, de Castro SS, Huang I-H, Al-Dilaimi A, Schröder J, et al. Pangenomic Study of Corynebacterium diphtheriae that provides insights into the genomic diversity of pathogenic isolates from cases of classical diphtheria, endocarditis, and pneumonia. Journal of Bacteriology. 2012;**194**(12):3199-3215

[43] Morvan A, Moubareck C, Leclercq A, Hervé-Bazin M, Bremont S, Lecuit M, et al. Antimicrobial resistance of *Listeria monocytogenes* strains isolated from humans in France. Antimicrobial Agents and Chemotherapy. 2010;**54**(6):2728-2731

[44] Reygaert W. An overview of the antimicrobial resistance mechanisms of bacteria. AIMS Microbiology. 2018;**4**(3): 482-501

[45] Munita JM, Bayer AS, Arias CA. Evolving resistance among gram-positive

pathogens. Clinical Infectious Diseases. 2015;**61**(suppl. 2):S48-S57

[46] Berger-Bächi B. Resistance mechanisms of Gram-positive bacteria. International Journal of Medical Microbiology. 2002;**292**(1):27-35

[47] Marshall KC, Stout R, Mitchell R. Mechanism of the initial events in the sorption of marine bacteria to surfaces. Journal of General Microbiology. 1971; **68**(3):337-348

[48] Mishra R, Panda AK, de Mandal S, Shakeel M, Bisht SS, Khan J. Natural anti-biofilm agents: Strategies to control biofilm-forming pathogens. Frontiers in Microbiology. 2020;**29**:11

[49] Eickhoff MJ, Bassler BL. SnapShot: Bacterial quorum sensing. Cell. 2018; **174**(5):1328

[50] Singh S, Singh SK, Chowdhury I, Singh R. Understanding the mechanism of bacterial biofilms resistance to antimicrobial agents. The Open Microbiology Journal. 2017;**11**(1):53-62

[51] Branda SS, Vik Å, Friedman L, Kolter R. Biofilms: The matrix revisited. Trends in Microbiology. 2005;**13**(1):20-26

[52] Renner LD, Weibel DB. Physicochemical regulation of biofilm formation. MRS Bulletin. 2011;**36**(5): 347-355

[53] Rabin N, Zheng Y, Opoku-Temeng C, Du Y, Bonsu E, Sintim HO. Biofilm formation mechanisms and targets for developing antibiofilm agents. Future Medicinal Chemistry. 2015;**7**(4):493-512

[54] Zhang J, Rui X, Wang L, Guan Y, Sun X, Dong M. Polyphenolic extract from *Rosa rugosa* tea inhibits bacterial quorum sensing and biofilm formation. Food Control. 2014;**42**:125-131

[55] Bodini SF, Manfredini S, Epp M, Valentini S, Santori F. Quorum sensing inhibition activity of garlic extract and *p*-coumaric acid. Letters in Applied Microbiology 2009;49(5):551-5

[56] Bjarnsholt T, Jensen PØ, Rasmussen TB, Christophersen L, Calum H, Hentzer M, et al. Garlic blocks quorum sensing and promotes rapid clearing of pulmonary Pseudomonas aeruginosa infections. Microbiology. 2005;**151**(12):3873-3880

[57] Rasmussen TB, Bjarnsholt T, Skindersoe ME, Hentzer M, Kristoffersen P, Köte M, et al. Screening for quorum-sensing inhibitors (QSI) by use of a novel genetic system, the QSI selector. Journal of Bacteriology. 2005; **187**(5):1799-1814

[58] Pattnaik A, Sarkar R, Sharma A, Yadav KK, Kumar A, Roy P, et al. Pharmacological studies on Buchanania lanzan Spreng.-A focus on wound healing with particular reference to anti-biofilm properties. Asian Pacific Journal of Tropical Biomedicine. 2013;**3**(12):967-974

[59] De R, Kundu P, Swarnakar S, Ramamurthy T, Chowdhury A, Nair GB, et al. Antimicrobial activity of Curcumin against *Helicobacter pylori* Isolates from India and during Infections in Mice. Antimicrobial Agents and Chemotherapy. 2009;**53**(4):1592-1597

[60] Hu P, Huang P, Chen MW. Curcumin reduces Streptococcus mutans biofilm formation by inhibiting sortase A activity. Archives of Oral Biology. 2013; **58**(10):1343-1348

[61] Mun S-H, Joung D-K, Kim Y-S, Kang O-H, Kim S-B, Seo Y-S, et al. Synergistic antibacterial effect of curcumin against methicillin-resistant *Staphylococcus aureus*. Phytomedicine. 2013;**20**(8–9):714-718

[62] Betts JW, Wareham DW. In vitro activity of curcumin in combination with epigallocatechin gallate (EGCG) versus multidrug-resistant *Acinetobacter baumannii*. BMC Microbiology. 2014; **14**(1):172

[63] Miyasaki Y, Rabenstein JD, Rhea J, Crouch M-L, Mocek UM, Kittell PE, et al. Isolation and characterization of antimicrobial compounds in plant extracts against multidrug-resistant *Acinetobacter baumannii*. PLoS One. 2013;**8**(4):e61594

[64] Khameneh B, Iranshahy M, Ghandadi M, Ghoochi Atashbeyk D, Fazly Bazzaz BS, Iranshahi M. Investigation of the antibacterial activity and efflux pump inhibitory effect of co-loaded piperine and gentamicin nanoliposomes in methicillin-resistant *Staphylococcus aureus*. Drug Development and Industrial Pharmacy. 2015;**41**(6):989-994

[65] Priya A, Pandian SK. Piperine impedes biofilm formation and hyphal morphogenesis of *Candida albicans*. Frontiers in Microbiology. 2020;**13**:11

[66] Obiang-Obounou B, Kang O-H, Choi J-G, Keum J-H, Kim S-B, Mun S-H, et al. The mechanism of action of sanguinarine against methicillin-resistant *Staphylococcus aureus*. The Journal of Toxicological Sciences. 2011; **36**(3):277-283

[67] Widyastuti N, Anjani G, Almira VG, Putri SE, Pratiwi AR. Effects of the administration of brewed Robusta coffee leaves on total antioxidant status in rats with high-fat, high-fructose diet-induced metabolic syndrome. Potravinarstvo Slovak Journal of Food Sciences. 2020;**14**:258-263

[68] Chakraborty P, Dastidar DG, Paul P, Dutta S, Basu D, Sharma SR, et al. Inhibition of biofilm formation of Pseudomonas aeruginosa by caffeine: A potential approach for sustainable management of biofilm. Archives of Microbiology. 2020;**202**(3):623-635

[69] Boberek JM, Stach J, Good L. Genetic evidence for inhibition of bacterial division protein FtsZ by berberine. PLoS One. 2010;**5**(10):e13745

[70] Roy SK, Kumari N, Pahwa S, Agrahari UC, Bhutani KK, Jachak SM, et al. NorA efflux pump inhibitory activity of coumarins from Mesua ferrea. Fitoterapia. 2013;**90**:140-150

[71] Tan N, Yazıcı-Tütüniş S, Bilgin M, Tan E, Miski M. Antibacterial activities of pyrenylated coumarins from the roots of *Prangos hulusii*. Molecules. 2017;**22**(7):1098

[72] Maxwell A. The interaction between coumarin drugs and DNA gyrase. Molecular Microbiology. 1993;**9**(4): 681-686

[73] Basile A, Sorbo S, Spadaro V, Bruno M, Maggio A, Faraone N, et al. Antimicrobial and antioxidant activities of coumarins from the roots of *Ferulago campestris* (Apiaceae). Molecules. 2009; **14**(3):939-952

[74] Sharifzadeh A, Khosravi AR, Shokri H, Shirzadi H. Potential effect of 2-isopropyl-5-methylphenol (thymol) alone and in combination with fluconazole against clinical isolates of *Candida albicans*, *C. glabrata* and *C. krusei*. Journal de Mycologie Médicale. 2018;**28**(2):294-299

[75] Togashi N, Hamashima H, Shiraishi A, Inoue Y, Takano A. Antibacterial activities against *Staphylococcus aureus* of Terpene alcohols with aliphatic carbon chains. Journal of Essential Oil Research. 2010; **22**(3):263-269

[76] Marinelli L, di Stefano A, Cacciatore I. Carvacrol and its derivatives as antibacterial agents. Phytochemistry Reviews. 2018;**17**(4): 903-921

[77] Khameneh B, Eskin NAM, Iranshahy M, Fazly Bazzaz BS. Phytochemicals: A promising weapon in the arsenal against antibiotic-resistant bacteria. Antibiotics. 2021;**10**(9):1044

[78] Bergquist SÅM, Gertsson UE, Knuthsen P, Olsson ME. Flavonoids in baby spinach (*Spinacia oleracea* L.): Changes during plant growth and storage. Journal of Agricultural and Food Chemistry. 2005;**53**(24):9459-9464

[79] Cushnie TPT, Lamb AJ. Antimicrobial activity of flavonoids. International Journal of Antimicrobial Agents. 2005;**26**(5):343-356

[80] Havsteen B. Flavonoids, a class of natural products of high pharmacological potency. Biochemical Pharmacology. 1983;**32**(7):1141-1148

[81] Wu Y, Bai J, Zhong K, Huang Y, Qi H, Jiang Y, et al. Antibacterial activity and membrane-disruptive mechanism of 3-p-trans-Coumaroyl-2-hydroxyquinic acid, a novel phenolic compound from pine needles of Cedrus deodara, against *Staphylococcus aureus*. Molecules. 2016; **21**(8):1084

[82] Mithöfer A, Boland W. Plant defense against herbivores: Chemical aspects. Annual Review of Plant Biology. 2012; **63**(1):431-450

[83] Zhang Q, Lyu Y, Huang J, Zhang X, Yu N, Wen Z, et al. Antibacterial activity and mechanism of sanguinarine against *Providencia rettgeri* in vitro. PeerJ. 2020; **8**:e9543

[84] Yan Y, Li X, Zhang C, Lv L, Gao B, Li M. Research progress on antibacterial activities and mechanisms of natural alkaloids: A review. Antibiotics. 2021; **10**(3):318

[85] Savoia D. Plant-derived antimicrobial compounds: Alternatives to antibiotics. Future Microbiology. 2012;**7**(8):979-990

[86] Bouyahya A, Bakri Y, Et-Touys A, Talbaoui A, Khouchlaa A, Charfi S, et al. Résistance aux antibiotiques et mécanismes d'action des huiles essentielles contre les bactéries. Phytothérapie. 2017

[87] Paduch R, Kandefer-Szerszeń M, Trytek M, Fiedurek J. Terpenes: Substances useful in human healthcare. Archivum Immunologiae et Therapiae Experimentalis. 2007;**55**(5):315

[88] Kimpen J. Polymicrobial diseases: Polymicrobial diseases. Clinical Infectious Diseases. 2003;**36**(2):238-239

[89] Bloom BR, Murray CJL. Tuberculosis: Commentary on a reemergent killer. Science. 1992; **257**(5073):1055-1064

[90] Wood AJJ, Iseman MD. Treatment of multidrug-resistant tuberculosis. The New England Journal of Medicine. 1993; **329**(11):784-791

[91] Khameneh B, Iranshahy M, Soheili V, Fazly Bazzaz BS. Review on plant antimicrobials: A mechanistic viewpoint. Antimicrobial Resistance and Infection Control. 2019;**8**(1):118

Chapter 9

The Mechanisms of Bacterial Biofilm Inhibition and Eradication: The Search for Alternative Antibiofilm Agents

Zeuko'O Menkem Elisabeth

Abstract

Biofilms are a community of microorganisms with accretions of their extracellular matrix that attach both to biological or non-biological surfaces, conferring a significant and incompletely understood mode of growth for bacteria. Biofilm formation represents a protected mode of growth of bacteria that allows cells to survive in hostile environments, facilitating the colonization of new areas. This biofilm formation appears to be produced by microorganisms to resist drug action, causing them to become resistant. Therefore, the search for alternative agents is necessary to counteract and reduce this production, creating suitable drugs against these biofilms. Natural products from medicinal plants possess an array of secondary metabolites and bioactive compounds that could have bioactive potentials that inhibit and eradicate biofilms.

Keywords: biofilms, inhibition, eradication

1. Introduction

Biofilms are complex communities of microbes found attached to a surface or may form aggregates without adhering to a surface. Biofilms also display unique properties, such as multidrug tolerance and resistance to both opsonization and phagocytosis, enabling them to survive in hostile environmental conditions by resisting selective pressures [1]. Sometimes, the host immune system is immunocompromised, making it ineffective in clearing biofilms with evidence that immune cells are paralyzed with disrupted phagocytosis capacities or decreased burst responses, lowering the production of reactive oxygen species [2, 3]. Moreso, these communities of microorganisms are unique since they involve several species in a cooperative. The biofilm thus constitutes a microbial society, with its own set of social rules and patterns of behavior, including altruism and cooperation, both of which favor the success of the group with task-sharing behavior. All of these characteristic patterns are orchestrated by chemical or genetic communication. The biofilm thus constitutes a unique way to stabilize interactions between species, inducing marked changes in the symbiotic

IntechOpen

relationships [3, 4]. Moreover, biofilms protect invading bacteria against the host's immune system via impaired activation of phagocytes and the complement system [5]. The use of antibiotics such as imipenem and colistin mostly reduces biofilms but does not eliminate the entire biofilm in most cases [6]. Due to their toxicity and side effects, it is not possible to reach the minimal concentration of antibiotics *in vivo*. This chapter describes the mechanisms of bacterial biofilm inhibition and eradication with the search for alternative antibiofilm agents.

2. Stages of biofilm formation

Bacteria form complex multicellular structures called biofilms. Biofilm formation is commonly considered to occur in four main stages [7]: (1) adhesion of planktonic cells, (2) microcolony formation, (3) biofilm maturation and (4) detachment (also termed dispersal) of bacteria, which may then colonize new areas (**Figure 1**). Sessile bacterial cells exist in the stationary or dormant growth phase, exhibiting phenotypes distinct from planktonic bacteria [8]. In biofilms, bacteria display exceptional resistance to environmental stresses, especially antibiotics. This makes biofilms a major public health problem, as they account for 60–80% of human microbial infections [9]. The different stages in biofilm formation involve different environments, as shown in **Figure 1**.

2.1 Attachment of planktonic cells

Biofilm formation starts with the attachment of microbial cells to abiotic or biotic surfaces. These biotic surfaces are living tissues such as endothelial lesions, mucosae, and nervous tissues, while abiotic surfaces are non-living cells including indwelling devices, prostheses, clinical environment surfaces, vascular and urinary catheters [10]. This initial attachment depends on the motility and adhesins expression

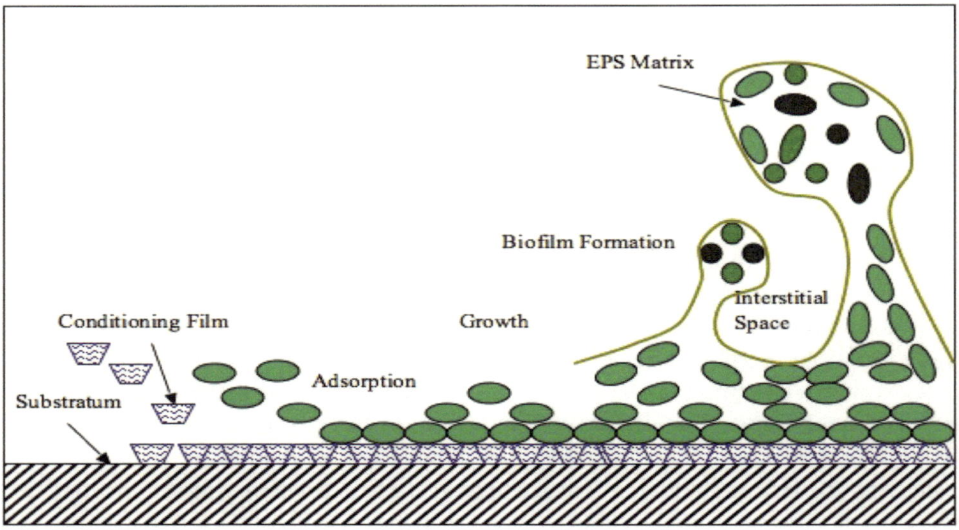

Figure 1.
Stages of biofilm development.

(microbial factors). The extension is influenced by the planktonic strains migrating to specific sites to either adhere to existing lesion or surface or directly cause tissue infection [11]. The physiology of the cell's changes affecting the surface membrane proteins making the removal of the attached cells laborious, necessitating the action of specific enzymes, sanitisers and detergent. The physicochemical properties of the surfaces (biotic and abiotic) controls microbial adherence making biofilms independent of surface extension [12].

2.2 The extracellular polymeric substance (EPS) matrix

The genes responsible for attachment and matrix assembly are activated when stimulated by factors such as population density and nutrient limitation [7]. The EPS matrix is composed of a mixture of biopolymers. The matrix produced is different and is surface- or medium-specific and differs between *in vivo* and *in vitro* conditions [11]. EPS is produced by planktonic cells, resulting in enhanced extension [13, 14].

2.3 Accumulation of multi-layered clusters of microbial cells

The microbial assembly development process results in simultaneous bacterial aggregation and growth. This disposition is entrenched as a distinct model with the aid of a confocal laser microscopy. The distinct model indicates that active metabolism is exhibited by the cells in the outer biofilm layers while those deeper inside the biofilm downregulate their metabolism, making them inactive in a persistent state [12, 15, 16]. This accumulation mostly involves intercellular adhesion. Specific genes and polysaccharide intercellular adhesin (PIA) are responsible for their accumulation on a polymer surface. However, the purification and structural analysis of these clustered microbial cells indicate the presence of two forms of that PIA, major polysaccharide I (>80%) and a minor polysaccharide II [17].

2.4 Biofilm maturation

In the biofilm maturation phase, the canals are created in the biofilm structure, allowing gradient-based passage of nutrients and signaling molecules based on their metabolic state, favoring the organized agglomeration and differentiation of cells [7, 12, 18]. These gradient passages are necessary for nutrients to enter the cells inside the biofilm layers. Biofilm structuring is a disruptive process causing the detachment of cell clusters controlling the biofilm invasion during *in vivo* biofilm infection leading to systemic dissemination [19].

2.5 The disentanglement and scattering of planktonic bacteria

The biofilms grow more thicker and compact on the interior, while external layers begin separating. The disentanglement and scattering occurs as a results of nutritional imbalance with insufficient carbon accessibility, increasing the synthesis of extracellular polymeric substances [20]. The scattered cells or clusters travel as septic emboli colonizing new sites, causing infection with possibly novel biofilms [2]. The dispersed cells form biofilms as a result of growth and may return quickly to their normal planktonic phenotype.

3. Bacterial biofilm structure, characteristics and chemical composition

3.1 Bacterial biofilm structure and characteristics

The basic structural units of a biofilm are microcolonies and separate communities of bacterial cells embedded into the EPS matrix. These microcolonies are in most cases mushroom-shaped or rod-like and can consist of one or more types of bacteria. The microcolonies consist of 10–25% cells and 79–90% EPS matrix depending on the bacterial type. This EPS matrix protects biofilm cells from various environmental conditions, such as UV radiation, changes in pH values, draining and temperature. There are channels through which water flows between microcolonies. These water channels function in distributing nutrients to microcolonies and receiving harmful metabolites as a simple circulatory system. Biofilms under different hydrodynamic conditions, such as laminar and turbulent flow, show changes in biofilm structure depending on the flow type. In laminar flow, bacterial microcolonies become round, and in turbulent flow, they extend in the downstream direction [21].

3.2 Chemical composition

The matrix of extracellular polymeric substances (EPSs) are self-secreted substances that keep bacterial cells in a compact structure attaching them to surfaces which makes the physical aspect of a biofilm [16]. The major constituent of the biomass of the biofilm is the hydrated EPS ranging between 2–15% of the total biofilm mass [4]. The EPS contains mostly extracellular DNA (eDNA), polysaccharides, proteins and lipids (**Table 1**) [22]. The EPS matrix exhibit three important characteristic features which are enhancing antimicrobial resistance, nutrient capture and social cooperation [14]. The tissues of higher organisms are similar to biofilms structures which are architecturally different and extremely heterogeneous in gene expression, all participating to the resistance mechanisms of biofilms [5, 23].

i. **Polysaccharides** is one of the major constituents of the EPS matrix adhering to cell surfaces forming a compact network. The majority of these molecules are heteropolysaccharides constituted of a mixture of neutral, charged sugar residues, organic and inorganic substituents contributing to their charged (polyanionic or polycationic) nature [24, 25]. The exopolysaccharide composition differ between microorganisms of the same species [26, 27]. These exopolysaccharides are indispensable to biofilm formation and constitute the protective barrier of the EPS matrix despite the heterogeneity among biofilms [21]. Additionally, they are also responsible for water retention within the biofilm. The high amount of water in the biofilm provides a highly hydrated environment that protects cells from fluctuations in water potential. The presence of water confers the biofilm to a nonrigid structure with different viscosities that allow movement of the cells within the matrix [28].

ii. **Extracellular proteins: structural proteins and enzymes**. These are also critical components of the matrix and are present in higher amounts than polysaccharides. The **structural proteins** are mainly involved in the stabilization of the biofilm architecture by connecting cells to the EPS [29]. The enzymes are essentially involved in the degradation of other matrix components, such as polysaccharides (dispersin B), matrix proteins (proteases), and eDNA (DNases).

Components	Percentage of matrix	Functions in biofilm	References
Microbial cells	2–5%	Cohesion of the structure	[5]
DNA and RNA	<1–2%	Cohesion of the structure Nutrient source Exchange of genetic information	[22]
Polysaccharides	1–2%	Cohesion of the structure Nutrient source Water retention Protective barrier Sorption of organic compounds and inorganic ions	[22]
Structural Proteins	<1–2%	Cohesion of the structure Nutrient source Protective barrier Sorption of organic compounds and inorganic ions Electron donor and acceptor	[5]
Enzymes	<1–2%	Enzymatic activity, Nutrient Source	[22]
Lipids and biosurfactants	<1–2%	Nutrient source	[22]
Water	Up to 97%	Lubricates the environment, simple circulatory system distributing nutrients to microcolonies	[22]

Table 1.
Chemical composition of biofilms.

Thus, the enzymatic activity within the biofilm provides nutrients to bacterial cells and promotes biofilm reorganization and dispersal [29]. In addition to polysaccharides and proteins, eDNA also contributes to the structural integrity of the matrix. The contribution of this component to the three-dimensional structure of the biofilm differs greatly among species [29]. The EPS matrix has an important role in biofilm formation, progression and durability as a result to its multiplex constitution and organization. It is also a protective barrier against external factors, a source of nutrients, enzymes and an intercellular connector. These unique features of the matrix participate in the high antimicrobial forbearance and/or recalcitrance of biofilms [15, 29].

4. Factors influencing bacterial biofilm formation and development

The formation of biofilms is a dynamic and complex process that includes the initial attachment of bacterial cells to the substratum, physiological changes within the microbe, multiplication of adhered cells to form microcolonies and finally biofilm maturation [30]. Biofilm-associated bacteria demonstrate distinct features from their free-living planktonic counterparts, such as different physiologies and high resistance to immune systems and antibiotics that render biofilms a source of chronic and persistent infections [2, 31]. It is known that the change in phenotype from planktonic to the sessile form occurs in response to changes in environmental conditions [3].

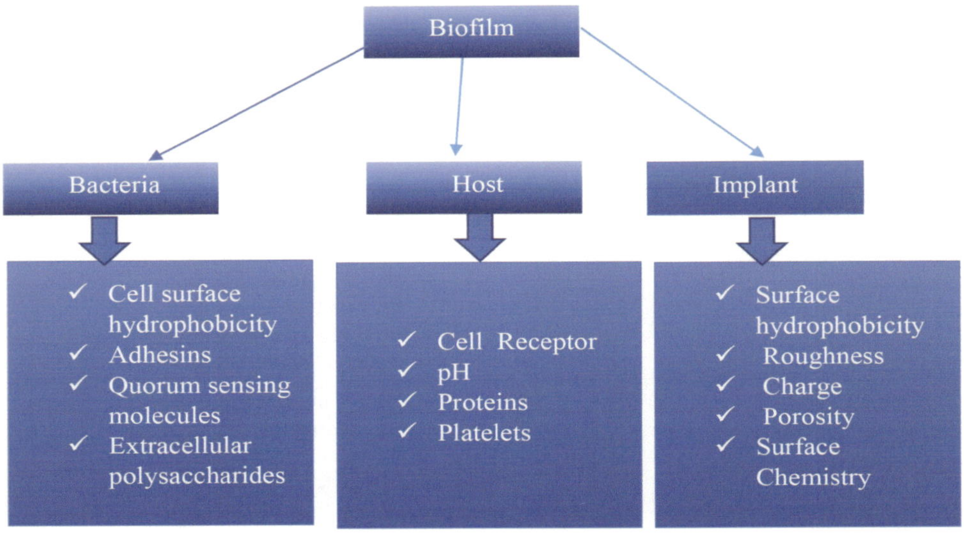

Figure 2.
Factors affecting biofilm formation.

Environmental factors, such as nutrient level, temperature, pH, and ionic strength, can influence biofilm formation, as shown in **Figure 2** [30]. These factors influence bacterial adhesion; cell surface properties, such as hydrophobicity, flagellation, and motility; surface properties, such as hydrophobicity and roughness; and environmental factors, such as temperature, pH, availability of nutrients and hydrodynamic conditions [21, 30, 32]. The cell surface properties, specifically the presence of extracellular appendages, such as fimbriae and flagella, the interactions involved in cell-to-cell communication and EPS production, such as surface-associated polysaccharides or proteins, possibly provide a competitive advantage for one organism in a mixed microbial community [3, 12]. Bacteria with hydrophobic properties are more likely to attach to surfaces than hydrophilic bacteria; however, the attachment of biofilms will occur readily on surfaces that are rough, hydrophobic, and coated by surface conditioning films.

The physicochemical properties of the substratum, such as texture (rough or smooth), hydrophobicity and charge, can also be modified by environmental conditions, such as pH, temperature, and nutrient levels [4, 10, 30]. In aquatic environments, the rate of microbial attachment can be increased by increasing the velocity of the flow, water temperature or nutrient concentration, providing that these factors do not exceed critical levels [6, 15].

Quorum Sensing: This is a bacterial cell–cell communication process that involves the production, detection, and response to extracellular signaling molecules called autoinducers (AIs) [33]. In Gram-positive bacteria, oligopeptides are used as signaling molecules to form biofilms, and quorum sensing is used for intraspecies communication. Quorum sensing controls processes such as bioluminescence, sporulation, competence, antibiotic production, biofilm formation, and virulence factor secretion [34]. Three main types of quorum sensing systems exist:

- Acyl-homoserine lactone quorum sensing system (AHL) in Gram-negative bacteria,

- Autoinducing peptide (AIP) quorum sensing system in Gram-positive bacteria

DOI: http://dx.doi.org/10.5772/intechopen.104772

- Autoinducer-2 (AI-2) system in both gram-negative and positive bacteria [34].

The acyl homoserine lactone-dependent QS system is a prominent cellular signaling molecules of homoserine lactones involved in quorum sensing regulation used primarily by Gram-negative bacteria. The AHL molecules have the homoserine lactone ring in common varying in length and substituents, synthesized by a specific AHL synthetase. The concentration of AHL contributes to bacterial growth. Autoinducing peptide (AIPs) are signal molecules secreted by membrane transporters and synthesized by Gram-positive bacteria. The AIPs bind to the histidine kinase sensor phosphorylating, consequently altering gene expression as the environmental concentration of AIPs augments [32, 35, 36]. These genes control the formation of innumerable toxins and decomposable exoenzymes [21, 36, 37]. The microorganisms can sense and translate the signals from distinct strains in AI-2 or autoinducer-2 interspecific signals, catalyzed by LuxS synthase as part of their cooperation and communication strategies [6, 25, 38]. Moreover, LuxS is involved in the activation of the methylation cycle and has been demonstrated to control the expression of hundreds of genes associated with the microbial processes of surface adhesion, detachment, and toxin production [24, 39, 40]. The QS system is a paramount target for the treatment of biofilm-associated infections [12].

5. Biofilm-producing bacteria and infections

Biofilm formation is present in approximately 65% of all bacterial infections and approximately 80% of all chronic infections according to the statistics of the National Institute of Health (NIH) (**Table 2**) [12]. Indwelling devices by bacteria settlement was associated with infections in 4% of the cases when pacemakers and inhaler were utilized and 2% in breast implant cases [35]. The device-related infections were estimated to be about 40% in ventricular-assisted devices, 2% in joint prostheses, 4% in mechanical heart valves and 6% in ventricular shunts [12, 25]. The heart infection (infective valve endocarditis) occurs as a result of the adherence of bacteria cells to the endothelium. The most frequent microbes being staphylococci and streptococci, members of the HACEK group, gram-negative bacteria and fungal strains [42]. The implanting of the endothelium generally occurs from colonization or the infection of different tracts (the genitourinary and gastrointestinal tract) or through the direct crossing of the skin barrier, either due to wounds or through injecting drugs [41]. Some biofilm-driven infections are chronic wounds, diabetic foot infections, and pulmonary infections in patients with cystic fibrosis and specific bacterial species (**Table 2**) [21, 37, 43].

6. Mechanisms of biofilm inhibition and eradication

i. Antibiofilm molecules and their mechanism of action:

The material matrix of implanted medical devices and biomaterials provides an ideal site for bacterial adhesion promoting mature biofilm formation [3]. Methods that prevent bacterial attachment to these materials represent a preventative strategy. The most common method for preventing bacterial extension is a surface modification (**Table 3**). The exterior surface of the implanted medical device or biomaterial

Bacterial strain	Gram stain	Types of infections	References
Staphylococcus aureus	Gram-positive	Chronic biofilm infections: chronic wound infection, right valve endocarditis, lung infections in patients with cystic fibrosis	[20]
Staphylococcus epidermidis	Gram-positive	Endocarditis: catheter-related infection, joint prosthesis infection	[20]
Streptococcus pneumoniae	Gram-positive	Lung infections, bacterial meningitis, acute or chronic otitis	[36]
Listeria monocytogenes	Gram-positive	Coculture interactions with *Pseudomonas*, Vibrio strains, listeriosis, contamination of food products	[36]
Burkholderia cepacia	Gram-negative	Opportunistic infections in patients with blood cancer	[41]
Escherichia coli	Gram-negative	Hemolytic uremic syndrome, acute diarrheic syndrome, urinary tract infections	[36]
Klebsiella pneumoniae	Gram-negative	Bacteremia, liver abscess, urinary tract infections	[20]
Pseudomonas putida	Gram-negative	Urinary tract infection	[36]
Pseudomonas aeruginosa	Gram-negative	Osteomyelitis, ventilator-associated pneumonia, lung infections in patients with cystic fibrosis, opportunistic infections in neutropenic patients, nosocomial infections.	[41]
Pseudomonas fluorescens	Gram-negative	Bioremediation, biocontrol- Pythium, Fusarium, antimicrobial properties –	[20]
Rhizobium leguminosarum	Gram-negative	Biocontrol properties – Pythium	[36]
Lactobacillus plantarum	Gram-positive	Salmonella infection	[41]
Lactococcus lactis	Gram-positive	Gastrointestinal tract infections	[20]

Table 2.
Examples of bacterial species involved in biofilm formation with their biological effects.

is altered, either directly or with the aid of a cover-producing barrier that is hostile to bacteria [45, 46]. This strategy has shown significant promise for preventing biofilm-related infections resulting from orthopedic implants. Thus, the area of surface modification to prevent biofilm formation is a large field [46–48]. The use of small molecule biofilm inhibitors is another approach used to prevent biofilm formation (**Figure 3**). The antibiofilm properties of a biofilm inhibitor are often employed to passivate the surface of an implanted medical device or biomaterial [41, 49, 50]. The use of biofilm inhibitors is one of the largest areas in biofilm remediation research, with a plethora of unique biofilm inhibitors currently described (phenols, imidazoles, furanone, indole, bromopyrrole) [51].

Anti-biofilm molecules are diverse compounds that inhibit biofilm formation. The identified anti-biofilm compounds are mainly isolated from natural sources, and some synthetic compounds, chelating agents, and antibiotics possess antibiofilm activity. The different antibiofilm molecules along with their target microorganisms are listed in **Table 2**. These antibiofilm molecules follow different mechanisms to inhibit biofilm formation in different bacteria, as listed in **Table 3**.

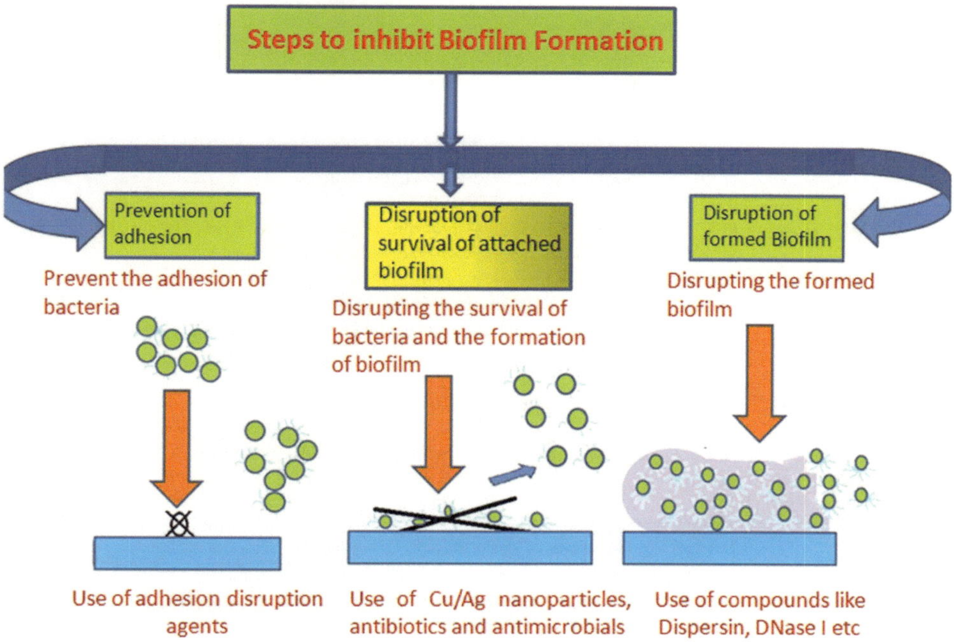

Figure 3.
The different steps in biofilm formation.

Resistance mechanism	Characteristics	References
Glycocalyx	The capsule is an important part of the biofilm in both Gram positive and negative bacteria. Its contribution to the maturation step relies on the electrostatic and hydrogen bonds established on the matrix and the abiotic surface. The composition in glycoprotein and polysaccharides varies with biofilm progression, permitting pathogens to live in difficult environment. The antimicrobial resistance is supported by the glycocalyx with the external layer acquiring antimicrobial compounds, serving as adherent for exoenzymes and protecting against antibacterial activity.	[6, 44]
Enzyme mediated resistance	The presence of heavy metals, such as cadmium, nickel, silver, zinc, copper, cobalt, and induces diversity of resistant phenotypes. This causes the enzymatic reduction of ionic particles mediating the transformation of toxic molecules to nontoxic or inactive.	[30]
Metabolism and growth rate heterogeneity	The bacterial metabolic activity and growth rate are influenced by the nutrients and oxygen concentrations within biofilms. This limits the metabolic activity inside the biofilm resulting in the reduction of the growing rate of strains. The enzymatic process inside biofilms is controlled by the changes in cell growth cycle regulating the metabolic and growth rate variations. These microbial communities increase the level of antimicrobial resistance inducing the expression of certain genes in different conditions.	[31, 32]
Cellular persistence	The infections' chronicity become tolerant to antibacterial agents with the persistent strains being responsible eliciting multidrug forbearance. The glycocalyx improves protection of the immune system inducing the growth of bacterial biofilm competing for antibiotic targets with multi-medicament resistance (MDR) protein synthesis.	[33–35]

Resistance mechanism	Characteristics	References
Metabolic state	The inaccessibility of nutrients due the exposition to bactericidal agent's inhibitory concentration affects the constitution of the prokaryotic envelope modifying it and conditioning the resistant cell population to exhibit phenotypic adjustment. **The genetic profile.** The mar operons are involved in the control of various genes' expression in *E. coli* assisting the MDR phenotype. The stress response cells display increase resistance to impaired factors within hours of exposure. The exposition of bacterial strains to molecular oxidants causes the diversified regulatory genes (oxyR and soxR) to exhibit persistence of the intracellular redox potential and the activation of stress response.	[38, 39]
Quorum sensing (QS)	QS regulates the heterogeneous organization with nutrient supply during the cell migration procedure. QS deficiency is linked with thinner microbial biofilm growth consequently lowering the EPS production.	[40, 42]
Stress response	The stress response acts as a preventive factor for cell damage more than repair. The causes of stress induction include starvation, decrease or increase temperature, high osmolality and low pH. The altered gene expression due to the stress response in immobilized strains result in increased resistance to antibiotics.	[41, 43, 45]
External membrane structure	The lipopolysaccharide layer prevents hydrophilic antimicrobials from entering through the outer membrane while the external membrane proteins reject hydrophobic molecules. Most antibacterial agents must penetrate the bacterial cells to target a specific site, modifying the cellular membrane that control antibiotic resistance.	[46, 47]
Efflux systems	The efflux pumps facilitate bacterial endurance under utmost environmental conditions exerting inherent and gained resistance to diverse antimicrobials of similar or divergent classes. The combination of similar recalcitrance processes leads to the overproduction of efflux pumps regulating the multi-medicament non-compliances. The efflux pumps are major player in the MDR of Gram-negative bacteria due to their clear mechanisms provided in drug discovery platforms of targeted bacterial pathogens.	[48–52]

Table 3.
Mechanism of biofilm-mediated antimicrobial resistance.

ii. Using Natural Products:

The formation and development of biofilms is a complicated procedure involving different stages that can be the target of natural antibiofilm agents for the prevention of biofilm development. Natural anti-biofilm agents either act solely or synergistically by diverse mechanisms.

There are five broad classes of natural compounds that have high antibiofilm properties, including phenolics, essential oils, terpenoids, lectins, alkaloids, polypeptides, and polyacetylenes [52]. Phenolics are a group of compounds. It has seven subclasses, which include phenolic acids, quinones, flavonoids, flavones, flavonols, tannins, and coumarins, out of which tannins, specifically condensed tannins, have anti-biofilm activity. These compounds act on biofilms by six main mechanisms, such as substrate deprivation, membrane disruption, binding to the adhesin complex and cell wall, binding to proteins, interacting with eukaryotic DNA, and blocking viral fusion [52, 53].

Many bioactive compounds from medicinal plants for the discovery of novel natural antibiofilm compounds are ongoing. The antibiofilm properties of Indian medicinal plants were studied with Cinnamomum *glaucescens (Nees) Hand.-Mazz, Syzygium praecox Roxb. Rathakr. & N. C. Nair, Bischofia javanica Blume, Elaeocarpus serratus L., Smilax zeylanica L., Acacia pennata (L.) Willd., Trema orientalis (L.) Blume, Acacia pennata (L.) Willd., Holigarna caustica (Dennst.) Oken, Murraya paniculata (L.) Jack, and Pterygota alata (Roxb.)* R. Br. extracts have promising antibiofilm activity against *S. aureus* [36, 53, 54]. Phytochemicals inhibit the quorum sensing mechanism mainly by blocking quorum sensing inducers such as AHL, autoinducers, and autoinducer type 2. Garlic extracts play a vital role in the inhibition of quorum sensing signaling molecules of *Pseudomonas* and *Vibrio* spp. Biofilms [5, 36, 52, 55]. Phytochemicals also play a significant role in inhibiting bacterial adhesion and suppressing genes related to biofilm formation. Biofilm development at the initial stages can be outlined by interfering with the forces (van der Waals force of attraction, electrostatic attraction, sedimentation and Brownian movements) that are responsible for the support of bacterial attachment to various surfaces [56]. Some phytocompounds have the potential to interfere with the extension along with the capability to stop the accessibility to nutrients essential for adhesion and bacterial growth. An alkaloid (norbgugaine) had a significant effect on *P. aeruginosa* biofilms by preventing adhesion due to loss of cell motility [9, 24, 55, 57]. A very recent study on *Adiantum philippense* L. crude extract showed a promising role in decreasing the content of biofilm exopolysaccharides [44, 58, 59]. It was reported that *A. philippense* L. crude extract restrained biofilms at the initial stages by targeting adhesin proteins, destroying the preformed biofilms inhibiting EPS assembly. Diverse group of phytocompounds especially polyphenols such as 7-epiclusianone, tannic acid, and casbane, have been identified and proved to protect cell surface. Members of Enterobacteriaceae express curli, an amyloid fiber on the cell surface that helps in attachment to characteristics and cell aggregation and enhances biofilm formation as well as a cellular invasion [41, 49, 60]. The phytocompounds of curlicide and pilicide nature can be exploited in therapeutic strategies of Enterobacteriaceae biofilm prevention [57, 61, 62]. These phytocompounds with fewer side effects are better therapeutic agents for biofilm-related infections, but recent reports suggest a combined approach that is always better than the individualistic approach [24, 44, 50, 51]. A few plant-based antimicrobials with the potential of anti-biofilm activity are summarized in **Table 4** [53].

7. Conclusion

Biofilm infections are highly resistant to antibiotics and physical treatments. Many strategies support biofilm antibiotic resistance and tolerance, such as persistent cells, adaptive responses, and limited antibiotic penetration. Thus, the underlying mechanisms of antibiotic forbearance and recalcitrance in biofilms are controlled by genes. In human infections, most organized bacterial cells gradually induce immune responses to form biofilms causing chronic infections leading to tissue destruction with permanent pathology. Therefore, biofilms arrangement is a vital perturbation in medical care environment. The exploration of alternative treatment procedures for biofilm-associated infections is of utmost importance. There are little novel and efficient antibiotic strategies which are scattering of biofilms, merging of antimicrobials with quorum sensing inhibitors, and a mixture of these procedures. Although the mentioned anti-biofilm strategies are key research areas, they are still in their infancy

Compound	Source	Pathogenic species	Experimental details	Molecular mechanism	References
Allicin	*Allium sativum* L.	*Pseudomonas aeruginosa*	*Invitro* (1pqsABCD knockout strain)	It decreases the bacterial attachment in the initial stages of biofilm formation as it reduces EPS formation It controls the expression of virulence factors hence interfere with the QS system	[52]
Ajoene		*P. aeruginosa* *P. aeruginosa* *Staphylococcus aureus*	*In vitro* (PMNs killing assays) and *in vivo* (Pulmonary infection mice model)	It downregulates rhamnolipid production It inhibits small regulatory RNA molecules (rsmY, rsmZ, and rnaIII) that operate in the later phase of QS signaling	[51]
Carvacrol (monoterpenoid)	*Origanum vulgare L.*	*P. aeruginosa*	*In vitro* (qPCR for relative expression of lasI/lasR genes) and ocking modeling of proteins LasI and LasR	It mainly acts on QS Machinery. The posttranslational inhibition against lasI, which effects AHL production.	[20]
Emodin (anthraquinone)	*Polygonum cuspidatum* Siebold & Zucc. *Rheum palmatum L.*	*S. aureus*	*In vitro* crystal violet biofilm assay and SEM analysis	It decreases the release of eDNA and downregulates the expression of biofilm-forming related genes like cidA, icaA, dltB, agrA, sortaseA, and sarA	[22]
Emodin (anthroquinone)	*R. palmatum L.*	*Candida albicans* *Candida krusei* *Candida parapsilosis* *Candida tropicalis*	*In vitro* (microdilution assay, kinase assay) and molecular docking for emodin in CK2 (Autodock Vina)	The biofilm formation is inhibited by targeting cellular kinase signaling It acts on planktonic cells by reducing hyphal formation. It acts as a competitive inhibitor of CK2.	[41]
Aloe-emodin	*Rheum officinale aill.*	*S. aureus*	*In vitro* (CLSM assays and Congo red assay)	It reduces the production of extracellular proteins and polysaccharide intercellular adhesin It inhibits biofilm formation on polyvinyl chloride surfaces	[52]

Compound	Source	Pathogenic species	Experimental details	Molecular mechanism	References
Hordenine	*Hordeum vulgare L.(sprouting)*	*P. aeruginosa*	*In vitro* (SEM and CLSM assays, qPCR for QS-related genes)	It decreases AHL production It reduces the exhibition of virulence factors (proteases, elastase, pyocyanin, rhamnolipid, alginate, and pyroviridine). It impedes the swimming and swarming activity It negatively regulates the expression of lasI, lasR, rhlI and rhlR genes	[52]
Pulverulentone A	*Callistemon citrinus* (Curtis) skeels leaves	*S. aureus*	*In vitro* broth microdilution assay, CLSM, TEM analysis Methicillin-resistant	It reduces styphyloxanthin production, thus inhibiting biofilm formation It disrupts the cell membrane.	[52]

Table 4.
Anti-biofilm activity of phytocompounds and their mechanism of action [53].

and has to be improved to upgrade and implement the strategies. The administration of a single antibiotic is often not enough to eradicate bacterial invasions, and a high concentration of the antibiotic can be extremely toxic. Also, some natural compounds as well as quorum sensing inhibitors, may be toxic and less effective. A possible solution might be the coadministration of antibiotics with antibiofilm peptides that allow the use of low antibiotic concentrations. New anti-biofilm molecules from natural substances with low or no harmful effects and synergistic effects with commonly used antibiotics are necessary. Moreso, natural products from medicinal plants and quorum sensing inhibiting compounds with little or no toxic effects will be of great importance in the fight against biofilms.

Author details

Zeuko'O Menkem Elisabeth[1,2]

1 Faculty of Health Sciences, Department of Biomedical Sciences, University of Buea, Buea, Cameroon

2 Faculty of Science, Antimicrobial and Biocontrol Agents Unit, Laboratory for Phytobiochemistry and Medicinal Plant Studies, Department of Biochemistry, University of Yaoundé I, Yaoundé, Cameroon

*Address all correspondence to: mllemenkem@gmail.com; zeukoo@yahoo.com; zeuko'o.elisabeth@ubuea.cm

References

[1] Rabin N, Zheng Y, Opoku-Temeng C, Du Y, Bonsu E, Sintim HO. Biofilm formation mechanisms and targets for developing antibiofilm agents. Future Medicinal Chemistry. 2015;**7**(4):493-512

[2] Kostakioti M, Hadjifrangiskou M, Hultgren SJ. Bacterial biofilms: Development, dispersal, and therapeutic strategies in the Dawn of the Postantibiotic era. Cold Spring Harbor Perspectives in Medicine. 2013;**3**(4):a010306-a010306

[3] Pugliese G, Favero MS. Biofilms: Survival mechanisms of clinically relevant microorganisms. Infection Control and Hospital Epidemiology. 2002;**23**(6):352-352

[4] Mari S, Vrane J. Characteristics and significance of microbial biofilm formation. Periodicum Biologorum. 2007;**109**(2):7

[5] Segev-Zarko L, Saar-Dover R, Brumfeld V, Mangoni ML, Shai Y. Mechanisms of biofilm inhibition and degradation by antimicrobial peptides. The Biochemical Journal. 2015;**468**(2):259-270

[6] Giaouris E, Simões M, Dubois-Brissonnet F. The role of biofilms in the development and dissemination of microbial resistance within the food industry. Food. 2020;**9**(6):816

[7] Crouzet M, Le Senechal C, Brözel VS, Costaglioli P, Barthe C, Bonneu M, et al. Exploring early steps in biofilm formation: Set-up of an experimental system for molecular studies. BMC Microbiology. 2014;**14**(1):253

[8] Percival SL, Malic S, Cruz H, Williams DW. Introduction to biofilms. In: Percival S, Knottenbelt D, Cochrane C, editors. Biofilms and Veterinary Medicine. Vol. 6. Berlin, Heidelberg: Springer Berlin Heidelberg; 2011. pp. 41-68. (Springer Series on Biofilms) Available from: http://link.springer.com/10.1007/978-3-642-21289-5_2

[9] Roy R, Tiwari M, Donelli G, Tiwari V. Strategies for combating bacterial biofilms: A focus on anti-biofilm agents and their mechanisms of action. Virulence. 2018;**9**(1):522-554

[10] Ponomareva AL, Buzoleva LS, Bogatyrenko EA. Abiotic environmental factors affecting the formation of microbial biofilms. The Biological Bulletin. 2018;**45**(5):490-496

[11] Bjarnsholt T, Alhede M, Alhede M, Eickhardt-Sørensen SR, Moser C, Kühl M, et al. The in vivo biofilm. Trends in Microbiology. 2013;**21**(9):466-474

[12] Georgiana PV, Săndulescu O. Communication is the key: Biofilms, quorum sensing, formation and prevention. Discover. 2019;**7**(3):e100

[13] Di Martino P. Groupe Biofilm et Comportement Microbien aux Interfaces, Laboratoire ERRMECe-EA1391, Université de Cergy-Pontoise, rue Descartes site de Neuville-sur-Oise 95031 Cergy-Pontoise, cedex France. Extracellular polymeric substances, a key element in understanding biofilm phenotype. AIMS Microbiol. 2018;**4**(2):274-288

[14] Tsuneda S, Aikawa H, Hayashi H, Yuasa A, Hirata A. Extracellular polymeric substances responsible for bacterial adhesion onto solid surface. FEMS Microbiology Letters. 2003;**223**(2):287-292

[15] Nandakumar V, Chittaranjan S, Kurian VM, Doble M. Characteristics of bacterial biofilm associated with implant material in clinical practice. Polymer Journal. 2013;**45**(2):137-152

[16] Jabbouri S, Sadovskaya I. Characteristics of the biofilm matrix and its role as a possible target for the detection and eradication of *Staphylococcus epidermidis* associated with medical implant infections. FEMS Immunology and Medical Microbiology. 2010;**59**(3):280-291

[17] von Eiff C, Peters G, Heilmann C. Pathogenesis of infections due to coagulasenegative staphylococci. The Lancet Infectious Diseases. 2002;**2**(11):677-685

[18] Liaqat I. Biofilm formation, maturation and prevention: A review. Journal of Bacteriology and Mycology. 2019;**6**(1):1092

[19] Otto M. Staphylococcal infections: Mechanisms of biofilm maturation and detachment as critical determinants of pathogenicity. Annual Review of Medicine. 2013;**64**(1):175-188

[20] Abdel-Aziz SM. Bacterial biofilm: Dispersal and inhibition strategies. SAJ Biotechnology. 2014;**1**(1):105. Available from: http://fulltext.scholarena.com/Bacterial-Biofilm-Dispersal-and-Inhibition-Strategies.php

[21] Wimpenny J, Manz W, Szewzyk U. Heterogeneity in biofilms: Table 1. FEMS Microbiology Reviews. 2000;**24**(5):661-671

[22] Guo X, Wang X, Liu J. Composition analysis of fractions of extracellular polymeric substances from an activated sludge culture and identification of dominant forces affecting microbial aggregation. Scientific Reports. 2016;**6**(1):28391

[23] Jin X, Marshall JS. Mechanics of biofilms formed of bacteria with fimbriae appendages. Nerenberg R, editor. PLoS One. 2020;15(12):e0243280.

[24] Pinto RM, Soares FA, Reis S, Nunes C, Van Dijck P. Innovative strategies toward the disassembly of the EPS matrix in bacterial biofilms. Frontiers in Microbiology. 2020;**26**(11):952

[25] Jamal M, Tasneem U, Hussain T, Andleeb S. Research & Reviews: Journal of Microbiology and Biotechnology. 2015;**4**(3):14

[26] Flemming H-C, Wingender J. The biofilm matrix. Nature Reviews. Microbiology. 2010;**8**(9):623-633

[27] Flemming H-C, Neu TR, Wozniak DJ. The EPS matrix: The "house of biofilm cells.". Journal of Bacteriology. 2007;**189**(22):7945-7947

[28] Flemming H-C, Wingender J, Szewzyk U, Steinberg P, Rice SA, Kjelleberg S. Biofilms: An emergent form of bacterial life. Nature Reviews. Microbiology. 2016;**14**(9):563-575

[29] Fong JNC, Yildiz FH. Biofilm matrix proteins. Ghannoum M, Parsek M, Whiteley M, Mukherjee P, editors. Microbiol Spectr. 2015;**3**(2):1-16. Available from: https://journals.asm.org/doi/10.1128/microbiolspec.MB-0004-2014. 2015 American Society for Microbiology. All rights reserved

[30] Alotaibi GF, Bukhari MA. Factors influencing bacterial biofilm formation and development. American Journal of Biomedical Science and Research. 2021;**12**(6):001820. DOI: 10.34297/AJBSR.2021.12.001820. ISSN: 2642-1747. www.biomedgrid.com.

[31] Reichhardt C, Fong JCN, Yildiz F, Cegelski L. Characterization of the vibrio cholerae extracellular matrix:

A top-down solid-state NMR approach. Biochim Biophys Acta BBA - Biomembr. 2015;**1848**(1):378-383

[32] Beitelshees M, Hill A, Jones C, Pfeifer B. Phenotypic variation during biofilm formation: Implications for anti-biofilm therapeutic design. Materials. 2018;**11**(7):1086

[33] Rutherford ST, Bassler BL. Bacterial quorum sensing: Its role in virulence and possibilities for its control. Cold Spring Harbor Perspectives in Medicine. 2012;**2**(11):a012427-a012427

[34] Mangwani N. Bacterial quorum sensing: Functional features and potential applications in biotechnology. Journal of Molecular Microbiology and Biotechnology. 2012;**22**:215-227

[35] Mombelli A, Décaillet F. The characteristics of biofilms in peri-implant disease: Biofilms in peri-implant disease. Journal of Clinical Periodontology. 2011;**38**:203-213

[36] Ribeiro SM, Felício MR, Boas EV, Gonçalves S, Costa FF, Samy RP, et al. New frontiers for anti-biofilm drug development. Pharmacology & Therapeutics. 2016;**160**:133-144

[37] Costerton JW, Lewandowski Z, Caldwell DE, Korber DR, Lappin-Scott HM. Microbial biofilms. Annu Rev Microbiol. 1995;**49**:711-745

[38] MacLeod FA, Guiot SR, Costerton JW. Layered structure of bacterial aggregates produced in an upflow anaerobic sludge bed and filter reactor. Applied and Environmental Microbiology. 1990;**56**(6):1598-1607

[39] Vickery K. Special issue: Microbial biofilms in healthcare: Formation. Prevention and Treatment. Materials. 2019;**12**(12):2001

[40] Melia CE, Bolla JR, Katharios-Lanwermeyer S, Mihaylov DB, Hoffmann PC, Huo J, et al. Architecture of cell–cell junctions in situ reveals a mechanism for bacterial biofilm inhibition. Proceedings of the National Academy of Sciences. 2021;**118**(31):e2109940118

[41] Srinivasan R, Santhakumari S, Poonguzhali P, Geetha M, Dyavaiah M, Xiangmin L. Bacterial biofilm inhibition: A focused review on recent therapeutic strategies for combating the biofilm mediated infections. Frontiers in Microbiology. 2021;**12**(12):676458

[42] Sen Yew H, Chambers ST, Roberts SA, Holland DJ, Julian KA, Raymond NJ, et al. Association between HACEK bacteraemia and endocarditis. Journal of Medical Microbiology. 2014;**63**(6):892-895

[43] Wilson M. Bacterial biofilms and human disease. Science Progress. 2001;**84**(3):235-254

[44] Harjai K, Kumar R, Singh S. Garlic blocks quorum sensing and attenuates the virulence of *Pseudomonas aeruginosa*. FEMS Immunology and Medical Microbiology. 2010;**58**(2):161-168

[45] Hobley L, Harkins C, MacPhee CE, Stanley-Wall NR. Giving structure to the biofilm matrix: An overview of individual strategies and emerging common themes. Albers S-V, editor. FEMS Microbiology Reviews. 2015;**39**(5):649-669

[46] Arciola CR, Campoccia D, Ravaioli S, Montanaro L. Polysaccharide intercellular adhesin in biofilm: Structural and regulatory aspects. Front Cell Infect Microbiol [Internet]. 2015;**5**:00007. Available from: http://journal.frontiersin.org/Article/10.3389/fcimb.2015.00007/abstract

[47] Bi Y, Xia G, Shi C, Wan J, Liu L, Chen Y, et al. Therapeutic strategies against bacterial biofilms. Fundam Res. 2021;**1**(2):193-212

[48] Hamzah H, Hertiani T, Utami Tunjung Pratiwi S, Nuryastuti T, Bayu Murti Y. The biofilm inhibition and eradication activity of curcumin againts polymicrobial biofilm. Susilowati H, Purwestri YA, Afiahayati, Gunadi, Dharmastiti R, Nuringtyas TR, et al., editors. BIO Web Conf. 2020;28:04001.

[49] Verderosa AD, Totsika M, Fairfull-Smith KE. Bacterial biofilm eradication agents: A current review. Frontiers in Chemistry. 2019;**28**(7):824

[50] Chen H, Wubbolts RW, Haagsman HP, Veldhuizen EJA. Inhibition and eradication of Pseudomonas aeruginosa biofilms by host Defence peptides. Scientific Reports. 2018;**8**(1):10446

[51] Di Somma A, Moretta A, Canè C, Cirillo A, Duilio A. Inhibition of bacterial biofilm formation. In: Dincer S, Sümengen Özdenefe M, Arkut A, editors. Bacterial Biofilms. IntechOpen; 2020 2022. Available from: https://www.intechopen.com/books/bacterial-biofilms/inhibition-of-bacterial-biofilm-formation DOI: 10.5772/intechopen.90614

[52] Cowan MM. Plant products as antimicrobial agents. Clinical Microbiology Reviews. 1999;**12**:19

[53] Mishra R, Panda AK, De Mandal S, Shakeel M, Bisht SS, Khan J. Natural anti-biofilm agents: Strategies to control biofilm-forming pathogens. Frontiers in Microbiology. 2020;**29**(11):566325

[54] Sandasi M, Leonard CM, Viljoen AM. The in vitro antibiofilm activity of selected culinary herbs and medicinal plants against listeria monocytogenes: Anti-biofilm activity. Letters in Applied Microbiology. 2010;**50**(1):30-35

[55] Vetrivel A, Ramasamy M, Vetrivel P, Natchimuthu S, Arunachalam S, Kim G-S, et al. Pseudomonas aeruginosa biofilm formation and its control. Biologics. 2021;**1**(3):312-336

[56] Ciofu O, Rojo-Molinero E, Macià MD, Oliver A. Antibiotic treatment of biofilm infections. APMIS. 2017;**125**(4):304-319

[57] Paluch E, Rewak-Soroczyńska J, Jędrusik I, Mazurkiewicz E, Jermakow K. Prevention of biofilm formation by quorum quenching. Applied Microbiology and Biotechnology. 2020;**104**(5):1871-1881

[58] Al-Daihan S, Al-Faham M, Al-shawi N, Almayman R, Brnawi A, zargar S, et al. Antibacterial activity and phytochemical screening of some medicinal plants commonly used in Saudi Arabia against selected pathogenic microorganisms. J King Saud Univ - Sci. 2013;25(2):115-120.

[59] Macia MD, Rojo-Molinero E, Oliver A. Antimicrobial susceptibility testing in biofilm-growing bacteria. Clinical Microbiology and Infection. 2014;**20**(10):981-990

[60] Lerche CJ, Schwartz F, Theut M, Fosbøl EL, Iversen K, Bundgaard H, et al. Anti-biofilm approach in infective endocarditis exposes new treatment strategies for improved outcome. Frontiers in Cell and Development Biology. 2021;**18**(9):643335

[61] Husain FM, Ahmad I, Al-thubiani AS, Abulreesh HH, AlHazza IM, Aqil F. Leaf extracts of Mangifera indica L. inhibit quorum sensing – Regulated production of virulence factors and

biofilm in test bacteria. Frontiers in Microbiology. 2017;**24**(8):727

[62] Murugaiyan J, Kumar PA, Rao GS, Iskandar K, Hawser S, Hays JP, et al. Progress in alternative strategies to combat antimicrobial resistance: Focus on antibiotics. Antibiotics. 2022;**11**(2):200

Chapter 10

Approaches to Enhance Therapeutic Activity of Drugs against Bacterial Biofilms

Sankar Veintramuthu and Selliamman Ravi Mahipriya

Abstract

Biofilm may be a consortium of microbial species where the cells of microbes attach to both life form and inanimate surfaces inside a self-made matrix of extracellular polymeric substance (EPS). Biofilm matrix surrounding the polymicrobial environment makes them highly resistant to harsh conditions and antibacterial treatments. The two significant factors that differentiate planktonic from biofilm resident microbes are EPS containing a variety of macromolecules and a diffusible molecule for transferring signals known as quorum sensing (QS). Against this backdrop of microbial resistance and cell signaling, different approaches have been developed to interfere with the specific mechanisms of intracellular and extracellular targets that include herbal active compounds and synthetic nanoparticles. This chapter outlines the features of biofilm development and the approaches with the evidence that can be incorporated into clinical usage.

Keywords: biofilm, antimicrobial resistance, quorum sensing, herbal compounds, nanoparticles

1. Introduction

In seventeenth century, Antonie von Leeuwenhoek saw microbial aggregates on the scrapings of the plaque from his teeth that was termed as "biofilm" by Bill Costerton in 1978 [1]. The biofilms were not characterized for their physical and chemical properties until the end of 1960 [2]. The evolution of scanning electron microscopy and transmission electron microscopy allowed for identifying the biofilm from wastewater treatment plant [3] after when Heukelekian and Heller identified the *"Bottle effect"* on marine microbes where there is a significant difference in the microbial population between *in situ* and *in vitro* due to environmental or man-made changes. Biofilm is an aggregation of microbially derived sessile communities having various bacterial colonies or individual cells in the group, which adheres to the surface. This group of cells attaches on an extracellular polymeric substance (EPS), a matrix that is mostly comprised of environmental DNA (eDNA), proteins, and polysaccharides, which provides significantly excessive resistance to antibiotics [4]. Bacterial biofilm can be formed in response to various factors such as high salt

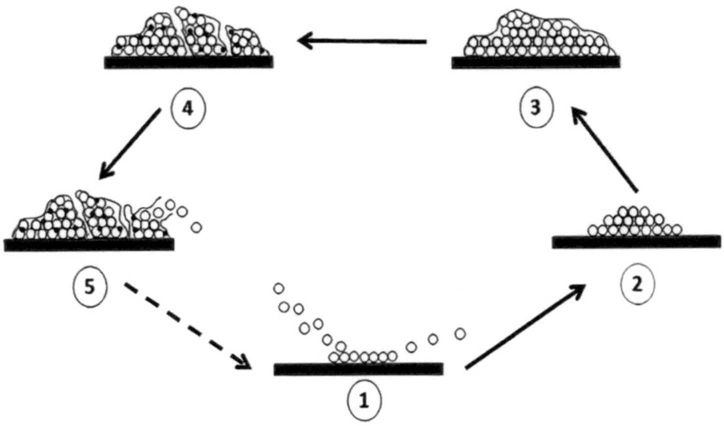

Figure 1.
Biofilm life process. (1) Planktonic bacteria attaches to the exterior face. (2) Adhesion, irreversible attachment occurs at this phase. (3) EPS is secreted and results in a matrix that forms the basis for biofilm's structure and initiates the onset of biofilm maturation. (4) The biofilm becomes completely matured, with the tower-like structures dispersed with water channels for the movement of oxygen, nutrients, and for discharging waste products.

concentration, restricted nutrients, high pH and pressure, and UV radiation. Biofilm life process is depicted in **Figure 1**.

The biofilm formation can be described in three steps:

A. Flexible attachment of bacteria to the surface subsequently irreversible attachment with the help of adhesive structures of bacteria.

B. Production of EPS and development of an organized structure entrapped inside an EPS matrix.

C. Finally, bacterial cell starts to break out from the biofilm and spread into the habitat through chemical signaling [5].

2. Biofilm: a threat to antibiotics and infections caused by biofilm

Around 80% of chronic and periodic microbial infections in the human bodies are caused by bacterial biofilm. Bacteria's present inside the biofilm aids to the chronic phase of infection, when released from the biofilm can cause an acute phase of infection [6]. The infections caused by bacterial biofilm can be placed in two broad categories such as device and non-device-associated infections. They can develop on or inside medical devices that are built in body such as central venous catheters, mechanical heart valves, pacemakers, urinary catheters, which cover both Gram-positive and Gram-negative bacteria or yeasts. These organisms on the medical devices may cause blood stream and urinary tract infections in the patient [7]. **Table 1** shows the microbial species that colonizes the devices based on the type of medical device and time taken for their action.

Microbial biofilm show 10–1000 times more antibiotic resistance than the planktonic species [12]. Bacterial biofilm offers huge evolutionary advantage for the

S. No	Medical device	Microbial organisms
1.	Contact lenses	*Escherichia coli, Staphylococcus epidermidis, Pseudomonas. aeruginosa,* species of *Candida, Serratia and Proteus, Staphylococcus aureus.* [8]
2.	Central venous catheters	*Klebsiella species, P. aeruginosa, and Enterobacter species* [9]
3.	Mechanical heart valve	*Enterococcus and Candida spp, Streptococcus species, S. aureus, S. epidermidis, Bacillus* [10]
4.	Urinary catheters	*E. coli, Enterococcus faecalis, S. epidermidis, P. aeruginosa, Proteus mirabilis, Klebsiella pneumonia* [11]

Table 1.
Medical devices and associated biofilm organisms.

bacteria including changes in environmental pH, resistance to antimicrobial agents, and phagocytic attack [13].

3. Quorum sensing (QS) and interaction

The bacterial cells have intercellular communication that is delivered through the extracellular signaling molecules known as autoinducers. The collection of signaling molecules enables individual bacterial cells to analyze the total number of bacteria, that is, cell density known as quorum sensing. In low-density planktonic populations, bacteria releases low-molecular-weight, highly diffusible, signal molecules (autoinducers, such as oligopeptides in Gram-positive bacteria and N-acyl-L-homoserine lactones in Gram-negative bacteria) at very low levels to produce changes in gene expression. When critical mass of bacterial population becomes high, the concentration of autoinducer molecules increases in the EPS followed by allowing individual bacteria to sense the presence of other bacterial species [14].

4. Conventional treatments and antimicrobial resistance

Biofilms are considered to be important owing to their potency in showing resistance toward antibiotics and antifungals. Once routed within the wound infection, biofilm shows enhanced tolerance to conventional treatments. Antibiotics work by deranging the cell wall of bacteria and affecting the DNA replication, repair, and protein synthesis. Apparently biofilm has various mechanisms through which they resist the effectiveness of antibiotics [15]. The primary defense mechanism involves EPS, which is capable of restricting the permeability of antibiotics into the cell thereby trapping them in the pores, followed by acidic internal environment and lack of oxygen. Ultimately, the lysis of genetic component can be carried between the cells to extend antimicrobial resistance.

The persister cells within the biofilm have the potency to restrict the effects of antibiotics targeting the cell division [16]. **Figure 2** illustrates the mechanism through which biofilm develops resistance to conventional antibiotics [17].

The resistance can be developed through persistent cells, phenotype of the biofilm, inhibition in antibiotics penetration, production of enzymes that resist the action of antimicrobial agents.

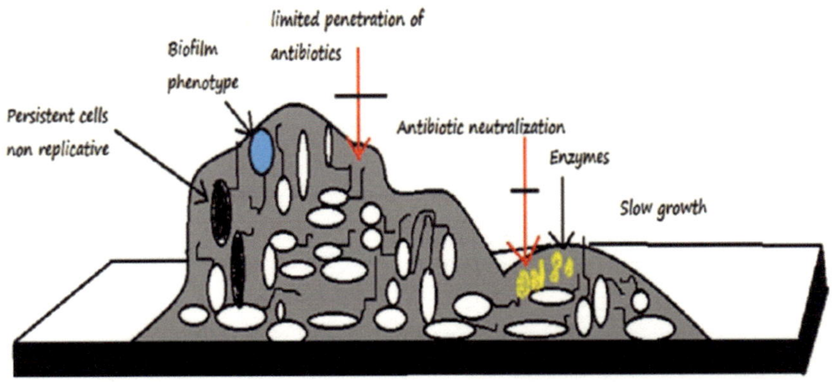

Figure 2.
Antibiotic resistance associated with biofilm.

5. Nanoparticles (NPs) as antibiofilm agents

Nanotechnology is fascinating, which likely benefited the field of biomedical and became widely conceded for the treatment of various diseases. Numerous resistance mechanisms set biofilm as one of the major disputes in infection treatment, which can be addressed by the strategy of using nanoparticles. NPs have two or three dimensions in the size range of 1 to 100 nm. They are of various types based on their size, shape, and composition [18]. Their higher surface area built them as suitable drug career, which has the capability to immobilize the compounds on their surface to increase their solubility and targeted delivery [19]. They can be of two types, polymer NPs and metallic NPs. Polymer NPs also possess the advantage of retaining the drug inside

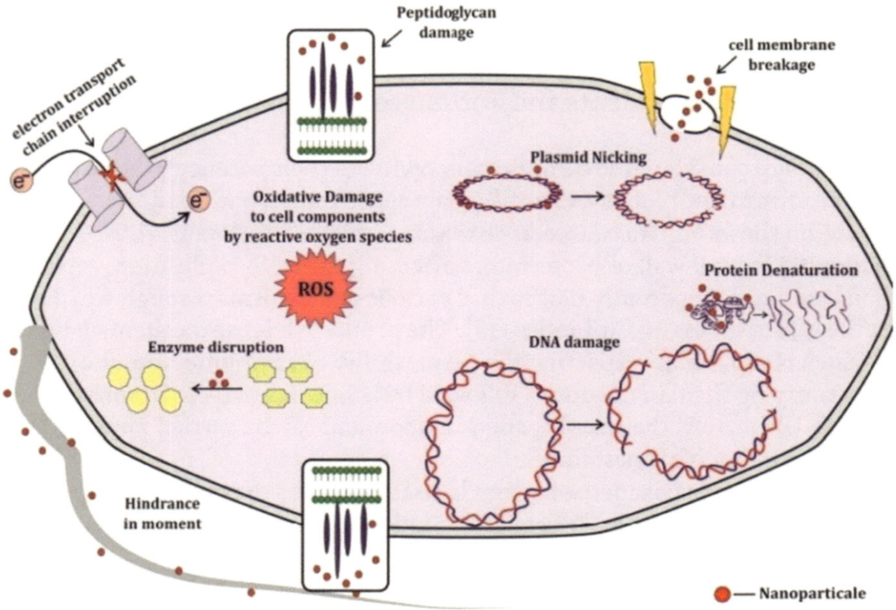

Figure 3.
Depiction of various mechanisms involved in combating biofilm through NPs.

the cavity and delivers the drug at the target area in either entangled or immobilized form. Reports suggest that NPs disrupt the integrity of biofilm by interacting with EPS, eDNA, proteins, lipids, and biofilm release reactive oxygen species (ROS) on interaction with NPs that can damage the cell envelope, cell membranes, cell structures, and biomolecules of the microbes. **Figure 3** represents the general mechanism involved in combating biofilm through NPs [20].

The nanoparticles can restrict biofilm by disruption electron transport between cell membrane, damaging the peptidoglycan layer, breaking through the cell membrane, denaturation of proteins, and DNA damage.

6. Synthesis of nanoparticles

Nanoparticles can be synthesized in laboratory broadly using two different approaches, that is, bottom-up and top-down techniques. The top-down approach implies breaking the bulk material into nanosized structures, which is based on miniaturizing the bulk substance through fabrication process and produces the NPs of appropriate properties. Bottom-up technique is an alternative approach because it creates less waste and involves building up of a material from the bottom [21].

7. Types of nanoparticles

Polymeric NPs can be engineered to release antibiotics, antibacterial agents, and bacteriostatic peptides or by modifying their chemical surface. The antibacterial activity of these organic NPs is due to polycationic groups accountable for cell damage through ion exchange interaction between bacteria and polymer surface with charges [22]. Metals are used in the synthesis of nanoparticles because of their antibacterial property broadly used in managing infections. Metallic nanoparticles can exert physical disruption to bacterial biofilms. **Table 2** enlists the types of metallic nanoparticles and their potential antimicrobial property [23].

The pH of micro-environment, magnetic field, or light can be used to turn on the nanomaterials or transform it to more active species enhancing their antibiofilm activity. These are often metallic nanoparticles due to their broad-spectrum

S. No	Metallic nanoparticles	Properties
1.	Zinc oxide	These NPs gets accumulated inside the cell releasing H_2O_2 and zinc ions thereby causing cell wall disruption.
2.	Titanium dioxide	Generation of reactive oxygen species.
3.	Copper oxide	Lipid oxidation takes place through reactive oxygen species and hydroxyl free radicals.
4.	Carbon nanotube	Reactive oxygen species results in cell wall disruptionthereby oxidizing lipids and proteins.
5.	Gold	They produce strong electrostatic effects and reacts with cell membrane.
6.	Silver	Releases silver ions and causes electron impairment DNA damage.

Table 2.
Metallic nanoparticles and their antimicrobial property.

antimicrobial activity and rich surface chemistry [24]. For negatively charged bacteria, the adhesion property rises because of the positively charged surface of NPs and the binding takes place through electrostatic interactions and Van der Waals interaction especially to cell membrane proteins [25].

8. Metal NPs against biofilm

CuO NPs inhibit formation of biofilm that was studied by *Agarwal et al.* that concluded eradication of biofilm formed by MRSA and *E. coli* with the exposure period of 4 days to CuO NPs at the concentration of 50 μg/ml [26]. ZnO NPs can exhibit antibacterial action between the concentration of 20–500 μg/ml for *E. coli* and *S. aureus* that can be enhanced by additional physical exposure and amplified by ultrasound [27]. MgO NPs can act against Gram-positive and Gram-negative bacteria, bacterial spores, and viruses at higher concentration of 100–1200 μg/ml. TiO_2 NPs can destroy biofilms of both Gram-positive and Gram-negative bacteria but the latter being more sensitive due to the sturdy layer of peptidoglycan that increases the absorption of reactive radicals [28]. However, their toxicity to humans and environment outweighs their advantages. Gold and silver NPs offer huge advantages such as higher surface area to volume ratio, small size, amenability, cheaper method of synthesis. Extensive research studies have been accomplished over the recent areas involving AgNPs and AuNPs. Three important steps involved in their antimicrobial action are a) interaction with biofilm when it comes into contact with the surface, b) subsequent penetration of NPs into the cell based on this interaction, and c) NPs as a whole or ions (Au^+ and Ag^+) reacts with cellular and biofilm components. The factor that plays significant role in penetration includes particle size, surface chemistry, surface charge, and concentration.

8.1 Silver nanoparticles: a biofilm buster

Silver has been used since remote time because of their therapeutic properties and their antibacterial activity and also explored through extensive research in medical field. Topical ointments and creams contain silver for treating burn wound infection. Several approaches are involved in synthesis of AgNPs, which include the use of microorganisms and plants but one of the easiest and convenient methods is through chemical synthesis [29]. **Table 3** enlists some of the sources that can be used for the synthesis of AgNP.

The synthesis of AgNPs can also be done by utilizing other physical methods such as evaporation-condensation and laser ablation, UV-initiated photo reduction, electrochemical synthetic method, irradiation methods [35]. Studies concluded that geometric mean diameter, shape, pH, and source for synthesis of AgNPs influence their efficiency. The synthesized AgNP can be characterized using UV-visible spectroscopy, X-ray diffraction, scanning electron microscopy, transmission electron microscopy for their structural properties. Although AgNPs were remarkably noted for their potential in pathogenic control, their effect on EPS has not been given sufficient attention [36].

8.2 Natural compounds as antibiofilm agents

Herbal compound aids the determination of novel constituents with interesting structures and biological activity. The antibiofilm properties of natural products rely on the inhibition of polymer matrix formation, resisting cell adhesion and attachment, breaking in ECM generation, and reducing virulence factors generation,

S.No	Approaches	Sources for synthesis
1.	Microbial approach	*Cedecea* sp., *Pseudomonas* sp., *Lactobacillus plantarum, Aspergillus fumigatus, Aeromonas sp., Klebsiella pneumonia, Corynbacterium sp., Enterobacter cloacae, Verticillium sp., Fusarium semitectum, Fusarium oxysporium.* [30–32]
2.	Plant synthesis	*Aloe vera* leaf extract, *Azadirachta indica, Cinnamomum camphora, Emblica officinalis, Pinus eldarica, Cassia auriculata* leaf extract, *Geranium* leaf extract, *Ficus benghalensis* leaf extract, Aqueous fruit extract of *Syzygium alternifolium,* fruit extract of *Sambucus nigra [33, 34]*
3.	Chemical reduction	DMF, $NaBH_4$, Trisodium citrate, ascorbic acid, dextrose, ethylene glycol, glucose.

Table 3.
Sources used for synthesis of AgNP.

thereby obstructing QS network and biofilm development [37]. The natural compounds that possess antibiofilm properties can be broadly classified into phenolics, essential oils, terpenoids, lectins, alkaloids, polypeptides, and polyacetylenes [38]. They either act merely or synergistically by different mechanisms. Various researches have been carried out with natural products that are discussed below:

8.2.1 Garlic

Allium sativum L has been extensively used in treating numerous diseases such as wound infection, malaria, common cold, sexually transmitted diseases [39]. Garlic possibly has a QS-interfering compound. DNA microarray analysis disclosed that Ajoene, a garlic-derived sulfur-containing compound, restricted QS-regulated gene expression in *P. aeruginosa*. Reasonable designing and biological screening of all compounds from garlic was carried out, resulting in the identification of a potent QS inhibitor N-(heptylsulfanylacetyl)-l-homoserine lactone. This element was found to disrupt the QS signaling by inhibiting transcriptional regulators LuxR and LasR. Recent studies have proved the antiswarming, anti-adherence, and antibiofilm activity of the aquatic extracts of garlic [40]. Ethanolic and methanolic extracts of garlic against six different bacterial species *(Escherichia coli, Staphylococcus aureus, Bacillus cereus, Streptococcus pneumoniae, Pseudomonas aeruginosa,* and *Klebsiella pneumonia*) show antibacterial activity at the concentration of 125–500 mg/mL through disc diffusion, and the *A. sativum* L extracts were potent enough restrict biofilm structures and the concentrations of each extract depend on the inhibitory effect [41].

8.2.2 Onion

Extracts of onion contains pharmaceutical properties that can be used as one of the promising therapies for the treatment of neoplastic, metabolic, and immunological diseases, which also involves bacterial, viral, and other fungal infections [42]. The anti-adherence, antibacterial, antibiofilm, and antimotility role of aqueous extracts of fresh or powdered onion and onion oil were studied from which the aqueous extracts of fresh and powdered onion showed more powerful inhibitory effects on biofilm than onion oil on the growth of both Gram-positive and Gram-negative bacteria [43]. Systematic assessment of quercetin, total phenolics, flavonoids, antioxidants, antibacterial, and antibiofilm or antibiofouling properties of methanolic extracts of fresh and aging onions of six varieties was studied by Kavitha *et al.,* which concluded that the onions that had been stored for 3 months showed the best antibiofilm effects. The red variety of *Allium cepa* extract was found to have higher antimicrobial activity when compared with the white and yellow varieties. At the range of about 50 μg mL^{-1}, the extracts were observed to reduce the biofilm growth of *P. aeruginosa* and *S. aureus* [44].

8.2.3 Rhubarb

Rhubarb is one of the most traditionally available medicinal materials included in Pharmacpoeia due it its bacteriostatic and anti-inflammatory properties. Emodin is the bioactive compound that has the ability to reverse multi-drug resistance. Natural emodin is obtained from Rheum palmatum L., Rheum tanguticum Maxim ex Balf, and Rheum ocinale [44]. Yan *et al.* studied the activity of emodin against *S. aureus* biofilm and confirmed the molecular mechanism that they decrease the release of eDNA and represses the biofilm-forming genes such as *cidA, icaA, dltB, agrA, sortaseA,* and *sarA* [45].

8.2.4 Banana

Studies concluded the antibacterial properties of banana in traditional medicine across the world. Generally, stem juice, flowers, and fruits of the banana plant are utilized for treating diarrhea and dysentery [46]. Vijayakumar *et al.* studied the antibiofilm properties of *Musa acuminata Colla.* against *P. aeruginosa* and described the mechanism of inhibiting the secretion of biofilm proteins and cell surface hydrophobicity productions [47].

8.2.5 Ginger

Ginger had been used in food and medicine for thousand years with the evidence demonstrating that it has antibacterial activity against the commercially available

S.No	Source	Active compound	Mechanism of action
1.	*Origanum vulgare* (oregano)	Carvacrol	Post-translational inhibition against*lasI*, which effects N- acyl-homoserine lactone secretion. It mostly acts on QS machinery against *P. aeruginosa* [51].
2.	*Apis mellifera* (Honey)	Defensin-1	Manuka and Honey dew significantly reduce cell viability of *S. aureus, P. aeruginosa, S. agalactia* [52]
3.	*Curcuma longa L* (Turmeric)	Curcumin	Restricts pellicle formation, Pilli motility and ring biofilm formation by interaction with biofilm response regulator *BfmR* [53].
4.	*Camellia sinesis (L)*	Epigallocatechin-3-gallate	Reduce the curli production and expression of curli-related proteins *csgA, csgB,* and *csgD* increases the degradation of sigma factor (RpoS) by ClpXP protease [54].
5.	*Capsicum annum* (Bell pepper)	Capsicum storage peptide 37 (CSP37)	Inhibited the formation and development of biofilm in common pathogenic strains at the concentration of 5 and 10 mg/ml through CSP [55].

Table 4.
Natural compounds with antibiofilm activity.

antibiotics by inhibiting QS signaling pathway [48]. Kim *et al.* initially investigated the inhibition of biofilm with ginger extract in *P. aeruginosa*. The biofilm assay demonstrated that the ginger extract decreased the biofilm development by 39–56% by reducing the formation of extracellular polymeric substances (EPS), which was associated with the suppression in secondary messenger, bis-(3c-5c)-cyclic dimeric-guagranosine [49]. Studies have shown that ginger essential oil at the biofilm inhibitory concentration (BIC) of 1.56 μL mL^{-1}, *S. aureus* had 94% inhibition of biofilm, and at BIC 0.78 μL mL^{-1} *Enterococcus faecalis, K. pneumonia, and E. coli* showed 91, 89, and 83% inhibition of biofilm [50].

Table 4 enlists the natural compounds with antibiofilm activity.

9. Conclusion

In recent times, the concept of biofilm has influenced almost every treatment step of infection due to high level of protection against antibiotics and antimicrobial agents, being the thrust to human medical management. Hence, there is a crucial demand to develop novel strategies to surpass the antibiotic resistance after understanding the clear mechanisms behind it. The plant compounds, phytochemicals, and nanoparticles can be fused with antimicrobial agents, which have substantial research evidence for their antibiofilm effects through their synergism. In spite of the clinical trials being done on such compounds, further study is required to prove their safety and effectiveness to support the clinical systems.

Acknowledgements

We would like to thank our principal Dr.M. Ramanathan D.Sc., for providing us with sufficient facilities to complete this book chapter.

Conflict of interest

The authors declare no conflict of interest.

Author details

Sankar Veintramuthu* and Selliamman Ravi Mahipriya
PSG College of Pharmacy, Coimbatore, Tamil Nadu, India

*Address all correspondence to: sankarv@psgpharma.ac.in

References

[1] Chandki R, Banthia P, Banthia R. Biofilms: A microbial home. Journal of Indian Society of Periodontology. 2011;**15**(2):111-114. DOI: 10.4103/0972-124X.84377

[2] Wyatt JE, Hesketh LM, Handley PS. Lack of correlation between fibrils, hydrophobicity and adhesion for strains of *streptococcus sanguis* biotypes 1 and 2. Microbios. 1987;**50**:7-15. DOI: 10.3109/08910609109140265

[3] Harty Derek WS, Handley PS. Expression of the surface properties of the fibrillar Streptococcus salivarius HB and its adhesion deficient mutants grown in continuous culture under glucose limitation. Journal of General Microbiology. 1989;**135**:11-21. DOI: 10.1099/00221287-135-10-2611

[4] Heukelekian H, Heller A. Relation between food concentration and surface for bacterial growth. Journal of Bacteriology. 1940;**40**:547-558. DOI: 10.1128/jb.40.4.547-558

[5] Jamal M, Tasneem U, Hussain T, Andleeb S. Bacterial biofilm: Its composition, formation and role in human infections. Journal of Microbiology and Biotechnology. 2015;**4**(3):1-14. DOI: 10.1016/j.jcma.2017.07.012

[6] Mah T-F. Biofilm-specific antibiotic resistance. Future Microbiology. 2012;**7**:1061-1072. DOI: 10.2217/fmb.12.76

[7] Donlan RM. Biofilms and device-associated infections. Emerging Infectious Disease. 2001;**7**:277-281. DOI: 10.3201/eid0702.010226

[8] Donlan RM, Costerton JW. Biofilms: Survival mechanisms of clinically relevant microorganisms. Clinical Microbiology Review. 2002;**15**:167-193. DOI: 10.1128/CMR.15.2.167-193.2002

[9] Maki DG, Mermel LA. Infections due to infusion therapy. In: Hospital Infections. 4th ed. Philadelphia: Lippincott-Raven; 1998. pp. 689-724. DOI: 10.1067/SO 196-6553(03)00703-X

[10] Braunwald E. Valvular heart disease. In: Heart Disease. 5th ed. Vol. 2. Philadelphia: W.B. Saunders Co; 1997. pp. 1007-1066. DOI: 10.4065%2Fmcp.2009.0706

[11] Kokare CR, Chakraborty S, Khopade AN, Mahadik KR. Biofilm importance and applications. Indian Journal of Biotechnology. 2009;**8**:159-168. Available from: http://hdl.handle.net/123456789/3883

[12] Sharma D, Misba L, Khan AU. Antibiotics versus biofilm: An emerging battleground in microbial communities. Antimicrobial Resistance and Infection Control. 2019;**8**(76):1-10. DOI: 10.1186/s13756-019-0533-3

[13] Wilkins M, Hall-Stoodley L, Allan RN, Faust SN. New approaches to the treatment of biofilm-related infections. Journal of Infection. 2014;**69**(1):47-52. DOI: 10.1016/j.jinf.2014.07.014

[14] Lazar V. Quorum sensing in biofilms: How to destroy the bacterial citadels or their cohesion/power? Anaerobe. 2011;**17**:280-285. DOI: 10.1016/j.anaerobe.2011.03.023

[15] Sharma D, Misba L, Khan AU. Antibiotics versus biofilm; an emerging battle ground in microbial communities. Antimicrobial Resistance and Infection

Control. 2019;**8**(1):76. DOI: 10.1186/s13756-019-0533-3

[16] Lewis K. Persister cells. Annual Review of Microbiology. 2010;**62**:357-372. DOI: 10.1146/annurev.micro.112408.134306

[17] Hall-Stoodley L et al. Bacterial biofilms: From the natural environment to infectious diseases. Nature Reviews Microbiology. 2004;**2**:95-108. DOI: 10.1038/nrmicro821

[18] Saleh T. Detection: From electrochemistry to spectroscopy with chromatographic techniques. Recent Trends with Nanotechnology Scientific Research. 2014;**2**:27-32. DOI: 10.4236/detection.2014.24005

[19] Ansari SA, Husain Q, Qayyum S, Azam A. Designing and surface modification of zinc oxide nanoparticles for biomedical applications. Food and Chemical Toxicology. 2011;**49**:2107-2115. DOI: 10.1016/j.fct.2011.05.025

[20] Qayyum S, Khan AU. Nanoparticles vs biofilm; a battle against another paradigm of antibioitic resistance. Medicinal Chemistry Communications. 2016;**7**:1479-1498. DOI: 10.1039/D1MA00639H

[21] Kammari R, Das NG, Das SK. Chapter 6 - Nanoparticulate systems for therapeutic and diagnostic applications. In: Micro and Nano Technologies, Emerging Nanotechnologies for Diagnostics, Drug Delivery and Medical Devices. O REILLY; 2017. pp. 105-144. DOI: 10.1016/B978-0-323-42978-8.00006-1

[22] Ramasamy MK, Lee J. Recent nanotechnology approaches for prevention and treatment of biofilm-associated infections on medical device. BioMedical Research International. 2016;**31**:1-18. DOI: 10.1155/2016/1851242

[23] Chaudhary S, Jyoti A, Shrivastava V, Tomar RS. Role of nanoparticles as antibiofilm agents: A comprehensive review. Current Trends in Biotechnology and Pharmacy. 2020;**2014**:97-110. DOI: 10.5530/ctbp.2020.1.10

[24] Tran HM, Tran H, Booth MA, Fox KE, Nguyen TH, Tran N, et al. Nanomaterials for treating bacterial biofilms on implantable medical devices. Nanomaterials. 2020;**10**:1-19. DOI: 10.3390/nano10112253

[25] Taylor EN, Webster TJ. The use of super paramagnetic nanoparticles for prosthetic biofilm prevention. International Journal of Nanomedicine. 2009;**4**:145-152. PMCID: PMC2747349

[26] Agarwala M, Choudhury B, Yadav RNS. Comparative study of antibiofilm activity of copper oxide and iron oxide nanoparticles against multidrug resistant biofilm forming uropathogens. Indian Journal of Microbiology. 2014;**54**:365-368. DOI: 10.1007%2Fs12088-014-0462-z

[27] Ikuma K, Madden AS, Decho AW, Lau BLT. Deposition of nanoparticles onto polysaccharide-coated surfaces: Implications for nanoparticle–biofilm interactions. Environmental Science. Nano. 2014;**1**:117-122. DOI: 10.3389/fmicb.2015.00677

[28] Shkodenko L, Kassirov I, Koshel E. Metal oxide nanoparticles against bacterial biofilms: Perspectives and limitations. Microorganisms. 2020;**8**(10):1545. DOI: 10.3390/microorganisms8101545

[29] Kishore MY, Kunal B, Kumar JS, Abeer H, Fathi AAE, Kumar MT. Anti-biofilm and antibacterial activities of silver nanoparticles synthesized by the reducing activity of phytoconstituents present in the indian medicinal plants.

Frontiers in Microbiology. 2020;**11**:1143. DOI: 10.3389/fmicb.2020.01143

[30] John MS, Nagoth JA, Ramasamy KP, Mancini A, Giuli G, et al. Synthesis of bioactive silver nanoparticles by a pseudomonas strain associated with the antarctic psychrophilic protozoon Euplotes focardii. Marine Drugs. 2020;**18**:38. DOI: 10.3390/md18010038

[31] Yusof HM, Rahman N'AA, Mohamad R, Zaidan UH. Microbial mediated synthesis of silver nanoparticles by lactobacillus plantarum TA4 and its antibacterial and antioxidant activity. Applied Sciences. 2020;**10**:6973. DOI: 10.3390/app10196973

[32] Othman AM, Elsayed MA, Al-BalakocyNG,HassanMM,ElshafeiAM. Biosynthesis and characterization of silver nanoparticles induced by fungal proteins and its application in different biological activities. Journal of Genetic and Engineering and Biotechnology. 2020;**17**(8):7. DOI: 10.1186/s43141-019-0008-1

[33] Iravani S, Korbekandi H, MirmohammadiSV,ZolfaghariB.Synthesis of silver nanoparticles: Chemical, physical and biological methods. Research in Pharmaceutical Sciences. 2014;**9**(6):385-406

[34] Alabdallah NM, Hasan MM. Plant-based green synthesis of silver nanoparticles and its effective role in abiotic stress tolerance in crop plants. Saudi Journal of Biological Sciences. 2021;**28**(10):5631-5639. DOI: 10.1016/j.sjbs.2021.05.081

[35] Kruis F, Fissan H, Rellinghaus B. Sintering and evaporation characteristics of gas-phase synthesis of size-selected PbS nanoparticles. Materials Science and Engineering. 2000;**69**:329-334. DOI: 10.1016/S0921-5107(99)00298-6

[36] Siddique MH, Aslam B, Imran M, Ashraf A, Nadeem H, Hayat S, et al. Effect of silver nanoparticles on biofilm formation and eps production of multidrug–resistant *Klebsiella pneumoniae*. Biomedical Research International. 2020;**19**:1-9. DOI: 10.1155/2020/6398165

[37] Hoyle BD, Jass J, Costerton JW. The biofilm glycocalyx as a resistance factor. Journal of Antimicrobial Chemotherapy. 1990;**26**(1):1-5. DOI: 10.1093/jac/26.1.1

[38] Mishra R, Panda A k, De Mandal S, Shakeel M, Bisht SS, Khan J. Natural anti-biofilm agents: Strategies to control biofilm – Forming pathogens. Frontiers in Microbiology. 2020;**11**:1-23. DOI: 10.3389/fmicb.2020.566325

[39] Tessema B, Mulu A, Kassu A, Yismaw G. An in vitro assessment of the antibacterial effect of garlic (Allium sativum) on bacterial isolates from wound infections. Ethiopian Medical Journal. 2006;**44**(4):385-389

[40] Hindi NKK, Al-Dabbagh NN, Dil Ghani Chabuck ZA. Anti-swarming, anti-adherence and anti-biofilm activities of garlic-related aquatic extracts: An *in vitro* study. Asian Journal of Microbiology Biotechnology and Environmental Sciences. 2018;**20**(2):137-147. DOI: 10.3390/molecules25235486

[41] Zeinab M, Mehdi H. The effects of extracts on biofilm formation and activities of six pathogenic bacteria, *Jundishapur*. Journal of Microbiology. 2015;**8**(8):1-7. DOI: 10.5812/jjm.18971v2

[42] Hamid NO, Mohammed Abdul-Hassan AL-Z, Nada KKH. An *invitro* study of anti-bacterial, anti-adherence, anti biofilm and anti-motility activities of the aqueous extracts of fresh and powdered onion and onion oil. International Journal of Pharmaceutical Science

and Research. 2018;**10**(6):1573-1578. DOI: 10.1111/j.1365-2672.2008.03882.x

[43] Kavita S, Neelima M, Yong RL. Systematic study on active compounds as antibacterial and antibiofilm agent in aging onions. Journal of Food and Drug Analysis. 2018;**26**:518-528. DOI: 10.1016/j.jfda.2017.06.009

[44] Yang T, Bai J, Yu Y, Bai X, Bello-Onaghise G, Xu Y, et al. Intervene effect on *Streptococcus suis* biofilm formation of emodin extracted from *Rheum ocinale baill*. Research Square. 2020:1-33. DOI: 10.3389/fphar.2018.00227

[45] Yan X, Gu S, Shi Y, Cui X, Wen S, Ge J. The effect of emodin on Staphylococcus aureus strains in planktonic form and biofilm formation in vitro. Archives of Microbiology. 2017;**199**(9):1267-1275. DOI: 10.1007/s00203-017-1396-8

[46] Jouneghani RS, Castro AHF, Panda SK, Swennen R, Luyten W. Antimicrobial activity of selected banana cultivars against important human pathogens, including *Candida* biofilm. Food. 2020;**435**(9):1-19. DOI: 10.3390/foods9040435

[47] Vijayakumar K, Ramanathan T. Musa acuminata and its bioactive metabolite 5-hydroxymethylfurfural mitigates quorum sensing (las and rhl) mediated biofilm and virulence production of nosocomial pathogen Pseudomonas aeruginosa *in vitro*. Journal of Ethnopharmacology. 2019;**246**:112242. DOI: 10.1016/j.jep.2019.112242

[48] Sebiomo A, Awofodu AD, Awosanya AO, Awotona FE, Ajayi AJ. Comparative studies of antibacterial effect of some antibiotics and ginger on two pathogenic bacteria. Journal of Microbiology and Antimicrobials. 2011;**3**:18-22. DOI: 10.1155/2014/562804

[49] Kim H-S, Park H-D. Ginger extract inhibits biofilm formation by *P.aeruginosa* PA14. PLoS One. 2013;**8**(9):1-16. DOI: 10.1371/journal.pone.0076106

[50] Aradhana D, Suchanda D, Rajesh KS, Saubhagini S, Enketeswara S. Antibiofilm and antibacterial activity of essential oil bearing *Zingiber offcinalis* Rosc. (ginger) rhizome against multi-drug resistant isolates. Journal of Essential Oil Bearing Plants. 2019;**22**(4):1163-1171. DOI: 10.1080/0972060X.2019.1683080

[51] Tapia-Rodriguez MR, Bernal-Mercado AT, Gutierrez-Pacheco MM, Vazquez-ArmentaFJ,Hernandez-MendozaA, Gonzalez-Aguilar GA, et al. Virulence of Pseudomonas aeruginosa exposed to carvacrol: Alterations of the quorum sensing at enzymatic and gene levels. Journal of Cell Communication and Signalling. 2019;**13**:531-537. DOI: 10.1007/s12079-019-00516-8

[52] Allen KL, Molan PC, Reid GM. The variability of the antibacterial activity of honey. Apiacta. 1991;**26**:114-121. DOI: 10.1371%2Fjournal.pone.0018229

[53] Raorane CJ, Lee JH, Kim YG, Rajasekharan SK, Garcia-Contreras R, Lee J. Antibiofilm and antivirulence efficacies of flavonoids and curcumin against Acinetobacter baumannii. Frontiers in Microbiology. 2019;**10**:990. DOI: 10.3389/fmicb.2019.00990

[54] Arita-Morioka KI, Yamanaka K, Mizunoe Y, Tanaka Y, Ogura T, Sugimoto S. Inhibitory effects of Myricetin derivatives on curli dependent biofilm formation in Escherichia coli. Scientific Reports. 2019;**8**:8452. DOI: 10.1038/s41598-018-26748-z

[55] Borowski RGV, Barros MP, Da Silva DB, Lopes NP, Zimmer KR, Staats CC, et al. Red peptide coatings control *S. epidermis* biofilm formation. International Journal of Pharmaceutics. 2020;**574**:1-11. DOI: 10.1016/j.ijpharm.2019.118872

Chapter 11

Development of Antibiofilm Substances by Endophytic Microorganisms with an Emphasis on Medicine

Saulo Henrique Rodrigues, Marcelo Assis, Camila Cristina de Foggi, Andréa Cristina Bogas, Mariana Ottaiano Gonçalves, Lavinia Cipriano, Elson Longo, Evandro Leite de Souza and Cristina Paiva de Sousa

Abstract

The growing antimicrobial resistance and persistence of pathogenic microorganisms in infections–particularly in nosocomial infections–have become a major problem for public health worldwide. One of the main causes of these issues is the formation of biofilms, which are microbial communities associated with extracellular polymeric substances (EPS) that form a slimy extracellular matrix, causing the bacteria to become more tolerant to usual drugs in these structures. Thus, the search for new antibiofilm compounds is part of a strategy to deal with this problem. Endophytic microorganisms such as bacteria and fungi, mutualistically associated with plants, are sources of compounds with biological properties, including antimicrobials, and can be important allies in the synthesis of antibiofilm. These secondary metabolites can interfere with cell-to-cell communication and cell adhesion ability, promoting the dispersal of bacterial colonies and affecting biofilm. Since endophytes are cultivable in laboratory conditions, these microorganisms are environmentally friendly, as they do not contribute to pollution, are easy to handle and are produced on a large scale. Furthermore, metabolites from endophytes are of natural origin and may contribute to the reduced use of synthetic drugs. Considering these aspects, this chapter will focus on the characterization of endophytic microorganisms as potential active sources of antibiofilm and antimicrobial compounds with applications in medicine.

Keywords: endophytes, biofilms, antimicrobial resistance, antibiofilm activity, anti-quorum sensing activity

1. Introduction

One of the most worrisome problems in public health nowadays is antimicrobial resistance and multi-resistance (AMR and MDR). This natural process has been accelerated by the unrestrained and irrational use of antimicrobials, such as antibiotics and antifungals [1]. One of the biggest challenges to overcome this problem is to equate the speed of development of new drugs with the adaptation of pathogens to current drugs, since the development of new compounds does not follow the growing resistance of microorganisms [2]. In addition, there is a large number of resistant pathogens involved in healthcare-related infections (HAI), making the treatment of diseases more difficult and expensive as well as increasing mortality and morbidity rates [3, 4]. Among the most common pathogens in nosocomial infections, bacteria from the ESKAPE group–an acronym used to refer to *Enterococcus faecium*, *Staphylococcus aureus, Klebsiella pneumoniae*, *Acinetobacter baumannii*, *Pseudomonas aeruginosa,* and *Enterobacter* are the most problematic, as they have mechanisms potentially involved in antimicrobial resistance [5]. Nevertheless, it is argued that the main cause of resistance may not be related to the classic mechanisms of microbial adaptation, but to the formation of a structure called biofilm [6].

Biofilms are organized, complex and dynamic communities of microorganisms adhered to a biotic or abiotic surface and protected by a polymeric extracellular matrix, which is composed of nucleic acids, polysaccharides, lipids, and proteins, generally called polymeric extracellular substances (EPS) [7]. This characteristic of adhering to different surfaces makes biofilms well disseminated in nature and easily found in different environments, including hospitals [8]. What makes biofilms so problematic for health is the fact that they allow the microorganisms inside them to thrive and persist in their environment. When related to infections, these structures tend to increase the tolerance of pathogens to treatments with conventional antimicrobial drugs, as they often prevent these compounds from reaching target cells [9]. Additionally, biofilms harbor different species of microorganisms that when acting together can lead to the development of chronic diseases [10] as well as to antimicrobial resistance due to horizontal gene transfer [11]. Another important point is the form of communication within biofilms. Through the so-called quorum sensing (QS), an intra and extracellular communication channel of microorganisms, they are able to coordinately regulate their activities in biofilms [12, 13]. Based on these considerations, the search for new compounds with antibiofilm activity becomes essential for combating resistant microorganisms.

A niche that has been gaining space because of its diversified production of biomolecules is endophytic microorganisms. By definition, endophytic microorganisms are bacteria and fungi that live symbiotically associated with healthy plant tissues without causing any apparent damage to their host [14]. Endophytes are a source of several secondary metabolites with, for example, antimicrobial [15], antitumor [16], enzymatic [17], anti-COVID [18], and antibiofilm activities. The main antibiofilm compounds currently sought are those capable of i) preventing or inhibiting microbial adhesion to avoid biofilm formation, ii) dispersing the already formed biofilm, and iii) interfering with intra/extracellular communication for biofilm formation (anti-QS) [19]. It is already known that natural products, such as those produced by endophytes, have advantages over synthetic compounds [20], for instance, rigidity, which provides better protein-protein interactions [21], and the possibility of being structurally shaped by evolution to be used by/in living beings [22]. Endophytic microorganisms can also be used in the synthesis of nanoparticles with antibiofilm activity. Nanoparticles can be defined as particles ranging from 1 to 100 nm and with size-related properties [23], being important allies in public health, as they can be applied in medicine [24]. Thus, the eco-friendliest method for the production of nanoparticles is precisely through the so-called green synthesis, which uses products from biological sources for the biosynthesis of nanoparticles [25].

This book chapter discusses the use of endophytic microorganisms and their compounds as potential tools for controlling and combating pathogenic biofilms, which are closely linked to antimicrobial resistance.

2. Natural antibiofilm and anti-quorum sensing products synthesized by endophytic microorganisms

Several studies have reported antibiofilm and anti-QS compounds produced by endophytes, reinforcing and highlighting the potential application of these microorganisms in various areas of health. Some of these studies are presented in **Table 1** and will be fully discussed throughout this chapter.

2.1 Natural antibiofilm agents from endophytic bacteria

Endophytic bacteria play a significant role in the production of a variety of secondary metabolites with potential applications in medicine [45], opening up new perspectives for the prospection of different bacterial species towards the discovery of novel antibiofilm agents against pathogenic microorganisms.

El-Gendy et al. [46] isolated 51 *Streptomyces* strains from the inner healthy tissue of *Sarcophyton convolutum* and determined the antibiofilm activity of ethyl acetate extracts of these endophytes onto 96-well polystyrene plates against seven methicillin-resistant *S. aureus* (MRSA) strains and nine multidrug-resistant *Pseudomonas* species (MRD). The *Streptomyces* strain MORSY 22 showed destructive activity of the biofilm produced by all *S. aureus* strains (MRSA1 to MRSA7), with values ranging from 87.46 to 95.75%, and all *Pseudomonas* species (MRD 1 to MRD9), with values ranging from 96.58 to 70.38%. These results revealed the potential of the strain MORSY 22 to prevent biofilm formation by bacterial pathogens and to develop antibiotic resistance.

Theodora et al. [47] screened the antibiofilm activity of endophytic bacteria against the pathogenic bacteria *Bacillus cereus* ATCC 14579, *S. aureus* ATCC 29213, *Enterococcus faecalis* ATCC 33186, *P. aeruginosa* ATCC 27853, *Salmonella typhimurium* and *Vibrio cholerae.* Crude extracts of isolates JB 19B and JB 18B showed the highest biofilm inhibition activity (90%) and biofilm destruction (76%), respectively, against *S. aureus*. Through scanning electron microscopy (SEM) analysis it was possible to verify a reduction in the extracellular matrix of the biofilms of *B. cereus* and *S. typhimurium* after treatment with extracts of isolates JB 18B and JB 19 B. The isolate JB 3B also showed inhibition activity against biofilm formation by all pathogenic bacteria. These findings confirmed the potential use of antibiofilm inhibitors from endophytic bacteria as a strategy for the control of bacterial infections.

Sabu et al. [48] isolated 14 endophytic actinomycetes from the rhizomes of *Zingiber officinale*. The crude extract of *Nocardiopsis* sp. ZoA1 at 200 μg/mL caused a reduction of more than 90% biofilm formation by multidrug-resistant coagulase-negative *Staphylococcus capitis* 267 and *Staphylococcus haemolyticus* 41 strains. GC-MS/MS analysis of *Nocardiopsis* sp. also revealed the presence of various compounds with antimicrobial activity, such as phenol, 2,4-bis (1,1-dimethylethyl), and trans-cinnamic acid. These results pointed to the inhibition of the synthesis of exopolysaccharide and proteinaceous factors by tested crude extracts and their potential to prevent biofilm formation by multidrug-resistant biofilm-forming strains.

Host Plant	Endophyte	Target	Compound	Reference
Datura metel	*Streptomyces californicus* ADR1	*S. aureus* ATCC 25923, *S. aureus* ATCC 29213, MR *S. aureus* ATCC 43300, MR *S. aureus* 562	Metabolite extract	[26]
Casuarina spp.	*Frankia casuarinae* DDNSF-02	*Candida* sp., *Pseudomonas* sp.	Ethyl Acetate Extract	[27]
Ventilago madraspatana	*Enterobacter aerogenes* VT66	*P. aeruginosa* PAO1	N-acyl homoserine lactone-lactonase	[28]
Coscinium fenestratum	*Enterobacter* sp.	*Aeromonas hydrophila*	Metallo-protein AHL-lactonase	[29]
Caulerpa racemosa	*Nocardiopsis* sp. DMS 2	*Klebsiella pneumoniae*	Crude Extract 1, 4-diaza-2, 5-dioxo-3-isobutyl bicyclo [4.3.0] nonane	[30]
Ocimum Sanctum	*Lasiodiplodiapseudotheobromae* IBRL OS-64	MR *S. aureus* ATCC 33591	Ethyl Acetate Extract	[31]
Acanthus ilicifolius	*Aspergillus flavipes* AIL8	*S. aureus*	Flavipesin A	[32]
Ammi majus L	*Aspergillus amstelodami* (MK215708)	*S. aureus, E. coli*	Dihydroauroglaucin	[33]
Ocimum Sanctum	*L. pseudotheobromae* IBRL OS-64	*Streptococcus mutans*	Ethyl Acetate Extract	[34]
Carica papaya	*Phomopsis tersa*	*P. aeruginosa* PAO1	Crude Extract	[35]
Delonix regia	*Talaromyces* sp.	*P. aeruginosa* PAO1	Dextranase	[36]
Ocimum Sanctum	*L. pseudotheobromae* IBRL OS-64	*P. aeruginosa* PAO1	Ethyl Acetate Extract	[37]
Anredera cordifolia	*P. aeruginosa* CP043328.1	*C. violacuem* ATCC 12472	Diisooctyl phthalate and [1, 2, 4] oxadiazole, 5-benzyl-3	[38]
Ventilago madraspatana	*Fusarium graminearum* *Lasidiplodia* sp	*C. violaceum* CV026	Crude Extract	[39]
Silybum marianum	*Penicillium restrictum*	MR *S. aureus*	Polyhydroxyanthraquinones	[40]
Diploria strigosa	*Fusarium* sp.	*Chromobacterium violaceum* CVO26	Crude Extract	[41]

DOI: http://dx.doi.org/10.5772/intechopen.104522

Host Plant	Endophyte	Target	Compound	Reference
Solanum nigrum	*Setosphaeria rostrata*	*P. aeruginosa*	Mycosilver nanoparticles	[42]
C. papaya	*Diaporthe phaseolorum* SSP12	*P. aeruginosa* PAO1	Crude Extract	[43]
Ipomoea carnea	*Aspergillus terreus* AH1	*P. aeruginosa*, *S. aureus*, *E. coli*, *B. subtilis*	Asterrelenin, Periplanamide, Butyrolacton, Pyranterron, Arenarin A	[44]

Table 1.
Antibiofilm and anti-QS activities of natural compounds produced by endophytic fungi and bacteria isolated from different host plants.

Biosurfactants are an important class of natural antibiofilm agents produced by microorganisms. They comprise a structural and heterogeneous group of amphipathic molecules, which include glycolipids, lipopeptides, phospholipids, fatty acids and neutral lipids, polymeric and particulate biosurfactants [49, 50]. These microbial molecules can interfere with cell-to-cell communication mediated by QS and cell adhesion ability, promoting the dispersal of bacterial colonies and affecting biofilm formation through distinct mechanisms, such as cell membrane damage, inhibition of electron transport chain and energy restriction [51, 52]. Additionally, microbial surfactants have been considered an eco-friendly alternative with low toxicity and high biodegradability, selectivity and compatibility when compared to chemically synthesized surfactants [53].

Recently, Ashitha et al. [54] studied the endophyte *Burkholderia* sp. WYAT7 was isolated from the medicinal plant *Artemisia nilagirica* (Clarke) Pamp. in order to evaluate its antibiofilm activity. The biosurfactant present in the culture supernatant was identified and characterized as a glycolipid, and an inhibitory effect on the *S. aureus* (MTCC 1430) biofilm formation was observed. The percentage of biofilm formation suppression by MTCC 1430 was 41.79% and 79.22% when treated with 1 mg/ml and 2 mg/ml, respectively. These results suggested that the surfactant produced by *Burkholderia* sp. WYAT7 could be explored as a therapeutic agent for the control of pathogenic bacteria.

Ceresa et al. [55] reported that lipopeptide biosurfactants produced by the endophytic *B. subtilis* AC7 (AC7BS) isolated from *Robinia pseudoacacia* efficiently reduced *Candida albicans* adhesion to and biofilm formation on medical-grade silicone elastomeric disks (SEDs) by 57–62% and 46–47%, respectively. Chemical analysis of the crude extract revealed the presence of surfactin and fengycin. Since the fungus *C. albicans* is considered responsible for colonizing medical implants and causing a high mortality rate, the authors suggested the potential use of these biosurfactants to coat silicone medical devices in order to limit colonization of the pathogen and prevent infections. Later, Ceresa et al. [56] studied the synergistic effect of lipopeptides of *B. subtilis* AC7 (AC7BS) combined with the QS molecule farnesol to counteract *C. albicans* biofilms on silicone elastomer in simulated physiological conditions. There was a significant reduction of up to 74% in the pathogen adhesion within 1.5 hours and up to 93% and 60% in the biofilm formation within 24 and 48 hours, respectively. These effects were confirmed by scanning electron microscopy (SEM) and confocal laser scanning microscopy (CLSM). According to the authors, these findings opened up new perspectives for the combination of biosurfactants and farnesol to counteract *C. albicans* adhesion to and biofilm formation on materials for medical use.

Cochis et al. [57] evaluated the preventive anti-adhesion activity of biosurfactants extracted from endophytes from *R. pseudoacacia* (AC5 and AC7) and *Nerium oleander* (OC5) against *C. albicans* biofilm on acrylic resin and disks of silicon. The effective concentrations for *C. candida* biofilm inhibition without cytotoxic effects on mouse fibroblasts (ATCC L929) and human keratinocytes (ATCC HeLa S3) were 156.3 g/ml and 78.1 g/ml, respectively. These results demonstrated the potential use of these biosurfactants for the prevention of *C. albicans* biofilm adhesion to catheter and prosthesis materials.

2.2 Natural antibiofilm agents from endophytic fungi

Several recent studies have shown the potential of endophytic fungi as producers of biomolecules with antimicrobial activity [58]. Historically, fungi are known for their diverse production, including penicillin–the first antibiotic discovered [59]. For such reason, over the years researchers have focused on the discovery of new fungal antimicrobials, such as clavatol, sordaricin, jesterone, and javanicin [60]. Based on this, it is evident how interesting endophytic fungi can be in terms of the production of antimicrobial compounds.

May Zin et al. [61] obtained several bioactive metabolites from the endophytic fungus *Eurotium chevalieri* KUFA 0006 isolated from *Rhizophora mucronata*. The new compounds were tested to verify their antibiofilm activity against *E. coli* ATCC 25922, *E. faecalis* ATCC 29212, and *S. aureus* ATCC 25923. Thirteen metabolites effectively inhibited the growth of biofilms, whereas eight inhibited the biofilm formation by *E. coli* ATCC 25922, six by *S. aureus* ATCC 25923 and only one by *E. faecalis* ATCC 29212. This work also highlighted compound 3, which showed antibiofilm activity against *E. coli* ATCC 25922 and *S. aureus* ATCC 25923, causing a reduction of about 80% in the staphylococcal biofilm. The authors also performed tests to evaluate the antibiotic activity of these metabolites against the same pathogenic strains and found a positive result in only one compound. This is a very interesting finding, because even though certain compounds did not present an inhibitory effect against the pathogen alone, they had an inhibitory activity against the biofilm.

Narmani et al. [62] isolated the fungus *Chaetosphaeronema achilleae* from *Taxus baccata* and reported the production of seven compounds from the endophyte. In general, the metabolites were tested at different concentrations against *S. aureus* DSM 1104 biofilms and all of them presented some inhibitory activity even at lower concentrations. Among them, compound 4 stood out, showing strong biofilm inhibitory activity of about 96.82% at a concentration of 256 μg/mL and approximately 91.95% at 128 μg/mL. In addition, compound 7 was able to inhibit about 96.18% at 256 μg/mL of the biofilm, which represents a quite positive result. In the same work, it was observed that not all compounds exhibited antimicrobial activity against *S. aureus* DSM 1104 alone, as only metabolites 2 and 7 were positive.

Kaur et al. [63] isolated the fungus *Alternaria destruens* (AKL-3) from *Calotropis gigantea* and observed antibiofilm activity of the active fractions AF1 and AF2 during biofilm formation and in the preformed biofilm. The test microorganisms were *P. aeruginosa*, *C. albicans*, *E. coli* and *Salmonella enterica*, and two different concentrations of each active fraction were tested. In the case of AF1, all biofilms had their formation relatively inhibited, in addition to having been moderately reduced in the preformed biofilm. With regard to AF2, the same could be observed, that is, all biofilms were inhibited in the initial phase and in the preformed biofilm. Nonetheless, according to the authors AF1 was more promising and showed significantly greater activity than AF2 in all tests with the pathogenic strains.

Kaur et al. [64] evaluated the antibiofilm activity of the chloroform extract of the endophytic *Aspergillus fumigatus* isolated from *Moringa oleifera* against *S. aureus* MTCC 740, *K. pneumoniae* MTCC 109, and *C. albicans* MTCC 227. In this study, the authors performed tests at different stages of the biofilm, namely, the initial cell fixation phase and the preformed biofilm. In the initial fixation tests, the fungal extract was able to inhibit the formation of *S. aureus*, *K. pneumoniae,* and *C. albicans* biofilms by 69.2%, 57.66%, and 55%, respectively, with the standard antimicrobials showing similar results. The authors also argued that the inhibition of the initial fixation of the *C. albicans* biofilm by the fungal extract was better than that of the standard antifungal (amphotericin B) since the value obtained was approximately 53.3%. Regarding the tests against preformed biofilms, the extract reduced by about 51%, 53.4% and 47.6% of the *S. aureus*, *K. pneumoniae* and *C. albicans* biofilms.

Elkhouly et al. [65] studied the metabolism of the endophytic fungus *Aspergillus Tubenginses* ASH4 isolated from *Hyoscyamus muticus* in order to understand the production of antibiofilm compounds. During the study, pathogenic biofilms of *S. aureus* ATCC6538-P, *Bacillus subtilis*, *P. aeruginosa* ATCC27853 and *E. coli* were bioindicators of the extract as well as of the pure compound. The endophytic extract was able to suppress the formation of the *S. aureus*, *B. subtilis*, *P. aeruginosa,* and *E. coli* biofilms by 60.8%, 50.06%, 28.44%, and 37.68%, respectively. Subsequently, the pure compound identified as anophinic acid was tested against the same strains, reaching an inhibition of 61.39%, 54.93%, 69.51%, and 34.45%, respectively. Based on these results, it is possible to observe that the values are similar between them, except in the case of *P. aeruginosa*.

Qader et al. [66] isolated the marine endophytic fungi *Epicoccum nigrum* M13 and *Alternaria alternata* 13A from *Thalassia hemprichii* and tested 16 pure compounds obtained from them. The bioindicators for the antibiofilm activity test were *E. coli*, *S. aureus*, *B. subtilis,* and *P. aeruginosa*, all clinically isolated from hospitals in Egypt. Among the tested compounds of *E. nigrum* M13, three showed antibiofilm activity against pathogenic strains ranging from moderate to weak. The authors pointed out that some compounds such as 1 exhibited moderate activity against *S. aureus* and *B. subtilis*, but weak activity against *E. coli* and *P. aeruginosa*. In addition, compounds 3 and 5 showed moderate activity against Gram-positive bacteria, but weak activity against Gram-negative ones. As for the compounds isolated from *A. alternata* 13A, five of them presented activity against the biofilms of the indicator strains. Unlike what was seen in *E. nigrum* M13, compounds 7, 8, 9 and 10 from *A. alternata* 13A inhibited by 70–80% the *S. aureus* and *B. subtilis* biofilms, indicating an excellent activity. The same compounds also showed moderate activity against *E. coli* and *P. aeruginosa* biofilms. On the other hand, compound 11 exhibited weak activity against *S. aureus*, *E. coli* and *P. aeruginosa* but moderate activity against *B. subtilis*.

2.3 Anti-quorum sensing activity of natural agents from endophytic microorganisms

Quorum sensing (QS) is a complex density-dependent microbial cell communication system that occurs in single or mixed populations through autoinducers (AIs) or QS molecules. It is a population-dependent signaling mechanism in which microorganisms activate some signaling molecules according to the cell density. This behavior can be observed in several species of fungi and bacteria [67–69], being considered an inter- and intraspecies communication behavior that leads to genetic responses to autoinducers. This allows the microbial community to perceive and respond to various factors, including the presence of threats. The QS activity is responsible for the regulation of several bacterial physiological activities, such as pathogenesis, biofilm formation, swarming motility, bioluminescence, pigment disposal, polysaccharide production, and virulence, transforming the QS molecules into an important target for alternative antimicrobial therapy and antibiofilm activity [70].

After their production, when AIs reach an optimal concentration they bind to receptors on microbial cells, causing an alteration in gene expression. This ability gives biofilms adaptability to the environment as well as greater resistance to elimination, which in turn increases their virulence [71, 72]. In addition, QS molecules are also considered responsible for inhibiting or delaying the growth of other bacteria or fungi that are not part of their biofilm.

It is known that QS molecules are different for each microbial species. Furthermore, the type of communication in mixed biofilms also differs, that is, it can be either inter or intraspecies. There are four main categories of AIs: AI-1, AI-2, AI-3, and AIP. According to Schauder et al. [73], the molecules AI-2 are responsible for interspecies communication, while Smith et al. [74] argue that the molecules AI-1, AI-3, and AIP are in charge of the intraspecies communication.

Figure 1 shows the QS mechanism in a fungal cell in a simplified way. AIs (named signal molecules) are synthesized by fungal cells and released to the outside of the cell. Signal receptor proteins detect AIs and stimulate the expression of various genes, such as virulence, growth, and morphogenesis regulators.

Since the QS mechanism is responsible for the survival and increased virulence of biofilms, the development of QS inhibition strategies has been of great importance. Most QS inhibition mechanisms use one of the following strategies: i) degradation and/or inactivation of AIs; ii) inhibition of AI synthesis; iii) inhibition of AI detector; and iv) antibiotics as QS inhibitors [76]. In the context of QS mechanisms of biofilms, endophytic microorganisms–considered to be synthesizers of QS inhibitors–have gained increasing attention. According to Mookherjee et al. [76], as endophytic microorganisms need to constantly produce defenses against competing microbial populations, they become an interesting source of QS inhibitors. QS inhibitor molecules can be produced by either endophytic fungi or bacteria [40, 77, 78].

Since QS can regulate the expression of virulence factors, QS inhibitors (QSIs) appear to be a promising antimicrobial strategy. As they act by imitating the QS autoinducers, they can be used to attenuate bacterial virulence, thus requiring lower doses, being more susceptible to the host immune system and reducing the use of antibiotics [39]. There are several studies reporting the QSI activity of biofilms.

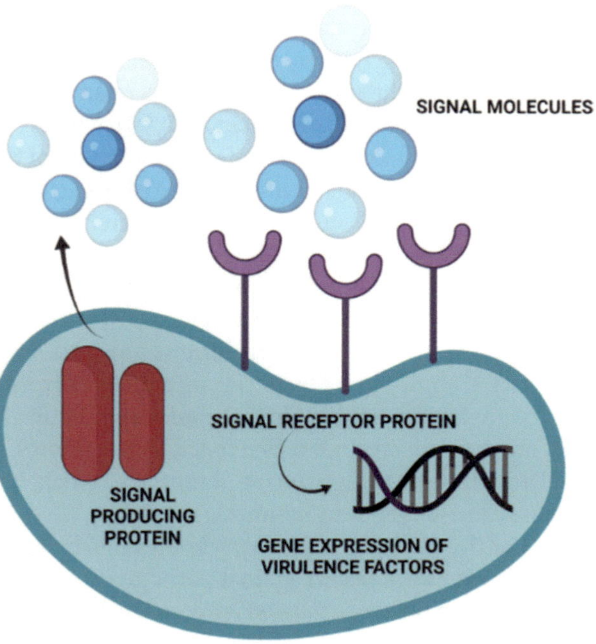

Figure 1.
QS mechanism scheme adapted from Sharma et al. [75].

2.3.1 QSIs produced by endophytes

It is known that endophytic fungi are responsible for the control and regulation of physiological activities of pathogens in animals and plants. Several studies have identified the production of QS inhibitors by endophytic fungi. Rajesh and Rai [39] isolated the endophytic fungus *Fusarium graminearum* from *Ventilago madraspatana* and measured the enzyme production using spectrophotometric and plate assay methods. Its anti-QS activity was analyzed against *Chromobacterium violaceum* CVO26, yielding strong positive results. Additionally, the extract of the endophytic fungus was able to inhibit the production of violacein pigment in the bacterium tested without any changes in bacterial growth. The authors then concluded that there was production of QS inhibitors by the endophytic fungus from *Ventilago madraspatana*, which in turn can be used for the development of anti-QS drugs–mainly against drug-resistant microorganisms.

Anti-QS molecules of *Lasiodiplodia sp.* from marine plants were also tested against *C. violaceum* CVO26 by Martín-Rodríguez et al. [41]. Four strains of the endophytic fungus stood out for their strong anti-QS activity. These strains were identified as belonging to four genera: *Sarocladium* (LAEE06), *Fusarium* (LAEE13), *Epicoccum* (LAEE14), and *Khuskia* (LAEE21). The authors reported that this was the first time that QS inhibitors were found in endophytic fungi extracted from marine plants.

Mishra et al. [70] showed that 2,4-di-tert-butylphenol (2,4-DBP), a component isolated from the endophytic fungus *Daldinia eschscholtzii*, is capable of inhibiting the QS activity of *P. aeruginosa*–one of the top three gram-negative bacteria considered a global threat due to its multiple drug resistance. They noticed that when exposed to 2,4-DBP, *P. aeruginosa* reduced the biofilm production and its virulence factors, as well as the expression of QS-related genes, confirming that 2,4-DBP can be used in combination with antibiotics to combat *P. aeruginosa.*

Zhou et al. [79] conducted a study that identified the QSI activity of 1-(4-amino-2-hydroxyphenyl) ethanone (AHE) isolated from the endophytic fungus *Phomopsis liquidambari* S47 from the leaves of *Punica granatum* against *P. aeruginosa* PAO1. The compound acted by suppressing the expression of genes related to QS, inhibiting the activity of antioxidant enzymes and enhancing oxidative stress. Pellissier et al. [80] explored the QSI activity of endophytic fungi extracted from the tropical palm *Astrocaryum sciophilum* against *P. aeruginosa.* Two pyran derivatives extracted from the endophytic strain *Laccophilus venezuelensis* showed activity affecting QS-regulated virulence factors.

Like endophytic fungi, bacteria are able to interact with each other (intra- and interspecies communication) through AIs. Kusari et al. [77] studied how endophytic bacteria from *Cannabis sativa* plants use QS inhibition as an antivirulence strategy in *C. violaceum*. A total of 13 endophytic bacteria were isolated from *C. sativa*, and their extracts were prepared and tested against *C. violaceum*. Four of them (*Bacillus* sp. strain B3, *Bacillus megaterium* strain B4, *Brevibacillus borstelensis* strain B8, and *Bacillus* sp. strain B11) exhibited the significant potential to weaken *C. violaceum* cell QS signals in a concentration-dependent manner.

Endophytic isolates of the phylum *Actinobacteria* previously isolated from common bean (*Phaseolus vulgaris*) were tested against pathogenic microorganisms by Lopes et al. [81]. Among them, *Microbacterium testaceum* BAC1065 and BAC1093 were found to inhibit QS of *C. violaceum* and *E. coli*. Kiarood et al. [82] found two strains (*Bacillus cereus* Si-Ps1 and *Pseudomonas nitrogenformans* La-Pot3–3) among 64 endophytic bacteria isolated from *Citrus sinensis* able to reduce the detection of QS molecules in *Pseudomonas syringae*. The *B. cereus* extract strongly inhibited *P. syringae* biofilm formation. An interesting fact reported by the authors was the increased number of cells in planktonic cultures treated with anti-QS molecules compared to control groups. This demonstrates that the molecules directly affect biofilm formation, but do not interfere with population growth.

3. Metal-based nanoparticles (NPs) synthesized from endophytic microorganisms as antibiofilm agents

The biosynthesis of metal-based NPs using endophytic microorganisms is a promising green synthetic route, considering the way to obtain these NPs and their final environmental impact [83]. These NPs can be used in many different technology sectors with emphasis on health [84, 85]. The biosynthesis of these NPs can occur intra- and/or extracellularly. The intracellular biosynthesis occurs through electrostatic interaction between positive charges from metal ions in a solution and negative charges from the bacterial/fungal cell wall [86]. In this process, microbial reductases dependent on NADH and NADPH are responsible for the transport of electrons, working as biocatalysts for redox reactions [87, 88]. In contrast, in extracellular synthesis, the culture supernatant, biomass, or cell-free extract is mixed with the metal ion solution, and the NPs are produced outside the microbial cell [89]. This process is performed by reductases produced and secreted into the culture medium by microbial cells and other cofactors [89, 90]. Therefore, biosynthesis through endophytic microorganisms can be used to obtain a series of different NPs, being the most common metallic/metallic oxides.

Noble metal NPs such as Ag has been widely used since ancient times for medicinal purposes due to their antimicrobial action [91]. Thus, it is natural that most of the works in the literature on the production of nanoparticles from endophytic microorganisms for microbial elimination are focused on Ag NPs. When these NPs are used for the inhibition of biofilms, the interaction between the NPs and the biofilm occurs in a succession of steps: first, the NPs are transferred to the biofilm surroundings; then, their superficial fixation occurs, followed by their migration to the biofilm interior [92]. Metal NPs can generate high local oxidative stress as a result of the production of reactive oxygen species (ROS), in addition to releasing M^+ ions, which can interact with various functional groups of microorganisms, such as proteins, lipids, and DNA [93]. Furthermore, they can bind to the cell membrane surface by electrostatic interactions and penetrate by endocytosis and direct diffusion [94]. Metal oxide NPs can generate a high concentration of ROS even in the dark, interacting similarly with metal NPs, and cause secondary effects due to both local contact of metal oxide NPs with microorganisms and ionic release (depending on the stability of the oxide in the reaction medium used) [95, 96]. **Figure 2** illustrates the mechanism of action of the nanoparticles on the biofilm.

Bakhtiari-Sardari et al. [97] biosynthesized Ag NPs from the inoculum of two strains of *Streptomyces* sp. (OSIP1 and OSNP14) using the cell-free supernatant from these cultures to inhibit *P. aeruginosa* ATCC 27853 biofilms, resulting in Ag NPs with a spherical shape and an average size of 8 and 15 nm, respectively. The growth of *P. aeruginosa* biofilms was inhibited by up to 85% at a minimum concentration of 125 μg/mL of Ag NPs. The highest activity of the Ag NPs synthesized by the strain of *Streptomyces* sp. OSIP1 was attributed to the smaller size of Ag NPs obtained. Ranjani et al. [98] used the same Ag NP biosynthesis strategy to inhibit the growth of *P. aeruginosa* ATCC

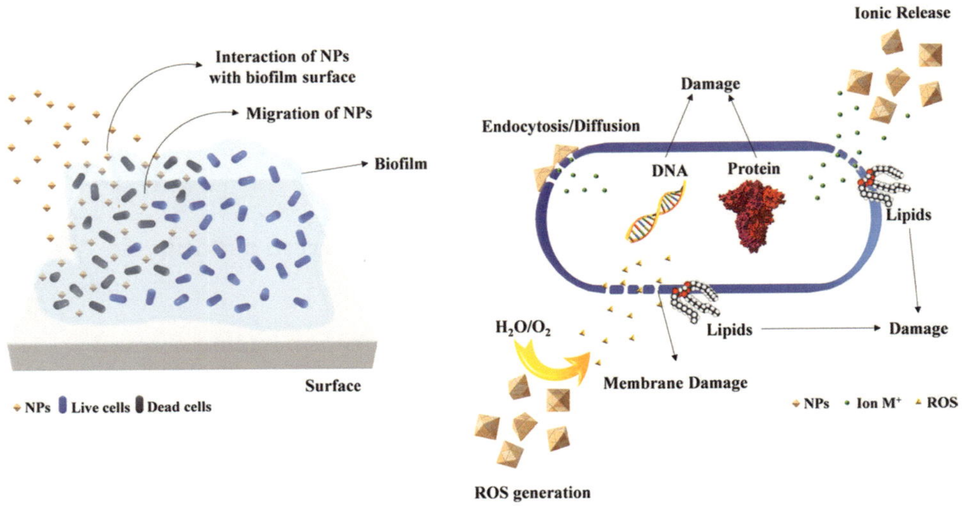

Figure 2.
Schematic illustration of antibiofilm effects of metal and metal oxide NPs.

27853. Using the cell extract of the fungus *L. theobromae* (MK942601), it was possible to obtain agglomerated Ag particles with an average size of 163.3 nm. The result of biofilm growth inhibition was 70% at a concentration of 50 μg/mL of Ag NPs. Bagur et al. [99] biosynthesized Ag NPs with an average size of 16.1 nm through a cell extract of the fungus *E. rostrata* due to its crucial role in the growth inhibition of *P. aeruginosa* and *S. aureus*. It was observed that there was a significant decrease in the growth of both pathogens at a concentration of 5 μg/mL of Ag NPs.

Neethu et al. showed in two different works the effectiveness of Ag NPs against the biofilm growth of the multidrug-resistant bacterium *A. baumanii* [100, 101]. In their first work, the biomass of the fungus *Peridinium polonicum* was used to synthesize spherical Ag NPs with sizes between 10 and 15 nm. It was observed that after 5 hours of exposure to the Ag NPs there was a reduction of more than 99.9% (3 log reduction) in the number of viable bacteria at a concentration of 15.6 μg/mL [100]. In their other work, the authors [101] produced a bionanocomposite coating with biosynthesized Ag NPs for a central venous catheter (CVC) using polydopamine as an adherent film of Ag NPs. Like in their previous work, it was observed that the CVC functionalized with Ag NPs eradicated the *A. baumanii* biofilm.

Ranjani et al. [102] synthesized Ag NPs nanocolloids and used them for the elimination of *E. coli* ATCC 25922 biofilms, commonly present in intensive care units (ICUs). The cell extract of the fungus *L. theobromae* (LtNc's) was able to produce Ag particles with an average size of 436.5 nm. At a concentration of 12.5 μg/mL of these Ag NPs, there was a 50% reduction in *E. coli* biofilm formation. In another work, Chandankere et al. [103] synthesized Ag NPs with sizes between 4 and 26 nm using the fungus *Colletotrichum* sp. DM16.3 to inhibit the growth of biofilms of bacteria *B. cereus* (Gram-positive) and *Vibrio cholerae* (gram-negative). At a concentration of 10 μg/mL of these Ag NPs, it was possible to observe an inhibition of biofilm growth of 45.6% for *B. cereus* and 85.1% for *V. cholerae*. Ibrahim et al. [104] used the cell extract of the bacterium *B. siamensis* to synthesize Ag NPs with sizes between 25 and 50 nm. It was observed that at a concentration of 20 μg/mL these Ag NPs were able to inhibit the growth of biofilms of *Xanthomonas oryzae* pv. *oryzae* LND0005 and *Acidovorax oryzae* RS-1 by 86.31 and 80.59%, respectively.

Metal oxide NPs can also be synthesized by endophytic microorganisms and used to inhibit biofilm growth. Dhandapani et al. [105] synthesized TiO_2 NPs (10–30 nm) from the biomass of the bacterium *B. subtilis* (FJ460362). Tests were performed using microorganisms present in local aquatic sources and in the presence of light so that TiO_2 produced more ROS, causing high oxidative stress to microorganisms. The Se and SeO_2 particles (75–225 nm) were synthesized from the extract of the bacterium *Bacillus* sp. MSh-1 and tested against the biofilms of *P. mirabilis*, *S. aureus,* and *P. aeruginosa*, resulting in inhibitions of 53.4, 48.1, and 55.1%, respectively [106]. Balaji et al. synthesized ZrO_2 particles using the bacterium *B. niancini* and used them to remove the biofilms of *E. coli* (91.5%), *Klebsiella aerogenes* (71%), *P. vulgaris* (83.25%), *S. aureus* (92.5%) and *S. mutant* (90.5%) at a concentration of 40 μg/ml [107].

4. Conclusions

Biofilms are known to be closely linked to the growing resistance of pathogens, posing a threat to public health. Based on this fact, endophytic microorganisms considered as potential and eco-friendly producers of compounds with antibiofilm activity may be a source for the discovery of new biomolecules to combat these pathogens since they can synthesize compounds with anti-adherent properties, being capable of dispersing pre-synthesized biofilms.

These microorganisms also produce QS inhibitors that can harm the communication between pathogens in biofilm and, consequently, interrupt its formation. There are several researches showing the capacity of endophytic production in the prevention and dispersion of biofilms of, for example, ESKAPE pathogens, and this is really relevant because these microorganisms had been causing such a considerable problem to public health.

In addition, the microbial products of endophytes can also be used in the biosynthesis of metal-based nanoparticles, which have been demonstrating an interesting activity against biofilms. Some studies showed that metal-based nanoparticles can allocate on the surface of biofilm and migration to its interior, interacting directly with the pathogens inside, causing their death in different ways.

Thus, endophytic microorganisms deserve a position in the discussion about the development of new antimicrobial and antibiofilm medicines, mainly because several researches described in this review showed the potential of endophytes against harmful pathogens and their biofilms.

Acknowledgements

This work was supported by grants from the São Paulo Research Foundation (FAPESP) No. 2020/16299-9 to SHR, Nos. 2016/13423-5 and 2017/12905-9 to CPS and No. 2013/07296-2 to EL, and from the Coordination for the Improvement of Higher Education Personnel (CAPES) under finance code (001) to MA.

Author details

Saulo Henrique Rodrigues[1,2], Marcelo Assis[3], Camila Cristina de Foggi[3], Andréa Cristina Bogas[1,4], Mariana Ottaiano Gonçalves[1,4], Lavinia Cipriano[1,2], Elson Longo[3], Evandro Leite de Souza[5] and Cristina Paiva de Sousa[1,4]*

1 Laboratory of Microbiology and Biomolecules - LaMiB, Biotechnology Graduate Program (PPGBiotec), Federal University of São Carlos, São Carlos, Brazil

2 Biological Sciences Undergraduate Program, Federal University of São Carlos, São Carlos, Brazil

3 Center for the Development of Functional Materials (CDMF), Federal University of São Carlos, São Carlos, Brazil

4 Biotechnology Graduate Program – PPGBiotec, Federal University of São Carlos, São Carlos, Brazil

5 Nutritional Sciences Graduate Program, Federal University of Paraíba, João Pessoa, Brazil

*Address all correspondence to: prokarya@ufscar.br

References

[1] Morgan DJ, Okeke IN, Laxminarayan R, Perencevich EN, Weisenberg S. Non-prescription antimicrobial use worldwide: A systematic review. The Lancet Infectious Diseases. 2011;**11**(9):692-701. DOI: 10.1016/s1473-3099(11)70054-8

[2] World Health Organization. Update from WHO and pew Charitable Trusts: Urgent Action Needed to Accelerate Antibiotic Development [Internet]. Geneva, Switzerland: World Health Organization; 2020. Available from: https://www.who.int/news/item/07-07-2020-update-from-who-and-pew-charitable-trusts-urgent-action-needed-to-accelerate-antibiotic-development

[3] Roca I, Akova M, Baquero F, Carlet J, Cavaleri M, Coenen S, et al. The global threat of antimicrobial resistance: Science for intervention. New Microbes and New Infections. 2015;**6**:22-29. DOI: 10.1016/j.nmni.2015.02.007

[4] Golkar Z, Bagasra O, Pace DG. Bacteriophage therapy: A potential solution for the antibiotic resistance crisis. The Journal of Infection in Developing Countries. 2014;**8**(02):129-136. DOI: 10.3855/jidc.3573

[5] Rice LB. Progress and challenges in implementing the research on ESKAPE pathogens. Infection Control & Hospital Epidemiology. 2010;**31**(S1):S7-S10. DOI: 10.1086/655995

[6] Santajit S, Indrawattana N. Mechanisms of antimicrobial resistance in Eskape pathogens. BioMed Research International. 2016;**2016**:1-8. DOI: 10.1155/2016/2475067

[7] Hall-Stoodley L, Stoodley P, Kathju S, Høiby N, Moser C, William Costerton J, et al. Towards diagnostic guidelines for biofilm-associated infections. FEMS Immunology & Medical Microbiology. 2012;**65**(2):127-145. DOI: 10.1111/j.1574-695X.2012.00968.x

[8] de la Fuente-Núñez C, Reffuveille F, Fernández L, Hancock REW. Bacterial biofilm development as a multicellular adaptation: Antibiotic resistance and new therapeutic strategies. Current Opinion in Microbiology. 2013;**16**(5):580-589. DOI: 10.1016/j.mib.2013.06.013

[9] Magalhães AP, Lopes SP, Pereira MO. Insights into cystic fibrosis polymicrobial consortia: The role of species interactions in biofilm development, phenotype, and response to in-use antibiotics. Frontiers in Microbiology. 2017;**7**:2146. DOI: 10.3389/fmicb.2016.02146

[10] Høiby N, Bjarnsholt T, Moser C, Bassi GL, Coenye T, Donelli G, et al. Escmid∗ guideline for the diagnosis and treatment of biofilm infections 2014. Clinical Microbiology and Infection. 2015;**21**:S1-S25. DOI: 10.1016/j.cmi.2014.10.024

[11] Jorge P, Magalhães AP, Grainha T, Alves D, Sousa AM, Lopes SP, et al. Antimicrobial resistance three ways: Healthcare crisis, major concepts and the relevance of biofilms. FEMS Microbiology Ecology. 2019;**95**(8):fiz115. DOI: 10.1093/femsec/fiz115

[12] Grandclément C, Tannières M, Moréra S, Dessaux Y, Faure D. Quorum quenching: Role in nature and applied developments. FEMS Microbiology Reviews. 2015;**40**(1):86-116. DOI: 10.1093/femsre/fuv038

[13] Li Y-H, Tian X. Quorum sensing and bacterial social interactions in biofilms.

Sensors. 2012;**12**(3):2519-2538. DOI: 10.3390/s120302519

[14] Hallmann J, Quadt-Hallmann A, Mahaffee WF, Kloepper JW. Bacterial endophytes in agricultural crops. Canadian Journal of Microbiology. 1997;**43**(10):895-914. DOI: 10.1139/m97-131

[15] Cao D, Sun P, Bhowmick S, Wei Y, Guo B, Wei Y, et al. Secondary metabolites of endophytic fungi isolated from Huperzia serrata. Fitoterapia. 2021;**155**:104970. DOI: 10.1016/j.fitote.2021.104970

[16] Tian L-L, Ren H, Xi J-M, Fang J, Zhang JZ, Wu Q-X. Diverse anti-inflammation and anti-cancer polyketides isolated from the endophytic fungi alternaria SP. MG1. Fitoterapia. 2021;**153**:105000. DOI: 10.1016/j.fitote.2021.105000

[17] Carla CM, Genoveva LF, Bianca S, Rafael C, Andréa CB, Pde C, et al. Leishmanicidal, cytotoxic, antimicrobial and enzymatic activities of Diaporthe species, a mangrove-isolated endophytic fungus. African Journal of Microbiology Research. 2020;**14**(9):516-524. DOI: 10.5897/AJMR2020.9397

[18] El-Hawary SS, Mohammed R, Bahr HS, Attia EZ, El-Katatny MH, Abelyan N, et al. Soybean-associated endophytic fungi as potential source for anti-Covid-19 metabolites supported by docking analysis. Journal of Applied Microbiology. 2021;**131**(3):1193-1211. DOI: 10.1111/jam.15031

[19] Parrino B, Schillaci D, Carnevale I, Giovannetti E, Diana P, Cirrincione G, et al. Synthetic small molecules as anti-biofilm agents in the struggle against antibiotic resistance. European Journal of Medicinal Chemistry. 2019;**161**:154-178. DOI: 10.1016/j.ejmech.2018.10.036

[20] Atanasov AG, Zotchev SB, Dirsch VM, Supuran CT. Natural products in drug discovery: Advances and opportunities. Nature Reviews Drug Discovery. 2021;**20**(3):200-216. DOI: 10.1038/s41573-020-00114-z

[21] Lawson AD, MacCoss M, Heer JP. Importance of rigidity in designing small molecule drugs to tackle protein–protein interactions (PPIs) through stabilization of desired conformers. Journal of Medicinal Chemistry. 2017;**61**(10):4283-4289. DOI: 10.1021/acs.jmedchem.7b01120

[22] Atanasov AG, Waltenberger B, Pferschy-Wenzig E-M, Linder T, Wawrosch C, Uhrin P, et al. Discovery and resupply of pharmacologically active plant-derived natural products: A review. Biotechnology Advances. 2015;**33**(8):1582-1614. DOI: 10.1016/j.biotechadv.2015.08.001

[23] Mehlhorn H. Nanoparticles – Definitions. In: Nanoparticles in the Fight against Parasites. Cham: Springer; 2016. pp. **1-14**. DOI: 10.1007/978-3-319-25292-6_1

[24] Mousavi SM, Hashemi SA, Ghasemi Y, Atapour A, Amani AM, Savar Dashtaki A, et al. Green synthesis of silver nanoparticles toward bio and medical applications: Review study. Artificial Cells, Nanomedicine, and Biotechnology. 2018;**46**(sup3):S855-S872. DOI: 10.1080/21691401.2018.1517769

[25] Hussain I, Singh NB, Singh A, Singh H, Singh SC. Green synthesis of nanoparticles and its potential application. Biotechnology Letters. 2015;**38**(4):545-560. DOI: 10.1007/s10529-015-2026-7

[26] Singh R, Dubey AK. Isolation and characterization of a new endophytic actinobacterium *Streptomyces*

californicus Strain ADR1 as a promising source of anti-bacterial, anti-biofilm and antioxidant metabolites. Microorganisms. 2020;**8**(6):929. DOI: 10.3390/microorganisms8060929

[27] Marappa N, Dharumadurai D, Nooruddin T, Mohammad AA. Morphological,molecularcharacterization and biofilm inhibition effect of endophytic Frankia sp. from root nodules of Actinorhizal plant Casuarina sp. South African Journal of Botany. 2020;**134**:72-83. DOI: 10.1016/j.sajb.2020.02.039

[28] Rajesh PS, Rai VR. Purification and antibiofilm activity of AHL-lactonase from endophytic Enterobacter aerogenes VT66. International Journal of Biological Macromolecules. 2015;**81**:1046-1052. DOI: 10.1016/j.ijbiomac.2015.09.048

[29] Shastry RP, Rekha PD, Rai VR. Biofilm inhibitory activity of metallo-protein AHL-lactonase from cell-free lysate of endophytic Enterobacter species isolated from Coscinium fenestratum Gaertn. Biocatalysis and Agricultural Biotechnology. 2019;**18**:101009. DOI: 10.1016/j.bcab.2019.01.047

[30] Nadar Rajivgandhi G, Ramachandran G, Chenthis Kanisha C, Li J-L, Yin L, Manoharan N, et al. Anti-biofilm compound of 1, 4-diaza-2, 5-dioxo-3-isobutyl bicyclo[4.3.0]nonane from marine Nocardiopsis sp. DMS 2 (MH900226) against biofilm forming K. pneumoniae. Journal of King Saud University-Science. 2020;**32**(8):3495-3502. DOI: 10.1016/j.jksus.2020.10.012

[31] Mat Jalil MT, Ibrahim D. Fungal extract of Lasiodiplodia pseudotheobromae IBRL OS-64 inhibits the growth of skin pathogenic bacterium and attenuates biofilms of methicillin-resistant Staphylococcus aureus. Malaysian Journal of Medical Sciences. 2021;**28**(4):24-36. DOI: 10.21315/mjms2021.28.4.4

[32] Bai Z-Q, Lin X, Wang Y, Wang J, Zhou X, Yang B, et al. New phenyl derivatives from endophytic fungus Aspergillus flavipes AIL8 derived of mangrove plant Acanthus ilicifolius. Fitoterapia. 2014;**95**:194-202. DOI: 10.1016/j.fitote.2014.03.021

[33] Fathallah N, Raafat MM, Issa MY, Abdel-Aziz MM, Bishr M, Abdelkawy MA, et al. Bio-guided fractionation of prenylated benzaldehyde derivatives as potent antimicrobial and antibiofilm from Ammi majus L fruits-associated Aspergillus amstelodami. Molecules. 2019;**24**(22):4118. DOI: 10.3390/molecules24224118

[34] Taufiq MMJ, Darah I. Study of ethyl acetate crude extract of Lasiodiplodia pseudotheobromae IBRL OS-64, an endophytic fungus isolated from the leaf of Ocimum sanctum for anti-biofilm agent (funded work). International Journal of pharma and Bio Sciences. 2020;**11**(2):7-18. DOI: 10.22376/ijpbs.2020.11.2.b7-18

[35] Meena H, Mishra R, Ranganathan S, Sarma VV, Ampasala DR, Kalia VC, et al. Phomopsis tersa as inhibitor of quorum sensing system and biofilm forming ability of Pseudomonas aeruginosa. Indian Journal of Microbiology. 2019;**60**(1):70-77. DOI: 10.1007/s12088-019-00840-y

[36] Ebaya MMA, El-Mowafy M, Adel El-Sokkary MM, Hassan R. Purification, characterization, and biocatalytic and antibiofilm activity of a novel dextranase from Talaromyces sp. Drlica K, editor. International Journal of Microbiology. 2020;**2020**:1-11. DOI: 10.1155/2020/9198048

[37] Taufiq MMJ, Darah I. Antibacterial and antibiofilm activities of crude extract

of Lasiodiplodia pseudotheobromae IBRL OS-64 against foodborne bacterium, Yersinia enterocolitica. Journal of Pharmaceutical Research International. 2020;**32**(14):87-102. DOI: 10.9734/jpri/2020/v32i1430609

[38] Ngidi LS, Nxumalo CI, Shandu JS, Maliehe TS, Rene K. Anti-quorum sensing and cytotoxic properties of the endophytic Pseudomonas aeruginosa CP043328.1's extract. Pharmacognosy Journal. 2020;**13**(2): 332-340. DOI: 10.5530/pj.2021.13.43

[39] Rajesh PS, Ravishankar RV. Hydrolytic enzymes and quorum sensing inhibitors from endophytic fungi of Ventilago madraspatana Gaertn. Biocatalysis and Agricultural Biotechnology. 2013;**2**(2):120-124. DOI: 10.1016/j.bcab.2013.01.002

[40] Figueroa M, Jarmusch AK, Raja HA, El-Elimat T, Kavanaugh JS, Horswill AR, et al. Polyhydroxyanthraquinones as quorum sensing inhibitors from the guttates of Penicillium restrictum and their analysis by desorption electrospray ionization mass spectrometry. Journal of Natural Products. 2014;**77**(6):1351-1358. DOI: 10.1021/np5000704

[41] Martín-Rodríguez A, Reyes F, Martín J, Pérez-Yépez J, León-Barrios M, Couttolenc A, et al. Inhibition of bacterial quorum sensing by extracts from aquatic fungi: First report from marine endophytes. Marine Drugs. 2014;**12**(11):5503-5526. DOI: 10.3390/md12115503

[42] Akther T, Khan MohdShahanbaj SH. Biosynthesis of silver nanoparticles via fungal cell filtrate and their anti-quorum sensing against Pseudomonas aeruginosa. Journal of Environmental Chemical Engineering. 2020;**8**(6):104365. DOI: 10.1016/j.jece.2020.104365

[43] Pattnaik SS, Ranganathan S, Ampasala DR, Syed A, Ameen F, Busi S. Attenuation of quorum sensing regulated virulence and biofilm development in Pseudomonas aeruginosa PAO1 by Diaporthe phaseolorum SSP12. Microbial Pathogenesis. 2018;**118**:177-189. DOI: 10.1016/j.micpath.2018.03.031

[44] Hamed A, Elkhouly H, Sidkey N, El hosainy A, Ghareeb M. Bioactive secondary metabolites from Endophytic Aspergillus terreus AH1 isolated from Ipomoea carnea growing in Egypt. Egyptian Journal of Chemistry. 2021;**64**(12):7511-7520. DOI: 10.21608/ejchem.2021.85908.4161

[45] Singh M, Kumar A, Singh R, Pandey KD. Endophytic bacteria: A new source of bioactive compounds. 3 Biotech. 2017;**7**(**5**):**315**. DOI: 10.1007/s13205-017-0942-Z

[46] El-Gendy MMAA, Yahya SMM, Hamed AR, El-Bondkly AMA. Assessment of the phylogenetic analysis and antimicrobial, antiviral, and anticancer activities of marine endophytic *Streptomyces* species of the soft coral *Sarcophyton convolutum*. International Microbiology. 2021;**25**:133-152. DOI: 10.1007/s10123-021-00204-x

[47] Theodora NA, Dominika V, Waturang DE. Screening and quantification of anti-quorum sensing and antibiofilm activities of phyllosphere bacteria against biofilm forming bacteria. BMC Research Notes. 2019;**12**:732. DOI: 10.1186/s13104-019-4775-1

[48] Sabu R, Soumya KR, RadhakrishnanEK.Endophytic*Nocardiopsis* sp. from *Zingiber officinale* with both antiphytopathogenic mechanisms and antibiofilm activity against clinical isolates. 3 Biotech. 2017;7:115. DOI: 10.007/s13205-017-0735-4

[49] Mukherjee AK, Das K. Microbial surfactants and their potential applications: An overview. In: Sen R, editor. Biosurfactants. Advances in Experimental Medicine and Biology. New York, NY: Springer; 2010. pp. 54-64. DOI: 10.1007/978-1-4419-5979-9_4

[50] Rosenberg E, Ron EZ. High- and low-molecular-mass microbial surfactants. Applied and Microbiololy and Biotechnology. 1999;**52**(2):154-162. DOI: 10.1007/s002530051502

[51] Barrantes K, Araya JJ, Chacón L, Procupez-Schtirbu R, Lugo F, Ibarra G, et al. Antiviral, antimicrobial, and antibiofilm properties of biosurfactants. In: In: Sarma H., Prasad M.N.V., editor. Biosurfactants for a Sustainable Future: Production and Applications in the Environment and Biomedicine. 1st ed. Chichester: Wiley Online Library; 2021. pp. 245-268. DOI: 10.1002/9781119671022.ch1

[52] Satpute SK, Banpurkar AG, Banat IM, Sangshetti JN, Patil RH, Gade WN. Multiple roles of biosurfactants in biofilms. Current Pharmaceutical Design. 2016;**22**:1429-1448. DOI: 10.2174/1381612822666160120152704

[53] Shekhar S, Sundaramanickam A, Balasubramanian T. Biosurfactant producing microbes and their potential applications: A review. Critical Reviews in Environmental Science and Technology. 2015;**45**:1522-1554. DOI: 10.1080/10643389.2014.955631

[54] Ashitha A, Radhakrishnan EK, Mathew J. Characterization of biosurfactant produced by the endophyte *Burkholderia* sp. WYAT7 and evaluation of its antibacterial and antibiofilm potentials. Journal of Biotechnology. 2020;**313**:1-10. DOI: 10.1016/j.jbiotec.2020.03.005

[55] Ceresa C, Rinaldi M, Chiono V, Carmagnola I, Allegrone G, Fracchia L. Lipopeptides from *Bacillus subtilis* AC7 inhibit adhesion and biofilm formation of *Candida albicans* on silicone. Antonie Van Leeuwenhoek. 2016;**109**(10):1375-1388. DOI: 10.1007/s10482-016-0736-z

[56] Ceresa C, Tessarolo F, Maniglio D, Coala I, Nollo G, Rinaldi M, et al. Inhibition of *Candida albicans* biofilm by lipopeptide AC7 coated medical-grade silicone in combination with farnesol. AIMS Bioengineering. 2018;**5**(3):192-208. DOI: 10.3934/bioeng.2018.3.192

[57] Cochis A, Fracchia L, Martinotti MG, Rimondini L. Biosurfactants prevent in vitro *Candida albicans* biofilm formation on resins and silicon materials for prosthetic devices. Oral Surgery, Oral Medicine, Oral Pathology and Oral Radiology. 2012;**113**(6):755-761. DOI: 10.1016/j.oooo.2011.11.004

[58] Terkar A, Borde M. Endophytic fungi: Novel source of bioactive fungal metabolites. In: New and Future Developments in Microbial Biotechnology and Bioengineering. Lucknow, India: Elsevier; 2021. pp. 95-105

[59] Link BL. Penicillin: Its discovery and early development. Seminars in Pediatric Infectious Diseases. 2004;**15**(1):52-57. DOI: 10.1053/j.spid.2004.02.001

[60] Gouda S, Das G, Sen SK, Shin HS, Patra JK. Endophytes: A treasure house of bioactive compounds of medicinal importance. Frontiers in Microbiology. 2016;**7**:1538. DOI: 10.3389/fmicb.2016.01538

[61] May Zin WW, Buttachon S, Dethoup T, Pereira JA, Wales L, Inácio Â, et al. Antibacterial and antibiofilm activities of the isolated metabolites

from the culture of the mangrove-derived endophytic fungus Eurotium chevalieri KUFA 0006. Phytochemistry. 2017;**141**:86-97. DOI: 10.1016/j.phytochem.2017.05.015

[62] Narmani A, Teponno RB, Helaly SE, Arzanlou M, Stadler M. Cytotoxic, anti-biofilm and antimicrobial polyketides from the plant associated fungus Chaetosphaeronema achilleae. Phytotherapy. 2019;**139**:104390. DOI: 10.1016/j.fitote.2019.104390

[63] Kaur J, Sharma P, Kaur R, Kaur S, Kaur A. Assessment of alpha glucosidase inhibitors produced from endophytic fungus Alternaria destruens as antimicrobial and antibiofilm agents. Molecular Biology Reports. 2020;**47**(1):423-432. DOI: 10.1007/s11033-019-05145-3

[64] Kaur N, Arora DS. Prospecting the antimicrobial and antibiofilm potential of Chaetomium globosum an endophytic fungus from Moringa oleifera. AMB Express. 2020;**10**(1):206. DOI: 10.1186/s13568-020-01143-y

[65] Elkhouly HI, Hamed AA, El Hosainy AM, Ghareeb MA, Sidkey NM. Bioactive secondary metabolite from Endophytic Aspergillus Tubenginses ASH4 isolated from Hyoscyamus muticus: Antimicrobial, antibiofilm. Antioxidant and Anticancer Activity. Pharmacognosy Journal. 2021;**13**(2):434-442

[66] Qader MM, Hamed AA, Soldatou S, Abdelraof M, Elawady ME, Hassane ASI, et al. Antimicrobial and antibiofilm activities of the fungal metabolites isolated from the marine endophytes Epicoccum nigrum M13 and Alternaria alternata 13A. Sea Drugs. 2021;**19**(4):232. DOI: 10.3390/md19040232

[67] Rutherford ST, Bassler BL. Bacterial quorum sensing: Its role in virulence and possibilities for its control. Cold Spring Harbor Perspectives in Medicine. 2012;**2**(11):a012427-a012427. DOI: 10.1101/cshperspect.a012427

[68] Deveau A, Hogan DA. Linking quorum sensing regulation and biofilm formation by Candida albicans. Methods in Molecular Biology (Clifton, NJ). 2011;**692**:219-233. DOI: 10.1007/978-1-60761-971-0_16

[69] Hogan DA. Talking to themselves: Autoregulation and quorum sensing in fungi. Eukaryotic Cell. 2006;**5**(4):613-619. DOI: 10.1128/ec.5.4.613-619.2006

[70] Mishra R, Kushveer JS, MohdIK K, Pagal S, Meena CK, Murali A, et al. 2,4-Di-tert-butylphenol isolated from an endophytic fungus, *Daldinia eschscholtzii*, reduces virulence and quorum sensing in *Pseudomonas aeruginosa*. Frontiers in Microbiology. 2020;**11**:1668. DOI: 10.3389/fmicb.2020.01668

[71] Antunes LCM, Ferreira RBR, Buckner MMC, Finlay BB. Quorum sensing in bacterial virulence. Microbiology. 2010;**156**(8):2271-2282. DOI: 10.1099/mic.0.038794-0

[72] Antunes LCM, Ferreira RBR. Intercellular communication in bacteria. Critical Reviews in Microbiology. 2009;**35**(2):69-80. DOI: 10.1080/10408410902733946

[73] Schauder S, Shokat K, Surette MG, Bassler BL. The LuxS family of bacterial autoinducers: Biosynthesis of a novel quorum-sensing signal molecule. Molecular Microbiology. 2001;**41**(2):463-476. DOI: 10.1046/j.1365-2958.2001.02532.x

[74] Smith JL, Fratamico PM, Novak JS. Quorum sensing: A primer for food microbiologists†. Journal of Food Protection. 2004;**67**(5):1053-1070. DOI: 10.4315/0362-028x-67.5.1053

[75] Sharma R, Jangid K. Fungal quorum sensing inhibitors. In: Quorum Sensing Vs Quorum Quenching: A Battle with No End in Sight. New Delhi: Springer; 2014. pp. 237-257. DOI: 10.1007/978-81-322-1982-8_20

[76] Mookherjee A, Singh S, Maiti MK. Quorum sensing inhibitors: Can endophytes be prospective sources? Archives of Microbiology. 2017;**200**(2):355-369. DOI: 10.1007/s00203-017-1437-3

[77] Kusari P, Kusari S, Lamshöft M, Sezgin S, Spiteller M, Kayser O. Quorum quenching is an antivirulence strategy employed by endophytic bacteria. Applied Microbiology and Biotechnology. 2014;**98**(16):7173-7183. DOI: 10.1007/s00253-014-5807-3

[78] Wang JW, Wu JH, Huang WY, Tan RX. Laccase production by *Monotospora* sp., an endophytic fungus in *Cynodon dactylon*. Bioresource Technology. 2006;**97**(5):786-789. DOI: 10.1016/j.biortech.2005.03.025

[79] Zhou J-W, Jia A-Q, Jiang H, Li P-L, Chen H, Tan X-J, et al. 1-(4-Amino-2-hydroxyphenyl)ethanone from *Phomopsis liquidambari* showed quorum sensing inhibitory activity against *Pseudomonas aeruginosa*. Applied Microbiology and Biotechnology. 2020;**105**(1):341-352. DOI: 10.1007/s00253-020-11013-z

[80] Pellissier L, Leoni S, Marcourt L, Ferreira Queiroz E, Lecoultre N, Quiros-Guerrero L-M, et al. Characterization of Pseudomonas aeruginosa quorum sensing inhibitors from the endophyte Lasiodiplodia venezuelensis and evaluation of their antivirulence effects by metabolomics. Microorganisms. 2021;**9**(9):1807. DOI: 10.3390/microorganisms9091807

[81] Lopes RBM, de Oliveira Costa LE, Vanetti MCD, de Araújo EF, de Queiroz MV. Endophytic bacteria isolated from common bean (*Phaseolus vulgaris*) exhibiting high variability showed antimicrobial activity and quorum sensing inhibition. Current Microbiology. 2015;**71**(4):509-516. DOI: 10.1007/s00284-015-0879-6

[82] Akbari Kiarood SL, Rahnama K, Golmohammadi M, Nasrollanejad S. Quorum-quenching endophytic bacteria inhibit disease caused by *Pseudomonas syringae* pv. Syringae in *citrus* cultivars. Journal of Basic Microbiology. 2020;**60**(9):746-757. DOI: 10.1002/jobm.202000038

[83] Messaoudi O, Bendahou M. Biological synthesis of nanoparticles using endophytic microorganisms: Current development. In: Nanotechnology and the Environment. London, UK: IntechOpen; 2020. DOI: 10.5772/intechopen.93734

[84] Khan I, Saeed K, Khan I. Nanoparticles: Properties, applications and toxicities. Arabian Journal of Chemistry. 2017;**12**(7):908. DOI: 10.1016/j.arabjc.2017.05.011

[85] Das P, Ghosh S, Nayak B. Phyto-fabricated nanoparticles and their anti-biofilm activity: Progress and current status. Frontiers in Nanotechnology. 2021;**3**:76. DOI: 10.3389/fnano.2021.739286

[86] Ovais M, Khalil A, Ayaz M, Ahmad I, Nethi S, Mukherjee S. Biosynthesis of metal nanoparticles via microbial enzymes: A mechanistic approach. International Journal of Molecular Sciences. 2018;**19**(12):4100. DOI: 10.3390/ijms19124100

[87] Blank LM, Ebert BE, Buehler K, Bühler B. Redox biocatalysis and metabolism: Molecular mechanisms and metabolic network analysis.

Antioxidants & Redox Signaling. 2010;**13**(3):349-394. DOI: 10.1089/ars.2009.2931

[88] Hulkoti NI, Taranath TC. Biosynthesis of nanoparticles using microbes—A review. Colloids and Surfaces B: Biointerfaces. 2014;**121**:474-483. DOI: 10.1016/j.colsurfb.2014.05.027

[89] Rahman S, Rahman L, Khalil AT, Ali N, Zia D, Ali M, et al. Endophyte-mediated synthesis of silver nanoparticles and their biological applications. Applied Microbiology and Biotechnology. 2019;**103**(6):2551-2569. DOI: 10.1007/s00253-019-09661-x

[90] Rana KL, Kour D, Yadav N, Yadav AN. Endophytic microbes in nanotechnology: Current development, and potential biotechnology applications. In: Microbial Endophytes. Udaipur, India: Woodhead Publishing; 2020. pp. 231-262. DOI: 10.1016/B978-0-12-818734-0.00010-3

[91] Ivanova N, Gugleva V, Dobreva M, Pehlivanov I, Stefanov S, Andonova V. Silver nanoparticles as multi-functional drug delivery systems. In: Nanomedicines. Rijeka, Croatia: IntechOpen; 2018. DOI: 10.5772/intechopen.80238

[92] Shkodenko L, Kassirov I, Koshel E. Metal oxide nanoparticles against bacterial biofilms: Perspectives and limitations. Microorganisms. 2020;**8**(10):1545. DOI: 10.3390/microorganisms8101545

[93] Slavin YN, Asnis J, Häfeli UO, Bach H. Metal nanoparticles: Understanding the mechanisms behind antibacterial activity. Journal of Nanobiotechnology. 2017;**15**(1):65. DOI: 10.1186/s12951-017-0308-z

[94] dos Santos CA, Ingle AP, Rai M. The emerging role of metallic nanoparticles in food. Applied Microbiology and Biotechnology. 2020;**104**(6):2373-2383. DOI: 10.1007/s00253-020-10372-x

[95] Wang L, Hu C, Shao L. The antimicrobial activity of nanoparticles: Present situation and prospects for the future. International Journal of Nanomedicine. 2017;**12**:1227-1249. DOI: 10.2147/IJN.S121956

[96] Długosz O, Szostak K, Staroń A, Pulit-Prociak J, Banach M. Methods for reducing the toxicity of metal and metal oxide NPs as biomedicine. Materials. 2020;**13**(2):279. DOI: 10.3390/ma13020279

[97] Bakhtiari-Sardari A, Mashreghi M, Eshghi H, Behnam-Rasouli F, Lashani E, Shahnavaz B. Comparative evaluation of silver nanoparticles biosynthesis by two cold-tolerant Streptomyces strains and their biological activities. Biotechnology Letters. 2020;**42**(10):1985-1999. DOI: 10.1007/s10529-020-02921-1

[98] Adnan M. Synthesis, characterization and applications of endophytic fungal nanoparticles. Inorganic and Nano-Metal Chemistry. 2020;**51**(2):280-287. DOI: 10.1080/24701556.2020.1784231

[99] Bagur H, Poojari CC, Melappa G, Rangappa R, Chandrasekhar N, Somu P. Biogenically synthesized silver nanoparticles using endophyte fungal extract of ocimum tenuiflorum and evaluation of biomedical properties. Journal of Cluster Science. 2019;**31**(6):1241-1255. DOI: 10.1007/s10876-019-01731-4

[100] Neethu S, Midhun SJ, Radhakrishnan EK, Jyothis M. Green synthesized silver nanoparticles by marine endophytic fungus Penicillium polonicum and its antibacterial efficacy against biofilm forming, multidrug-resistant Acinetobacter baumanii.

Microbial Pathogenesis. 2018;**116**:263-272. DOI: 10.1016/j.micpath.2018.01.033

[101] Neethu S, Midhun SJ, Radhakrishnan EK, Jyothis M. Surface functionalization of central venous catheter with mycofabricated silver nanoparticles and its antibiofilm activity on multidrug resistant Acinetobacter baumannii. Microbial Pathogenesis. 2020;**138**:103832. DOI: 10.1016/j.micpath.2019.103832

[102] Das R, Ahmed M, Lalnunmawii E, Nachimuthu S. Myco-nanocolloids manipulate growth, biofilm formation and virulence genes in UTI causing E. coli. Inorganic and Nano-metal Chemistry. 2020;**51**(12):1725-1734. DOI: 10.1080/24701556.2020.1852426

[103] Chandankere R, Chelliah J, Subban K, Shanadrahalli VC, Parvez A, Zabed HM, et al. Pleiotropic functions and biological potentials of silver nanoparticles synthesized by an endophytic fungus. Frontiers in Bioengineering and Biotechnology. 2020;**8**. DOI: 10.3389/fbioe.2020.00095

[104] Ibrahim E, Fouad H, Zhang M, Zhang Y, Qiu W, Yan C, et al. Biosynthesis of silver nanoparticles using endophytic bacteria and their role in inhibition of rice pathogenic bacteria and plant growth promotion. RSC Advances. 2019;**9**(50):29293-29299. DOI: 10.1039/c9ra04246f

[105] Dhandapani P, Maruthamuthu S, Rajagopal G. Bio-mediated synthesis of TiO2 nanoparticles and its photocatalytic effect on aquatic biofilm. Journal of Photochemistry and Photobiology B: Biology. 2012;**110**:43-49. DOI: 10.1016/j.jphotobiol.2012.03.003

[106] Shakibaie M, Forootanfar H, Golkari Y, Mohammadi-Khorsand T, Shakibaie MR. Anti-biofilm activity of biogenic selenium nanoparticles and selenium dioxide against clinical isolates of Staphylococcus aureus, Pseudomonas aeruginosa, and Proteus mirabilis. Journal of Trace Elements in Medicine and Biology. 2015;**29**:235-241. DOI: 10.1016/j.jtemb.2014.07.020

[107] Balaji S, Shanmugam VK. Enhanced antibiofilm activity of endophytic bacteria mediated zirconium nanoparticles. Biointerface Research in Applied Chemistry. 2021;**12**(1):74-82. DOI: 10.33263/BRIAC121.074082

Chapter 12

Natural Products as Antibiofilm Agents

Cynthia Amaning Danquah, Prince Amankwah Baffour Minkah, Theresa A. Agana, Phanankosi Moyo, Michael Tetteh, Isaiah Osei Duah Junior, Kofi Bonsu Amankwah, Samuel Owusu Somuah, Michael Ofori and Vinesh J. Maharaj

Abstract

Biofilms, are vastly structured surface-associated communities of microorganisms, enclosed within a self-produced extracellular matrix. Microorganisms, especially bacteria are able to form complex structures known as biofilms. The presence of biofilms especially in health care settings increases resistance to antimicrobial agents which poses a major health problem. This is because biofilm-associated persistent infections are difficult to treat due to the presence of multidrug-resistant microorganisms. This chapter will give an idea about documented agents including isolated compounds, crude extracts, decoctions, fractions, etc. obtained from natural sources such as plants, bacteria, fungi, sponge and algae with antibiofilm activities. Furthermore, we have done phylogenetic analysis to identify plant families most prolific in producing plant species and compounds with good antibiofilm properties so as to aid in prioritizing plant species to investigate in future studies. The data in this chapter will help serve as valuable information and guidance for future antimicrobial development.

Keywords: biofilm, natural products, quorum sensing, anti-biofilm agents, antimicrobials

1. Introduction

The empirical approach to antimicrobial therapy among health care professionals and the concurrent patronage of over-the-counter antibiotics by patients have together caused an exponential rise in multidrug resistance among clinically relevant antimicrobials and with increasing trends for the past two decades [1]. Different mechanisms of antimicrobial resistance have been proposed, including the (i) alteration of the antibiotic target by genetic mutations or post-translational modification, (ii) deactivation of the antibiotic through hydrolysis or modification, such as phosphorylation by an enzyme, (iii) increased efflux of the antibiotic out of the cell by efflux pumps and porins, (iv) decreased influx/penetration of the antibiotic into the cell, through changes in cell wall structure; and overproduction of the antibiotic target through gene amplification [2]. However, one of bacteria's preferred and commonly deployed strategies to overcome the effect of antimicrobials is the formation of biofilms. Over 90% of pathogenic bacterial species, including *Staphylococcus aureus* (*S. aureus*) and *Pseudomonas aeruginosa* (*P. aeruginosa*), possess an inherent ability to produce biofilms, making biofilms the leading cause of multidrug resistance among microorganisms [3–5].

Biofilm is a complex community of sessile microbial communities embedded in a self-producing polymeric matrix comprising exopolysaccharides, proteins, nucleic acids, and cell surface proteins [6–8]. As a community of microorganisms, biofilms constitute either a single microbial species or a combination of a different class of bacteria, fungi, protozoa, archaea, and yeast, with a unique ability to colonize almost any environmental niche, biotic or inert surfaces [9–13]. Biofilm enables microorganisms to withstand harsh environmental conditions such as nutrient deficiencies, high osmotic pressure, the low potential of hydrogen, oxidative stress, and antimicrobial insults [14]. The increased resistance of biofilms to antimicrobials arise from phenotypic cell variation and gene transcription. In particular, there is an exponential growth of microorganisms and genetic transfer of extrachromosomal elements via cell-to-cell communication system called quorum sensing [14–17]. Quorum sensing is critical in the development and survival of biofilms; thus, it regulates the nutritional demands of microorganisms within the biofilm to meet the external supply of resources [18, 19]. In addition quorum sensing is essential for the biosynthesis and secretion of small molecule signals that activate a range of downstream processes including virulence and drug resistance mechanisms as seen in biofilms [20, 21].

The health risks of biofilms are enormous, which underscore their utilization in plant protection, bioremediation, wastewater treatment, and corrosion prevention in agricultural and industrial settings [22–24]. In particular, the biofilm grows on living human tissues such as the lungs and teeth and the surfaces of implanted biomedical devices, including contact lenses, central venous catheters [8, 25], prosthetic joints, pacemakers, and intrauterine devices [7]. Unlike single bacterial plankton cells, the treatment of biofilm-mediated infections is challenging owing to the decreased susceptibility to antimicrobial agents and other chemotherapeutics. The availability of qualitative (such as Congo red agar, microtitre plate, tube methods) and quantitative (including polymerase chain reaction (PCR)) techniques have enabled the detection and measurement of biofilms [26]. Conversely, the evaluation and screening of antimicrobials against biofilms are of great challenge. In particular, standard microdilution testing cannot evaluate the susceptibility of biofilms to antimicrobial drugs because these tests focus on planktonic (suspended) organisms rather than biofilm (surface-associated) organisms [7]. Instead, susceptibility must be determined directly against biofilm-associated organisms, preferably under conditions that mimic *in vitro* and/or *in vivo* conditions. In this light, several biofilm models systems have been developed to permit accurate screening and evaluation of novel agents for their antibiofilm activity [27, 28].

Although nature has provided a plethora of natural products with varying chemotherapeutic properties to fight human infectious diseases, discovering new and effective antimicrobials has been slow. The decline in the efficacy of existing chemotherapy and the surge in drug resistance has triggered an expedient exploration of natural products, especially from plants and microbial origin, for their antibiofilm activity against biofilm-mediated human infections. Plant extracts and plant-derived chemical products, such as essential oils, flavonoids, terpenoids, have been shown in vitro to have antimicrobial and antibiofilm activity [27–31]. Secondary metabolites and other peptidic compounds from microorganisms also exhibit antagonistic effects against biofilms [6, 32]. These chemical constituents exert their action by inhibiting critical elements within a biofilm and/or terminating biofilm formation processes [33]. Given the unique nature of plants and microbes, natural products derived from these sources could provide an avenue for developing newly efficacious and clinically desirable chemotherapies against biofilms-mediated infections and their associated health consequences.

This chapter aims to provide a comprehensive summary of natural products from plants and microbial sources as potential sources of antibiofilm agents. Again, it highlights the strategies and model organisms used to identify and evaluate the antibiofilm capacity of these naturally isolated chemical compounds.

2. Biofilm formation

Biofilm formation represents a survival mechanism deployed by microorganisms in response to unfavorable environmental conditions [34]. Structurally, biofilms are a collection of adherent microorganisms in a milieu of an extracellular matrix consisting of polysaccharides, proteins, nucleic acids, and lipids. This unique architecture enables biofilms to cling firmly to surfaces of implanted body organs and biomedical devices and, more importantly, increase their resistance to antimicrobial therapy. The presence of bacterial secreted glycocalyx and degrading matrix enzymes reduces the antimicrobial concentration of which individual plankton cells within the biofilm are exposed [35, 36].

The morphogenesis of biofilms constitutes five distinct stages; namely, reversible attachment, irreversible adhesion, production of extracellular polymeric substances, biofilm maturation, and dispersal/detachment. As the initial step in biofilm formation, reversible attachment is characterized by the interaction between plankton cells and the conditioned surface. Fewer plankton cells move to the surface of the substrate by convection, pedesis, or sedimentation [37]. Consequently, chemotaxis directs bacterial cells along a nutrient gradient [38]. Upon reaching the surface of the substratum, the interaction between the cell surfaces and the substratum is dependent on the net sum of repulsive or attractive forces generated by the two characters [39, 40]. The presence of fimbriae, flagella, pili, and glycocalyx enables the microorganisms to overcome the repulsive forces (such as electrostatic, hydrophobic, Van der Waals, and hydration interactions) from the substratum and subsequently cling [39, 41, 42]. The rate of biofilms formation is influenced by the substrate's physicochemical properties, including the surface roughness, hydrophobicity, surface charge, and the presence of conditioning films [41, 43, 44].

Furthermore, bacterial cells transition into an irreversible adhesion phase. Irreversible attachment occurs through the combined effect of short-range forces of the substrate (such as dipole-dipole, hydrogen, ionic and covalent interactions) and adhesive structures of the bacterial cells. The flagella and pili, for instance, are critically important in the attachment process of various strains of microorganisms [45–48]. For example, Vatanyoopaisarn et al. demonstrated the firm clinging ability of wild-type *Listeria monocytogens* (*L. monocytogens*) compared to the non-flagellated mutant type [45]. Similarly, Di Martino and colleagues showed the distinctive role of type one and type three fimbriae in initiating the attachment of *Klebsiella pneumonia* (*K. pneumonia*) to abiotic surfaces [46]. Alarcon and coworkers also observed the critical role of pilus in the twitching substrate movement of *P. aeruginosa* [48].

Moreover, the resident plankton cells produce extracellular polymeric substances (EPS), an essential biofilm component. Quorum sensing and cyclic-di-GMP mediated

EPS formation [49–52]. The formation of EPS promotes cohesion among bacteria and the adhesion of biofilms via hydrophobic and ionic interactions [49, 53, 54]. In addition, EPS is vital in constructing biofilms, maintaining biofilm architecture, quorum sensing, and genetic transfer among individual organisms within the biofilm [49, 55].

The resident bacterial cells proliferate into microcolonies mediated by autoinducers (AIs). AIs are chemical signaling molecules that permit intra-species and inter-species bacterial cell-to-cell communication [56, 57]. The surge in AIs activates critical enzymatic machinery in bacterial species for regulating the formation of microcolonies and the maturation of biofilms [52]. For example, the increase in AIs causes synchronous activation of the 15 gene-long epsA-O in *Bacillus subtilis (B. subtilis)* that causes an increased production of EPS. The proliferation of microcolonies and the increased accumulation of EPS trigger gene expression [52]. This alteration in gene expression reversibly stimulates additional EPS as adhesive molecules to bind individual plankton cells. In addition to EPS production, water channels are created to facilitate the inflow of nutrients to the individual cells within the biofilm [58]. During the maturation stage of biofilm formation, there is restricted motility of the bacterial cells together with characteristic variation in gene and protein expression between biofilm and plankton cells [59, 60].

The terminal phase of biofilm formation, delineated as detachment or dispersal, is regulated by a complex mechanism constituting signal transduction, effector, and environmental factors [61]. Detachment/dispersal represents a unique phase in the life cycle, where plankton cells segregate and escape from biofilms to establish microcolonies on fresh surfaces [62, 63]. Of note, the dispersal phase of a biofilm is characterized by the detachment of plankton cells from hitherto biofilm, seeding or passive movement of plankton to new uncolonized surfaces, and clinging or attachment to substrates [61, 64, 65].

3. Models for assessing antibiofilm activity

Several methods have been developed to study the antibiofilm activities of various compounds *in vitro*. However, only a few *in vivo* strategies for studying biofilms have been described. Given the importance of bacterial biofilm infections worldwide, we describie some models for assessing the efficacy of antibiofilm compounds *in vivo.*

3.1 The human organoid model

The human epidermis organoid model has a tough methicillin-resistant *S. aureus* (MRSA) USA300 and *P. aeruginosa* PAO1 biofilm system for studying host-microbe interplay and enable the screening of novel antibiofilm agents. This model allows the screening of synthetic host peptides to reveal their superior antibiofilm activity against MRSA compared to the antibiotic mupirocin. This model provides an exciting tool for elucidating disease pathology and testing novel drugs toxicities and efficacies. It also has the added advantage of reducing the use of animals in pre-clinical testing and replacing in vivo infection models with an ethical alternative that better reflects human disease [27].

This method involves establishing bacterial biofilm by seeding the center of the skin model with 5 μL of 2 × 10^8 CFU/ml of MRSA or *P. aeruginosa* PA01 or fluorescently-tagged MRSA or PA01-mCherry or luminescent MRSA-lux or PA01-lux resuspended in PBS and cultured at 37 °C and 7.3% CO_2. 30 μL of 1–4 mg/ml DJK-5 peptide was then added on top of the biofilm for 4 h, 1–3 days post inoculation. Luminisense signal are monitored daily after the establishment of infection until luminescence are observed in the culture medium underneath the skin. This is to study how long the skin could endure biofilm growth. ChemiDoc imaging system is used to visualize biofilms and bacterial counts quantified by sonicating, votexing and serially diluting excised skin samples on agar plates [27].

3.2 Wound models

Among the most widely used models to investigate antibiofilm compounds is the skin wound model. It involves either causing damage to the skin (abrasion, burns or surgical excisions) and subsequently infecting the injured region with biofilm-forming bacteria, or inducing the formation of absess or wounds by seeding high-density biofilm forming bacteria subcutaneously. The commonly used clinically relevant organisms are *S. aureus, Stapylococcus epidermidis (Staphylococcus epidermidis)* and *P. aeuruginosa* [66]. The inoculum can differ depending on the expected severity of the infection ranging from acute to chronic, with chronic infections mimicking biofilm infection in human more accurately. Recovery and/or healing of the infected wound therefore indicates antibiofilm activity. Effectiveness of antibiofilm compounds can also be assessed by (a) examining the infectious process and recovery via real-time imaging with an in vivo imaging system as well as wound size measurement using calipers and photographs, (b) tissue analysis to assess tissue regeneration process, (c) assessment of genetic fingerprints associated with the formation of biofilms such as *pslD, mucC* and quorum sensing related genes (d) analysis of inflammatory patterns (e) assessmet of underlying organs [67, 68].

3.3 Oral infections model

Various biofilms from disease and non-disease causing microorganisms results in the formation of dental caries. Dental caries results from the interation between diet and microbiota-matrix that occur on the oral surface [69]. This is mostly replicated in animal models using newly weaned rats. Prior treatmet with antbiotics is essential to elimintate existing microbiome. Subsequently, the animals are fed with cariogenic diet while also receiving the bacteria (e.g. *Streptococcus mutans* (*S. mutans*)) orally daily for period of 5–7 days. The infection is ascertained by sowing oral samples. The topical application of the compounds is carried out on the teeth, daily for 30–45 days and the mandibles and molars excised at the end of the study to evalauate the carious lesions [70].

Periodontitis can as well be replicated in animal models using its associated bacteria (e.g. *Streptococcus gordoni* (*S. gordoni*) and *Porphyromonas gingivalis* (*P. gingivalis*)) and confirmed by oral sowing or PCR analysis [71]. The treatment can be perfomed topically either to prevent or eradicate already formed biofilm infection. The animals are euthanized at the end of the experiment, and the skull excised for alveolar bone loss assay of the maxilla [71, 72].

3.4 Respiratory tract chronic infections model

The primary organism associated with biofilm lung infection in cystic fibrosis (CF) has been identified to be *P. aeruginosa.* In the cystic fibrosis murine model, bacteria are inoculated either intrathecally, intranasally or by instillation [73]. The inoculum and the frequency of inoculation underscores the severity of infection.

Bacteria carriers such as alginate formed by the bacteria strain itself or by bacteria incorporation on agar beads can be used to establish chronic pulmonary infection. Intrathecal instillation is however, the most preferred route for inoculation of bacteria in this scenario [74].

Clinical isolates of *P. aeuruginosa* has also been used in some models. This model has an advantage of having a shorter time between establishment of infection and end of treatment than that described above. Since the bacteria is directly inoculated, it can result in severe acute respiratory distress (SARS) and eventually death even before treatment has been effective [67].

3.5 Foreign body infection model

The ability of biofilm forming bacteria to grow and multiply on the surfaces of certain medical devices [75] has led to the discovery of this model. The preformation of biofilm on these surgically implanted foreign bodies affect the activity of defense cells [25]. This model can be executed using two (2) approaches. These are Site Specific Device Model where biofilm forming bacteria are introduced at the injection site after devices are placed in particular organ or region in humans for evaluation of antibiofilm activity, and Subcutaneous Device Model where deliberately colonized foreign bodies are inserted in the subcutaneous layer, mostly at the back of the animals [76]. In Site Specific Device Model, antibiofim activity is measured at the part of the device that made contact with bacteria or measured by bacterial recovery at injection site [75]. In Subcutaneous Device Model, the mobility of antibiofilm peptides can be restricted with the aim of preventing bacterial contact and eventually biofilm development [75]. However other modes of assessment like histological analysis, imaging by IVIS, scanning microscopy, and inflammatory response detection can also be employed in evaluating antibiofilm activity in test organisms [75].

4. Methods used to determine anti-biofilm effects of natural products

Bacteria undergo an evolutionary mechanism to withstand harsh environmental conditions. The antibacterial agents derived from natural sources may serve as an effective alternative due to the presence of secondary metabolites, which possess selectional advantages against the biofilm-forming microorganisms [77–79]. Several methods have been reported as reliable protocols to investigate the anti-biofilm effects of natural products (**Table 1**) [88, 89]. Crystal violet assay is the widely accepted assay used to identify the anti-biofilm potentials of natural products despite the limitation, including the repeated washing that could lead to loss of cells and biofilm disruption [77, 88, 90, 91]. Other methods used to determine the antibiofilm effects of natural products are the Tissue Culture Plate (TCP) method [82], which exists as the most typical use standard method and is a comparatively reliable method to Congo Red Agar method (CRA) and Tube method [80]. Tube method and Congo red agar methods qualitatively detect biofilm formed, whiles the tissue culture plate method quantitatively determines the amount of biofilm formed [76]. Real time, conventional and multiplex PCR are other techniques used at molecular level to detect biofilm genes [92–94].

In measuring the anti-biofilm activities of natural products, viability and matrix biomass can be assessed, where resazurin and crystal violet staining are performed sequentially in the same plate. Wheat germ agglutinin-Alexa Fluor 488 fluorescent

Method of biofilm detection	Principle	Aim
Tissue culture plate	It involves the staining of cells with crystal violet dye [77, 80, 81]	Biofilm detected quantitatively
Tube method	Crystal violet staining where visible lining forms at the bottom and wall of the tube [80]	Biofilm detected qualitatively
Congo red agar	Congo red staining formed black colonies crystals [81–84]	Biofilm detected qualitatively
Crystal violet assay	Quantifies the dye bound to biofilm [77, 85]	Quantitative determination of biofilm
Real-time PCR, Multiplex PCR and conventional PCR	Amplification of DNA to the generation of fluorescence which can simply be detected [86, 87]	Detection of biofilm genes

Table 1.
Methods to determine anti-biofilm effects of natural products.

conjugate is mainly used to stain the matrix, which is essential to measure the biofilm matrix, biomass, and viability to investigate the potencies of anti-biofilm effects of natural products [95, 96].

5. Antibiofilm agents from nature

5.1 Plant-derived antibiofilm agents

Plants have since time immemorial served as a source of therapeutics for the treatment and prevention of a plethora of diseases. This practice continues today, with more than 80% of people globally reportedly using various herbal remedies as

Plant species	Comment
Aralia spinosa (Araliaceae)	$MBIC_{50}$ = 2 μg/ml against *S. aureus* [101]
Juglans regia (Juglandaceae)	$MBIC_{50}$ = 7.21 μg/ml and MBEC50 = 57.71 μg/ml against *S. epidermis* [102]
Liriodendron tulipifera (Magnoliaceae)	$MBIC_{50}$ = 32 μg/ml against *S. aureus* [101]
Citrus bergamia (Rutaceae)	Inhibited *P. aeruginosa* biofilm formation 79% at 1.56 μg/ml [103]
Gymnopodium floribundum (Polygonaceae)	IC_{50} = 53.6 μg/ml against *S. aureus* [104]
Zygophyllum coccineum (Zygophyllaceae)	MBEC = 15.63, 3.9, 15.63 and 15.63 μg/ml against *Streptococcus pneumoniae*, *S. aureus*, *P. aeruginosa*, and *E. coli*, respectively [105]
Ziziphus jujuba (Rhamnaceae)	50% inhibition at 1.41 μg/ml against *S. aureus* [106]
Matayba oppositifolia (Sapindaceae)	IC_{50} = 10.4 μg/ml against *S. aureus* [104]
Schoepfia schreberi (Schoepfiaceae)	IC50 = 17.7 μg/ml against *S. aureus* [104]

IC, inhibitory concentration; MBIC, minimum biofilm inhibitory concentration; MBEC, minimum biofilm eradication concentration.

Table 2.
Potent antibiofilm plant species.

Compound and plant source	Comment
Xanthohumol (*Humulus lupulus*)	100% inhibition of *S. aureus* biofilm formation at 9.8 μg/ml [107]
5-Hydroxymethylfurfural (*Musa acuminata*)	83% inhibition at 10 μg/ml against *P. aeruginosa* [108]
Lupulone (*H. lupulus*)	100% inhibition of *S. aureus* biofilm formation at 1.2 μg/ml [107]
Cyanidin 3-O-glucoside (*Lonicera caerulea*)	$MICB_{50}$ = 3.3 μg/ml against *Porphyromonas gingivalis* [109]
Hodiendiol I (*P. artemisioides*)	78, 75 and 13% inhibition of *Listeria monocytogenes*, *Pseudomonas aeruginosa,* and *Staphylococcus aureus* biofilms at 4 μg/ml [110]
Negletein (*S. oblonga*)	72–88% reduction of biofilms of *S. aureus*, *B. subtilis*, *P. aeruginosa* and *E. coli* at 12 μg/ml [111]
Syringopicroside (*Syringa oblata*)	92% inhibition at 1,28 μg/ml against *S. aureus* [112]
Quercitin-3-glucoside (*S. oblonga*)	92–98% reduction of biofilms of *S. aureus*, *B. subtilis*, *P. aeruginosa* and *E. coli* at 12 μg/ml [111]
Panduratin A (*Kaempferia pandurate*)	Prevented *S. mutans* and *S. sanguis* biofilm growth by >50% at 8 μg/ml, and reduced the biofilms by >70% at 10 μg/ml [113]

Table 3.
Potent antibiofilm plant-derived compounds.

a source of primary healthcare [97]. In mainstream medicine, plants have proven to be a prolific source of novel chemical matter from which essential drugs used to treat various diseases have been developed [98]. Galvanized by the emergence and spread of the antimicrobial drug resistance phenomena, numerous plant species have been

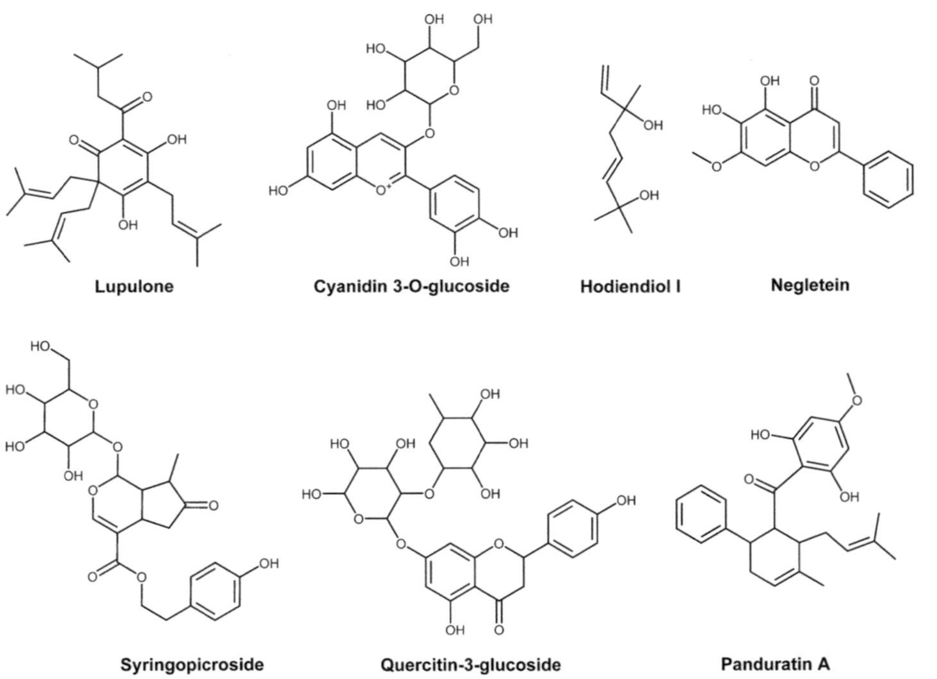

Figure 1.
Chemical structures of some active plant derived antibiofilm compounds.

thoroughly investigated as novel sources of antibacterial agents. To complement these strategies, the search for agents that can reverse resistance (resistance breakers) or target alternative mechanisms of overcoming antibacterial resistance, including biofilms, is being pursued [99, 100]. Plants have been identified as a potential oasis of such agents, prompting many studies in the last decade inspired towards the search for antibiofilm agents from plants. This section summarizes current studies on the investigation of antibiofilm agents, including crude extracts, fractions thereof, and pure compounds from plants (**Tables 2** and **3**; **Figure 1**).

5.1.1 Apiaceae

Despite being one of the least investigated, the Apiaceae plant family has produced some of the most prolific antibiofilm plant species. Among them is the annual herb *Trachyspermum ammi* popularly called bishop's weed [114]. Investigations on its seed led the isolation of a potent novel naphthalene compound, (4aS, 5R, 8aS) 5, 8a-di-1-propyl-octahydronaphthalen-1-(2H)-one, which remarkably inhibited both adherence (IC_{50} = 39.06 μg/ml) and formation of *S. mutans* biofilms (~60% inhibition at 78.13 μg/ml) *in vitro* (**Figure 2**). This activity was strikingly more pronounced than its parent compound's bacteriostatic and bactericidal properties (MIC = 156.25 μg/ml; MBC = 312.5 μg/ml) against *S. mutans* [114]. Thymol, a monoterpenoid isolated from *Carum copticum*, showed good activity against three bacterial species, namely *Klebsiella pneumoniae*, *Escherichia coli* (*E. coli*), and *Enterobacter cloacae* (*E. cloacea*), at sub-MIC levels, reducing biofilm formation by 80, 78, and 83%, respectively at 50 μg/ml (**Figure 2**). The compound was approximately fourfold more potent than its parent species [115].

5.1.2 Asteraceae

The Asteraceae is one of the most prominent species-rich plant families that produce highly active terpenoid compounds. A study on *Helichrysum italicum* led to the isolation of 21 compounds demonstrating varied activity of either inhibiting the formation or eradication of preformed *P. aeruginosa* biofilms. From the 21 compounds screened, chlorogenic acid emerged as the most active inhibiting biofilm formation (45% inhibition at 128 μg/ml). In contrast, biofilm eradication for all compounds was weak (<30%) [116]. Chondrillasterol, a terpenoid isolated from the plant *Vernonia adoensis,* has shown an intriguing activity profile being more potent in disrupting

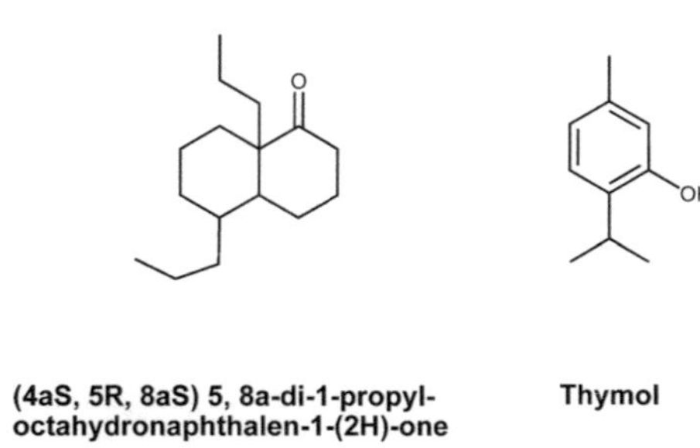

Figure 2.
Chemical structures of (4aS, 5R, 8aS) 5, 8a-di-1-propyl-octahydronaphthalen-1-(2H)-one and thymol.

P. aeruginosa biofilms (complete disruption at 1.6 μg/ml) in comparison to inhibiting biofilm formation (wholly inhibited at 100 μg/ml) (**Figure 3**) [117].

5.1.3 Burseraceae

An aqueous extract of *Commiphora leptophloeos* showed promising inhibition of *Staphylococcus epidermidis* biofilm formation on different surfaces. At a concentration of 4 mg/ml, an aqueous stem bark extract of *C. leptophloeos* showed equipotent activity on inhibiting *S. epidermis* biofilms on a polystyrene (84% inhibition) and glass surface (82% inhibition) [118]. *Boswellia papyrifera (B. papyrifera)* is a deciduous tree 12 m high with a rounded crown, a white to pale brown bark that peels off in large flakes and exudes a fragrant resin [119]. Traditionally, as therapeutics, its leaves and roots are used to manage lymphadenopathy, while the resin serves as a febrifuge. The burnt leaves of *B. papyrifera* act as a mosquito repellent [120]. Essential oils obtained from *B. papyrifera* resin inhibited preformed *S. epidermidis* and *S. aureus* biofilms by 99–71%, and 95.3–59.1% at 217.3–6.8 μg/ml, respectively [121]. At a sub-MIC concentration of 0.27 μg/ml, the essential oil of *B. papyrifera* observed, under fluorescence microscopy, showed to inhibit the adhesion of stained *S. epidermidis* cells [122].

5.1.4 Combretaceae

The medicinal plant *Terminalia bellerica* (*T. bellerica*) is found predominantly in India, Sri Lanka, Bangladesh, and South-East Asia. Its fruits are traditionally used as a laxative, astringent, and antipyretic in treating menstrual disorder, piles, and leprosy. An investigation by Ahmed et al. [122] showed that the dried fruits of *T. bellerica* ethanol extracts could inhibit *S. mutans* biofilm formation *in vitro* on a glass surface by 92.2% at 250 μg/ml. Another *Terminalia* species, *T. fagifolia*, has been shown to have good antibiofilm properties. The ethanol stem bark extract of *T. fagifolia* inhibited the formation of preformed *S. epidermis* and *S. aureus* strains *in vitro*. It was particularly active against S. *epidermis* by inhibiting biofilm formation by ~70% at a sub-MIC concentration of 12.5 μg/ml compared to ~85% inhibition at 50 μg/ml against *S. aureus* [123]. Similarly, a water fraction of *Combretum elaeagnoides* showed potency against multiple species being able to reduce biofilm formation of *S. aureus*, *Salmonella typhimurium* (*S. typhimurium*), *Salmonella*

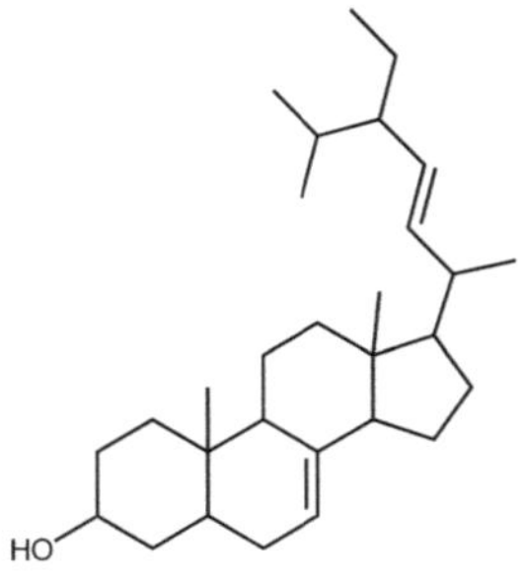

Chondrillasterol

Figure 3.
Chemical structure of chondrillasterol.

enteritidis (*S. enteriditis*), *Klebsiella pneumoniae,* and *Enterobacter cloacae* by 80, 73, 63, 54, and 66%, respectively, at 1 mg/ml [124].

5.1.5 Fabaceae

Along with the Asteraceae, the Fabaceae family is one plant species that has received substantial interest as a source of antibiofilm agents. *Copaifera pauper C. paupera*) is a medicinal tree commonly found in South America that exhibits activity against monospecies and multispecies formed biofilms [125]. For the monospecies (*Aggregatibacter actinomycetemcomitans* and *Porphyromonas gingivalis*) *produced biofilms*, *C. paupera* oleoresins showed marked activity against the individual strains and with IC_{50} (eradication of biofilm) values of 58.66 μg/ml and 104.9 μg/ml, respectively. Activity against the multispecies biofilms was marginally lower with a measured IC_{50} (eradication of biofilm) of 594.5 μg/ml. *Copaifera pubiflora* oleoresins have shown a similar pattern of activity against individual *A. actinomycetemcomitans* [IC_{50} (eradication of biofilm) = 189.4 μg/ml)] and *P. gingivalis* [IC_{50} (eradication of biofilm) = 94.02 μg/ml)] strains and their combined multispecies biofilm [IC_{50} (eradication of biofilm) = 556.8 μg/ml)]. Three compounds, namely polylactic acid, hardwikiic acid, and kaurenoic acid, have been isolated from a *Copaifera* spp. and also shown to have potency against both the monospecies and multispecies biofilms of *A. actinomycetemcomitans* and *P. gingivalis* [IC_{50} (eradication of biofilm) ranging from 55.79 to 462 μg/ml)] [125]. Other species that have shown marked activity against multispecies biofilms include *Pityrocarpa moniliformis*, *Anadenanthera colubrina,* and *Dioclea grandiflora* [125].

Trigonella foenum-graceum (*T. foenum-graceum*), commonly called fenugreek, is an annual legume and a traditional spice crop native to the eastern Mediterranean. It has been known for its medicinal properties in the Mediterranean and Asian cultures for many years. Fenugreek seeds are traditionally used as laxative, expectorant, carminative, and demulcent [126]. The methanol extracts of *T. foenum-graceum* seeds inhibited *P. aeruginosa* biofilms in a dose-dependent pattern (24.1–68.7% at 125–1000 μg/ml) without affecting bacterial proliferation [127]. The extract caused a reduction to the exopolysaccharide (EPS) produced by *P. aeruginosa* biofilms. In addition to *P. aeruginosa*, *T. foenum-graceum* showed activity against the aquatic pathogen *Aeromonas hydrophila* reducing EPS production and biofilm formation by 46 and 76.9%, respectively, at 800 μg/ml [127].

5.1.6 Lamiaceae

The Lamiaceae is a family of flowering plants commonly known as the mint family with a cosmopolitan distribution containing about 236 genera and about 6900–7200 species. Many plants in this family are aromatic and include widely used culinary herbs like basil, mint, rosemary, and sage [128]. Several Lamiaceae species have been interrogated for their antibiofilm activity and have shown pronounced activity against different biofilm stages of various microorganisms. One such species is the plant *Marrubium vulgare* (*M. vulgare*), a perennial herb found right across the globe. The plant is well renowned for its medicinal properties and serves as a therapeutic agent for several ailments, including gastrointestinal disorders, asthma, pulmonary infections, and ulcers. The aqueous decoctions of *M. vulgare* inhibited adherence of methicillin-resistant *S. aureus* biofilms with IC_{50} of 8 μg/ml and IC_{90} of 128 μg/ml [129].

However, the plant was less effective in inhibiting *S. aureus* biofilm growth on a plastic surface (31% inhibition at 128 μg/ml). Surprisingly, at the highest test concentration of 128 μg/ml, *M. vulgare* showed no bacteriostatic activity suggesting the species is selectively more potent against biofilm mechanisms. Aqueous extract prepared from the aerial parts of *Ballota nigra*, mirrored this bioactivity profile. Specifically, inhibiting methicillin-resistant *S. aureus* biofilm formation and adherence by 45–90% at 8–128 μg/ml while demonstrating limited bacteriostatic activity at the highest test concentration [129].

The genus *Salvia* is well documented for its bacteriostatic and bactericidal properties. Various species within this genus possess dual antibiofilm properties. Hexane-soluble and dichloromethane soluble fractions and sub-fractions of *Salvia officinalis* (*S. officinalis*) have shown impeccable antibiofilm and bacteriostatic properties with an $MBIC_{50}$ and MIC values ranging from 3.668 to 200 μg/ml and 25 to 400 μg/ml, respectively, against *P. gingivalis*, *F. nucleatum*, *P. melaninogenica,* and *A. actinomycetemcomitans*. The labdane diterpenoid manool has been isolated and identified as the active principle from *S. officinalis*, showing pronounced activity with $MBIC_{50}$ and MIC values of 12.5 μg/ml and 3.12 μg/ml, respectively against *A. actinomycetemcomitans* (**Figure 4**) [130].

While *Mentha piperita* oil is considerably active against *Chromobacterium violaceum* (Inhibited biofilm formation by 72.5% at 0.049 μg/ml), it is inactive against *P. aeruginosa* at reasonably higher test concentrations of 6.25, 3.125 and 1.56 μg/ml. In the same study, *Thymus vulgare* essential oil showed marked potency against both species inhibiting their biofilm formation by 70% at 0.049 μg/ml (against *C. violaceum*) and 65% at 3.125 μg/ml (against *P. aeruginosa*) [103]. Equally impressive is the species *Perovskia artemisioides,* which has inhibited biofilm formation of *L. monocytogenes*, *P. aeruginosa, S. aureus, Acinetobacter baumanii (A. baumanii), and Pectobacterium carotovorum* by 92, 95, 71, 35, and 94% at 4 μg/ml. Subsequent work led to the identification of numerous antibiofilm compounds from *P. artemisioides* [110].

5.1.7 Malvaceae

Alcea longipedicellata (*Aulonemia longipedicellata*) is a member of the *Alcea* genus with over 80 flowering plants in the family Malvaceae, commonly known as the hollyhocks and native to Asia and Europe. The compound, malvin, isolated from the flowers of *A. longipedicellata* flower, exhibited about 55% inhibition of *S. mutans* biofilm adherence at 0.1% v/v (**Figure 5**) [131]. *Hibiscus rosa-sinensis* a tropical shrub used in folk medicine to treat respiratory disorders and diarrhea, among other ailments, has shown remarkable activity against drug-resistant strains of *Helicobacter pylori* (*H. pylori*).

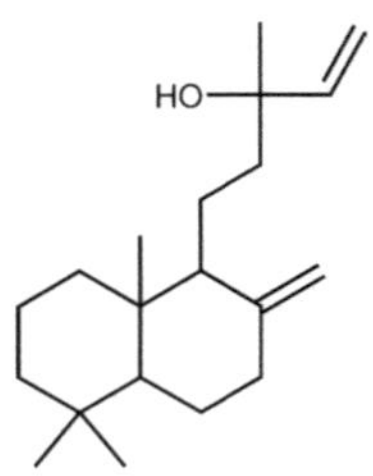

Manool

Figure 4.
Chemical structure of manool.

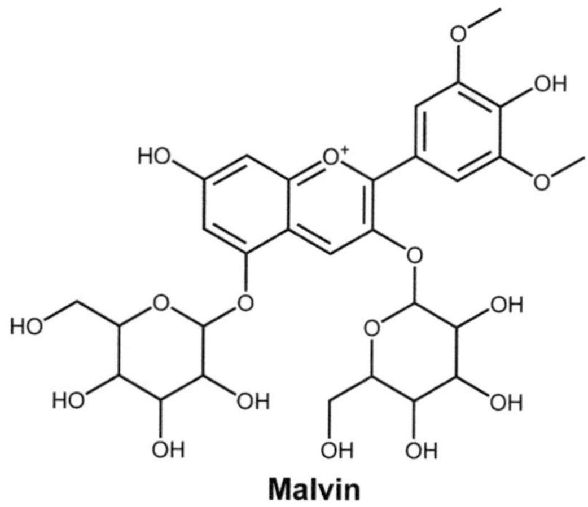

Figure 5.
Chemical structure of malvin.

An ethyl acetate fraction of *H. rosa-sinensis* demonstrated strong biofilm formation inhibition against *H. pylori* at sub-MIC concentrations (79% inhibition at 125 μg/ml) [132].

5.1.8 Myristicaceae

The Myristicaceae are flowering plants native to Africa, Asia, Pacific islands, and the Americas. The family consists of 20 genera and at least 500 species. Fruit of the Myristicaceae, particularly the lipid-rich aril surrounding the seed in some species, are essential as food for birds and mammals of tropical forests [133]. Plants in the

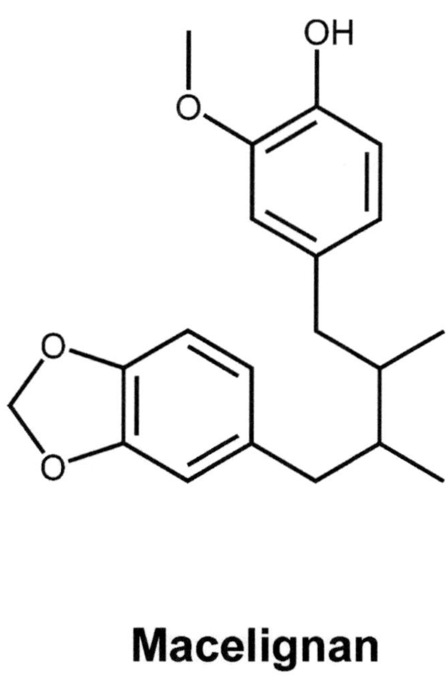

Figure 6.
Chemical structure of macelignan.

family Myristicaceae with reported antibiofilm activities include *Myristica fragrans* (*M. fragrans*), *Syzygium aromaticum,* and *Syzygium cumini*. *M. fragrans* has been shown to inhibit *Salmonella enterica* biofilm formation by 88% at 50 μg/ml. Biosynthesised silver nanoparticles of *M. fragrans* showed marginally improved activity inhibiting the formation of *S. enterica* biofilm by 99.1% at 50 μg/ml [134]. Another study on *M. fragrans* led to the isolation of the compound macelignan, which reduced the formation of *S. mutans* and *S. sanguis* biofilm by >50% at 10 μg/ml (**Figure 6**) [113]. The methanol fruit extract of *S. cumini* disrupted *Klebsiella pneumoniae* biofilm biomass in a dose-dependent manner by 35.85, 64.03, and 79.94% at test concentrations of 0.1, 0.5, and 1 mg/ml, respectively [135]. Essential oils from the aerial parts *of S. aromaticum* reduced *Staphylococcus epidermidis* biofilm biomass by 50.3% at 20 μg/ml [136].

5.1.9 Amaryllidaceae

Extracts of *Crinum asciaticum*, a member of the family Amaryllidaceae, was investigated for its anti-tuberculosis, anti-efflux pump and antibiofilm activity. This study reealed the anti-infective activity of the extracts against *Mycobacterium smegmatis* (*M. smegmatis*) (NCTC 8159) and *Mycobacterium aurum* (*M. aurum*) (NCTC 10437) at MICs of 125 μg/ml and 250 μg/ml respectively. Also, efflux pump inhibition was observed for both *M. smegmatis* and *M. aurum.* Of great importance is the *in vitro* inhibition of *M. smegmatis* and *M. aurum* biofilms which was very significant at $p < 0.005$ [77].

5.2 Antibiofilm agents obtained from mushrooms

Research has shown that some species of macrofungi have various chemical components with antibacterial, antifungal, antiviral, antioxidant, anticancer and antiprotozoal properties [137]. The extracts of some species, including *Laetiporus sulphureus*, *Ganoderma lucidum*, and *Lentinus edodes* have demonstrated antibacterial activity [138]. *Fistulina hepatica*, *Ramaria botrytis*, and *Russula delica* extracts had promising antibacterial activity against multi-resistant microorganisms namely MRSA, *E. coli* and *Proteus mirabilis.*

In addition, some of these compounds were found to inhibit biofilm formation [137].

Studies on the aqueous extracts of *Macrolepiota procera*, *Pleurotus ostreatus*, *Auricularia auricula-judae*, *Armillaria mellea*, and *Laetiporus sulphureus* were shown to inhibit *Staphylococcal spp* biofilm formation. These extracts reduced biofilm formation by 47.72–70.87% without affecting bacterial growth [139].

A study by Borges et al demonstrated that ferulic and gallic acid inhibited biofilm formation in *P. aeruginosa* by interfering with cell motility and physico-chemical features on the cell surface. It also inhibited biofilm formation by *E. coli* due to phenolic compounds present therein [140]. Again, wild mushroom extracts had antibiofilm activity against *E. coli*, *Leucopaxillus gigantes* and *Mycenus rosea.* From this same study, extracts from *Sarcodon imbricants,* and *Russula delica* inhibited biofilm formation of *P. mirabilis* that is resistant to fluoroquinolones, ampicillin, and cephalosporins [138].

Extracts from *Lentinus edodes*, one of the mostly cultivated edible mushrooms, reacted negatively to biofilm proliferation by some bacteria in a study conducted by Lingström and colleagues [141]. Upon further fractionation and isolation, the compounds; oxalic acid, quinic acid, inosine and uridine (**Figure 7**) were discovered to be responsible for the various levels of antibiofilm activity against *S. mutans*, *Actinomyces naeslundii*, and *Prevotella intermedia* strains [141].

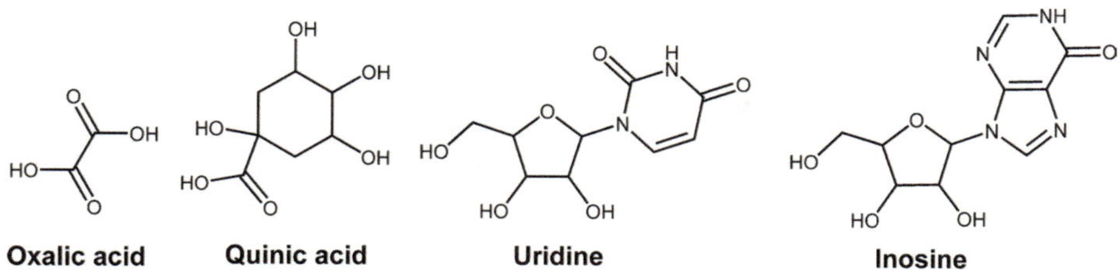

Figure 7.
Structures of compounds isolated from mushrooms with antibiofilm activities.

Melanin obtained from *Auricularia aricula*, an edible mushroom, has established antibiofilm properties [142]. This pigment exhibited significant antibiofilm inhibitory activity against *E. coli* K-12, *P. aeruginosa* PA01, and *Pseudomonas fluorescens* P-3 [142].

5.3 Sponges as antibiofilm agents

Marine sponges produce an array of secondary metabolites such as enzymes, enzyme inhibitors, and antibiotics and represent an untapped reservoir of bioactive compounds [143]. These compounds serve as defense against environmental threats like microbial infection, competition for space, or overgrowth by fouling organisms [144].

Phorbaketals isolated from the Korean marine sponge *Phorbas spp*. had antibiofilm activity against *S. aureus* [143]. Moreover, all six phorbaketals (phorbaketal A, phorbaketal B, phorbaketal C, phorbaketal A acetate, phorbaketal B acetate, phorbaketal C acetate, **Figure 8**) assessed for their antibiofilm activities revealed a minimum inhibitory concentration against *S. aureus* 6538 higher than 200 μg/ml. All six compounds significantly inhibited biofilm formation of methicillin-sensitive *S. aureus* in a dose-dependent manner, with Phorbaketal B and Phorbaketal C having the highest inhibitory effects, probably due to the presence of two hydroxyl groups in its structure. Phorbaketal B and C exerts their action via reduction of the expression of alpha-hemolysin (*hla*) and nuclease (*nuc1*). Phorbaketal C further reduced the expression of RNAIII (a regulatory molecule) which stimulates *hla* translation, thereby repressing the expression of *hla* [143].

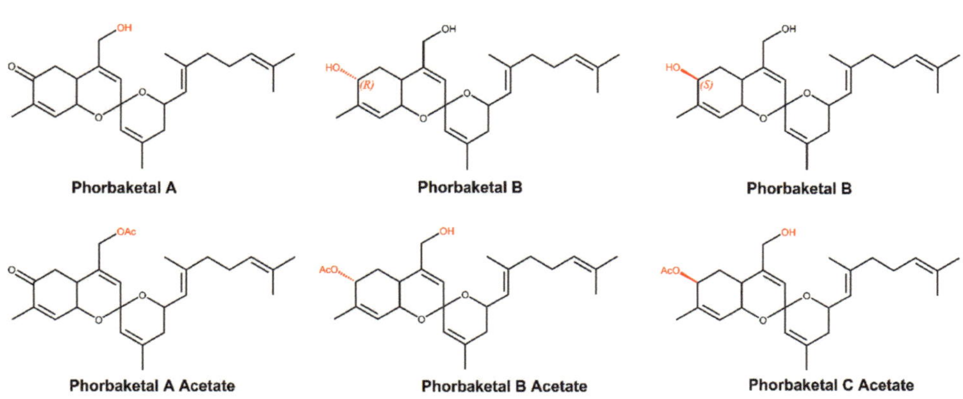

Figure 8.
Chemical structures of phorbiketals isolated from Phorbas sp.

In addition, natural compounds such as collismycin, hydroxyl flavonoids, hydroxylbipyridine, and hydroxyl anthraquinones exhibited antibiofilm activity depending on the number and positions of hydroxyl groups in the backbone structures [145]. The planktonic cell growth of *S. aureus* was relatively unaffected by the six phorbaketals at <100 μg/ml [143].

In another study by Paul and Puglisi, cell-free supernatants (CFSs) isolated from the sponge-associated bacteria belonging to the genera *Colwellia, Pseudoalteromonas, Shewanella* and *Winogradskyella* were evaluated for antibiofilm activity at 4°C and 25°C against Antarctic strains of *P. aeruginosa* ATCC27853 and *S. aureus* ATCC29213. Inhibition of biofilm formation was observed differently among strains which was dependent on the incubation temperature. Significant antibiofilm activity was observed by CFSs at 4 °C and 25 °C respectively against *S. aureus* and *P. auruginosa* without exhibiting cidal activity on bacterial growth [146]. The different physico-chemical nature of exopolymers produced by the *Colwellia sp.* GW185, *Shewanella sp*. CAL 606 and *Winogradsyella* CAL396 is responsible for their antibiofilm activity (**Table 4**).

In another study, marine sponge-derived *Strepomyces sp*. SBT343 extracts were investigated for their antibiofilm activity on *Staphylococcal* biofilm formation. Results from *in vitro* biofilm assay of an organic extract showed inhibition of biofilm formation on polysterene, glass and contact lens surfaces. This same extract inhibited biofilm formation of *Staphylococcus epidermidis* and *S. aureus* with no antibiofilm activity against *Pseudomonas* biofilms [147].

5.4 Algal sources of antibiofilm agents

Existing literature proves the existence of compounds obtained from algae that possess antibiofilm properties against human pathogenic microbes. The scientific

Species and strain	Major constituents	Antibiofilm activity against organisms
Colwellia spp. GW185	Glucose, mannose, galactose, galactosamine	*P. aeruginosa, S. aureus*
Shewanella spp. CAL606	Glucose, mannose, galactose, galactosamine	
Winogradskyella spp. CAL396	Mannose, arabinose, galacturonic acid	

Table 4.
Bacterial exopolysaccharide with antibiofilm activity against pathogenic bacteria [143].

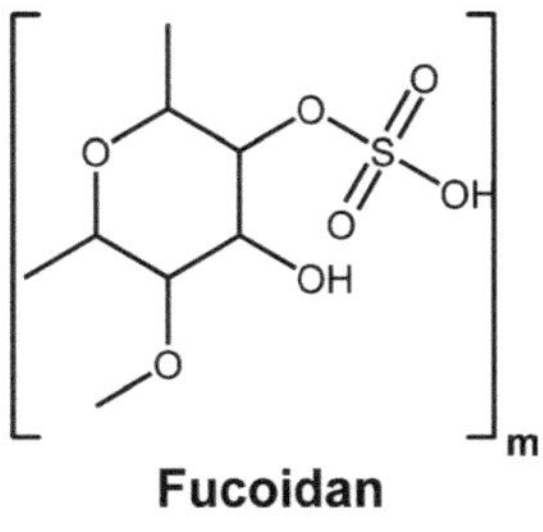

Figure 9.
Structure of fucoidan.

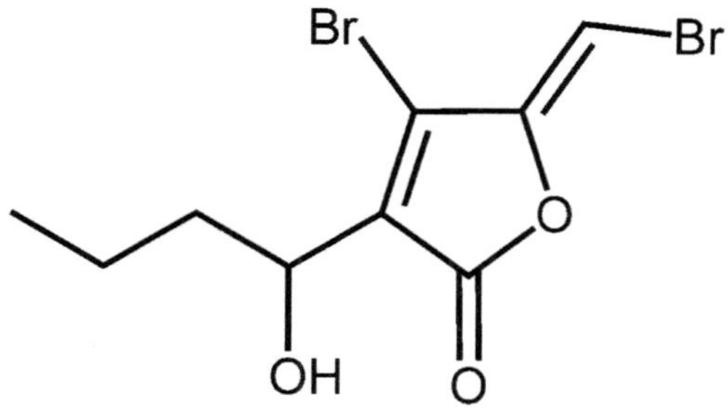

Figure 10.
Structure of a halogenated furanones.

research community however, continues to discover such natural antibiofilm agents. These compounds do not exist in their pure forms but are isolated from crude extracts through a series of processes [148].

Marine algae produce certain sulfated polysaccharides that exhibit antimicrobial and antibiofilm activities [149]. Fucoidan F85 (**Figure 9**), a sulfated polysaccharide extracted from *Fucus vesiculosus* upon observation was found to possess antimicrobial and antibiofilm properties against some dental plaque bacteria [149]. Fucoidans are made up of L-fucose and sulfate esters with other different molecules [150] and are normally extracted from brown algae using acid, solvent or water at a high temperature and a long reaction [151]. According to Yunhai and colleagues, Icelandic local seaweed species (*Ascophyllum nodosum* and *Laminaria digitate*), are sources of fucoidans with antibacterial activity [152].

A study conducted by Maggs et al proves that marine brown algae, *Halidrys siliquosa* produces compounds with antibiofilm activity against *Staphylococcus sp*, *Streptococcus sp*, *Enterococcus sp*, *Pseudomonas sp*, *Stenotrophomonas sp*, and *Chromobacterium sp*. *Halidrys siliquosa* can be found in rock pools and sometimes forests in the shallow subtidal zone [148].

Delisea pulchra red alga, produces halogenated furanones which show antibiofilm effects against *B. subtilis*, *E. coli* [153] and *P. aueroginosa* [154]. These furanones oppose the transmission of intracellular signals and speed up LuxR transcription turnover (**Figure 10**) [155].

The algal fronds of *Plocamium magga* has been reported to produce an isolate, KS8 from the *Pseudoalteromonas* genus that shows antibiofilm activity against acyl homoserine lactone base reporter strains (*Chromobacterium violaceum* (*CV*) ATCC 12472 and CV026) [156].

Ethanolic extracts of *Chlorella vulgaris* and *Dunaliella salina* can inhibit biofilm formation by *S. mutans* and *P. aueroginosa* [157]. This antibiofilm characteristic may be associated with the activity of glucotransferases [157].

Methanol extract of *Oscillatoria sp*., green algae containing silver nanoparticles also showed strong antibiofilm activity against all test pathogens in an experiment conducted by Adebayo-Tayo and associates [158].

Silver nanoparticles associated with aqueous extract of *Turbinaria conoides* have been reported to possess antibiofilm activity via adherence inhibition against *Salmonella typhi*, *E. coli* and *Serratialique faciens* [159].

6. Miscellaneous agents with antibiofilm activities

Several agents from natural products such as essential oils, honey etc. have shown great potential as bacterial biofilm inhibitors. These have been described below;

6.1 Essential oil

Essential oils from medicinal plants have received attention in recent times for their potential exploitations. This is as a result of the increasing reports of their composition and biochemicals to possess medicinal properties. A number of *in vitro* evidences indicates that essential oils can act as antibacterial and antibiofilm agents against a large spectrum of pathogenic bacterial strains.

The effect of *Lippia alba* (*L. alba*) and *Cymbopogon citratus* (*C. citratus*) (lemon grass) essential oils on biofilms of *S. mutans* was tested by Tofiño-Rivera et al. in an attempt to find new compounds against dental caries using the MBEC-high-throughput (MBEC-HTP) assay. The *L. alba* essential oils demonstrated significant eradication activity against *S. mutans* biofilms of 95.8% in 0.01 mg/dL concentration, and *C. citratus* essential oils showed eradication activity of 95.4% at 0.1 and 0.01 mg/dL concentrations and of 93.1% in the 0.001 mg/dL concentration [160]. Further, geraniol and citral were later identified as the major components of the essential oils. A similar investigation by Ortega-Cuadros et al., showed 93.0% growth inhibition of *S. mutans* biofilms at a concentration of 1.00 μg/ml of *C. citratus* essential oil [161].

In an investigation to access the ability of *Allium sativum* fermented extract and cannabinol oil extract to inhibit and remove *P. aeruginosa* biofilms on soft contact lenses, the cannabinol oil extract inhibited biofilm formation by about 70% and eradicated preformed biofilms in both *P. aeruginosa* (ATCC 9027 strain) and *P. aeruginosa* clinical isolates from the ocular swabs tested [162]. Cannabigerol, a non-psychoactive cannabinoid which is also naturally present in trace amounts in the Cannabis plant was able to reduced the QS-regulated bioluminescence and biofilm formation of *Vibrio harveyi* (a marine quorum-sensing and biofilm-producing bacterial species) at concentrations not affecting the planktonic bacterial growth [163].

Essential oils from *Cyclamen coam* (*C. coam*) and *Zataria multiflora* (*Zinnia multiflora*) extracts inhibited biofilm formation on *P. aeruginosa* 214, a strong biofilm producing clinical strain [164]. *C. coam* and *Z. multiflora* essential oils inhibited biofilm formation completely at concentrations <0.062 mg/ml and 4 μl/ml, respectively. It is reported that carvacrol, a major constituent of *Z. multiflora* essential oil inhibits biofilm formation by preventing the initial adhesion of biofilm cells to the surface [165, 166].

6.2 Lectin

A study by Moura et al. reported the antibiofilm activity of a lectin extracted from *Moringa oleifera* (*M. oleifera*) seed. The lectin from this plant exhibited antibiofilm activity against *Bacillus* spp. and *Serratia marcescens* at concentrations of 20.8–41.6 μg/ml and 0.325–1.3 μg/ml respectively [167]. The antibiofilm activity of the *M. oleifera* seed lectin might be due to the ability of these lectins to damage the cell wall and cell membranes through its interactions with glycoconjugates and polysaccharides constituents within the bacterial cell wall [168].

Solanum tuberosum lectins had a varying biofilm inhibitory effect when evaluated against an isolate of *P. aeruginosa* PA01. At a concentration between 2.5 and 15 μg/ml, the lectins inhibited the biofilm formation by 5–20% [169].

Plant lectins are reported to also exhibit antibiofilm activities against pathogenic microorganisms. A typical example are, lectins extracted from *Canavalia ensiformis*, *Calliandra surinamensis*, *Canavalia marítima* and *Alpinia purpurata* [170].

6.3 Chitosan

Chitosan is a polysaccharide composed of units of glucosamine (2-amino-2-deoxy-D-glucose) and *N*-acetyl glucosamine (2-acetamido-2-deoxy-D-glucose) linked by β (1 → 4) bonds. Chitosan is produced as a result of partial deacetylation of chitin leads. Chitin is found on the shells of crustaceans, arthropods and fungal cell wall [171].

The antibiofilm activity of chitosan from crab and shrimp species indigenous to the Philippines was investigated against *P. aeruginosa* and *S. aureus*. Biofilm inhibitory activity for both crab and shrimp chitosan were not observed against *S. aureus* at the concentration used, but activity was observed for shrimp chitosan at a concentration of 2.5 g/L. A 2.5 g/L mixed (1:1) chitosan solution of the two extracts had the highest percentage antibiofilm formation inhibition in *P. aeruginosa* biofilms. *S. aureus* biofilm formation was sensitive to the 10 g/L mixed (1:1) solution. The same mixed solution produced an inhibition against *P. aeruginosa* [172].

Costa et al. also reported that chitosan demonstrated antibiofilm and biofilm eradication activity against the fungus *Candida albicans* [171].

6.4 Honey

The exploration of new antibiotics to combat biofilm formation in resistant microbes has led to an increase interest evaluating the antibiofilm properties of honey. Manuka honey have demonstrated good antibiofilm forming activity against a range of bacteria, including *Streptococcus and Staphylococcus* species, *P. mirabilis, A. baumannii, E. coli, E. cloacae* and *P. aeruginosa* [173, 174].

Lu and colleagues studied the antibiofilm properties of four New Zealand based honeys; monofloral manuka honey, Medihoney (a manuka-based medical-grade honey), manuka-kanuka blend, and a clover honey on two *P. aeruginosa* strains PAO1 and PA14 with different biofilm forming abilities. All the different types of honey used in the study were effective at inhibiting both the planktonic cell growth and biofilm formation of both strains. In the study of the biofilm eradication properties of the honey, they concluded that honey used at clinically obtainable concentrations completely eradicated established *P. aeruginosa* biofilms [175]. Similar results were obtained using different strains of *S. aureus*, including methicillin-resistant *S. aureus* (MRSA) strains. In this study, they demonstrated that honey is able to reduce biofilm mass and also to kill cells that remain embedded in the biofilm matrix; and planktonic cells released from biofilms following honey treatment do not have elevated resistance to honey [176].

The biofilm inhibitory effect of Costa Rican Meliponini stingless bee honeys has also been reported against *S. aureus* and *P. aeruginosa* biofilm formation. The meliponini stingless bee honeys in a concentration-dependent manner inhibited the planktonic growth and biofilm formation, and also caused the destruction of *S. aureus* biofilm [177].

Australian honey has also been reported to possess antibacterial and biofilm inhibitory activities. Sindi A and colleagues in their investigation reported that Western Australian honeys from *Eucalyptus marginata* (Jarrah) and *Corymbia calophylla* (Marri) trees exhibited antimicrobial activity against Gram-negative and

Gram-positive pathogens. They reduced both the formation of biofilms and the production of bacterial pigments, which are both regulated by quorum sensing. The Western Australian honey when applied to preformed biofilms had biofilm eradication activity by reducing metabolic activity in the biofilms [178].

6.5 Peptides

Peptides are small molecules made of 10–100 amino acids that are part of the innate immune response, and found among all classes of life contributing to the first line of defense against infections. In the search for an effective agent that can treat chronic infections, antimicrobial peptides (AMPs) have been shown to demonstrate antimicrobial, antibiofilm and biofilm eradication properties. Although there has not been much studies on the biofilm inhibitory action of AMP compared to its antibacterial activity, some naturally occurring AMP's have been reported to exhibit strong antibiofilm activities against multidrug resistant as well as clinically isolated bacterial biofilms [179].

Cathelicidin peptides are one of the most important classes of AMP. Investigation of cathelicidin AMP, indicates that SMAP-29, BMAP-28, and BMAP-27 have antimicrobial activity and are able to significantly reduce biofilm formation by multidrug-resistant (MDR) *P. aeruginosa* strains isolated from patients with cystic fibrosis. In addition, they were bactericidal in preformed biofilms [180]. Blower et al. also demonstrated that the SMAP-29 peptide is able to inhibit biofilm production in *Burkholderia thailandensis* by about 50% at peptide concentrations at or above 3 μg/ml [181].

Hepcidin 20 alters the biofilm architecture of *Staphylococcus epidermidis* by targeting the polysaccharide intercellular adhesin after it has reduced the extracellular matrix mass [182].

The peptides lactoferrin, conjugated lactoferricin, melimine and citropin 1.1 have all shown good anti-biofilm activity against *S. aureus* and *P. aeruginosa* infection in medical devices [183].

7. Conclusion

Microorganisms, though form biofilms as a defense mechanism for survival, this action poses a threat to the healthcare system by compromising the therapeutic efficacy of antimicrobial agents and causing ascendancies in antimicrobial resistance. Natural products from plants and microorganisms provide a plethora of chemical compounds with antibiofilm properties capable of disrupting pre-formed biofilms or inhibiting the formation of new biofilms. Identifying novel antibiofilm compounds from these sources is essential to mitigate biofilm-mediated infections. Similarly, the exploration of model systems is critical for evaluating the antibiofilm properties of newly identified medicinal compounds. Altogether, understanding the antibiofilm potential of these natural products could serve as an impetus in antimicrobial drug discovery.

Author details

Cynthia Amaning Danquah[1*], Prince Amankwah Baffour Minkah[1,2], Theresa A. Agana[3], Phanankosi Moyo[4], Michael Tetteh[1], Isaiah Osei Duah Junior[5], Kofi Bonsu Amankwah[6], Samuel Owusu Somuah[7], Michael Ofori[1] and Vinesh J. Maharaj[4]

1 Department of Pharmacology, Faculty of Pharmacy and Pharmaceutical Sciences, College of Health Sciences, Kwame Nkrumah University of Science and Technology, Kumasi, Ghana

2 Global Health and Infectious Disease Research Group, Kumasi Centre for Collaborative Research in Tropical Medicine, Kwame Nkrumah University of Science and Technology, Kumasi, Ghana

3 Department of Pharmaceutics, Faculty of Pharmacy and Pharmaceutical Sciences, College of Health Sciences, Kwame Nkrumah University of Science and Technology, Kumasi, Ghana

4 Department of Chemistry, University of Pretoria, Pretoria, South Africa

5 Department of Optometry and Visual Science, College of Science, Kwame Nkrumah University of Science and Technology, Kumasi, Ghana

6 Department of Biomedical Sciences, University of Cape Coast, Cape Coast, Ghana

7 Department of Pharmacy Practice, School of Pharmacy, University of Health and Allied Sciences, Ho, Ghana

*Address all correspondence to: cadanq@yahoo.com;
cadanquah.pharm@knust,edu.gh

References

[1] Strathdee SA, Davies SC, Marcelin JR. Confronting antimicrobial resistance beyond the COVID-19 pandemic and the 2020 US election. Lancet. 2020;**396**(10257):1050-1053. DOI: 10.1016/S0140-6736(20)32063-8

[2] Blair JM, Webber MA, Baylay AJ, Ogbolu DO, Piddock LJ. Molecular mechanisms of antibiotic resistance. Nature Reviews. Microbiology. 2015;**13**(1):42-51. DOI: 10.1038/nrmicro3380

[3] Jamal M et al. Bacterial biofilm and associated infections. Journal of the Chinese Medical Association. 2018;**81**(1):7-11. DOI: 10.1016/j.jcma.2017.07.012

[4] Romling U, Balsalobre C. Biofilm infections, their resilience to therapy and innovative treatment strategies. Journal of Internal Medicine. 2012;**272**(6):541-561. DOI: 10.1111/joim.12004

[5] Attinger C, Wolcott R. Clinically addressing biofilm in chronic wounds. Advances in Wound Care. 2012;**1**(3):127-132. DOI: 10.1089/wound.2011.0333

[6] Lemos ASO et al. Antibacterial and antibiofilm activities of psychorubrin, a pyranonaphthoquinone isolated from *Mitracarpus frigidus* (Rubiaceae). Frontiers in Microbiology. 2018;**9**:724. DOI: 10.3389/fmicb.2018.00724

[7] Donlan RM. Biofilm formation: A clinically relevant microbiological process. Clinical Infectious Diseases. 2001;**33**(8):1387-1392. DOI: 10.1086/322972

[8] Donlan RM, Costerton JW. Biofilms: Survival mechanisms of clinically relevant microorganisms. Clinical Microbiology Reviews. 2002;**15**(2):167-119. DOI: 10.1128/CMR.15.2.167

[9] Chmit M, Kanaan H, Habib J, Abbass M, McHeik A, Chokr A. Antibacterial and antibiofilm activities of polysaccharides, essential oil, and fatty oil extracted from *Laurus nobilis* growing in Lebanon. Asian Pacific Journal of Tropical Medicine. 2014;**7S1**:S546-S552. DOI: 10.1016/S1995-7645(14)60288-1

[10] Baker BJ, Banfield JF. Microbial communities in acid mine drainage. FEMS Microbiology Ecology. 2003;**44**(2):139-152. DOI: 10.1016/S0168-6496(03)00028-X

[11] Sweet MJ, Croquer A, Bythell JC. Development of bacterial biofilms on artificial corals in comparison to surface-associated microbes of hard corals. PLoS One. 2011;**6**(6):e21195. DOI: 10.1371/journal.pone.0021195

[12] Raghupathi PK, Liu W, Sabbe K, Houf K, Burmolle M, Sorensen SJ. Synergistic interactions within a multispecies biofilm enhance individual species protection against grazing by a pelagic protozoan. Frontiers in Microbiology. 2017;**8**:2649. DOI: 10.3389/fmicb.2017.02649

[13] Costa-Orlandi CB et al. Fungal biofilms and polymicrobial diseases. Journal of Fungi. 2017;**3**(2):22. DOI: 10.3390/jof3020022

[14] Costerton JW, Lewandowski Z, DeBeer D, Caldwell D, Korber D, James G. Biofilms, the customized microniche. Journal of Bacteriology. 1994;**176**(8):2137-2142. DOI: 10.1128/jb.176.8.2137-2142.1994

[15] Flemming HC, Wingender J, Szewzyk U, Steinberg P, Rice SA,

Kjelleberg S. Biofilms: An emergent form of bacterial life. Nature Reviews. Microbiology. 2016;**14**(9):563-575. DOI: 10.1038/nrmicro.2016.94

[16] Wright GD. Molecular mechanisms of antibiotic resistance. Chemical Communications. 2011;**47**(14):4055-4061. DOI: 10.1039/c0cc05111j

[17] Taylor PK, Yeung AT, Hancock RE. Antibiotic resistance in Pseudomonas aeruginosa biofilms: Towards the development of novel anti-biofilm therapies. Journal of Biotechnology. 2014;**191**:121-130. DOI: 10.1016/j.jbiotec.2014.09.003

[18] An AY, Choi KG, Baghela AS, Hancock REW. An overview of biological and computational methods for designing mechanism-informed anti-biofilm agents. Frontiers in Microbiology. 2021;**12**:640787. DOI: 10.3389/fmicb.2021.640787

[19] Abraham WR. Going beyond the control of quorum-sensing to combat biofilm infections. Antibiotics (Basel). 2016;**5**(1):3. DOI: 10.3390/antibiotics5010003

[20] Zhao X, Yu Z, Ding T. Quorum-sensing regulation of antimicrobial resistance in bacteria. Microorganisms. 2020;**8**(3):425. DOI: 10.3390/microorganisms8030425

[21] Rutherford ST, Bassler BL. Bacterial quorum sensing : Its role in virulence and possibilities for its control. Cold Spring Harbor Perspectives in Medicine. 2012;**2**(11):p.a012427. DOI: 10.1101/cshperspect.a012427

[22] Edwards SJ, Kjellerup BV. Applications of biofilms in bioremediation and biotransformation of persistent organic pollutants, pharmaceuticals/personal care products, and heavy metals. Applied Microbiology and Biotechnology. 2013;**97**(23):9909-9921. DOI: 10.1007/s00253-013-5216-z

[23] Berlanga M, Guerrero R. Living together in biofilms: The microbial cell factory and its biotechnological implications. Microbial Cell Factories. 2016;**15**(1):165. DOI: 10.1186/s12934-016-0569-5

[24] Bogino PC, Oliva Mde L, Sorroche FG, Giordano W. The role of bacterial biofilms and surface components in plant-bacterial associations. International Journal of Molecular Sciences. 2013;**14**(8):15838-15859. DOI: 10.3390/ijms140815838

[25] Wu H, Moser C, Wang HZ, Hoiby N, Song ZJ. Strategies for combating bacterial biofilm infections. International Journal of Oral Science. 2015;7(1):1-7. DOI: 10.1038/ijos.2014.65

[26] Kırmusaoğlu S. The Methods for Detection of Biofilm and Screening antibiofilm Activity of Agents. London, UK: IntechOpen; 2019. DOI: 10.5772/intechopen.84411. Available from: https://www.intechopen.com/chapters/65613

[27] Wu BC et al. Human organoid biofilm model for assessing antibiofilm activity of novel agents. NPJ Biofilms Microbiomes. 2021;**7**(1):8. DOI: 10.1038/s41522-020-00182-4

[28] Thomsen K, Trostrup H, Moser C. Animal models to evaluate bacterial biofilm development. Methods in Molecular Biology. 2014;**1147**:127-139. DOI: 10.1007/978-1-4939-0467-9_9

[29] Chung PY. Plant-derived compounds as potential source of novel anti-biofilm agents against *Pseudomonas aeruginosa*. Current Drug Targets. 2017;**18**(4):414-420. DOI: 10.2174/1389450117666161019102025

[30] Guimaraes R et al. Antibiofilm potential of medicinal plants against Candida spp. oral biofilms: A review. Antibiotics (Basel). 2021;**10**(9). DOI: 10.3390/antibiotics10091142

[31] Lahiri D, Dash S, Dutta R, Nag M. Elucidating the effect of anti-biofilm activity of bioactive compounds extracted from plants. Journal of Biosciences. 2019;**44**(2):52

[32] Alasil SM, Omar R, Ismail S, Yusof MY. Antibiofilm activity, compound characterization, and acute toxicity of extract from a novel bacterial species of paenibacillus. International Journal of Microbiology. 2014;**2014**:649420. DOI: 10.1155/2014/649420

[33] Roy R, Tiwari M, Donelli G, Tiwari V. Strategies for combating bacterial biofilms: A focus on anti-biofilm agents and their mechanisms of action. Virulence. 2018;**9**(1):522-554. DOI: 10.1080/21505594.2017.1313372

[34] Karatan E, Watnick P. Signals, regulatory networks, and materials that build and break bacterial biofilms. Microbiology and Molecular Biology Reviews. 2009;**73**(2):310-347. DOI: 10.1128/MMBR.00041-08

[35] Arciola CR, Campoccia D, Speziale P, Montanaro L, Costerton JW. Biofilm formation in Staphylococcus implant infections. A review of molecular mechanisms and implications for biofilm-resistant materials. Biomaterials. 2012;**33**(26):5967-5982. DOI: 10.1016/j.biomaterials.2012.05.031

[36] Hoyle BD, Jass J, Costerton JW. The biofilm glycocalyx as a resistance factor. The Journal of Antimicrobial Chemotherapy. 1990;**26**(1):1-5. DOI: 10.1093/jac/26.1.1

[37] Palmer J, Flint S, Brooks J. Bacterial cell attachment, the beginning of a biofilm. Journal of Industrial Microbiology & Biotechnology. 2007;**34**(9):577-588. DOI: 10.1007/s10295-007-0234-4

[38] Porter SL, Wadhams GH, Armitage JP. Signal processing in complex chemotaxis pathways. Nature Reviews. Microbiology. 2011;**9**(3):153-165. DOI: 10.1038/nrmicro2505

[39] Carniello V, Peterson BW, van der Mei HC, Busscher HJ. Physico-chemistry from initial bacterial adhesion to surface-programmed biofilm growth. Advances in Colloid and Interface Science. 2018;**261**:1-14. DOI: 10.1016/j.cis.2018.10.005

[40] Morra M, Cassinelli C. Bacterial adhesion to polymer surfaces: A critical review of surface thermodynamic approaches. Journal of Biomaterials Science. Polymer Edition. 1997;**9**(1):55-74. DOI: 10.1163/156856297x00263

[41] Donlan RM. Biofilms: Microbial life on surfaces. Emerging Infectious Diseases. 2002;**8**(9):881-890. DOI: 10.3201/eid0809.020063

[42] Vacheethasanee K et al. Bacterial surface properties of clinically isolated *Staphylococcus epidermidis* strains determine adhesion on polyethylene. Journal of Biomedical Materials Research. 1998;**42**(3):425-432. DOI: 10.1002/(sici)1097-4636(19981205)42:3<425::aid-jbm12>3.0.co;2-f

[43] Fletcher M, Loeb GI. Influence of substratum characteristics on the attachment of a marine pseudomonad to solid surfaces. Applied and Environmental Microbiology. 1979;**37**(1):67-72. DOI: 10.1128/aem.37.1.67-72.1979

[44] Pringle JH, Fletcher M. Influence of substratum wettability on attachment of freshwater bacteria to solid

surfaces. Applied and Environmental Microbiology. 1983;**45**(3):811-817. DOI: 10.1128/aem.45.3.811-817.1983

[45] Vatanyoopaisarn S, Nazli A, Dodd CE, Rees CE, Waites WM. Effect of flagella on initial attachment of listeria monocytogenes to stainless steel. Applied and Environmental Microbiology. 2000;**66**(2):860-863. DOI: 10.1128/AEM.66.2.860-863.2000

[46] Di Martino P, Cafferini N, Joly B, Darfeuille-Michaud A. Klebsiella pneumoniae type 3 pili facilitate adherence and biofilm formation on abiotic surfaces. Research in Microbiology. 2003;**154**(1):9-16. DOI: 10.1016/s0923-2508(02)00004-9

[47] Jagnow J, Clegg S. Klebsiella pneumoniae MrkD-mediated biofilm formation on extracellular matrix- and collagen-coated surfaces. Microbiology. 2003;**149**(Pt 9):2397-2405. DOI: 10.1099/mic.0.26434-0

[48] Alarcon I, Evans DJ, Fleiszig SM. The role of twitching motility in *Pseudomonas aeruginosa* exit from and translocation of corneal epithelial cells. Investigative Ophthalmology & Visual Science. 2009;**50**(5):2237-2244. DOI: 10.1167/iovs.08-2785

[49] Costa OYA, Raaijmakers JM, Kuramae EE. Microbial extracellular polymeric substances: Ecological function and impact on soil aggregation. Frontiers in Microbiology. 2018;**9**:1636. DOI: 10.3389/fmicb.2018.01636

[50] Fong JNC, Yildiz FH. Biofilm matrix proteins. Microbiology Spectrum. 2015;**3**(2). DOI: 10.1128/microbiolspec.MB-0004-2014

[51] Ramirez-Mata A, Lopez-Lara LI, Xiqui-Vazquez ML, Jijon-Moreno S, Romero-Osorio A, Baca BE. The cyclic-di-GMP diguanylate cyclase CdgA has a role in biofilm formation and exopolysaccharide production in Azospirillum brasilense. Research in Microbiology. 2016;**167**(3):190-201. DOI: 10.1016/j.resmic.2015.12.004

[52] Toyofuku M, Inaba T, Kiyokawa T, Obana N, Yawata Y, Nomura N. Environmental factors that shape biofilm formation. Bioscience, Biotechnology, and Biochemistry. 2016;**80**(1):7-12. DOI: 10.1080/09168451.2015.1058701

[53] Bacosa HP et al. Extracellular polymeric substances (EPS) producing and oil degrading bacteria isolated from the northern Gulf of Mexico. PLoS One. 2018;**13**(12):e0208406. DOI: 10.1371/journal.pone.0208406

[54] Jayathilake PG et al. Extracellular polymeric substance production and aggregated bacteria colonization influence the competition of microbes in biofilms. Frontiers in Microbiology. 2017;**8**:1865. DOI: 10.3389/fmicb.2017.01865

[55] Limoli DH, Jones CJ, Wozniak DJ. Bacterial extracellular polysaccharides in biofilm formation and function. Microbiology Spectrum. 2015;**3**(3). DOI: 10.1128/microbiolspec.MB-0011-2014

[56] Bassler BL, Greenberg EP, Stevens AM. Cross-species induction of luminescence in the quorum-sensing bacterium Vibrio harveyi. Journal of Bacteriology. 1997;**179**(12):4043-4045. DOI: 10.1128/jb.179.12.4043-4045.1997

[57] Chen X et al. Structural identification of a bacterial quorum-sensing signal containing boron. Nature. 2002;**415**(6871):545-549. DOI: 10.1038/415545a

[58] Garnett JA, Matthews S. Interactions in bacterial biofilm development:

A structural perspective. Current Protein & Peptide Science. 2012;**13**(8):739-755. DOI: 10.2174/138920312804871166

[59] Hall-Stoodley L, Stoodley P. Developmental regulation of microbial biofilms. Current Opinion in Biotechnology. 2002;**13**(3):228-233. DOI: 10.1016/s0958-1669(02)00318-x

[60] He X, Ahn J. Differential gene expression in planktonic and biofilm cells of multiple antibiotic-resistant Salmonella Typhimurium and Staphylococcus aureus. FEMS Microbiology Letters. 2011;**325**(2):180-188. DOI: 10.1111/j.1574-6968.2011.02429.x

[61] Kaplan JB. Biofilm dispersal: Mechanisms, clinical implications, and potential therapeutic uses. Journal of Dental Research. 2010;**89**(3):205-218. DOI: 10.1177/0022034509359403

[62] Diaz-Salazar C et al. The stringent response promotes biofilm dispersal in pseudomonas putida. Scientific Reports. 2017;**7**(1):18055. DOI: 10.1038/s41598-017-18518-0

[63] Singh PK et al. Vibrio cholerae combines individual and collective sensing to trigger biofilm dispersal. Current Biology. 2017;**27**(21):3359-3366 e7. DOI: 10.1016/j.cub.2017.09.041

[64] Shen D, Langenheder S, Jurgens K. Dispersal modifies the diversity and composition of active bacterial communities in response to a salinity disturbance. Frontiers in Microbiology. 2018;**9**:2188. DOI: 10.3389/fmicb.2018.02188

[65] Lee K, Yoon SS. Pseudomonas aeruginosa biofilm, a programmed bacterial life for fitness. Journal of Microbiology and Biotechnology. 2017;**27**(6):1053-1064. DOI: 10.4014/jmb.1611.11056

[66] Chung EMC, Dean SN, Propst CN, Bishop BM, Van Hoek ML. Komodo dragon-inspired synthetic peptide DRGN-1 promotes wound-healing of a mixed-biofilm infected wound. npj Biofilms Microbiomes. 2017;**3**(1):1-13. DOI: 10.1038/s41522-017-0017-2

[67] Silveira GGOS et al. Antibiofilm peptides: Relevant preclinical animal infection models and translational potential. ACS Pharmacology & Translational Science. 2021;**4**(1):55-73. DOI: 10.1021/ACSPTSCI.0C00191

[68] Roberts AEL, Kragh KN, Bjarnsholt T, Diggle SP. The limitations of in vitro experimentation in understanding biofilms and chronic infection. Journal of Molecular Biology. 2015;**427**(23):3646-3661. DOI: 10.1016/J.JMB.2015.09.002

[69] Zhang T, Wang Z, Hancock REW, De La Fuente-Núñez C, Haapasalo M. Treatment of oral biofilms by a D-enantiomeric peptide. PLoS One. 2016;**11**(11):e0166997. DOI: 10.1371/JOURNAL.PONE.0166997

[70] Bowen WH. Rodent model in caries research. Odontology. 2013;**101**(1):9-14. DOI: 10.1007/S10266-012-0091-0

[71] Daep CA, James DAM, Lamont RJ, Demuth DR. Structural characterization of peptide-mediated inhibition of Porphyromonas gingivalis biofilm formation. Infection and Immunity. 2006;**74**(10):5756-5762. DOI: 10.1128/IAI.00813-06

[72] Daep CA, Novak EA, Lamont RJ, Demuth DR. Structural dissection and in vivo effectiveness of a peptide inhibitor of Porphyromonas gingivalis adherence to Streptococcus gordonii. Infection and Immunity. 2011;**79**(1):67-74. DOI: 10.1128/IAI.00361-10

[73] Harrington NE, Sweeney E, Harrison F. Building a better biofilm-formation of in vivo-like biofilm

structures by *Pseudomonas aeruginosa* in a porcine model of cystic fibrosis lung infection. Biofilms. 2020;**2**:100024. DOI: 10.1016/J.BIOFLM.2020.100024

[74] Song Z et al. Effects of Intratracheal administration of Novispirin G10 on a rat model of mucoid *Pseudomonas aeruginosa* lung infection. Antimicrobial Agents and Chemotherapy. 2005;**49**(9):3868. DOI: 10.1128/AAC.49.9.3868-3874.2005

[75] Franco L, Cardoso MH, Carvalho CME, De Fuente-nunez C. Antibio film peptides: Relevant preclinical animal infection models and translational potential. ACS Pharmacology & Translational Science. 2021;**4**(1):55-73. DOI: 10.1021/acsptsci.0c00191

[76] Lebeaux D, Chauhan A, Rendueles O, Beloin C. From in vitro to in vivo models of bacterial biofilm-related infections. Pathogens. 2013;**2**(2):288-356. DOI: 10.3390/pathogens2020288

[77] Ofori M, Danquah CA, Ativui S, Doe P, Asamoah WA. In-vitro anti-tuberculosis, anti-efflux pumps and anti-biofilm effects of crinum asiaticum bulbs. 2021;**14**(December):1905-1915

[78] Sen T, Karmakar S, Sarkar R. Evaluation of Natural Products against Biofilm-Mediated Bacterial Resistance. Elsevier Inc.; 2015

[79] Butler MS, Buss AD. Natural products—The future scaffolds for novel antibiotics? Biochemical Pharmacology. 2006;**71**(7):919-929. DOI: 10.1016/j.bcp.2005.10.012

[80] Hassan A, Usman J, Kaleem F, Omair M, Khalid A, Iqbal M. Evaluation of different detection methods of biofilm formation in the clinical isolates. Brazilian Journal of Infectious Diseases is the official publication of the Brazilian Society of Infectious Diseases. 2011;**15**(4):305-311

[81] Abdel Halim RM, Kassem NN, Mahmoud BS. Detection of biofilm producing Staphylococci among different clinical isolates and its relation to methicillin susceptibility. Open Access Macedonian Journal of Medical Sciences. 2018;**6**(8):1335-1341. DOI: 10.3889/oamjms.2018.246

[82] Freeman DJ, Falkiner FR, Keane CT. New method for detecting slime production by coagulase negative staphylococci. Journal of Clinical Pathology. 1989;**42**(8):872-874. DOI: 10.1136/jcp.42.8.872

[83] Darwish SF, Asfour HAE. Investigation of biofilm forming ability in Staphylococci causing bovine mastitis using phenotypic and genotypic assays. Scientific World Journal. 2013;**2013**:378492. DOI: 10.1155/2013/378492

[84] Chakraborty P, Bajeli S, Kaushal D, Radotra BD, Kumar A. Tuberculosis. Nature Communications. 2021. DOI: 10.1038/s41467-021-21748-6

[85] Djordjevic D, Wiedmann M, McLandsborough LA. Microtiter plate assay for assessment of Listeria monocytogenes biofilm formation. Applied and Environmental Microbiology. 2002;**68**(6):2950-2958. DOI: 10.1128/AEM.68.6.2950-2958.2002

[86] Zegaer BH et al. Detection of bacteria bearing resistant biofilm forms, by using the universal and specific PCR is still unhelpful in the diagnosis of periprosthetic joint infections. Frontiers in Medicine. 2014;**1**:30. DOI: 10.3389/fmed.2014.00030

[87] Shahmoradi M, Faridifar P, Shapouri R, Mousavi SF, Ezzedin M, Mirzaei B. Determining the biofilm forming gene profile of *Staphylococcus aureus* clinical isolates via multiplex colony

PCR method. Reports of Biochemistry and Molecular Biology. 2019;7(2):181-188

[88] Roy R, Tiwari M, Donelli G, Tiwari V. Strategies for combating bacterial bio films : A focus on anti-biofilm agents and their mechanisms of action. Virulence. 2018;**9**(1):522-554

[89] Pathogens I. Efficacy and mechanism of traditional medicinal plants and bioactive compounds against clinically important pathogens. Antibiotics (Basel). 2019;**8**(4):257

[90] Danquah CA et al. Analogues of disulfides from allium stipitatum demonstrate potent anti-tubercular activities through drug efflux pump and biofilm inhibition. Scientific Reports. 2018. DOI: 10.1038/s41598-017-18948-w

[91] Hall-Stoodley L, Costerton JW, Stoodley P. Bacterial biofilms: From the natural environment to infectious diseases. Nature Reviews. Microbiology. 2004;**2**(2):95-108. DOI: 10.1038/nrmicro821

[92] Chen J-Q, Healey S, Regan P, Laksanalamai P, Hu Z. PCR-based methodologies for detection and characterization of Listeria monocytogenes and Listeria ivanovii in foods and environmental sources. Food Science and Human Wellness. 2017;**6**(2):39-59. DOI: 10.1016/j.fshw.2017.03.001

[93] Maheaswari R, Kshirsagar JT, Lavanya N. Polymerase chain reaction: A molecular diagnostic tool in periodontology. Journal of Indian Society of Periodontology. 2016;**20**(2):128-135. DOI: 10.4103/0972-124X.176391

[94] Yang Q, Rui Y. Two multiplex real-time PCR assays to detect and differentiate *Acinetobacter baumannii* and non-baumannii bla OXA-58-like genes. 2016;**11**(7):e0158958. DOI: 10.1371/journal.pone.0158958

[95] Hiltunen A, Skogman M. Exploration of microbial communities using the thermo scientific. BioTechniques. 2017;**63**:236-237. DOI: 10.2144/000114613

[96] Emde B, Heinen A, Gödecke A, Bottermann K, Düsseldorf H. Wheat germ agglutinin staining as a suitable method for detection and quantification of fibrosis in cardiac tissue after myocardial infarction. 2014;**58**(4):2448. DOI: 10.4081/ejh.2014.2448

[97] Ozioma E-OJ, Antoinette O, Chinwe N. Herbal medicines in African traditional medicine. In: Builders PF editors. Herbal Medicine. 2019. DOI: 10.5772/INTECHOPEN.80348

[98] Newman DJ, Cragg GM. Natural products as sources of new drugs over the nearly four decades from 01/1981 to 09/2019. Journal of Natural Products. 2020;**83**(3):770-803. DOI: 10.1021/ACS.JNATPROD.9B01285/SUPPL_FILE/NP9B01285_SI_009.PDF

[99] Ahmad I, Husain FM, Maheshwari M, Zahin M. Medicinal plants and phytocompounds: A potential source of novel antibiofilm agents. Biology. 2014:205-232. DOI: 10.1007/978-3-642-53833-9_10

[100] Melander RJ, Basak AK, Melander C. Natural products as inspiration for the development of bacterial antibiofilm agents. Natural Product Reports. 2020;**37**(11):1454-1477. DOI: 10.1039/D0NP00022A

[101] Dettweiler M et al. American civil war plant medicines inhibit growth, biofilm formation, and quorum sensing by multidrug-resistant bacteria. Scientific Reports. 2019;**9**(1):7692. DOI: 10.1038/S41598-019-44242-Y

[102] Acquaviva R et al. Antibacterial and anti-biofilm activities of walnut

pellicle extract (*Juglans regia* L.) against coagulase-negative staphylococci. Natural Product Research. 2021;**35**(12):2076-2081. DOI: 10.1080/14786419.2019.1650352

[103] Ahmed SO, Zedan HH, Ibrahim YM. Quorum sensing inhibitory effect of bergamot oil and aspidosperma extract against *Chromobacterium violaceum* and *Pseudomonas aeruginosa*. Archives of Microbiology. 2021;**203**(7):4663-4675. DOI: 10.1007/S00203-021-02455-8

[104] Uc-Cachón AH et al. Antibacterial and antibiofilm activities of Mayan medicinal plants against Methicillin-susceptible and -resistant strains of *Staphylococcus aureus*. Journal of Ethnopharmacology. 2021;**279**:114369. DOI: 10.1016/j.jep.2021.114369

[105] Mohammed HA et al. Phytochemical profiling, in vitro and in silico anti-microbial and anti-cancer activity evaluations and staph GyraseB and h-TOP-IIβ receptor-docking studies of major constituents of *Zygophyllum coccineum* L. aqueous-ethanolic extract and its subsequent fractions: An approach to validate traditional phytomedicinal knowledge. Molecules. 2021;**26**(3):557. DOI: 10.3390/MOLECULES26030577

[106] Miao W et al. The impact of flavonoids-rich *Ziziphus jujuba* Mill. extract on *Staphylococcus aureus* biofilm formation. BMC Complementary Medicine and Therapies. 2020;**20**(1):187. DOI: 10.1186/S12906-020-2833-9/FIGURES/5

[107] Bocquet L et al. Phenolic compounds from *Humulus lupulus* as natural antimicrobial products: New weapons in the fight against methicillin resistant *Staphylococcus aureus*, *Leishmania mexicana* and *Trypanosoma brucei* strains. Molecules. 2019;**24**(6):1024. DOI: 10.3390/MOLECULES24061024

[108] Vijayakumar K, Ramanathan T. Musa acuminata and its bioactive metabolite 5-hydroxymethylfurfural mitigates quorum sensing (las and rhl) mediated biofilm and virulence production of nosocomial pathogen *Pseudomonas aeruginosa* in vitro. Journal of Ethnopharmacology. 2020;**246**:112242. DOI: 10.1016/J.JEP.2019.112242

[109] Minami M, Takase H, Nakamura M, Makino T. Effect of *Lonicera caerulea* var. *emphyllocalyx* fruit on biofilm formed by *Porphyromonas gingivalis*. BioMed Research International. 2019;**2019**. DOI: 10.1155/2019/3547858

[110] Sadeghi Z, Masullo M, Cerulli A, Nazzaro F, Farimani MM, Piacente S. Terpenoid constituents of Perovskia artemisioides aerial parts with inhibitory effects on bacterial biofilm growth. Journal of Natural Products. 2021;**84**(1):26-36. DOI: 10.1021/ACS.JNATPROD.0C00832

[111] Rajendran N et al. Antimicrobial flavonoids isolated from Indian medicinal plant Scutellaria oblonga inhibit biofilms formed by common food pathogens. Natural Product Research. 2016;**30**(17):2002-2006. DOI: 10.1080/14786419.2015.1104673

[112] Tang Y et al. Effect of syringopicroside extracted from *Syringa oblata* lindl on the biofilm formation of *Streptococcus suis*. Molecules. 2021;**26**(5):1295. DOI: 10.3390/MOLECULES26051295

[113] Yanti Y, Rukayadi K, Lee H, Hwang JK. Activity of panduratin A isolated from Kaempferia pandurata Roxb. against multi-species oral biofilms in vitro. Journal of Oral Science. 2009;**51**(1):87-95. DOI: 10.2334/JOSNUSD.51.87

[114] Khan R, Zakir M, Khanam Z, Shakil S, Khan AU. Novel compound from Trachyspermum ammi (Ajowan

caraway) seeds with antibiofilm and antiadherence activities against Streptococcus mutans: A potential chemotherapeutic agent against dental caries. Journal of Applied Microbiology. 2010;**109**(6):2151-2159. DOI: 10.1111/J.1365-2672.2010.04847.X

[115] Maheshwari M, Abul Qais F, Althubiani AS, Abulreesh HH, Ahmad I. Bioactive extracts of Carum copticum and thymol inhibit biofilm development by multidrug-resistant extended spectrum β-lactamase producing enteric bacteria. Biofouling. 2019;**35**(9):1026-1039. DOI: 10.1080/08927014.2019.1688305

[116] D'Abrosca B et al. Spectroscopic identification and anti-biofilm properties of polar metabolites from the medicinal plant Helichrysum italicum against *Pseudomonas aeruginosa*. Bioorganic & Medicinal Chemistry. 2013;**21**(22):7038-7046. DOI: 10.1016/J.BMC.2013.09.019

[117] Mozirandi W, Tagwireyi D, Mukanganyama S. Evaluation of antimicrobial activity of chondrillasterol isolated from Vernonia adoensis (Asteraceae). BMC Complementary and Alternative Medicine. 2019;**19**(1):1-11. DOI: 10.1186/S12906-019-2657-7/FIGURES/7

[118] Trentin DDS et al. Potential of medicinal plants from the Brazilian semi-arid region (Caatinga) against Staphylococcus epidermidis planktonic and biofilm lifestyles. Journal of Ethnopharmacology. 2011;**137**(1):327-335. DOI: 10.1016/J.JEP.2011.05.030

[119] Verghese J. Olibanum in Focus | Perfumer & Flavorist. 1988;**13**(1):1-12

[120] Gebremedhin T. *Boswellia papyrifera* (Del.) hochst. from Western Tigray: Opportunities, constraints and seed germiniation responses. 1997. DOI: 10.3/JQUERY-UI.JS

[121] Schillaci D, Arizza V, Dayton T, Camarda L, Di Stefano V. In vitro anti-biofilm activity of Boswellia spp. oleogum resin essential oils. Letters in Applied Microbiology. 2008;**47**(5):433-438. DOI: 10.1111/J.1472-765X.2008.02469.X

[122] Yadav S. antibiofilm formation activity of terminalia bellerica plant extract against clinical isolates of *Streptococcus mutans* and *Streptococcus sobrinus*: Implication in oral hygiene. International Journal of Pharmaceutical and Biological Science Archive. 2012;**3**(4)

[123] de Araujo AR et al. Antibacterial, antibiofilm and cytotoxic activities of Terminalia fagifolia Mart. extract and fractions. Annals of Clinical Microbiology and Antimicrobials. 2015;**14**(1):1-10. DOI: 10.1186/S12941-015-0084-2/FIGURES/9

[124] Erhabor RC, Aderogba MA, Erhabor JO, Nkadimeng SM, McGaw LJ. In vitro bioactivity of the fractions and isolated compound from *Combretum elaeagnoides* leaf extract against selected foodborne pathogens. Journal of Ethnopharmacology. 2021;**273**:113981. DOI: 10.1016/J.JEP.2021.113981

[125] Abrão F et al. Oleoresins and naturally occurring compounds of Copaifera genus as antibacterial and antivirulence agents against periodontal pathogens. Scientific Reports. 2021;**11**(1):4953. DOI: 10.1038/S41598-021-84480-7

[126] Yoshikawa M, Murakami T, Komatsu H, Murakami N, Yamahara J, Matsuda H. Medicinal foodstuffs. IV. Fenugreek seed. (1): Structures of trigoneosides Ia, Ib, IIa, IIb, IIIa, and IIIb, new furostanol saponins from the seeds

of Indian *Trigonella foenum-graecum* L. Chemical and Pharmaceutical Bulletin. 1997;**45**(1):81-87. DOI: 10.1248/CPB.45.81

[127] Husain FM, Ahmad I, Khan MS, Al-Shabib NA. Trigonella foenum-graceum (seed) extract interferes with quorum sensing regulated traits and biofilm formation in the strains of *Pseudomonas aeruginosa* and *Aeromonas hydrophila*. Evidence-Based Complementary and Alternative Medicine. 2015;**2015**:879540. DOI: 10.1155/2015/879540

[128] Kuete V. Medicinal Spices and Vegetables from Africa: Therapeutic Potential against Metabolic, Inflammatory, Infectious and Systemic Diseases. Medicinal Spices and Vegetables from Africa. Vol. 43. Elsevier Inc.; 2017

[129] Quave CL, Plano LRW, Pantuso T, Bennett BC. Effects of extracts from Italian medicinal plants on planktonic growth, biofilm formation and adherence of methicillin-resistant *Staphylococcus aureus*. Journal of Ethnopharmacology. 2008;**118**(3):418. DOI: 10.1016/J.JEP.2008.05.005

[130] Mendes FSF et al. Antibacterial activity of *Salvia officinalis* L. against periodontopathogens: An in vitro study. Anaerobe. 2020;**63**:102194. DOI: 10.1016/J.ANAEROBE.2020.102194

[131] Esmaeelian B, Kamrani YY, Amoozegar MA, Rahmani S, Rahimi M, Amanlou M. Anti-cariogenic properties of malvidin-3,5-diglucoside isolated from Alcea longipedicellata against oral bacteria. International Journal of Pharmacology. 2007;**3**(6):468-474. DOI: 10.3923/IJP.2007.468.474

[132] Ngan LTM et al. Antibacterial activity of *Hibiscus rosa-sinensis* L. red flower against antibiotic-resistant strains of *Helicobacter pylori* and identification of the flower constituents. Brazilian Journal of Medical and Biological Research. 2021;**54**(7):e10889. DOI: 10.1590/1414-431X2020E10889

[133] Christenhusz MJM, Byng JW. The number of known plants species in the world and its annual increase. Phytotaxa. 2016;**261**(3):201-217. DOI: 10.11646/PHYTOTAXA.261.3.1

[134] Balakrishnan S et al. Antiquorum sensing and antibiofilm potential of biosynthesized silver nanoparticles of Myristica fragrans seed extract against MDR Salmonella enterica serovar Typhi isolates from asymptomatic typhoid carriers and typhoid patients. Environmental Science and Pollution Research International. 2020;**27**(3):2844-2856. DOI: 10.1007/S11356-019-07169-5

[135] Gopu V, Kothandapani S, Shetty PH. Quorum quenching activity of *Syzygium cumini* (L.) Skeels and its anthocyanin malvidin against *Klebsiella pneumoniae*. Microbial Pathogenesis. 2015;**79**:61-69. DOI: 10.1016/J.MICPATH.2015.01.010

[136] Chemsa AE et al. Chemical constituents of essential oil of endemic Rhanterium suaveolens Desf. growing in Algerian Sahara with antibiofilm, antioxidant and anticholinesterase activities. Natural Product Research. 2015;**30**(18):2120-2124. DOI: 10.1080/14786419.2015.1110705

[137] Karaca B, Çöleri Cihan A, Akata I, Altuner EM. Anti-biofilm and antimicrobial activities of five edible and medicinal macrofungi samples on some biofilm producing multi drug resistant Enterococcus strains. Turkish Journal of Agriculture—Food Science and Technology. 2020;**8**(1):69. DOI: 10.24925/turjaf.v8i1.69-80.2723

[138] Alves MJ, Ferreira ICFR, Lourenço I, Costa E, Martins A, Pintado M. Wild mushroom extracts as inhibitors of bacterial biofilm formation. Pathogens. 2014;**3**(3):667-679. DOI: 10.3390/pathogens3030667

[139] Čuvalová A, Strapáč I, Handrová L, Kmeť V. Antibiofilm activity of mushroom extracts against *Staphylococcus aureus*. Annales Universitatis Paedagogicae Cracoviensis Studia Naturae. 2018;**3**:17-23. DOI: 10.24917/25438832.3supp.2

[140] Borges A, Saavedra MJ, Simões M. The activity of ferulic and gallic acids in biofilm prevention and control of pathogenic bacteria. Biofouling. 2012;**28**(7):755-767. DOI: 10.1080/08927014.2012.706751

[141] Lingström P, Zaura E, Ofek I, Wilson M. Function Components in Lentinus edodes mushroom with anti-bio fi lm activity directed against bacteria involved in caries and gingivitis. 2018. DOI: 10.1039/c7fo01727h

[142] Bin L, Wei L, Xiaohong C, Mei J, Mingsheng D. In vitro antibiofilm activity of the melanin from *Auricularia auricula*, an edible jelly mushroom. Annales de Microbiologie. 2012;**62**(4):1523-1530. DOI: 10.1007/s13213-011-0406-3

[143] Kim YG, Lee JH, Lee S, Lee YK, Hwang BS, Lee J. Antibiofilm activity of phorbaketals from the marine sponge phorbas sp. against *Staphylococcus aureus*. Marine Drugs. 2021;**19**(6):1-9. DOI: 10.3390/md19060301

[144] Paul VJ, Puglisi MP. Chemical mediation of interactions among marine organisms. Natural Product Reports. 2004;**21**(1):189-209. DOI: 10.1039/b302334f

[145] Lee JH, Kim E, Choi H, Lee J. Collismycin C from the micronesian marine bacterium Streptomyces sp. MC025 Inhibits *Staphylococcus aureus* biofilm formation. Marine Drugs. 2017;**15**(12):387. DOI: 10.3390/md15120387

[146] Rizzo C et al. Antibiofilm activity of antarctic sponge-associated bacteria against *Pseudomonas aeruginosa* and *Staphylococcus aureus*. Journal of Marine Science and Engineering. 2021;**9**(3):1-16. DOI: 10.3390/jmse9030243

[147] Balasubramanian S et al. Marine sponge-derived Streptomyces sp. SBT343 extract inhibits staphylococcal biofilm formation. Frontiers in Microbiology. 2017;**8**(Feb):1-14. DOI: 10.3389/fmicb.2017.00236

[148] Maggs CA, Gilmore BF. Against Clinically Relevant Human Pathogens. 2015. pp. 3581-3605. DOI: 10.3390/md13063581

[149] Jun JY, Jung MJ, Jeong IH, Yamazaki K, Kawai Y, Kim BM. Antimicrobial and antibiofilm activities of sulfated polysaccharides from marine algae against dental plaque bacteria. Marine Drugs. 2018;**16**(9):301. DOI: 10.3390/md16090301

[150] Achmad H, Huldani, Ramadhany YF. Antimicrobial activity and sulfated polysaccharides antibiofilms in marine algae against dental plaque bacteria: A literature review. Systematic Reviews in Pharmacy. 2020;**11**(6):459-465. DOI: 10.31838/srp.2020.6.72

[151] Huang CY, Wu SJ, Yang WN, Kuan AW, Chen CY. Antioxidant activities of crude extracts of fucoidan extracted from *Sargassum glaucescens* by a compressional-puffing-hydrothermal extraction process. Food Chemistry. 2016;**197**:1121-1129. DOI: 10.1016/j.foodchem.2015.11.100

[152] Yunhai H, Eyþórsdóttir A, Scully SM. In vitro antibacterial activity of fucoidan isolated from *Ascophyllum nodosum* and *Laminaria digitata*. Fisheries Training Programmes. 2016

[153] Ren D, Sims JJ, Wood TK. Ren, 2002 Inhibition of biofilm formation.PDF. 2002;**2**(Fujikawa 1994):293-299

[154] Hentzer M et al. Inhibition of quorum sensing in Pseudomonas aeruginosa biofilm bacteria by a halogenated furanone compound. Microbiology. 2002;**148**(1):87-102. DOI: 10.1099/00221287-148-1-87

[155] Manefield M et al. Halogenated furanones inhibit quorum sensing through accelerated LuxR turnover. Microbiology. 2002;**148**(4):1119-1127. DOI: 10.1099/00221287-148-4-1119

[156] Busetti A, Shaw G, Megaw J, Gorman SP, Maggs CA, Gilmore BF. Marine-derived quorum-sensing inhibitory activities enhance the antibacterial efficacy of tobramycin against *Pseudomonas aeruginosa*. Marine Drugs. 2015;**13**(1):1-28. DOI: 10.3390/md13010001

[157] López Y, Soto SM, Sara M. Antibiotics the usefulness of microalgae compounds for of microalgae preventing biofilm infections compounds for preventing biofilm infections. 2020;(i)

[158] Adebayo-Tayo B, Salaam A, Ajibade A. Green synthesis of silver nanoparticle using Oscillatoria sp. extract, its antibacterial, antibiofilm potential and cytotoxicity activity. Heliyon. 2019;**5**(10):e02502. DOI: 10.1016/j.heliyon.2019.e02502

[159] Vijayan SR, Santhiyagu P, Singamuthu M, Kumari Ahila N, Jayaraman R, Ethiraj K. Synthesis and characterization of silver and gold nanoparticles using aqueous extract of seaweed, turbinaria conoides, and their antimicrofouling activity. Scientific World Journal. 2014;**2014**. DOI: 10.1155/2014/938272

[160] Tofiño-Rivera A, Ortega-Cuadros M, Galvis-Pareja D, Jiménez-Rios H, Merini LJ, Martínez-Pabón MC. Effect of Lippia alba and Cymbopogon citratus essential oils on biofilms of *Streptococcus mutans* and cytotoxicity in CHO cells. Journal of Ethnopharmacology. 2016;**194**(December 2015):749-754. DOI: 10.1016/j.jep.2016.10.044

[161] Ortega-Cuadros M, Tofiño-Rivera AP, Merini LJ, Martínez-Pabón MC. Antimicrobial activity of Cymbopogon citratus (Poaceae) on Streptococcus mutans biofilm and its cytotoxic effects. Revista de Biología Tropical. 2018;**66**(4):1519-1529. DOI: 10.15517/rbt.v66i4.33140

[162] E. Cbd et al., Prevention of *Pseudomonas aeruginosa* biofilm formation on soft contact lenses by *Allium sativum* fermented extract (BGE) and cannabinol oil. pp. 1-12

[163] Aqawi M, Gallily R, Sionov RV, Zaks B, Friedman M, Steinberg D. Cannabigerol prevents quorum sensing and biofilm formation of *Vibrio harveyi*. Frontiers in Microbiology. 2020;**11**(May):1-13. DOI: 10.3389/fmicb.2020.00858

[164] Abdi-Ali A, Mohammadi-Mehr M, Agha Alaei Y. Bactericidal activity of various antibiotics against biofilm-producing *Pseudomonas aeruginosa*. International Journal of Antimicrobial Agents. 2006;**27**(3):196-200. DOI: 10.1016/J.IJANTIMICAG.2005.10.007

[165] Liu F et al. Carvacrol oil inhibits biofilm formation and exopolysaccharide

production of Enterobacter cloacae. Food Control. 2021;**119**:107473. DOI: 10.1016/J.FOODCONT.2020.107473

[166] Gutierrez-Pacheco MM et al. Carvacrol inhibits biofilm formation and production of extracellular polymeric substances of Pectobacterium carotovorum subsp. carotovorum. Food Control. 2018;**89**:210-218. DOI: 10.1016/J.FOODCONT.2018.02.007

[167] Moura MC et al. Multi-effect of the water-soluble *Moringa oleifera* lectin against *Serratia marcescens* and Bacillus sp.: Antibacterial, antibiofilm and anti-adhesive properties. Journal of Applied Microbiology. 2017;**123**(4):861-874. DOI: 10.1111/jam.13556

[168] Moura MC, Napoleão TH, Coriolano MC, Paiva PMG, Figueiredo RCBQ, Coelho LCBB. Water-soluble Moringa oleifera lectin interferes with growth, survival and cell permeability of corrosive and pathogenic bacteria. Journal of Applied Microbiology. 2015;**119**(3):666-676. DOI: 10.1111/jam.12882

[169] Feng Y, Song J, Zhao Z, Zhao F, Yang l, Jiao C. A rapid and effective method for purification of a heat-resistant lectin from potato (*Solanum tuberosum*) tubers. DOI: 10.1007/s10719-018-9836-5

[170] Roberto W, Larissa Y, Ferreira A, Macário I, Cavalcanti F. Antibacterial and antibiofilm lectins from plants—a review Lectinas antibacterianas e antibiofilmes de plantas—uma revisão Lectinas antibacterianas y antibiofilms vegetales—una revision. 2021;**2021**:1-14

[171] Costa E, Silva S, Tavaria F, Pintado M. Antimicrobial and antibiofilm activity of chitosan on the oral pathogen *Candida albicans*. 2014;**3**(4):908-919. DOI: 10.3390/pathogens3040908

[172] Aurestila BJ, Villaver EAM, Tan EY. Anti-biofilm activity of chitosan from crab and shrimp species indigenous to the Philippines on established biofilms of *Pseudomonas aeruginosa* and *Staphylococcus aureus*. Journal of Pharmacognosy & Natural Products. 2018;**04**(01):1-5. DOI: 10.4172/2472-0992.1000149

[173] Maddocks SE, Jenkins RE, Rowlands RS, Purdy KJ, Cooper RA. Manuka honey inhibits adhesion and invasion of medically important wound bacteria in vitro. Future Microbiology. 2013;**8**(12):1523-1536. DOI: 10.2217/FMB.13.126/ASSET/IMAGES/LARGE/FIGURE3.JPEG

[174] Majtan J, Bohova J, Horniackova M, Klaudiny J, Majtan V. Anti-biofilm effects of honey against wound pathogens Proteus mirabilis and Enterobacter cloacae. Phytotherapy Research. 2014;**28**(1):69-75. DOI: 10.1002/PTR.4957

[175] Lu J et al. Honey can inhibit and eliminate biofilms produced by Pseudomonas aeruginosa. Scientific Reports. 2019;**9**(1):1-13. DOI: 10.1038/s41598-019-54576-2

[176] Lu J et al. Manuka-type honeys can eradicate biofilms produced by Staphylococcus aureus strains with different biofilm-forming abilities. PeerJ. 2014;**2014**(1):1-25. DOI: 10.7717/peerj.326

[177] Zamora LG et al. An insight into the antibiofilm properties of Costa Rican stingless bee honeys. Journal of Wound Care. 2017;**26**(4):168-177. DOI: 10.12968/jowc.2017.26.4.168

[178] Sindi A et al. Anti-biofilm effects and characterisation of the hydrogen peroxide activity of a range of Western Australian honeys compared to Manuka and multifloral honeys. Scientific Reports. 2019;**9**(1):1-17. DOI: 10.1038/s41598-019-54217-8

[179] Brogden KA. Antimicrobial peptides: Pore formers or metabolic inhibitors in bacteria? Nature Reviews Microbiology. 2005;**3**(3):238-250. DOI: 10.1038/nrmicro1098

[180] Pompilio A et al. Antibacterial and anti-biofilm effects of cathelicidin peptides against pathogens isolated from cystic fibrosis patients. Peptides. 2011;**32**(9):1807-1814. DOI: 10.1016/J.PEPTIDES.2011.08.002

[181] Blower RJ, Barksdale SM, van Hoek ML. Snake cathelicidin NA-CATH and smaller helical antimicrobial peptides are effective against *Burkholderia thailandensis*. PLoS Neglected Tropical Diseases. 2015;**9**(7):e0003862. DOI: 10.1371/JOURNAL.PNTD.0003862

[182] Brancatisano FL et al. Inhibitory effect of the human liver-derived antimicrobial peptide hepcidin 20 on biofilms of polysaccharide intercellular adhesin (PIA)-positive and PIA-negative strains of *Staphylococcus epidermidis*. Undefined. 2014;**30**(4):435-446. DOI: 10.1080/08927014.2014.888062

[183] Yoshinari M, Kato T, Matsuzaka K, Hayakawa T, Shiba K. Prevention of biofilm formation on titanium surfaces modified with conjugated molecules comprised of antimicrobial and titanium-binding peptides. Journal of Bioadhesion and Biofilm Research. 2009;**26**(1):103-110. DOI: 10.1080/08927010903216572

Chapter 13

Efficacy of Radiations against Bacterial Biofilms

Salma Kloula Ben Ghorbal, Rim Werhani and Abdelwaheb Chatti

Abstract

A biofilm has been defined as a community of bacteria living in organized structures at a liquid interface. Biofilms can colonize a wide range of domains, including essentially industrial sectors, different natural environments, and also biomedical environments. Bacteria in biofilms are generally well protected against environmental stresses and, as a consequence, are extremely difficult to eradicate. The current study was to investigate the efficacy of different radiations against bacterial biofilms on different surfaces. It was established that the majority of available treatments have proven less effective against pathogenic biofilms, compared to planktonic bacteria. Therefore, new biofilm treatment strategies are needed, including physical treatments such as radiations. UV LEDs offer new solutions to prevent biofilm formation on inaccessible surfaces, such as medical and food equipment and, potentially, sanitary facilities, to limit nosocomial infections, compared to continuous UV irradiation treatment. Moreover, the antimicrobial effectiveness of gamma irradiation is therefore guaranteed in the treatment of bacteria associated with a biofilm, compared to planktonic bacteria. However, limited studies have been conducted to evaluate the inactivation effect of low-energy X-rays on more resistant biofilm pathogens on food-contact surfaces.

Keywords: biofilm, bacteria, UV, X-rays, gamma irradiation, efficacy

1. Introduction

Biofilms consist of structured communities of bacteria, embedded in a self-produced polymeric matrix and adherent to inert or living surfaces [1–3]. Biofilm mode of growth is an approach in microorganisms to survive harsh growth conditions. Most microorganisms such as *Pseudomonas aeruginosa* [4], *Staphylococcus aureus* [5], and *Escherichia coli* [6] favor a way of life where the bacterial population is attached to a support, named sessile state, rather than free and isolated in the environment, named planktonic state. The attachment to a surface is a "survival strategy" that allows the bacteria to settle and colonize an environment. This structure represents the normal way of life of a bacterium [7]. This way of life is of great interest for the bacteria since it gives them a resistance to different sources of stress to which planktonic bacteria are sensitive [1]. In effect, bacteria in biofilms are generally well protected against environmental stresses, antibiotics [8], disinfectants, and the host immune system [9] and as a consequence are extremely difficult to eradicate [10]. Therefore, biofilms constitute a protected mode of growth that allows survival in a hostile environment. This strength is essentially due to the biofilm matrix composed of numerous polysaccharides, proteins, and extracellular DNA (eDNA) which is crucial in biofilm structural integrity [11]. Although it is widely accepted that eDNA is released primarily by cell lysis, several studies have shown that other mechanisms of active secretion may coexist [12]. This implies that eDNA is an interesting target in the control of biofilms. Numerous studies have demonstrated that biofilm formation can be prevented by enzymatic degradation of eDNA by DNase [11]. It was reported for *Campylobacter. jejuni* biofilm-attached to stainless steel surfaces that degradation of eDNA by exogenous addition of DNase led to rapid biofilm removal and is likely to potentiate the activity of antimicrobial treatments and thus synergistically aid disinfection treatments, like radiations, antibiotics [13]. For UVC radiation, they target genomic DNA by forming thymidine dimers in RNA and DNA, which can interfere with transcription and replication and thus induce bacterial death [14]. For the extracellular DNA, the formation of thymidine dimers, following exposure to UVC, has no consequence on bacterial multiplication. Therefore, the presence of eDNA in the matrix can only increase the viscosity of the matrix and therefore continues to block the passage of radiation through the biofilm which limits the effectiveness of the radiations.

2. Surfaces colonized by biofilms

Biofilms can colonize a wide range of domains, including essentially industrial sectors, different natural environments (soil, sediment, etc.), and biomedical environments [15]. Many bacteria form clumps at the bottom of the containers. Then, they reach the surface of the liquid-type media. However, some bacteria such as *Salmonella* [16], *E. coli, P. fluorescens,* and *Vibrio cholera* produce rigid or fragile pellicle structures at air-liquid interfaces [17]. Biofilm production by the colonization of the air interface can facilitate and contribute to gas exchange while enabling the acquisition of nutrients and water from the liquid phase. The biofilms at air-liquid interfaces can cause severe problems in industrial water systems [18].

In the medical sector, microbial adhesion resulting in biofilm formation on implanted medical devices is a common occurrence and can lead to serious illness and death [19]. Implanted medical devices like intravascular catheters, urinary catheters, pacemakers, heart valves, stents, and orthopedic implants, normally used for therapeutic purposes, can also be the source of real infectious risks when colonized by bacterial biofilms [20].

3. Biofilms treatment

The majority of available antibacterial treatments have shown their effectiveness against planktonic bacteria. However, these treatments have proven to be ineffective against pathogenic biofilms [21, 22], which are thousands of times more resistant to this type of treatment [23–25]. It is therefore difficult to eradicate biofilms effectively because of the phenomenon of biofilm recalcitrance [22]. Despite the importance of biofilm treatment either in the medical or environmental sectors, studies into the effectiveness of irradiation on biofilm-associated cells are lacking. Therefore, new biofilm treatment strategies are needed, including physical treatments such as radiations. This review presents an overview of bacterial biofilm development and seeks to highlight the efficacy of radiations against bacterial biofilms.

3.1 Continuous UVC irradiation treatment efficiency on biofilms

Though germicidal UV radiation is widely applied for disinfection of water and food from planktonic bacteria, it may also be used to prevent bacterial growth and colonization on surfaces, as biofilms, within engineered systems [26]. Moreover, the UVC-based method is to be of practical use for disinfection of catheters in the clinic, as they are the major sources of infection [27]. However, higher UV doses would be required to inactivate biofilm-bound bacteria than planktonic bacteria because the biofilm would provide some degree of protection from the effects of UVC irradiation [28].

Torkzadeh et al. [26] have developed an experimental device and method to ensure the growth of biofilms in the presence of UV radiation and to measure the resulting reduction in surface biofilm growth. Under optimal growth conditions and after 48 h of growth, the reduction of the bio-volume of the *E. coli* surface is about 95% by a UV intensity of 50.5 $\mu W/cm^2$ at 254 nm, compared to the control biofilms. The UV intensity required for biofilm prevention was greater than that expected due to the UV dose–response of tested bacteria and the cumulative doses applied to the tested surfaces. This results indicate that biofilms can form even under irradiation conditions that should inactivate planktonic cells completely. This is probably due to the protective effects of colloidal material and microbial exudates, that form biofilm matrix.

In water and wastewater infrastructure, biofilms pose a real problem for disinfection. Until now, the majority of ultraviolet (UV) disinfection studies focus on planktonic bacteria, with limited attention given to UV irradiation of biofilms. Among the few outstanding studies, the study of Myriam et al. [29] focused on the study of UVC dose/biofilm production relationship for five *P.aeruginosa* strains, isolated from wastewater. The aim was to evaluate the impact of incremental UVC doses, up to 100 $mJ.cm^{-2}$, on the ability of *Pseudomonas* strains to produce biofilm, knowing that the UV dose equal to 40 $mJ.cm^{-2}$ is the dose recommended for the disinfection of water in Europe and America. The results of this study showed that biofilm production presents a progressive increase in function of an increasing of exposure UVC dose until a threshold UV dose. Moreover, the values of threshold UV doses were different in relation with the response of each bacteria strain to UVC dose (dose/response). This may be explained by the fact that intraspecific difference showed in the UV dose/response relationship is probably dependent on several factors: the degree of DNA damage induced by UV, the speed of induction of DNA repair mechanisms for each tested bacteria. On the other hand, beyond the threshold, a progressive decrease in the production of biofilm correlated with the increase of UV dose was noticed. This decrease in biofilm production can be explained by the fact that the bacterial strains have received a lethal UV dose reducing bacterial sustainability by the accumulation of photoproducts surpassing the capability of bacteria DNA repair mechanisms allowing for consequent, a decrease of biofilm formation and the weakening of this resistant structure.

The UV treatment has evolved a lot since the development of UV light sources from the conventional mercury lamp to the light-emitting diode (LED). It was established that pulsed UV can be more effective than a continuous emitting mode to control biofilms. Moreover, adaptable UV LED is promising to control biofilms in the water distribution system, according to the review of [30]. Luo et al. [30] have, recently, demonstrated that pulsed UV can be more effective than a continuous emitting mode to control biofilms, on one side and that a selective combination of UV LED wavelengths allows targeting damaged biofilm components, on the other hand.

In the medical sector, an application of radiation treatment on catheters looks promising. In this context, the study of Jimmy Bak et al. [31], who proposed a method for disinfecting the inner surface of catheters biofilm, has demonstrated that mean killing rates were 89.6% for 0.5 min exposure, 98% for 2 min exposure, and 99% for 60 min exposure. About 99% of the cells were killed with a UVC dose of 15 kJ m^{-2}. This dose, which is 100 to 1000 times higher than the lethal dose required for planktonic cells, is assumed to be the maximum dose necessary to avoid contamination of newly inserted catheters. The need for high doses to kill mature biofilm and the limited effect of currently available UVC light sources result in a relatively long treatment time of about 60 minutes, hence the need for new UV sources like UV LED.

Recently, Jimmy Bak et al. [31] have tested a newly developed UVC disinfection device, which can be connected to a Luer catheter hub, on polymer tubes contaminated with a wide range of either bacterium, including *S. aureus*, *E. coli,* and *P. aeruginosa* and fungi like *Candida albicans*. Their results have shown no viable counts after 2 min of radiation for bacteria. Whereas, Killing of *C. albicans* needs more than 20 minutes to be obtained in a UVC absorbing suspension.

On any type of surface contaminated by biofilm, the effectiveness of UVC light in inactivating biofilm-forming microorganisms is mainly due to the ability of DNA molecules to absorb UV photons between 200 and 300 nm, with an absorption peak at 260 nm, at first. Then, this uptake causes damage to the DNA by altering the pairing of nucleotide bases, creating new bonds between adjacent nucleotides on the same DNA strand. This damage occurs particularly between pyrimidine bases [32].

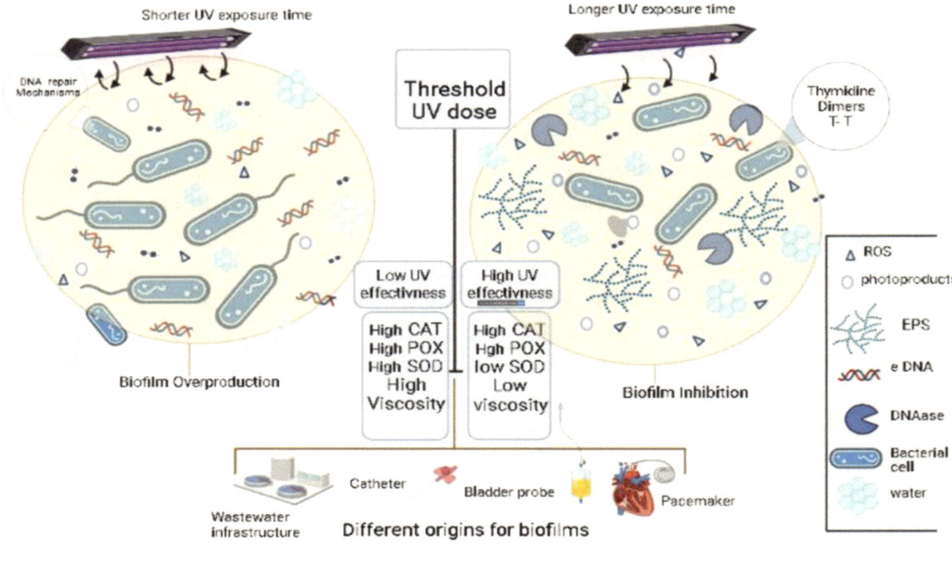

Figure 1.
UVC time/biofilm production relationship and DNA repair mechanisms.

Therefore, to limit UV damages, bacteria generally possess molecular mechanisms to restore DNA lesions [33], which preserve the irradiated biofilm, from damage due to UVC exposure. This repair mechanism has been shown to be effective up to a threshold dose-related to a maximum accumulation of photoproducts and of reactive oxygen species, which can no longer be managed by this mechanism [29]. Our study in 2016, confirmed the oxidative stress through ROS accumulation, following UVC exposure, and has demonstrated that, in the enzymatic ROS-scavenging pathways, catalase and peroxidase enhancement improved the resistance of *P. aeruginosa* treated with incremental UV-C doses. However, longer exposure to UV-C rays inhibited SOD activity. This result confirms that SOD cannot efficiently remove superoxide radicals that accumulated in cells of *P. aeruginosa* at longer irradiation time and further confirms the inability of the repair system besides the ROS-scavenging pathways to deal with photoproducts and ROS accumulation, respectively [34, 35].

We can then conclude that the resistance of bacteria to UVC treatment remains at dose limits. Beyond these doses, there is an exhaustion of the repair system and a sure bacterial death. Hence the need to exceed the dose limits in order to escape bacterial resistance (**Figure 1**).

3.2 UV LED irradiation treatment efficiency on biofilms

UV LEDs are emerging as competitive light sources because of advantages such as the possible selection of combined-wavelength UV LED [30], adjustable emitting mode, and the designable configuration that facilitate their incorporation into confined spaces. Therefore, UV LEDs offer new solutions to prevent biofilm formation on inaccessible surfaces, such as medical and food equipment and, potentially, sanitary facilities, to limit nosocomial infections. These results imply that surfaces more exposed to bacterial colonization require adequate UVC irradiation to prevent biofilm establishment. Furthermore, continuous surface irradiation may be insufficient as a sole source for biofilm prevention in many circumstances [26]. However, problems with low wall plugs and reliable power supplies still limit the effectiveness of UV LEDs, which further enlightens the prospective of UV in dealing with the biofilm issue in water infrastructure and also in the medical sector.

In this context, the study of Aikaterini et al. [36] on *P. aeruginosa* biofilms at different growth stages, within 24, 48, and 72 h of growth, was conducted to judge the effectiveness of ultraviolet B (UVB), at 296 nm and ultraviolet C (UVC) irradiation, with central wavelength at 266 nm, two different light-based treatments. The effectiveness of the UVB and UVC irradiations was quantified by counting colony-forming units. For UV exposure, a type of AlGaN light-emitting diodes (LEDs) was used to distribute UV irradiation on the biofilms. For *P.aeruginosa* biofilms, it appears that UVB irradiation is much more effective than UVC radiation for the inactivation of mature biofilms. The fact that UVB at 296 nm is present in daylight and has such a disinfecting capacity on biofilms opens the way to the treatment of infectious pathologies [36].

In parallel, the study of Gora et al. [37] has demonstrated that UV LED irradiation at 265 nm achieved 1.3 log inactivation of biofilm-bound *P. aeruginosa* at a UV dose of 8 mJ/cm^2. This inactivation level is lower than those that have been reported by researchers using UVC LEDs to inactivate planktonic *P. aeruginosa*, a finding that can be explained by the higher resistance of biofilm-bound bacteria to UV inactivation.

Moreover, the combination of UV LED and Blue laser was tested on *S. aureus* biofilm and gave the highest biofilm reduction of about 80.57%. It was then demonstrated that it to be the best choice to eradicate more biofilm [38].

Concerning the effect of radiations on biofilm matrix, it is well established that bacteria enclosed in a layer of exopolysaccharides are protected by 13% from UVC radiation. It was also confirmed that absorption of UV light by the alginate, an important matrix molecule, translated into a higher survival rate than observed with planktonic cells, for the same UV dose [39]. In effect, alginate water retention seems to be at the origin of the obvious ability to survive severe environments, like UVC exposure. On the other hand, the effect of UV LED on exopolysaccharides (EPS) has not been extensively studied, but it is predicted to be similar to the effect of continuous UVC on EPS. It is then assumed that following the prolonged exposure to UVC radiation, the production of EPS is stimulated [34]. Moreover, in the framework of the development of a profitable strategy to improve the EPS yield, UV irradiation mutagenesis of *Bacillus licheniformis* significantly improved the EPS yield. Significantly enhanced yield (>3-folds) of EPS after UVC exposure can only confirm the stimulating effect of UVC radiation on the production of EPS, to ensure better protection against UVC rays and then bacterial survival [40] (**Table 1**).

3.3 Ionizing radiation treatment efficiency on biofilms

Ionizing radiation is a non-thermal destruction technique that inactivates pathogens that may contaminate certain food products, by exposing them to irradiation sources such as high-energy X-rays at about 5 MeV, gamma rays at about 2.5 MeV, or electron beams at about 10 MeV [41]. Compared to these conventional high-energy irradiation techniques, low-energy X-rays have a higher linear energy transfer (LET) value, resulting in a greater relative biological effect (RBE) [42]. Some previous studies have shown that low-energy X-rays is effective in destroying certain planktonic germs such as *E. coli* O157:H7, *Salmonella*, *Listeria monocytogenes*, and *Shigella flexneri* [43–45]. However, few studies have investigated the effect of low-energy X-rays on more resistant pathogens in mono-microbial or poly-microbial cultured biofilms and on food contact surfaces.

Despite of this, we could not simply conclude that low-energy X-rays destroyed EPS in biofilm. Therefore, we could at least postulate that low-energy X-rays irradiation weakened EPS structure in biofilm. Typical EPS mainly comprises

Light sources	Microorganisms	UV dose	Inactivation	Reference
UV LED	Biofilm-bound *P.aeruginosa*	8 mJ/cm^2 (265 nm)	1.3 log reduction	Gora et al. [37]
	Less mature *P. aeruginosa* biofilms (24 h grown)	72–10,000 J/m^2	1 log reduction	Aikaterini et al. [36]
	Mature biofilms (48 and 72 h grown)	20 000 J/m^2	0.8 _ 0.3 log reductions	
Continuous low-intensity UVC irradiation	*E. coli*	50.5 μW/cm^2 (254 nm)	95%	Torkzadeh et al. [26]
	Catheter biofilm	15 kJ/m^2	99%	Bak et al. [27, 31]
	Wastewater	40 mJ/cm^2		

Table 1.
UV doses required for the treatment of biofilms for different microorganisms.

homopolymers like cellulose and dextran and heteropolymers of alginate, emulsan, gellan, and xanthan, which maintain the stability of the biofilm matrix [46]. Ionizing irradiation can break down glycosidic bonds and consequently degrade polysaccharides and destabilize the biofilm [47].

Similarly, some in vitro studies also showed that the direct effect of radiation on oral *C. albicans* cells leads to a rapid proliferation ability, increase of virulent factors, and resistance to drugs [48]. Moreover, irradiated *Klebsiella oxytoca* strains of oral origin were more virulent than non-irradiated ones [49]. All of these results indicated that direct exposure of X-rays can affect the virulence of oral bacteria microbes even at therapeutic doses [50].

Concerning gamma irradiation, it is an established technology of well-documented safety and efficacy for the inactivation of pathogenic microorganisms such as *Salmonella* [51, 52]. Recently, gamma-ray sterilization was proven to be a viable method of sterilization of conducting polymer-based biomaterials for biomedical applications [53].

The study of [54] has demonstrated that in bacterial biofilms attached to stainless steel, gamma irradiation at a dose of 10.0 kGy reduced the counts of *S. aureus* attached for 1 hr. and overnight by ≥5.1 and 5.0 log CFU/cm^2, respectively. Gamma irradiation at a dose of 1.0 kGy reduced the counts of *P. aeruginosa* counts to below the limit of detection (<2logCFU/cm^2).

Concerning food sterilization, *Salmonella* is a problematic bacterium due to its biofilm resistance to chemical sanitizing treatments. Ionizing radiation is known to be used to inactivate *Salmonella* on a variety of foods and contact surfaces in the food industry. The relative efficacy of the process against biofilm-associated cells versus free-living planktonic cells was tested for three food-borne-illness-associated isolates of *Salmonella*, by by Niemira and Solomonet [55]. They demonstrated that the dose of radiation required to reduce 90% (D10 values) of *Salmonella enterica* serovar *Anatum* was not significantly different between biofilm-forming bacteria (0.645 kGy) and planktonic cells (0.677 kGy). In contrast, biofilm-forming cells of *S. enterica* serovar *Stanley* were significantly more sensitive to ionizing radiation, with a D10 of 0.531, than planktonic cells, with a D10 of 0.591 kGy. D10 values of *S. enterica* serovar *Enteritidis* were similarly decreased for biofilm-associated cells (0.436 kGy) in comparison to planktonic cells (0.535 kGy). The anti-microbial efficiency of ionizing

Light sources	Microorganisms	Dose	Inactivation	Reference
Gamma irradiation	Biofilm-bound *S. aureus* attached for 1 h	10.0 kGy	≥5.1 log CFU/cm^2 5.0 log	[54]
	Biofilm-bound *S. aureus* attached overnight		5.0 log CFU/cm^2	
	Biofilm-bound *P. aeruginosa*	1.0 kGy	<2 log CFU/cm^2	
	Salmonella enterica serovar *A*	0.645 kGy	1 log	[31]
	S. enterica serovar Stanley	0.531 kGy	1 log	
	S. enterica serovar Enteritidis	0.436 kGy	1 log	

Table 2.
Gamma irradiation and doses required for the treatment of biofilms for different microorganisms.

radiation is therefore guaranteed in the treatment of bacteria associated with a biofilm. Ben Miloud YahiaYahia [52] proposed that the biofilm-forming abilities could be reduced with temperature decrease and increasing gamma radiation doses (**Table 2**).

4. Conclusion

This study has demonstrated that ionizing and non-ionizing radiation effectively reduces the populations of both planktonic and biofilm-associated bacteria. However, biofilms are confirmed to be more difficult to eradicate and require enhanced doses for their eradication. It was also confirmed that radiation sensitivity is microorganism specific. Likewise, the influence on radiation sensitivity of the cultured state of the organism, between planktonic and biofilm-associated, is also isolate specific, confirmed for gamma-treated *Salmonella*. But also, the stage of biofilm growth seems to affect the effectiveness of radiations treatment, as confirmed for *Pseudomonas* and *Staphylococcus* biofilms. In general, these results show that, in contrast to chemical antimicrobial treatments, the antimicrobial efficacy of radiation is preserved or enhanced when treating biofilm-associated bacteria, compared to planktonic cells.

Author details

Salma Kloula Ben Ghorbal[1]*, Rim Werhani[1] and Abdelwaheb Chatti[2]

1 Laboratoire de Traitement et Valorisation de Rejets Hydriques, Center des Recherches et des Technologies des Eaux (CERTE), Université de Carthage, Tunis, Tunisia

2 Faculté des Sciences de Bizerte, Unité de Biochimie des Lipides et interaction des macromolécules en Biologie, Laboratoire de Biochimie et Biologie Moléculaire, Université de Carthage, Tunis, Tunisia

*Address all correspondence to: salmakloula@yahoo.fr

References

[1] Costerton JW, Stewart PS, Greenberg EP. Bacterial biofilms: A common cause of persistent infections. Science. 1999;**284**(5418):1318-1322

[2] Donlan RM, Costerton JW. Biofilms: Survival mechanisms of clinically relevant microorganisms. Clinical Microbiology Reviews. 2002;**15**(2):167-193

[3] Hall-Stoodley L, Hu FZ, Gieseke A, Nistico L, Nguyen D, Hayes J, et al. Direct detection of bacterial biofilms on the middle-ear mucosa of children with chronic otitis media. JAMA. 2006;**296**(2):202-211

[4] Gellatly SL, Hancock RE. Pseudomonas aeruginosa: New insights into pathogenesis and host defenses. Pathogenic Diseases. 2013;**67**:159-173. DOI: 10.1111/2049-632X.12033

[5] Gordon RJ, Lowy FD. Pathogenesis of methicillinresistant *Staphylococcus aureus* infection. Clinical Infectious Diseases. 2008;**46**:S350-S359. DOI: 10.1086/533591

[6] Beloin C, Roux A, Ghigo JM. *Escherichia coli* biofilms. Current Topical Microbiology. 2008;**322**:249-289. DOI: 10.1007/978-3-540-75418-3_12

[7] Klinger E, Bouchard S, Légeron P, Roy S, Lauer F, Chemin I, et al. Virtual reality therapy versus cognitive behavior therapy for social phobia: A preliminary controlled study. Cyberpsychology & Behavior. 2005;**8**(1):76-88

[8] Høiby N, Bjarnsholt T, Givskov M, Molin S, Ciofu O. Antibiotic resistance of bacterial biofilms. International journal of antimicrobial agents. 2010;**35**(4):322-332

[9] Jensen PO, Givskov M, Bjarnsholt T, Moser C. The immune system vs.*Pseudomonas aeruginosa* biofilms. FEMS Immunology and Medical Microbiology. 2010;**59**(3):292-305

[10] Burmølle M, Thomsen TR, Fazli M, Dige I, Christensen L, Homøe P, et al. Biofilms in chronic infections–a matter of opportunity–monospecies biofilms in multispecies infections. FEMS Immunology & Medical Microbiology. 2010;**59**(3):324-336

[11] Okshevsky M, Regina VR, Meyer RL. Extracellular DNA as a target for biofilm control. Current Opinion in Biotechnology. 2015;**33**:73-80

[12] Wu J, Xi C. Evaluation of different methods for extracting extracellular DNA from the biofilm matrix. Applied and Environmental Microbiology. 2009;**75**(16):5390-5395

[13] Brown HL, Hanman K, Reuter M, Betts RP, Van Vliet AH. *Campylobacter jejuni* biofilms contain extracellular DNA and are sensitive to DNase I treatment. Frontiers in Microbiology. 2015;**10**(6):699

[14] Pereira RV, Bicalho ML, Machado VS, et al. Evaluation of the effects of ultraviolet light on bacterial contaminants inoculated into whole milk and colostrum, and on colostrum immunoglobulin G. Journal of Dairy Science. 2014;**97**(5):2866-2875. DOI: 10.3168/jds.2013-7601

[15] Bryers JD. Medical biofilms. Biotechnology and Bioengineering. 2008;**100**(1):1-8

[16] Scher K, Romling U, Yaron S. Effect of heat, acidification, and chlorination on *S. enterica* serovar Typhimurium

cells in a biofilm formed at the air-liquid interface. Applied and environmental microbiology. 2005;**71**(3):1163-1168

[17] Armitano J, Méjean V, Jourlin-Castelli C. G ram-negative bacteria can also form pellicles. Environmental Microbiology Reports. 2014;**6**(6):534-544

[18] Rühs PA, Böcker L, Inglis RF, Fischer P. Studying bacterial hydrophobicity and biofilm formation at liquid–liquid interfaces through interfacial rheology and pendant drop tensiometry. Colloids and Surfaces B: Biointerfaces. 2014;**117**:174-184

[19] Habash M, Reid G. Microbial biofilms: Their development and significance for medical device-related infections. Journal of Clinical Pharmacology. 1999;**39**:887-898. DOI: 10.1177/00912709922008506

[20] Francolini I, Donelli G. Prevention and control of biofilmbased medical-device-related infections. FEMS Immunology Medicine and Microbiology. 2010;**59**:227-238. DOI: 10.1111/j.1574-695X.2010.00665.x

[21] Costerton JW, Cheng KJ, Geesey GG, Ladd TI, Nickel JC, Dasgupta M, et al. Bacterial biofilms in nature and disease. Annual Review of Microbiology. 1987;**41**:435-464. DOI: 10.1146/annurev.mi.41.100187.002251

[22] Lebeaux D, Ghigo J-M, Beloin C. Biofilm-related infections: Bridging the gap between clinical management and fundamental aspectsof recalcitrance toward antibiotics. Microbiology and Molecular Biology Reviews. 2014;**78**:510-543. DOI: 10.1128/MMBR.00013-14

[23] Davies D. Understanding biofilm resistance to antibacterial agents. Natural Review Drug Discovery. 2003;**2**:114-122. DOI: 10.1038/nrd1008

[24] Luppens SBI, Reij MW, van der Heijden RWL, Rombouts FM, Abee T. Development of a standard test to assess the resistance of *S. aureus* biofilm cells to disinfectants. Applied Environmental Microbiology. 2002;**68**:4194-4200. DOI: 10.1128/AEM.68.9.4194-4200.2002

[25] Stewart PS, andWilliam Costerton, J. Antibiotic resistance of bacteria in biofilms. Lancet. 2001;**358**:135-138. DOI: 10.1016/S0140-6736(01)05321-1

[26] Torkzadeh H, Zodrow KR, Bridges WC, Cates EL. Quantification and modeling of the response of surface biofilm growth to continuous low intensity UVC irradiation. Water Research. 2021;**1**(193):116895

[27] Bak J, Ladefoged SD, Tvede M, Begovic T, Gregersen A. Dose requirements for UVC disinfection of catheter biofilms. Biofouling. 2009;**25**(4):289-296

[28] Da Silva Araújo, Paula Alexandra. Biofilm Control with Antimicrobial Agents: The Role of the Exopolymeric Matrix. Diss. Universidade do Porto (Portugal), 2014.

[29] Myriam BE, Khefacha S, Maalej L, Imen DA, Hassen A. Effect of ultraviolet, electromagnetic radiation subtype C (UV-C) dose on biofilm formation by *Pseudomonas aeruginosa*. African Journal of Microbiology Research. 2011;**5**(25):4353-4358

[30] Luo X, Zhang B, Lu Y, Mei Y, Shen L. Advances in application of ultraviolet irradiation for biofilm control in water and wastewater infrastructure. Journal of Hazardous Materials. 2022;**421**:126682

[31] Bak J, Begovic T, Bjarnsholt T, Nielsen A. A UVC device for intra-luminal disinfection of catheters: In vitro tests on soft polymer tubes contaminated

with *Pseudomonas aeruginosa*, *Staphylococcus aureus*, *Escherichia coli* and *Candida albicans*. Photochemistry and Photobiology. 2011;**87**(5):1123-1128

[32] Myriam BS, Otaki M, Shinobu K, Abdennaceur H. Detection of active E. coli after irradiation by pulsed UV light using a Q phage. African Journal of Microbiology Research. 2010;**4**(11):1128-1134

[33] Rastogi RP, Kumar A, Tyagi MB, Sinha RP. Molecular mechanisms of ultraviolet radiation-induced DNA damage and repair. Journal of Nucleic Acids. 2010;**2010**:1-32

[34] Kloula Ben Ghorbal S, Rim W, Abdelwaheb C. Role of Mn-cofactored superoxide dismutase in the aptitude of P. aeruginosa to form biofilm, under UV-C radiations. Journal of New Sciences, Agriculture and Biotechnology. 2022;**88**; article in press

[35] Salma KB, Lobna M, Sana K, Kalthoum C, Imene O, Abdelwaheb C. Antioxidant enzymes expression in *P. aeruginosa* exposed to UV-C radiation. Journal of Basic Microbiology. 2016;**56**(7):736-740

[36] Argyraki A, Markvart M, Bjørndal L, Bjarnsholt T, Petersen PM. Inactivation of Pseudomonas aeruginosa biofilm after ultraviolet light-emitting diode treatment: A comparative study between ultraviolet C and ultraviolet B. Journal of Biomedical Optics. 2017;**22**(6):065004

[37] Gora SL, Rauch KD, Ontiveros CC, Stoddart AK, Gagnon GA. Inactivation of biofilm-bound Pseudomonas aeruginosa bacteria using UVC light emitting diodes (UVC LEDs). Water research. 2019;**15**(151):193-202

[38] Astuti SD, Rasheed A, Rahmawati I, Puspita PS, Kholimatussa'diah S, Putra AP. Combination of blue laser exposure with UV-LED to improve antimicrobial effects on *Staphylococcus aureus* biofilm. Malaysian Journal of Medicine and Health Sciences. 2020;**16**(S4):6-11

[39] Elasri MO, Miller RV. Study of the response of a biofilm bacterial community to UV radiation. Applied and Environmental Microbiology. 1999;**65**(5):2025-2031

[40] Asgher M, Urooj Y, Qamar SA, Khalid N. Improved exopolysaccharide production from Bacillus licheniformis MS3: Optimization and structural/functional characterization. International Journal of Biological Macromolecules. 2020;**151**:984-992

[41] Zhang J, Xie Y, Li Y, Shen C, Xia Y. Covid-19 screening on chest x-ray images using deep learning based anomaly detection. 2020;**27**(27). arXiv preprint arXiv:2003.12338

[42] Hill MA. The variation in biological effectiveness of X-rays and gamma rays with energy. Radiation Protection Dosimetry. 2004;**112**(4):471-481

[43] Aleid SM, Dolan K, Siddiq M, Jeong S, Marks B. Effect of low-energy X-ray irradiation on physical, chemical, textural and sensory properties of Dates. International journal of food science & technology. 2013;**48**(7):1453-1459

[44] Jeong H, Kim SH, Han SS, Kim MH, Lee KC. Changes in membrane fatty acid composition through proton-induced fabF mutation enhancing 1-butanol tolerance in E. coli. Journal of the Korean Physical Society. 2012;**61**(2):227-233

[45] Moosekian SR, Jeong S, Ryser ET. Inactivation of sanitizer-injured *Escherichia coli* O157: H7 on baby spinach using X-ray irradiation. Food Control. 2014;**36**(1):243-247

[46] Svenningsen NB et al. The biofilm matrix polysaccharides cellulose and alginate both protect Pseudomonas putida mt-2 against reactive oxygen species generated under matric stress and copper exposure. Microbiology. 2018;**164**(6):883-888

[47] Lembre P, Lorentz C, Di Martino P. Exopolysaccharides of the biofilm matrix: A complex biophysical world. The complex world of polysaccharides. 2012;**31**:371-392

[48] Ueta E, Tanida T, Osaki T. A novel bovine lactoferrin peptide, FKCRRWQWRM, suppresses Candida cell growth and activates neutrophils. The Journal of Peptide Research. 2001;**57**(3):240-249

[49] Vanhoecke BW, De Ryck TR, De Boel K, Wiles S, Boterberg T, Van de Wiele T, et al. Low-dose irradiation affects the functional behavior of oral microbiota in the context of mucositis. Experimental Biology and Medicine. 2016;**241**(1):60-70

[50] Wang Z, Zhou Y, Han Q, Ye X, Chen Y, Sun Y, et al. Synonymous point mutation of gtfB gene caused by therapeutic X-rays exposure reduced the biofilm formation and cariogenic abilities of *Streptococcus mutans*. Cell & Bioscience. 2021;**11**(1):1-3

[51] Rezende AC, Igarashi MC, Destro MT, Franco BD, Landgraf M. Effect of gamma radiation on the reduction of Salmonella strains, Listeria monocytogenes, and Shiga toxin–producing *Escherichia coli* and sensory evaluation of minimally processed spinach (Tetragonia expansa). Journal of Food Protection. 2014;**77**(10):1768-1772

[52] Ben Miloud Yahia N, Ghorbal SKB, Maalej L, Chatti A, Elmay A, Chihib NE, et al. Effect of temperature and gamma radiation on Salmonella Hadar biofilm production on different food contact surfaces. Journal of Food Quality. 2018

[53] Kim S, Jeong JO, Lee S, Park JS, Gwon HJ, Jeong SI, et al. Effective gamma-ray sterilization and characterization of conductive polypyrrole biomaterials. Scientific Reports. 2018;**8**(1):1-10

[54] Kim JH, Jo CR, Rho YT, Lee CB, Byun MW. Comparison of gamma irradiation and sodium hypochlorite treatments to inactivate *Staphylococcus aureus* and *Pseudomonas aeruginosa* biofilms on stainless steel surfaces. Food Science and Biotechnology. 2007;**16**(2):315-319

[55] Niemira BA, Solomon EB. Sensitivity of planktonic and biofilm-associated Salmonella spp. to ionizing radiation. Applied and Environmental Microbiology. 2005;**71**(5):2732-2736

Chapter 14

Antifouling Strategies-Interference with Bacterial Adhesion

Zhen Jia

Abstract

Biofilm refers to a viable bacterial community wrapped in self-produced extracellular polymeric substances (EPS) matrix. As bacteria shielded by EPS are viable and can resist broad hostile environments and antimicrobial agents, biofilm poses a massive challenge to industries and human health. Currently, biofilm has accounted for widespread and severe safety issues, infections, and economic loss. Various antifouling strategies have been designed and developed to prevent biofilm formation. As bacterial biofilm is perceived as a dynamic multistage process in which bacterial attachment on solid surfaces is the prerequisite for biofilm formation, the interference with the attachment is the most promising environmentally benign option to antifouling. The chapter summarizes and discusses the antifouling strategies that interfere with the adhesion between bacteria and substrate surfaces. These strategies primarily focus on modifying the substrate surface's topographical and physicochemical properties.

Keywords: biofilm, antifouling, modification, topography, physicochemical property

1. Introduction

Bacterial biofilm is a structured community of bacterial cells within a self-generated hydrated extracellular polymetric substance (EPS) matrix anchored to a surface [1]. The physical channels formed during biofilm formation facilitate nutrients, air, and water to penetrate and distribute to cells [2], promoting microbial reproduction, metabolism, and EPS secretion. EPS is a biopolymer produced by bacterial cells following surface attachment, serving as a house or shelter for cells [3, 4]. It mainly consists of a wide variety of exopolysaccharides (40–95%), proteins (1–60%), nucleic acids (1–10%), and lipids (1–40%) [2, 5], which are critical factors to enhance bacterial adhesion behavior. On the one hand, EPS possesses mechanical stability, protecting cells from mechanical damages and shear and providing a functional microenvironment for bacterial growth [6]. On the other hand, EPS creates a physical barrier that enables bacteria inside to survive under harsh conditions and to resist antibiotics and antimicrobial agents [7].

Biofilm, different from planktonic cells, is a self-protection growth pattern of bacteria. Over 99% of the world's bacteria present as a form of biofilm [8], broadly distributing on broad infrastructure elements, systems, and devices. Due to strong self-protection ability and resistance to harsh conditions, the unwanted biofilms pose severe threats and challenges to human health and industries, such as the transmissions of disease and infections and interferences of system functions and decreases in the endurance of surfaces and devices [9]. In the medical system, bacteria can form biofilm in healthcare settings (such as sinks, drains, and showers) and medical devices (such as surgical instruments and implantable biomedical devices). Up to 80% of hospital-acquired infections (HAIs) contribute to biofilm infections [10]. Such HAIs affect about 10% of all hospital patients in the United States and lead to nearly 100,000 deaths annually [11, 12]. In the food industry, biofilms have been widely reported on food surfaces, food contact surfaces, and processing systems, leading to product contamination, cross-contamination, food withdrawal, and disease outbreaks [13–15]. In the marine system, biofilm accumulation accelerates corrosion on marine vehicles, resulting in equipment clogging, damage, and roughness [16]. In addition, biofilm increases hydrodynamic drag, which adversely interferes with equipment performance and increases fuel expenditure up to 45% [17]. The economic losses caused due to biofilm are also enormous. For US Navy alone, the estimated fuel cost per annum is around $500 million, of which $75–100 million account for drag induced by fouling organisms [18]. Therefore, it is critical and urgent to prevent biofilm formation.

Biofilm formation is a dynamic process, typically containing five stages: initial reversible attachment, irreversible attachment, micro-colony formation, biofilm formation and maturation, and dispersion. Among them, initial reversible attachment is critical. In this stage, bacteria actively seek and anchor to surfaces relying on the motility of planktonic cells using extracellular organelles and proteins (such as pili, curli fibers, flagella, and outer membrane proteins), cells' gravitational transportation, physical forces between cells and surfaces (such as van der Waals forces, steric interactions, and electrostatic interactions), and hydrodynamic forces of the surrounding environment [19, 20]. Additionally, other forces include acid-base interactions at a very short range, around 5 nm range, responsible for bond formation and hydrophobic forces [21] and divalent cations responsible for crosslinking between bacterial surface polymers that aid in matrix stabilization [22]. The attachment of a microbial cell to a surface is called adhesion [23]. The adhesion is reversible as bacteria are loosely attached. The attached cells still exhibit Brownian motion and can easily dissociate back to planktonic forms. The adhesion of bacteria is primarily influenced by various factors, including surface properties, environmental conditions (like pressure and temperature), and bacterial orientation [24].

Based on the process of biofilm formation, it is worth noting that bacterial adhesion is an initial prerequisite for biofilm formation. After being attached to surfaces, bacterial cells initiate to reproduce and ultimately grow into a biofilm, demonstrating that bacterial adhesion is the fundamental and critical step responsible for biofilm formation. Therefore, inhibiting bacterial adhesion is the desirable and key antifouling approach to prevent biofilm formation. The adhesion of bacteria is mainly affected by various factors, including surface properties of substrates, physicochemical properties of microbes, and environmental conditions [25]. As the properties of substrate surfaces are changeable and can be manipulated depending on the purpose, antifouling approaches to control biofilm formation mostly focus on modifying surface properties, including surface topography and physicochemistry (**Table 1**).

The purpose of this chapter is to provide insights into antifouling strategies related to the topographical and physicochemical properties of substrate surfaces in the prevention of cell adhesion and to elucidate corresponding theoretical mechanisms. This chapter also covers the main challenges and future trends of antifouling materials.

Strategies	Properties	Coating
Physicochemical modification	Hydrophilic surface	Polymers: PEO, PEG, OEG, poly-HEMA, dextran, phosphatidylcholines, poly (acrylic acid), etc.
		Nanoparticles: TiO_2, SiO_2, ZnO, Fe_3O_4, silver nanoparticles, etc.
	Hydrophobic surface	Silicone-based coatings
		Fluorine-based coatings
		Sol-gel
		Biosurfactants: surfactin & pseudofactin
		Organic materials: polydimethylsiloxane, polyethylene, polystyrene, polyalkylpyrrole, etc.
		Inorganic materials: ZnO and TiO_2 etc.
	Chemical properties	Metal ions and their compounds: Zn, ZnO, Cu, CuO, Mg, MgO, TiO_2, etc.
		Biomacromolecules: proteins/peptides, polysaccharides, antibodies, etc.
Topographic modification	Micro-scale topography	
	Nano-scale topography	

Table 1.
Surface modification techniques.

2. Physicochemical modification strategy

It is well documented that bacterial adhesion can be effectively tuned and reduced by altering surface physicochemical properties using chemically active antifouling coatings [26, 27]. Currently, various coatings have been extensively reported for their effectiveness in preventing bacterial initial adhesion.

2.1 Surface energy

Surface energy is the binding or interfacial attractive force between materials and solid substrates [28]. It is an essential physicochemical property of a solid surface. Many studies have demonstrated that changing surface energy was related to affecting bacterial adhesion [29, 30]. Baier analyzed the relationship between surface energy and bacterial adhesion, known as the Baier curve depicted in **Figure 1** [31]. According to the curve, bacterial adhesion is minimized when the surface energy of a substrate is in the range of 20–30 mN/m (the lowest values), while antifouling occurs when surface energy is higher than 70 mN/m [32, 33].

Surface energy represents the degree to which water can bind on the surface [34] and can be determined by contact angle (θ) [35]. θ characterizes the ability of water to maintain contact with a solid surface. 'Hydrophilic surface ($\theta < 90°$)' and 'hydrophobic surface ($\theta \geq 90°$)' are two common terms to describe the incongruous behavior of water on solid surfaces [36]. Hydrophilic surfaces are surfaces with high surface energy, while hydrophobic surfaces are surfaces with low surface energy [37].

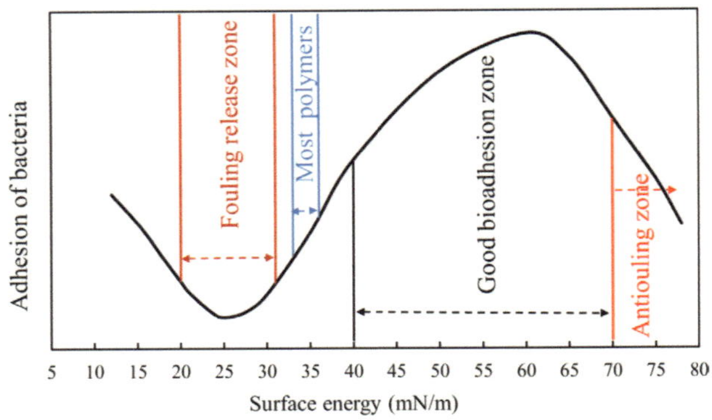

Figure 1.
Correlation between bacterial adhesion and surface energy (Baier curve).

2.1.1 Hydrophilic surfaces

Hydrophilic surfaces can be successfully fabricated by functionalizing with polymers or nanoparticles. Polymers include poly (ethylene oxide) (PEO), poly (ethylene glycol) (PEG), oligo (ethylene glycol) (OEG), dextran, phosphatidylcholines, poly (acrylic acid), and poly-(2-hydroxyethyl methacrylate) (poly-HEMA), etc. [38]. Nanoparticles cover TiO_2, SiO_2, ZnO, Fe_3O_4, and silver nanoparticles [39–42]. The presence of −CH2− CH2−O− structure and C−C−C linkage enable PEG to be highly water-soluble [43].

Many researchers have reported the antiadhesion ability of hydrophilic surfaces. An increase in surface hydrophilicity can reduce bacterial adhesion [44]. Dong et al. indicated that PEG-modified SS exhibited higher hydrophilicity than bare stainless steel (SS), leading to a 96% reduction in *Listeria monocytogenes* attachment [45]. ZnO nanoparticles, composed of hydrophilic groups like −OH, $-SO_3H$, and −COOH, possess strong hydrophilicity [46]. The increased hydrophilicity derived from ZnO nanoparticles promoted antifouling properties of poly (ether sulfone) surface towards *Escherichia coli* (*E. coli*) and *Staphylococcus aureus* (*S. aureus*) [47].

Superhydrophilic surfaces hold near-zero water contact angles ($\theta < 5°$) and exhibit outstanding antifouling properties. Superhydrophilic surfaces can be developed by hydrophilic functionalities, such as metal oxides including TiO_2, ZnO, SiO_2, SnO_2, CuO, and WO3, by applying various fabrication methods (e. g. UV irradiation, plasma, sol-gel self-assembly, etching, and spay/spin/dip-coating) [48–50]. TiO_2 and ZnO are the primary metal oxides to create superhydrophilic film due to their photoinduced self-cleaning property [51, 52]. SiO_2 is also frequently used due to its low price and easy to reach [53]. The adhesion number of *E. coli* cells on superhydrophilic TiO2 coated surface was approximately 45% lower than the surface without coating [54]. Qian et al. prepared superhydrophilic film on the 316L stainless steel surface using methoxy-polyethylene-glycol thiol [55]. The surface showed excellent superhydrophilicity with a water contact angle of zero and exhibited enhanced and more durable antibacterial performances against *E. coli* and *S. aureus* [55].

The antiadhesion mechanism of the (super)hydrophilic surface contributes to forming a highly hydrated layer. Hydrophilic compounds on substrate surfaces, such as PEG or OEG, can strongly bond water molecules, connecting each chain through ether oxygen and generating a thin water film (a highly hydrated layer) between bacteria and surface, which physically blocks bacterial adhesion (as shown in **Figure 2A**) [56, 57]. In addition, the number of anchor sites can be effectively diminished by the water layer [58]. The more hydrophilic surface is, the more resistant it is to the adhesion of bacteria [59].

2.1.2 Hydrophobic surfaces

Hydrophobic coatings, such as silicone- or fluorine-based coatings, polydimethylsiloxane (PDMS), and sol-gel, enable the surface to be more hydrophobic [60]. Besides, some biosurfactants, like surfactin secreted by genus *Bacillus* strains and pseudofactin produced by *Pseudomonas fluorescens*, have also been verified to successfully promote surface hydrophobicity [61, 62]. Extensive studies demonstrated that hydrophobicity was closely associated with the antiadhesive ability of surfaces. The adhesion-resistant ability of hydrophobic surfaces is attributed to low surface energy. Microbial adhesion is less to low-energy surfaces and more accessible to clean because of weaker binding at the interface [63]. Zhao et al. compared bacterial adhesion behavior on hydrophobic surfaces with various surface energy, indicating that the number of *E. coli* attachments was significantly reduced when surface energy ranged between 20 and 30 mJ/m^2 [64]. By spraying hydrophobic perfluoroalkoxy/nano-silver coatings onto aluminum substrates, Zhai et al. found that besides contact killing of silver ions, the hydrophobic surface property could synergistically prevent the adhesion of *E. coli* [65]. With the presence of surfactin coating (surface energy is roughly 27 mN/m), stainless steel, polypropylene, and polyvinyl chloride could effectively prohibit adhesion of *Enterobacter sakazakii*, *Listeria monocytogenes* (*L. monocytogenes*), and *Salmonella* Typhimurium [66, 67].

A superhydrophobic surface is a surface having a water contact angle greater than 150°, a sliding grade lower than 5°, and high stability of the Cassie model state [68, 69]. In general, superhydrophobic surfaces can be acquired by rendering with fluorocarbon materials containing $-CF_3$ and $-CF_2-$ groups, silicones, organic materials (for example, polyethylene, polystyrene, and polyalkylpyrrole), and inorganic materials (like ZnO and TiO_2) [68, 70–72]. The remarkable and well-known property of superhydrophobic surfaces is that an air layer known as air plastron is physically entrapped between liquid and surface (as shown in **Figure 2B**) when a substrate is immersed in liquid or bacterial suspension [73]. The air plastron exhibits a great potency in antifouling and corrosive resistance [74]. The contact area between bacteria and the superhydrophobic surface is reduced by the air plastron, resulting in significant mitigation of adherent bacteria [74, 75]. In addition, due to the higher contact angle and low sliding angle of a superhydrophobic surface, droplets cannot stay on the superhydrophobic surface and roll off immediately, known as the 'lotus effect,' accounting for the low-adhesion or self-cleaning property of the superhydrophobic surface [76, 77]. An approximately 80% reduction in the adhesion

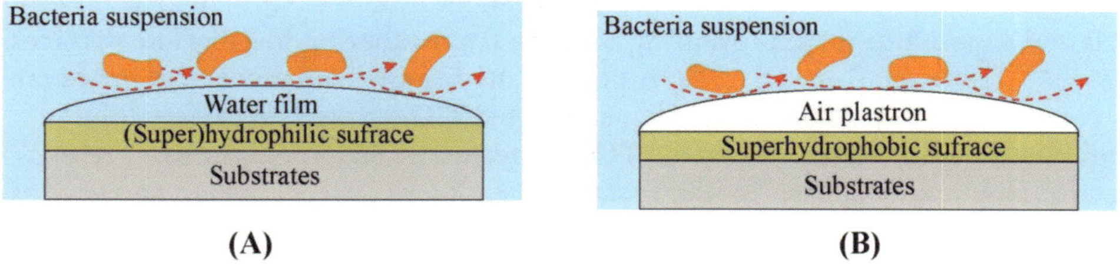

Figure 2.
Mechanism of superhydrophilic (A) and superhydrophobic (B) surfaces.

of *E. coli* K-12 was achieved on a superhydrophobic surface [54]. Freschauf et al. demonstrated low initial concentration (~2%) of *E. coli* could attach to the superhydrophobic polystyrene, polycarbonate, and polyethylene surfaces [78]. Compared to bare glass, poly-pyrene-F6 coated glass showed a significant impact against bacterial attachment: bacterial adhesion could be diminished by about 65% for *Pseudomonas aeruginosa* (*P. aeruginosa*) and *S. aureus* [79].

2.2 Chemical properties

2.2.1 Metal ions and their compounds

Metals in various forms, coated on substrate surfaces, are well known for their antibacterial effects [80, 81]. The main metals applied include silver, gold, copper, zinc, magnesium, calcium, cerium, strontium, nickel, titanium, europium, yttrium ions, and anions (such as selenium and fluoride) [82]. Silver can deactivate protein activities by interacting with thiol groups in proteins and interfere with transmembrane energy generation and ion transport by generating stable S-Ag bonds in the cell membrane [81]. Moreover, the silver ion can bind to nucleic acid, affecting replication ability and denaturing them [81, 83]. The antibacterial capability of silver has been utilized to prevent bacterial infection for decades, and nearly 650 types of bacteria are associated [84, 85]. Copper exhibits contact-killing properties by damaging cell membranes, inducing the formation of reactive oxygen species (ROS), inhibiting enzymes' activities, and denaturing nucleic acid [81]. Estimated 90 types of bacteria have been reported to be killed using contacting copper [81]. In hospitals, copper alloys, used in doorknobs and other surfaces, exerted an antimicrobial effect against *E. coli* O157, methicillin-resistant *S. aureus* (MRSA), and *Clostridium difficile* while equivalent stainless-steel surfaces did not [86].

Metal oxides such as zinc oxide (ZnO), copper oxide (CuO), Fe_2O_3, MgO, and titanium oxide (TiO_2) have been implemented to prevent biofilm formation in recent years since they are stable under harsh conditions and generally safe for humans and animals [87]. Among metal oxide antibacterial agents, ZnO and TiO_2 aroused increasing attention due to their efficient antibacterial activities on a broad spectrum of bacteria [88, 89]. The antibacterial ability of ZnO may contribute to its destruction of bacterial cell integrity and the formation of ROS [90]. ZnO is a photocatalytic material that can respond to UV light and induce ROS creation [91]. TiO_2, also known as a photocatalyst, has received more attention because of its strong antiadhesion and antibacterial properties [92]. Moreover, TiO_2 is abundant in nature, biologically and chemically stable, non-toxic, corrosion-resistive, and inexpensive [93]. When illuminated by ultraviolet light with paper energy under aerobic conditions, TiO_2 can induce the generation of electrons and holes that react with organic substance and dioxygen molecules to form hydroxyl radicals and superoxide ions, preventing bacteria from adhering to substrate surfaces [94–96] by penetrating cell walls, rupturing membrane, and discomposing organic substances [97, 98]. Many studies have reported TiO_2-coated surfaces exhibited antiadhesion properties against both Gram-negative and Gram-positive bacteria, such as *S. aureus* and *Streptococcus mutants* [99], *E. coli* [93], *L. monocytogenes* [100], and *Salmonella* [101].

Besides, metal/metal oxide nanoparticles and metal-organic frameworks (MOFs) are porous materials with nanostructures, acting as reservoirs of metal ions. They also possessed significant antibacterial ability, inhibiting biofilm formation and acting as antiadhesion agents [102]. The mechanisms of their actions are similar to those at the molecular level [103].

2.2.2 Biomacromolecules

Surfaces modified by natural/synthetic proteins/peptides exhibit effective ability to prevent/reduce bacterial adhesion [104]. Proteins/peptides are low toxicity, assembly, and biocompatibility and can be coated on the surfaces of various materials, such as metals, oxides, and polymers [105]. Proteins/peptides avoid bacterial attachment by shifting the hydrophobicity of surfaces and providing hydration [106]. Binding between proteins and bacterial cells is also responsible for adhesion resistance [107]. In addition, due to zwitterionic charges and high hydrogen bond-donor/acceptor abilities of polar functional groups, proteins can interact with negative charged groups on the bacterial cell membrane, destructing cells' integrity [108–110] and exhibiting non-fouling characteristics [111]. Albumins, such as human serum albumin (HSA) and bovine serum albumin, are remarkable proteins that can prevent bacterial adherence to implant surfaces. Eighty-two to ninety-five percent of *S. aureus* was significantly inhibited from binding to HAS-coated titanium surfaces [112]. The antibody is a 'Y-shaped' protein. Its opsonization can impede the adhesion of bacterial cells to implant surfaces by blocking the way of cell-surface attachment and phagocytizing cells [113]. With the presence of antibodies, the adhesion of *E. coli* was markedly reduced on polymer substrates [114].

Probiotic microorganisms, such as *Lactobacillus* and Lactic acid bacteria (LAB), play an important role in antiadhesion. Due to their high adherence capability, probiotics exhibited vigorous antiadhesion activity by competing with bacteria for attachment sites [115]. In addition, antimicrobial substances (such as bacteriocins and hydrogen peroxide) produced by probiotics can also inhibit bacterial adhesion [116]. Studies on the antiadhesion ability of LAB and *Lactobacillus* strains have been largely reported, including *Lactobacillus fermentum (L. fermentum)* in the prevent adhesion of *S. aureus* [117], antiadhesion effects of *L. Plantarum*, *L. crustorum*, *L. coryniformis*, and *L. rhamnosus* on *E. coli* [118], and antiadhesion activity of *L. crispatus* against *Enterococcus faecalis* [115].

Bioactive materials present effective possibilities of resisting biofilm formation. Polysaccharides are a crucial bioactive substance [119], like chitosan, hyaluronic, and alginic acid. The mechanism of the antiadhesion capability of polysaccharides might be that polysaccharides could dissolve biofilms by interacting with the EPS layer and distort biofilm formation and kill cells by inhibiting the metabolic activity of bacterial cells [120]. Chitosan possesses significant antibacterial and antibiofilm activities, making it widely used in medical and food fields, such as food preservation, scaffolds, and bandages [121–123]. The positive-charge property of chitosan enables it to bind with negatively charged cell membranes, inducing the leakage of proteinaceous and other intracellular constituents [124]. Moreover, chitosan can cross through the membrane, bind with DNA, and interfere with the synthesis of mRNA and protein [113]. It was found that chitosan with quaternary ammonium groups could eradicate biofilm formation of *Staphylococcus aureus* [125], and carboxymethyl chitosan could restrain *S. aureus* or *P. aeruginosa* from adhering to surfaces with an efficiency of >90% [126].

3. Topographic modification strategy

Topographical features of substratum surfaces can modulate bacterial attachment and biofilm formation as surface morphology dominates surface roughness and wettability [127]. Typically, the topographical surface can be classified into three different scales: macro-, micro-, and nano-scale [128]. Roughness is a critical factor affecting bacterial attachment by reducing the attachment area between a particle and a surface [129]. Since most microbes are approximately 0.2–2 μm in diameter [130] which is much smaller than the groove distance of macro-roughness, cells can swim and entrap into the grooves of macro-roughness surfaces, suggesting that macro-scale roughness surfaces are not related to antifouling [127]. Therefore, micro-and nano-scale topography surfaces are crucial for preventing bacterial adhesion. Many studies have investigated how micro/nano-scale topographies affect bacterial adhesion. Discrete, ordered, and hierarchical surface structures from nano-scale to micro-scale were self-assembled, designed, or bioinspired by mimicking natural surfaces (such as skins of marine mammals and sharks, shells of mollusks and crabs, wings of insects and birds, and leaves of plants) [131, 132].

3.1 Micro-scale topography surfaces

Micro-structure can be fabricated on surfaces of metals, plastics, and polymer films, like stainless steel [133], polyethylene terephthalate (PET) [134], and PDMS [135]. The micro-patterned topographies exhibited positive influences on preventing the adhesion of various bacteria strains while being non-toxic [135]. Wang et al. designed and fabricated micro-patterned PET surfaces, which simultaneously include curved and straight edges, flat plateaus (top of pillars), and flat surfaces between pillars [134]. The results indicated that PET surfaces with pillars could significantly reduce the attachment of *E. coli* cells under both static and dynamic (shaking at 200 r/min) conditions in nutritious media and oligotrophic solution at 37°C. The Sharklet diamond-shaped micropattern, inspired by shark surface architecture, was widely reported due to its impressive ability to prevent colonization and biofilm formation of various bacteria strains, including *Mycobacterium abscessus* [136], *E. coli* [137], *S. aureus* [138], and *P. aeruginosa* [139].

Features of micropatterns, including pattern shape, size, and groove distance, affect antifouling efficiency [140]. Varied topographical pattern shapes have been created and presented antiadhesion ability. Pattern shapes cover ordered geometric shapes (i.e. line [26], pyramid [141], and cross [142]), pillar [143], pit [144], brush [145], wrinkle [17], and biomimetic shapes (like Sharklet diamond shape [136], lotus-like shape [146], rice leaf [147], rose petals [148], and mytilid shells [149]). In general, with the increase in pattern size, the antiadhesion ability of micropatterns decreased. Lu et al. studied the adhesion of *E. coli*, *P. aeruginosa*, and *S. aureus* on micro-patterned PDMS films with three different pattern sizes [135]. It was found that when pattern size was smaller than bacteria size, the surface was effective in preventing bacteria adhesion; however, as the pattern size was comparable to or larger than bacteria size, the antiadhesion capability of the surface decreased markedly, with more bacteria attachment but still less compared with the flat surface. Similar results were reported by other researchers [150]. This phenomenon might be attributed to the contact area between microorganisms and the surfaces. The available cell-surface contact area reduces with a smaller pattern size than bacterial cell size [151]. The groove between patterns provides anchor sites for cell contact, creates vortices under dynamic conditions, and acts as dead zones for cells sheltered from sanitation treatment [152, 153]. It was also reported that bacteria prefer to distribute in the grooves rather than the top of protruding patterns [135]. As groove distance is smaller than bacteria size, less bacterial cells are entrapped [154]. Similar results were obtained by Lu et al. and Romero et al. [135, 155]. However, the attachment of bacteria can be enhanced when the groove distance is equal to bacteria size because microorganism cells can fit between grooves, and binding energy can be increased [135].

Besides, the effectiveness of surface microstructures on antifouling is also affected by surface energy and hydrophobicity [156]. According to Wenzel and Cassie and Baxter, surface topography can alter the surface to be hydrophobic and superhydrophobic [157]. Carman et al. demonstrated that hexagons could increase the hydrophobicity of the polydimethylsiloxane elastomer [158]. Micro-scale structure could enhance surface hydrophobic ability, allowing more air bubbles to effectively form between surface and liquid [159]. Since a large portion of surfaces was occupied by air, the contact area between bacteria and surfaces was significantly reduced, leading to less cell attachment [160]. Additionally, due to the effect of surface tension, bacteria cannot cross the air-water interface, thereby inhibiting bacterial adhesion [157].

3.2 Nano-scale topography surfaces

Nano-topography provides an effective way to repel bacterial adhesion and prohibit biofouling. Like the micro-scale patterns, the topographical features such as shape, size, density, and groove width can markedly affect cell adhesion onto surfaces [161, 162]. Compared to low-density patterns, nanostructures with highly dense patterns greatly improve the reduction rate of bacterial attachment [163, 164]. Adhesion numbers of *E. coli* and *S. aureus* were significantly reduced by 55.6 and 40.5% on a nanoscale (6 nm) titanium surface with a low density of 213 peaks/μm^2 compared to 2 nm with a high density of 2240 peaks/μm^2 [165].

Numerous shapes of nano-patterns with varying size, depth, and groove width have been reported as excellent impeders of bacterial adhesion and biofilm formation [166–169]. A topographical surface characterized by nanometer-size pores (approximately 0.20 μm^2) surrounded by nano ridges, mimicking the pilot whale skin, exhibited antifouling activity based on reduced available space for bacterial attachment [170]. The more the topography resembled the size and shape of features on bio-skins, the better the antifouling activity was [16, 171]. Bhadra et al. fabricated a nanowire array (average size is approximately 40.2 nm) on titanium and estimated its antifouling ability [172]. It was revealed that the nanowire arrays could render titanium as a moderately effective bactericidal surface, with more excellent bactericidal activity, eliminating almost 50% of *P. aeruginosa* cells and about 20% of *S. aureus* cells. The surfaces of cicada and dragonfly wings exhibit bactericidal properties towards some bacteria strains due to their nano-scale pillar structure [173, 174]. Cicada-inspired fluoridated hydroxyapatite with nanopillars has been successfully fabricated using electrochemical additive manufacturing (ECAM) by Ge et al. [175]. Different types of nanopillar array were obtained: with diameters, heights, and aspect ratios of ~65–95 nm, ~380–510 nm, and ~4.5–7.5 nm, respectively. It was demonstrated that the nanopillars with diameters of ~80 nm were lethal to both Gram-negative and Gram-positive bacteria when the nanopillar density is proper [176, 177].

The cell-nanostructure adhesion mechanisms are still poorly understood. Currently, there are three mechanisms proposed to elucidate the antifouling behavior of nano-textured surfaces. (1) nanostructures induce the formation of the superhydrophobic surface [178]. As explained by the Cassie-Baxter state, the nanostructure can promote air pockets generating in the solid/liquid interface and increase the surface contact angle [179]. As a result, the available contact area for bacteria on the surface is reduced, thus preventing bacterial adhesion [180]. (2) Bacterial membrane can be ruptured and stretched by the nanostructure, leading to cell disruption and eventually cell death, known as the biophysical model, developed by Pogodin et al. [181]. This occurs because the size of most bacterial cells is in the micrometer range, while the structured surfaces are in the nanometer range [182]. Based on the model, bacterial cells absorbed on pattern surface may lead to a drastic increase of contact area, accompanied by stretching the cell membrane between the pillars, which induces membrane disruption and cell death. Furthermore, in terms of the model, the rigidity of cell membranes plays a crucial role in bacterial attachment behavior: the more rigid cells are, the more resistant they are. This may be the reason why nano-pillared surfaces were less effective against gram-positive bacteria strains (*Bacillus subtilis*, *Planococcus maritimus*, and *S. aureus*) when compared to less rigid gram-negative bacteria strains (*P. aeruginosa*) [173, 183–185]. (3) Since the nano-structured topography is unfavorable for bacterial cells, the immobilized cells push and pull the structure while attempting to move away, imposing fatal shear force on the membrane, which initiates bacterial membrane damage [174]. In addition, the solid adhesive force between bacteria and nanostructure also facilitates membrane deformation and cell membrane rupture [174].

4. Conclusions, challenges, and future trends

Bacterial biofilm is a universal and ubiquitous phenomenon. It can directly cause severe problems on public health, the environment, and industries and subsequently lead to economic losses. Consequently, various strategies have been developed and implemented to control biofilm formation. As bacterial adhesion on a surface is the prerequisite for biofilm formation, much attention has been paid to the antifouling strategies that utilize topography and physicochemistry modification to prevent bacterial adhesion to surfaces. This chapter only summarizes the positive effect of surface topographical and physicochemical properties on preventing bacterial adhesion. However, inconsistent and even conflicting impacts could be found in various reported studies. No one particular surface structure or physicochemical property has demonstrated universal antiadhesion ability against all types of microorganisms. Therefore, it is needed to continue the development of strategies that are truly and broadly effective. Furthermore, though surface topographical and physicochemical properties exhibited significant and effective ability to resist the adhesion of specific bacteria strains, the surface structures and physicochemical properties are easily destroyed by various forces, thus decreasing their antifouling capabilities. Therefore, developing a long-term and durable surface with effective antifouling properties remains a huge challenge for the future.

Author details

Zhen Jia
Department of Food Science and Human Nutrition, University of Florida, Gainesville, Florida, United States

*Address all correspondence to: janejia1988@gmail.com

References

[1] Sauer K, Rickard AH, Davies DG. Biofilms and biocomplexity. Microbe. 2007;**2**(7):347. DOI: https://ci.nii.ac.jp/naid/10029122639/en/

[2] Rabin N, Zheng Y, Opoku-Temeng C, Du Y, Bonsu E, Sintim HO. Biofilm formation mechanisms and targets for developing antibiofilm agents. Future Medicinal Chemistry. 2015;**7**(4):493-512. DOI: 10.4155/fmc.15.6

[3] Vickery K, Hu H, Jacombs AS, Bradshaw DA, Deva AK. A review of bacterial biofilms and their role in device-associated infection. Healthcare Infection. 2013;**18**(2):61-66. DOI: 10.1071/HI12059

[4] Flemming HC, Neu TR, Wozniak DJ. The EPS matrix: The "house of biofilm cells". Journal of Bacteriology. 2007; **189**(22):7945-7947. DOI: 10.1128/JB.00858-07

[5] Kumar MA, Anandapandian KTK, Parthiban K. Production and characterization of exopolysaccharides (EPS) from biofilm forming marine bacterium. Brazilian Archives of Biology and Technology. 2011;**54**:259-265. DOI: 10.1590/S1516-89132011000200006

[6] Karygianni L, Ren Z, Koo H, Thurnheer T. Biofilm matrixome: Extracellular components in structured microbial communities. Trends in Microbiology. 2020;**28**(8):668-681. DOI: 10.1016/j.tim.2020.03.016

[7] Dzianach PA, Dykes GA, Strachan NJ, Forbes KJ, Pérez-Reche FJ. Challenges of biofilm control and utilization: Lessons from mathematical modeling. The Royal Society Publishing. 2019;**16**(155):20190042. DOI: 10.1098/rsif.2019.0042

[8] Donlan RM. New approaches for the characterization of prosthetic joint biofilms. Clinical Orthopaedics and Related Research. 2005;**437**:12-19. DOI: 10.1097/01.blo.0000175120.66051.29

[9] Oh JK, Yegin Y, Yang F, Zhang M, Li J, Huang S, et al. The influence of surface chemistry on the kinetics and thermodynamics of bacterial adhesion. Scientific Reports. 2018;**8**(1):1-13. DOI: 10.1038/s41598-018-35343-1

[10] Bryers JD. Medical biofilms. Biotechnology and Bioengineering. 2008;**100**(1):1-18. DOI: 10.1002/bit.21838

[11] Davies D. Understanding biofilm resistance to antibacterial agents. Nature Reviews Drug Discovery. 2003;**2**(2):114-122. DOI: 10.1038/nrd1008

[12] Klevens RM, Edwards JR, Richards CL, Horan TC, Gaynes RP, Pollock DA, et al. Estimating health care-associated infections and deaths in US hospitals, 2002. Public Health Reports. 2007;**122**(2):160-166. DOI: 10.1177/003335490712200205

[13] Srey S, Jahid IK, Ha S-D. Biofilm formation in food industries: A food safety concern. Food Control. 2013;**31**(2):572-585. DOI: 10.1016/j.foodcont.2012.12.001

[14] Shi X, Zhu X. Biofilm formation and food safety in food industries. Trends in Food Science & Technology. 2009;**20**(9):407-413. DOI: 10.1016/j.tifs.2009.01.054

[15] Zhao X, Zhao F, Wang J, Zhong N. Biofilm formation and control strategies of foodborne pathogens: Food

safety perspectives. RSC Advances. 2017;**7**(58):36670-36683. DOI: 10.1039/C7RA02497E

[16] Kirschner CM, Brennan AB. Bio-inspired antifouling strategies. Annual Review of Materials Research. 2012;**42**:211-229. DOI: 10.1146/annurev-matsci-070511-155012

[17] Epstein AK, Hong D, Kim P, Aizenberg J. Biofilm attachment reduction on bioinspired, dynamic, micro-wrinkling surfaces. New Journal of Physics. 2013;**15**(9):095018. DOI: 10.1088/1367-2630/15/9/095018

[18] Lewis JA. Marine biofouling and its prevention on underwater surfaces. Materials Forum. 1998;**22**:41-61

[19] Thomas WE, Nilsson LM, Forero M, Sokurenko EV, Vogel V. Shear-dependent 'stick-and-roll'adhesion of type 1 fimbriated *Escherichia coli*. Molecular Microbiology. 2004;**53**(5):1545-1557. DOI: 10.1111/j.1365-2958.2004.04226.x

[20] Kumar A, Alam A, Rani M, Ehtesham NZ, Hasnain SE. Biofilms: Survival and defense strategy for pathogens. International Journal of Medical Microbiology. 2017;**307**(8):481-489. DOI: ttps://doi.org/10.1016/j.ijmm.2017.09.016

[21] Vera CV, Peterson BW, Mei HC, Busscher HJ. Physico-chemistry from initial bacterial adhesion to surface-programmed biofilm growth. Advances in Colloid and Interface Science. 2018;**261**:1-14. DOI: 10.1016/j.cis.2018.10.005

[22] Bos R, Busscher HJ. Role of acid-base interactions on the adhesion of oral streptococci and actinomyces to hexadecane and chloroform-influence of divalent cations and comparison between free energies of partitioning and free energies obtained by extended DLVO analysis. Colloids and Surfaces B: Biointerfaces. 1999;**14**:169-177. DOI: 10.1016/S0927-7765(99)00034-X

[23] Gupta P, Sarkar S, Das B, Bhattacharjee S, Tribedi P. Biofilm, pathogenesis and prevention—A journey to break the wall: A review. Archives of Microbiology. 2016;**198**(1):1-15. DOI: 10.1007/s00203-015-1148-6

[24] Medilanski E, Kaufmann K, Wick LY, Wanner O, Harms H. Influence of the surface topography of stainless steel on bacterial adhesion. The Journal of Bioadhesion and Biofilm Research. 2002;**18**(3):193-203. DOI: 10.1080/08927010290011370

[25] Chmielewski RAN, Frank JF. Biofilm formation and control in food processing facilities. Comprehensive Reviews in Food Science and Food Safety. 2003;**2**(1):22-32. DOI: 10.1111/j.1541-4337.2003.tb00012.x

[26] Graham MV, Mosier AP, Kiehl TR, Kaloyeros AE, Cady NC. Development of antifouling surfaces to reduce bacterial attachment. Soft Matter. 2013;**9**(27):6235-6244. DOI: 10.1039/C3SM50584G

[27] Chavant P, Martinie B, Meylheuc T, Bellon-Fontaine MN, Hebraud M. *Listeria monocytogenes* LO28: Surface physicochemical properties and ability to form biofilms at different temperatures and growth phases. Applied and Environmental Microbiology. 2002;**68**(2):728-737. DOI: 10.1128/AEM.68.2.728-737.2002

[28] Liu Y, Zhao Q. Influence of surface energy of modified surfaces on bacterial adhesion. Biophysical Chemistry. 2005;**117**(1):39-45. DOI: 10.1016/j.bpc.2005.04.015

[29] Zhao Q, Wang S, Müller-Steinhagen H. Tailored surface free energy of membrane diffusers to minimize microbial adhesion. Applied Surface Science. 2004;**230**(1-4):371-378. DOI: 10.1016/j.apsusc.2004.02.052

[30] Cheng G, Zhang Z, Chen S, Bryers JD, Jiang S. Inhibition of bacterial adhesion and biofilm formation on zwitterionic surfaces. Applied Surface Science. 2007;**28**(29):4192-4199. DOI: 10.1016/j.biomaterials.2007.05.041

[31] Baier RE. Surface behaviour of biomaterials: The theta surface for biocompatibility. Journal of Materials Science: Materials in Medicine. 2006;**17**(11):1057-1062. DOI: 10.1007/s10856-006-0444-8

[32] Pistone A, Scolaro C, Visco A. Mechanical properties of protective coatings against marine fouling: A review. Polymers. 2021;**13**(2):173. DOI: 10.3390/polym13020173

[33] Liu W, Fryer P, Zhang Z, Zhao Q, Liu Y. Identification of cohesive and adhesive effects in the cleaning of food fouling deposits. Innovative Food Science & Emerging Technologies. 2006;**7**(4):263-269. DOI: 10.1016/j.ifset.2006.02.006

[34] Callow ME, Fletcher RL. The influence of low surface energy materials on bioadhesion—A review. International Biodeterioration & Biodegradation. 1994;**34**(3-4):333-348. DOI: 10.1016/0964-8305(94)90092-2

[35] Baier R. Influence of the initial surface condition of materials on bioadhesion. In: Proceedings of the Third International Congress on Marine Corrosion and Fouling. Evanston, Illinois, USA: Northwestern University Press; 1973

[36] Drelich J, Chibowski E, Meng DD, Terpilowski K. Hydrophilic and superhydrophilic surfaces and materials. Soft Matter. 2011;**7**(21):9804-9828. DOI: 10.1039/C1SM05849E

[37] Genzer J, Efimenko K. Recent developments in superhydrophobic surfaces and their relevance to marine fouling: A review. The Journal of Bioadhesion and Biofilm Research. 2006;**22**(5):339-360. DOI: 10.1080/08927010600980223

[38] Kobayashi M, Terayama Y, Yamaguchi H, Terada M, Murakami D, Ishihara K, et al. Wettability and antifouling behavior on the surfaces of superhydrophilic polymer brushes. Langmuir. 2012;**28**(18):7212-7222. DOI: 10.1021/la301033h

[39] Moghimifar V, Raisi A, Aroujalian A. Surface modification of polyethersulfone ultrafiltration membranes by corona plasma-assisted coating TiO_2 nanoparticles. Journal of Membrane Science. 2014;**461**:69-80. DOI: 10.1016/j.memsci.2014.02.012

[40] Vatanpour V, Shahsavarifar S, Khorshidi S, Masteri-Farahani M. A novel antifouling ultrafiltration membranes prepared from percarboxylic acid functionalized SiO_2 bound Fe_3O_4 nanoparticle (SCMNP-COOOH)/polyethersulfone nanocomposite for BSA separation and dye removal. Journal of Chemical Technology & Biotechnology. 2019;**94**(4):1341-1353. DOI: 10.1002/jctb.5894

[41] Hosseini S, Afshari M, Fazlali A, Farahani SK, Bandehali S, Van der Bruggen B, et al. Mixed matrix PES-based nanofiltration membrane decorated by (Fe_3O_4–polyvinylpyrrolidone) composite nanoparticles with intensified antifouling and separation characteristics. Chemical Engineering Research and Design. 2019;**147**:390-398. DOI: 10.1016/j.cherd.2019.05.025

[42] Laohaprapanon S, Vanderlipe AD, Doma BT Jr, You SJ. Self-cleaning and antifouling properties of plasma-grafted poly (vinylidene fluoride) membrane coated with ZnO for water treatment. Journal of the Taiwan Institute of Chemical Engineers. 2017;**70**:15-22. DOI: 10.1016/j.jtice.2016.10.019

[43] Dong B, Manolache S, Wong AC, Denes FS. Antifouling ability of polyethylene glycol of different molecular weights grafted onto polyester surfaces by cold plasma. Polymer Bulletin. 2011;**66**(4):517-528. DOI: 10.1007/s00289-010-0358-y

[44] Katsikogianni M, Amanatides E, Mataras D, Missirlis YF. *Staphylococcus epidermidis* adhesion to He, He/O_2 plasma treated PET films and aged materials: Contributions of surface free energy and shear rate. Colloids and Surfaces B: Biointerfaces. 2008;**65**(2):257-268. DOI: 10.1016/j.colsurfb.2008.04.017

[45] Dong B, Manolache S, Somers EB, Lee Wong AC, Denes FS. Generation of antifouling layers on stainless steel surfaces by plasma-enhanced crosslinking of polyethylene glycol. Journal of Applied Polymer Science. 2005;**97**(2):485-497. DOI: 10.1002/app.21766

[46] Shen L, Wang X, Li R, Yu H, Hong H, Lin H, et al. Physicochemical correlations between membrane surface hydrophilicity and adhesive fouling in membrane bioreactors. Journal of Colloid and Interface Science. 2017;**505**:900-909. DOI: 10.1016/j.jcis.2017.06.090

[47] Jo Y, Choi E, Choi N, Kim CK. Antibacterial and hydrophilic characteristics of poly (ether sulfone) composite membranes containing zinc oxide nanoparticles grafted with hydrophilic polymers. Industrial & Engineering Chemistry Research. 2016;**55**(28):7801-7809. DOI: 10.1021/acs.iecr.6b01510

[48] Otitoju T, Ahmad A, Ooi BS. Superhydrophilic (superwetting) surfaces: A review on fabrication and application. Journal of Industrial and Engineering Chemistry. 2017;**47**:19-40. DOI: 10.1016/j.jiec.2016.12.016

[49] Zhu W, Feng X, Feng L, Jiang L. UV-manipulated wettability between superhydrophobicity and superhydrophilicity on a transparent and conductive SnO_2 nanorod film. Chemical Communications. 2006;**26**:2753-2755. DOI: 10.1039/B603634A

[50] Tang K, Wang X, Yan W, Yu J, Xu R. Fabrication of superhydrophilic Cu2O and CuO membranes. Journal of Membrane Science. 2006;**286**(1-2):279-284. DOI: 10.1016/j.memsci.2006.10.005

[51] Sorcar S, Razzaq A, Tian H, Grimes CA, In S. Facile electrochemical synthesis of anatase nano-architectured titanium dioxide films with reversible superhydrophilic behavior. Journal of Industrial and Engineering Chemistry. 2017;**46**:203-211. DOI: 10.1016/j.jiec.2016.10.032

[52] Du X, Huang X, Li X, Meng X, Yao L, He J, et al. Wettability behavior of special microscale ZnO nail-coated mesh films for oil–water separation. Journal of Colloid and Interface Science. 2015;**458**:79-86. DOI: 10.1016/j.jcis.2015.07.040

[53] Cho H, Kim M, Lee S, Hwang W. Stable and conformable superhydrophilic surface fabricated via surface-initiated silicification on polyaniline nanofibers. Applied Surface Science. 2016;**367**:432-437. DOI: 10.1016/j.apsusc.2016.01.215

[54] Yoon SH, Rungraeng N, Song W, Jun S. Superhydrophobic and

superhydrophilic nanocomposite coatings for preventing *Escherichia coli* K-12 adhesion on food contact surface. Journal of Food Engineering. 2014;**131**:135-141. DOI: 10.1016/j.jfoodeng.2014.01.031

[55] Qian H, Yang J, Lou Y, ur Rahman O, Li Z, Ding X, et al. Mussel-inspired superhydrophilic surface with enhanced antimicrobial properties under immersed and atmospheric conditions. Applied Surface Science. 2019;**465**:267-278. DOI: 10.1016/j.apsusc.2018.09.173

[56] Gu Y, Yu L, Mou J, Wu D, Xu M, Zhou P, et al. Research strategies to develop environmentally friendly marine antifouling coatings. Marine Drugs. 2020;**18**(7):371. DOI: 10.3390/md18070371

[57] Lin PH, Li BR. Antifouling strategies in advanced electrochemical sensors and biosensors. The Analyst. 2020;**145**(4):1110-1120. DOI: 10.1039/C9AN02017A

[58] Zhao X, Xuan H, Chen Y, He C. Preparation and characterization of superior antifouling PVDF membrane with extremely ordered and hydrophilic surface layer. Journal of Membrane Science. 2015;**494**:48-56. DOI: 10.1016/j.memsci.2015.07.052

[59] Marshall A, Munro P, Trägårdh G. The effect of protein fouling in microfiltration and ultrafiltration on permeate flux, protein retention and selectivity: A literature review. Desalination. 1993;**91**(1):65-108. DOI: 10.1016/0011-9164(93)80047-Q

[60] Buskens P, Wouters M, Rentrop C, Vroon Z. A brief review of environmentally benign antifouling and foul-release coatings for marine applications. Journal of Coatings Technology and Research. 2013;**10**(1):29-36. DOI: 10.1007/s11998-012-9456-0

[61] Seydlová G, Svobodová J. Review of surfactin chemical properties and the potential biomedical applications. Central European Journal of Medicine. 2008;**3**(2):123-133. DOI: 10.2478/s11536-008-0002-5

[62] Biniarz P, Baranowska G, Feder-Kubis J, Krasowska A. The lipopeptides pseudofactin II and surfactin effectively decrease *Candida albicans* adhesion and hydrophobicity. Antonie Van Leeuwenhoek. 2015;**108**(2):343-353. DOI: 10.1007/s10482-015-0486-3

[63] Hamza A, Pham V, Matsuura T, Santerre JP. Development of membranes with low surface energy to reduce the fouling in ultrafiltration applications. Journal of Membrane Science. 1997;**131**(1-2):217-227. DOI: 10.1016/S0376-7388(97)00050-1

[64] Zhao Q, Wang C, Liu Y, Wang SJ. Bacterial adhesion on the metal-polymer composite coatings. Journal of Membrane Science. 2007;**27**(2):85-91. DOI: 10.1016/S0376-7388(97)00050-1

[65] Zhai M, Gong Y, Chen X, Xiao T, Zhang G, Xu L, et al. Mass-producible hydrophobic perfluoroalkoxy/nano-silver coatings by suspension flame spraying for antifouling and drag reduction applications. Surface and Coatings Technology. 2017;**328**:115-120. DOI: 10.1016/j.surfcoat.2017.08.049

[66] Nitschke M, Araújo L, Costa S, Pires R, Zeraik A, Fernandes A, et al. Surfactin reduces the adhesion of food-borne pathogenic bacteria to solid surfaces. Surface and Coatings Technology. 2009;**49**(2):241-247. DOI: 10.1016/j.surfcoat.2017.08.049

[67] Mireles JR, Toguchi A, Harshey RM. *Salmonella enterica* serovar Typhimurium

swarming mutants with altered biofilm-forming abilities: Surfactin inhibits biofilm formation. Journal of Bacteriology. 2001;**183**(20):5848-5854. DOI: 10.1128/JB.183.20.5848-5854. 2001

[68] Yin X, Yu S, Wang K, Cheng R, Lv Z. Fluorine-free preparation of self-healing and anti-fouling superhydrophobic Ni3S2 coating on 304 stainless steel. Chemical Engineering Journal. 2020;**394**:124925. DOI: 10.1016/j.cej.2020.124925

[69] Jeevahan J, Chandrasekaran M, Joseph GB, Durairaj R, Mageshwaran G. Superhydrophobic surfaces: A review on fundamentals, applications, and challenges. Journal of Coatings Technology and Research. 2018;**15**(2): 231-250. DOI: 10.1007/s11998-017-0011-x

[70] Lu X, Peng Y, Qiu H, Liu X, Ge L. Anti-fouling membranes by manipulating surface wettability and their anti-fouling mechanism. Desalination. 2017;**413**:127-135. DOI: 10.1016/j.desal.2017.02.022

[71] Ma M, Hill RM. Superhydrophobic surfaces. Current Opinion in Colloid & Interface Science. 2006;**11**(4):193-202. DOI: 10.1016/j.cocis.2006.06.002

[72] Latthe SS, Gurav AB, Maruti CS, Vhatkar RS. Recent progress in preparation of superhydrophobic surfaces: A review. Journal of Surface Engineered Materials and Advanced Technology. 2012;**2**:76-94. DOI: 10.4236/jsemat.2012.22014

[73] Simovich T, Rosenhahn A, Lamb RN. Thermoregeneration of plastrons on superhydrophobic coatings for sustained antifouling properties. Advanced Engineering Materials. 2020;**22**(3):1900806. DOI: 10.1002/adem.201900806

[74] Hwang GB, Page K, Patir A, Nair SP, Allan E, Parkin IP. The anti-biofouling properties of superhydrophobic surfaces are short-lived. ACS Nano. 2018;**12**(6):6050-6058. DOI: 10.1021/acsnano.8b02293

[75] Huang Q, Lin L, Yang Y, Hu R, Vogler EA, Lin C. Role of trapped air in the formation of cell-and-protein micropatterns on superhydrophobic/superhydrophilic microtemplated surfaces. Biomaterials. 2012;**33**(33): 8213-8220. DOI: 10.1016/j.biomaterials.2012.08.017

[76] Zhu H, Guo Z, Liu W. Adhesion behaviors on superhydrophobic surfaces. Chemical Communications. 2014;**50**(30):3900-3913. DOI: 10.1039/C3CC47818A

[77] Solga A, Cerman Z, Striffler BF, Spaeth M, Barthlott W. The dream of staying clean: Lotus and biomimetic surfaces. Bioinspiration & Biomimetics. 2007;**2**(4):S126. DOI: 10.1088/1748-3182/2/4/S02

[78] Freschauf LR, McLane J, Sharma H, Khine M. Shrink-induced superhydrophobic and antibacterial surfaces in consumer plastics. PLoS One. 2012;**7**(8):e40987. DOI: 10.1371/journal.pone.0040987

[79] Morán G, Ramos-Chagas G, Hugelier S, Xie X, Boudjemaa R, Ruckebusch C, et al. Superhydrophobic polypyrene films to prevent *Staphylococcus aureus* and *Pseudomonas aeruginosa* biofilm adhesion on surfaces: High efficiency deciphered by fluorescence microscopy. Photochemical & Photobiological Sciences. 2018;**17**(8):1023-1035. DOI: 10.1039/c8pp00043c

[80] Matsumoto N, Sato K, Yoshida K, Hashimoto K, Toda Y. Preparation and

characterization of β-tricalcium phosphate co-doped with monovalent and divalent antibacterial metal ions. Acta Biomaterialia. 2009;**5**(8):3157-3164. DOI: 10.1016/j.actbio.2009.04.010

[81] Mittapally S, Taranum R, Parveen S. Metal ions as antibacterial agents. Journal of Drug Delivery & Therapeutics. 2018;**8**(6-s):411-419. DOI: 10.22270/jddt.v8i6-s.2063

[82] Goldmann WH. Biosensitive and antibacterial coatings on metallic material for medical applications. Cell Biology Biologyiology. 2021;**45**:1624-1632. DOI: 10.1002/cbin.11604

[83] Jeon HJ, Yi SC, Oh SG. Preparation and antibacterial effects of Ag–SiO_2 thin films by sol–gel method. Biomaterials. 2003;**24**(27):4921-4928. DOI: 10.1016/S0142-9612(03)00415-0

[84] Li SM, Jia N, Ma MG, Zhang Z, Liu QH, Sun RC. Cellulose–silver nanocomposites: Microwave-assisted synthesis, characterization, their thermal stability, and antimicrobial property. Carbohydrate Polymers. 2011;**86**(2):441-447. DOI: 10.1016/j.carbpol.2011.04.060

[85] Chernousova S, Epple M. Silver as antibacterial agent: Ion, nanoparticle, and metal. Angewandte Chemie International Edition. 2013;**52**(6):1636-1653. DOI: 10.1002/anie.201205923

[86] Jaiswal S, McHale P, Duffy B. Preparation and rapid analysis of antibacterial silver, copper and zinc doped sol–gel surfaces. Colloids and Surfaces B: Biointerfaces. 2012;**94**:170-176. DOI: 10.1016/j.colsurfb.2012.01.035

[87] Liu Y, He L, Mustapha A, Li H, Hu Z, Lin M. Antibacterial activities of zinc oxide nanoparticles against *Escherichia coli* O157: H7. Journal of Applied Microbiology. 2009;**107**(4):1193-1201. DOI: 10.1111/j.1365-2672.2009.04303.x

[88] Tayel AA, El-tras WF, Moussa S, El-baz AF, Mahrous H, Salem MF, et al. Antibacterial action of zinc oxide nanoparticles against foodborne pathogens. Journal of Food Safety. 2011;**31**(2):211-218. DOI: 10.1111/j.1745-4565.2010.00287.x

[89] Sun T, Hao H, Hao W, Yi S, Li X, Li J. Preparation and antibacterial properties of titanium-doped ZnO from different zinc salts. Nanoscale Research Letters. 2014;**9**(1):1-11. DOI: 10.1186/1556-276X-9-98

[90] Sirelkhatim A, Mahmud S, Seeni A, Kaus NHM, Ann LC, Bakhori SKM, et al. Review on zinc oxide nanoparticles: Antibacterial activity and toxicity mechanism. Nano-Micro Letters. 2015;**7**(3):219-242. DOI: 10.1007/s40820-015-0040-x

[91] Zhang H, Chen B, Jiang H, Wang C, Wang H, Wang XJB. A strategy for ZnO nanorod mediated multi-mode cancer treatment. Biomaterials. 2011;**32**(7):1906-1914. DOI: 10.1016/j.biomaterials.2010.11.027

[92] Petrini P, Arciola C, Pezzali I, Bozzini S, Montanaro L, Tanzi MC, et al. Antibacterial activity of zinc modified titanium oxide surface. Biomaterials. 2006;**29**(4):434-442. DOI: 10.1016/j.biomaterials.2010.11.027

[93] Alhadrami H, Al-Hazmi F. Antibacterial activities of titanium oxide nanoparticles. Journal of Bioelectronics and Nanotechnology. 2017;**2**(1):1-5

[94] Zhang S, Liang X, Gadd GM, Zhao Q. Advanced titanium dioxide-polytetrafluorethylene (TiO2-PTFE) nanocomposite coatings on stainless steel surfaces with antibacterial and

anti-corrosion properties. Applied Surface Science. 2019;**490**:231-241. DOI: 10.1016/j.apsusc.2019.06.070

[95] Visai L, De Nardo L, Punta C, Melone L, Cigada A, Imbriani M, et al. Titanium oxide antibacterial surfaces in biomedical devices. The International Journal of Artificial Organs. 2011;**34**(9): 929-946. DOI: 10.5301/ijao.5000050

[96] Chhattani S, Shetty PC, Laxmikant S, Ramachandra CS. In vitro assessment of photocatalytic titanium oxide surface-modified stainless steel and nickel titanium orthodontic wires for its antiadherent and antibacterial properties against *Streptococcus* mutans. Journal of Indian Orthodontic Society. 2014;**48**(2):82-87. DOI: 10.5005/jp-journals-10021-1223

[97] Dalrymple OK, Stefanakos E, Trotz MA, Goswami DY. A review of the mechanisms and modeling of photocatalytic disinfection. Applied Catalysis B: Environmental. 2010;**98**(1-2):27-38. DOI: 10.1016/j.apcatb.2010.05.001

[98] Sunada K, Watanabe T, Hashimoto K. Studies on photokilling of bacteria on TiO2 thin film. Journal of Photochemistry and Photobiology A: Chemistry. 2003;**156**(1-3):227-233. DOI: 10.1016/S1010-6030(02)00434-3

[99] Tavassoli HS, Mahdavi M, Hamzeh F, Rezvani MB, Ahmadian BF, Saderi H, et al. Effect of nano titanium dioxide on self-cleaning and antibacterial properties of dental mirrors. Journal of Dental School Shahid Beheshti University of Medical Sciences. 2013;**98**:104-112. DOI: https://www.sid.ir/en/journal/ViewPaper.aspx?id=502407

[100] Chorianopoulos N, Tsoukleris D, Panagou E, Falaras P, Nychas GJE. Use of titanium dioxide (TiO2) photocatalysts as alternative means for *Listeria monocytogenes* biofilm disinfection in food processing. Food Microbiology. 2011;**28**(1):164-170. DOI: 10.1016/j.fm.2010.07.025

[101] Szczawiński J, Tomaszewski H, Jackowska-Tracz A, Szczawińska ME. Effect of UV radiation on survival of *Salmonella* Enteritidis on the surface of ceramic tiles coated with TiO2. Bulletin of the Veterinary Institute in Pulawy. 2010;**54**:479-483. Available from http://www.piwet.pulawy.pl

[102] De Faria AF, Martinez DST, Meira SMM, de Moraes ACM, Brandelli A, Souza Filho AG, et al. Anti-adhesion and antibacterial activity of silver nanoparticles supported on graphene oxide sheets. Colloids and Surfaces B: Biointerfaces. 2014;**113**:115-124. DOI: 10.1016/j.colsurfb.2013.08.006

[103] Wyszogrodzka G, Marszałek B, Gil B, Dorożyński P. Metal-organic frameworks: Mechanisms of antibacterial action and potential applications. Drug Discovery Today. 2016;**21**(6):1009-1018. DOI: 10.1016/j.drudis.2016.04.009

[104] Dolid A, Gomes LC, Mergulhão FJ, Reches M. Combining chemistry and topography to fight biofilm formation: Fabrication of micropatterned surfaces with a peptide-based coating. Colloids and Surfaces B: Biointerfaces. 2020;**196**:111365. DOI: 10.1016/j.colsurfb.2020.111365

[105] Maity S, Nir S, Zada T, Reches M. Self-assembly of a tripeptide into a functional coating that resists fouling. Chemical Communications. 2014;**50**(76):11154-11157. DOI: 10.1039/C4CC03578J

[106] Fletcher M, Marshall KC. Bubble contact angle method for evaluating substratum interfacial characteristics

and its relevance to bacterial attachment. Applied and Environmental Microbiology. 1982;**44**(1):184-192. DOI: 10.1128/aem.44.1.184-192.1982

[107] Brokke P, Dankert J, Carballo J, Feijen J. Adherence of coagulase-negative staphylococci onto polyethylene catheters in vitro and in vivo: A study on the influence of various plasma proteins. Journal of Biomaterials Applications. 1991;**5**(3):204-226. DOI: 10.1177/088532829100500305

[108] Ederth T, Lerm M, Orihuela B, Rittschof D. Resistance of zwitterionic peptide monolayers to biofouling. Langmuir. 2018;**35**(5):1818-1827. DOI: 10.1021/acs.langmuir.8b01625

[109] Jiang C, Wang G, Hein R, Liu N, Luo X, Davis JJ. Antifouling strategies for selective in vitro and in vivo sensing. Chemical Reviews. 2020;**120**(8):3852-3889. DOI: 10.1021/acs.chemrev.9b00739

[110] De La Fuente-Núñez C, Korolik V, Bains M, Nguyen U, Breidenstein EB, Horsman S, et al. Inhibition of bacterial biofilm formation and swarming motility by a small synthetic cationic peptide. Antimicrobial Agents and Chemotherapy. 2012;**56**(5):2696-2704. DOI: 10.1128/AAC.00064-12

[111] Ye H, Wang L, Huang R, Su R, Liu B, Qi W, et al. Superior antifouling performance of a zwitterionic peptide compared to an amphiphilic, non-ionic peptide. ACS Applied Materials & Interfaces. 2015;**7**(40):22448-22457. DOI: 10.1021/acsami.5b06500

[112] Kinnari TJ, Peltonen LI, Kuusela P, Kivilahti J, Könönen M, Jero J, et al. Bacterial adherence to titanium surface coated with human serum albumin. Otology & Neurotology. 2005;**26**(3):380-384. DOI: 10.1097/01.mao.0000169767.85549.87

[113] Qiu Y, Zhang N, An Y, Wen X. Biomaterial strategies to reduce implant-associated infections. The International Journal of Artificial Organs. 2007;**30**(9):828-841. DOI: 10.1177/039139880703000913

[114] Rojas IA, Slunt JB, Grainger DW. Polyurethane coatings release bioactive antibodies to reduce bacterial adhesion. Journal of Controlled Release. 2000;**63**(1-2):175-189. DOI: 10.1016/S0168-3659(99)00195-9

[115] Todoriki K, Mukai T, Sato S, Toba T. Inhibition of adhesion of food-borne pathogens to Caco-2 cells by *Lactobacillus* strains. Journal of Applied Microbiology. 2001;**91**(1):154-159. DOI: 10.1046/j.1365-2672.2001.01371.x

[116] Reid G, Burton J. Use of *Lactobacillus* to prevent infection by pathogenic bacteria. Microbes and Infection. 2002;**4**(3):319-324. DOI: 10.1016/S1286-4579(02)01544-7

[117] Jayashree S, Karthikeyan R, Nithyalakshmi S, Ranjani J, Gunasekaran P, Rajendhran J. Anti-adhesion property of the potential probiotic strain *Lactobacillus fermentum* 8711 against methicillin-resistant *Staphylococcus aureus* (MRSA). Frontiers in Microbiology. 2018;**9**:411. DOI: 10.3389/fmicb.2018.00411

[118] Wang T, Sun H, Chen J, Luo L, Gu Y, Wang X, et al. Anti-adhesion effects of *Lactobacillus* strains on caco-2 cells against *Escherichia coli* and their application in ameliorating the symptoms of dextran sulfate sodium-induced colitis in mice. Probiotics and Antimicrobial Proteins. 2021;**13**:1632-1643. DOI: 10.1007/s12602-021-09774-8

[119] Vishwakarma J, Sirisha VL. Unraveling the anti-biofilm potential of green algal sulfated polysaccharides

against *Salmonella* enterica and *Vibrio harveyi*. Applied Microbiology and Biotechnology. 2020;**104**:6299-6314. DOI: 10.1007/s00253-020-10653-5

[120] Vishwakarma J, Vavilala SL. Evaluating the antibacterial and antibiofilm potential of sulphated polysaccharides extracted from green algae *Chlamydomonas reinhardtii*. Journal of Applied Microbiology. 2019;**127**(4):1004-1017. DOI: 10.1111/jam.14364

[121] Kurita K. Chitin and chitosan: Functional biopolymers from marine crustaceans. Marine Biotechnology. 2006;**8**(3):203-226. DOI: 10.1007/s10126-005-0097-5

[122] Di Martino A, Sittinger M, Risbud MV. Chitosan: A versatile biopolymer for orthopaedic tissue-engineering. Biomaterials. 2005;**26**(30):5983-5990. DOI: 10.1016/j.biomaterials.2005.03.016

[123] Liu Y, Jiang Y, Zhu J, Huang J, Zhang H. Inhibition of bacterial adhesion and biofilm formation of sulfonated chitosan against *Pseudomonas aeruginosa*. Carbohydrate Polymers. 2019;**206**:412-419. DOI: 10.1016/j.carbpol.2018.11.015

[124] Rabea EI, Badawy ME-T, Stevens CV, Smagghe G, Steurbaut W. Chitosan as antimicrobial agent: Applications and mode of action. Biomacromolecules. 2003;**4**(6):1457-1465. DOI: 10.1021/bm034130m

[125] Sahariah P, Mr M, Meyer RL. Quaternary ammoniumyl chitosan derivatives for eradication of *staphylococcus aureus* biofilms. Biomacromolecules. 2018;**19**(9):3649-3658. DOI: 10.1021/acs.biomac.8b00718

[126] Tan Y, Han F, Ma S, Yu W. Carboxymethyl chitosan prevents formation of broad-spectrum biofilm. Carbohydrate Polymers. 2011;**84**(4):1365-1370. DOI: 10.1016/j.carbpol.2011.01.036

[127] Nurioglu AG, Esteves ACC, With G. Non-toxic, non-biocide-release antifouling coatings based on molecular structure design for marine applications. Journal of Materials Chemistry B. 2015;**3**(32):6547-6570. DOI: 10.1039/C5TB00232J

[128] Verran J, Boyd RD. The relationship between substratum surface roughness and microbiological and organic soiling: A review. The Journal of Bioadhesion and Biofilm Research. 2001;**17**(1):59-71. DOI: 10.1080/08927010109378465

[129] Barthlott W, Neinhuis C. Purity of the sacred lotus, or escape from contamination in biological surfaces. Planta. 1997;**202**(1):1-8. DOI: 10.1007/s004250050096

[130] Maniloff J. Nannobacteria: Size limits and evidence. Science. 1997;**276**(5320):1773-1776. DOI: 10.1126/science.276.5320.1773e

[131] Bixler GD, Theiss A, Bhushan B, Lee SC. Anti-fouling properties of microstructured surfaces bio-inspired by rice leaves and butterfly wings. Journal of Colloid and Interface Science. 2014;**419**:114-133. DOI: 10.1016/j.jcis.2013.12.019

[132] Ivanova EP, Linklater DP, Aburto-Medina A, Le P, Baulin VA, Nguyen HKD, et al. Antifungal versus antibacterial defence of insect wings. Journal of Colloid and Interface Science. 2021;**603**:886-897. DOI: 10.1016/j.jcis.2021.06.093

[133] George R, Muraleedharan P, Sreekumari K, Khatak HS. Influence of surface characteristics and

microstructure on adhesion of bacterial cells onto a type 304 stainless steel. The Journal of Bioadhesion and Biofilm Research. 2003;**19**(1):1-8. DOI: 10.1080/08927010290031017

[134] Wang L, Chen W, Terentjev E. Effect of micro-patterning on bacterial adhesion on polyethylene terephthalate surface. Journal of Biomaterials Applications. 2015;**29**(10):1351-1362. DOI: 10.1177/0885328214563998

[135] Lu N, Zhang W, Weng Y, Chen X, Cheng Y, Zhou P. Fabrication of PDMS surfaces with micro patterns and the effect of pattern sizes on bacteria adhesion. Food Control. 2016;**68**:344-351. DOI: 10.1016/j.foodcont.2016.04.014

[136] Kim E, Kinney WH, Ovrutsky AR, Vo D, Bai X, Honda JR, et al. A surface with a biomimetic micropattern reduces colonization of *Mycobacterium abscessus*. FEMS Microbiology Letters. 2014;**360**(1):17-22. DOI: 10.1111/1574-6968.12587

[137] Reddy ST, Chung KK, McDaniel CJ, Darouiche RO, Landman J, Brennan AB. Micropatterned surfaces for reducing the risk of catheter-associated urinary tract infection: An in vitro study on the effect of sharklet micropatterned surfaces to inhibit bacterial colonization and migration of uropathogenic *Escherichia coli*. Journal of Endourology. 2011;**25**(9):1547-1552. DOI: 10.1089/end.2010.0611

[138] Chung KK, Schumacher JF, Sampson EM, Burne RA, Antonelli PJ, Brennan AB. Impact of engineered surface microtopography on biofilm formation of *Staphylococcus aureus*. Biointerphases. 2007;**2**(2):89-94. DOI: 10.1116/1.2751405

[139] Sakamoto A, Terui Y, Horie C, Fukui T, Masuzawa T, Sugawara S, et al. Antibacterial effects of protruding and recessed shark skin micropatterned surfaces of polyacrylate plate with a shallow groove. FEMS Microbiology Letters. 2014;**361**(1):10-16. DOI: 10.1111/1574-6968.12604

[140] Lee SW, Phillips KS, Gu H, Kazemzadeh-Narbat M, Ren D. How microbes read the map: Effects of implant topography on bacterial adhesion and biofilm formation. Biomaterials. 2021;**268**:120595. DOI: 10.1016/j.biomaterials.2020.120595

[141] Choi DC, Jung SY, Won YJ, Jang JH, Lee JW, Chae HR, et al. Effect of pattern shape on the initial deposition of particles in the aqueous phase on patterned membranes during crossflow filtration. Environmental Science & Technology Letters. 2017;**4**(2):66-70. DOI: 10.1021/acs.estlett.6b00468

[142] Flavel BS, Sweetman MJ, Shearer CJ, Shapter JG, Voelcker NH. Micropatterned arrays of porous silicon: Toward sensory biointerfaces. ACS Applied Materials & Interfaces. 2011;**3**(7):2463-2471. DOI: 10.1021/am2003526

[143] Atthi N, Sripumkhai W, Pattamang P, Thongsook O, Srihapat A, Meananeatra R, et al. Fabrication of robust PDMS micro-structure with hydrophobic and antifouling properties. Microelectronic Engineering. 2020;**224**:111255. DOI: 10.1016/j.mee.2020.111255

[144] Vasudevan R, Kennedy AJ, Merritt M, Crocker FH, Baney RH. Microscale patterned surfaces reduce bacterial fouling-microscopic and theoretical analysis. Colloids and Surfaces B: Biointerfaces. 2014;**117**:225-232. DOI: 10.1016/j.colsurfb.2014.02.037

[145] Du T, Ma S, Pei X, Wang S, Zhou F. Bio-inspired design and fabrication

of micro/nano-brush dual structural surfaces for switchable oil adhesion and antifouling. Small. 2017;**13**(4):1602020. DOI: 10.1002/smll.201602020

[146] Jiang R, Hao L, Song L, Tian L, Fan Y, Zhao J, et al. Lotus-leaf-inspired hierarchical structured surface with non-fouling and mechanical bactericidal performances. Chemical Engineering Journal. 2020;**398**:125609. DOI: 10.1016/j.cej.2020.125609

[147] Ayazi M, Golshan Ebrahimi NG. High stable self-cleaning surface developed by monolithic hierarchical roughness. Surface Engineering. 2020;**36**(6):628-635. DOI: 10.1080/02670844.2019.1687970

[148] Malshe AP, Bapat S, Rajurkar KP, Haitjema H. Bio-inspired textures for functional applications. CIRP Annals. 2018;**67**(2):627-650. DOI: 10.1016/j.cirp.2018.05.001

[149] Bers A, Díaz E, Da Gama B, Vieira-Silva F, Dobretsov S, Valdivia N, et al. Relevance of mytilid shell microtopographies for fouling defence-a global comparison. The Journal of Bioadhesion and Biofilm Research. 2010;**26**(3):367-377. DOI: 10.1080/08927011003605888

[150] Whitehead KA, Verran J. The effect of surface topography on the retention of microorganisms. Food and Bioproducts Processing. 2006;**84**(4):253-259. DOI: 10.1205/fbp06035

[151] Scardino A, Guenther J, De Nys R. Attachment point theory revisited: The fouling response to a microtextured matrix. The Journal of Bioadhesion and Biofilm Research. 2008;**24**(1):45-53. DOI: 10.1080/08927010701784391

[152] Jurusik AK. Factors Influencing the Transport and Deposition of *Salmonella* and Colloids from Evaporating Sessile Droplets on Polydimethylsiloxane Surfaces of Fresh Produce and Simple Micropatterns. Newark, Delaware, USA: University of Delaware; 2019

[153] Mok JH, Niu Y, Yousef A, Zhao Y, Sastry SK. Spatial persistence of *Escherichia coli* O157: H7 flowing on micropatterned structures inspired by stomata and microgrooves of leafy greens. Innovative Food Science & Emerging Technologies. 2021;**75**:102889. DOI: 10.1016/j.ifset.2021.102889

[154] Graham MV, Cady NC. Nano and microscale topographies for the prevention of bacterial surface fouling. Coatings. 2014;**4**(1):37-59. DOI: 10.3390/coatings4010037

[155] Romero M, Luckett J, Figueredo G, Carabelli AM, Carlier A, Vasilevich A, et al. Topographical biomaterials instruct bacterial surface attachment and the in vivo host-pathogen response. BioRxiv. 2020;**10**(10):328146. DOI: 10.1101/2020.10.10.328146

[156] Choi CH, Kim CJ. Advanced nanostructured surfaces for the control of biofouling: Cell adhesions to three-dimensional nanostructures. Green Tribology. Berlin, Heidelberg: Springer; 2012. p. 79-103

[157] Truong V, Webb H, Fadeeva E, Chichkov B, Wu A, Lamb R, et al. Air-directed attachment of coccoid bacteria to the surface of superhydrophobic lotus-like titanium. The Journal of Bioadhesion and Biofilm Research. 2012;**28**(6):539-550. DOI: 10.1080/08927014.2012.694426

[158] Carman ML, Estes TG, Feinberg AW, Schumacher JF, Wilkerson W, Wilson LH, et al. Engineered antifouling microtopographies-correlating wettability with cell

attachment. The Journal of Bioadhesion and Biofilm Research. 2006;**22**(1):11-21. DOI: 10.1080/08927010500484854

[159] Echeverria C, Torres MDT, Fernández-García M, Fuente-Nunez C, Muñoz-Bonilla A. Physical methods for controlling bacterial colonization on polymer surfaces. Biotechnology Advances. 2020;**43**:107586. DOI: 10.1016/j.biotechadv.2020.107586

[160] Cassie ABD, Baxter S. Wettability of porous surfaces. Transactions of the Faraday society. 1944;**40**:546-551

[161] Anselme K, Davidson P, Popa A, Giazzon M, Liley M, Ploux L. The interaction of cells and bacteria with surfaces structured at the nanometre scale. Acta Biomaterialia. 2010;**6**(10):3824-3846. DOI: 10.1016/j.actbio.2010.04.001

[162] Jeon H, Simon CG Jr, Kim GH. A mini-review: Cell response to microscale, nanoscale, and hierarchical patterning of surface structure. Applied Biomaterials. 2014;**102**(7):1580-1594. DOI: 10.1002/jbm.b.33158

[163] Ma J, Sun Y, Gleichauf K, Lou J, Li Q. Nanostructure on taro leaves resists fouling by colloids and bacteria under submerged conditions. Langmuir. 2011;**27**(16):10035-10040. DOI: 10.1021/la2010024

[164] Damodaran VB, Murthy NS. Bio-inspired strategies for designing antifouling biomaterials. Biomaterials Research. 2016;**20**(1):1-11. DOI: 10.1186/s40824-016-0064-4

[165] Lüdecke C, Roth M, Yu W, Horn U, Bossert J, Jandt KDJC, et al. Nanorough titanium surfaces reduce adhesion of *Escherichia coli* and *Staphylococcus aureus* via nano adhesion points. Colloids and Surfaces B: Biointerfaces. 2016;**145**:617-625. DOI: 10.1016/j.colsurfb.2016.05.049

[166] Lipovsky A, Gedanken A, Lubart R. Visible light-induced antibacterial activity of metaloxide nanoparticles. Photomedicine and Laser Surgery. 2013;**31**(11):526-530. DOI: 10.1089/pho.2012.3339

[167] Hsu LC, Fang J, Borca-Tasciuc DA, Worobo RW, Moraru CI. Effect of micro-and nanoscale topography on the adhesion of bacterial cells to solid surfaces. Applied and Environmental Microbiology. 2013;**79**(8):2703-2712. DOI: 10.1128/AEM.03436-12

[168] Epstein A, Hochbaum A, Kim P, Aizenberg J. Control of bacterial biofilm growth on surfaces by nanostructural mechanics and geometry. Nanotechnology. 2011;**22**(49):494007. DOI: 10.1088/0957-4484/22/49/494007

[169] Díaz C, Cortizo MC, Schilardi PL, Saravia SGG, Mele MAFL. Influence of the nano-micro structure of the surface on bacterial adhesion. Materials Research. 2007;**10**:11-14. DOI: 10.1590/S1516-14392007000100004

[170] Baum C, Meyer W, Stelzer R, Fleischer LG, Siebers D. Average nanorough skin surface of the pilot whale (Globicephalamelas, Delphinidae): Considerations on the self-cleaning abilities based on nanoroughness. Marine Biology. 2002;**140**(3):653-657. DOI: 10.1007/s00227-001-0710-8

[171] Nir S, Reches M. Bio-inspired antifouling approaches: The quest towards non-toxic and non-biocidal materials. Current Opinion in Biotechnology. 2016;**39**:48-55. DOI: 10.1016/j.copbio.2015.12.012

[172] Bhadra CM, Truong VK, Pham VT, Al Kobaisi M, Seniutinas G, Wang JY, et al. Antibacterial titanium nano-patterned arrays inspired by dragonfly wings. Scientific Reports. 2015;**5**(1):1-12. DOI: 10.1038/srep16817

[173] Hasan J, Webb HK, Truong VK, Pogodin S, Baulin VA, Watson GS, et al. Selective bactericidal activity of nanopatterned superhydrophobic cicada *Psaltoda claripennis* wing surfaces. Applied Microbiology and Biotechnology. 2013;**97**(20):9257-9262. DOI: 10.1007/s00253-012-4628-5

[174] Bandara CD, Singh S, Afara IO, Wolff A, Tesfamichael T, Ostrikov K, et al. Bactericidal effects of natural nanotopography of dragonfly wing on *Escherichia coli*. ACS Applied Materials & Interfaces. 2017;**9**(8):6746-6760. DOI: 10.1021/acsami.6b13666

[175] Ge X, Zhao J, Esmeryan KD, Lu X, Li Z, Wang K, et al. Cicada-inspired fluoridated hydroxyapatite nanostructured surfaces synthesized by electrochemical additive manufacturing. Materials & Design. 2020;**193**:108790. DOI: 10.1016/j.matdes.2020.108790

[176] Viela F, Navarro-Baena I, Hernández JJ, Osorio MR, Rodríguez I. Motheye mimetic cytocompatible bactericidal nanotopography: A convergent design. Bioinspiration & Biomimetics. 2018;**13**(2):026011. DOI: 10.1088/1748-3190/aaa903

[177] Wu S, Zuber F, Maniura-Weber K, Brugger J, Ren Q. Nanostructured surface topographies have an effect on bactericidal activity. Journal of Nanobiotechnology. 2018;**16**(1):1-9. DOI: 10.1186/s12951-018-0347-0

[178] Barish JA, Goddard JM. Anti-fouling surface modified stainless steel for food processing. Food and Bioproducts Processing. 2013;**91**(4):352-361. DOI: 10.1016/j.fbp.2013.01.003

[179] Martines E, Seunarine K, Morgan H, Gadegaard N, Wilkinson CD, Riehle MO. Superhydrophobicity and superhydrophilicity of regular nanopatterns. Nano Letters. 2005;**5**(10): 2097-2103. DOI: 10.1021/nl051435t

[180] An K, Long C, Sui Y, Qing Y, Zhao G, An Z, et al. Large-scale preparation of superhydrophobic cerium dioxide nanocomposite coating with UV resistance, mechanical robustness, and anti-corrosion properties. Surface and Coatings Technology. 2020;**384**:125312. DOI: 10.1016/j.surfcoat.2019.125312

[181] Pogodin S, Hasan J, Baulin VA, Webb HK, Truong VK, Nguyen THP, et al. Biophysical model of bacterial cell interactions with nanopatterned cicada wing surfaces. Biophysical Journal. 2013;**104**(4):835-840. DOI: 10.1016/j.bpj.2012.12.046

[182] Jaggessar A, Shahali H, Mathew A, Yarlagadda PKDV. Bio-mimicking nano and micro-structured surface fabrication for antibacterial properties in medical implants. Journal of Nanobiotechnology. 2017;**15**(1):1-20. DOI: 10.1186/s12951-017-0306-1

[183] Yao X, Walter J, Burke S, Stewart S, Jericho M, Pink D, et al. Atomic force microscopy and theoretical considerations of surface properties and turgor pressures of bacteria. Colloids and Surfaces B: Biointerfaces. 2002;**23**(2-3):213-230. DOI: 10.1016/S0927-7765(01)00249-1

[184] Arnoldi M, Fritz M, Bäuerlein E, Radmacher M, Sackmann E, Boulbitch A. Bacterial turgor pressure can be measured by atomic force microscopy. Physical Review. 2000;**62**(1):1034. DOI: 10.1103/PhysRevE.62.1034

[185] Ivanova EP, Hasan J, Webb HK, Gervinskas G, Juodkazis S, Truong VK, et al. Bactericidal activity of black silicon. Nature Communications. 2013;**4**(1):1-7. DOI: 10.1038/ncomms3838

Chapter 15

Curcuma Xanthorrhiza Roxb. An Indonesia Native Medicinal Plant with Potential Antioral Biofilm Effect

Dewi F. Suniarti, Ria Puspitawati, Rezon Yanuar and Ranny R. Herdiantoputri

Abstract

Most common oral diseases are directly related to oral biofilm, a complex community of microorganisms inhibiting the oral cavity. Recent studies provide deeper knowledge on how free-floating bacteria form a structurally organized micro-ecosystem and on its pathogenicity and its self-defense mechanisms; thus, creating an understanding of the challenges in eliminating oral biofilm and maintaining the balance of oral ecosystem. Chlorhexidine has been the standard oral antimicrobial agent for decades. However, studies showed that it is less effective against bacteria in the form of biofilm that leads to an ongoing search of another method to fight against biofilm, including the use of plant-derived compounds. Medicinal plants are known to contain secondary metabolites, which are not only important in protecting the plant from any harmful environment but also potential as antimicroorganism and antioral biofilm for humans. *Curcuma xanthorrhiza* Roxb., containing xanthorrhizol (XNT), an essential bioactive compound, is an Indonesian native medicinal plant proven to have antibacterial and antibiofilm activities by several in vitro studies. The understanding of biofilm formation, its resistance to common drugs, and the potential role of *C. xanthorrhiza*-derived compounds as antibacterial and antibiofilm may contribute to developing *C. xanthorrhiza* into the alternative weapon against oral biofilm-related diseases.

Keywords: *Curcuma xanthorrhiza* Roxb., xanthorrhizol, oral biofilm, antibacterial

1. Introduction

Oral biofilm or dental plaque is the complex community of microorganisms that can be found on the surfaces of various orodental tissues, especially on tooth surfaces. It had become a common knowledge that oral biofilm directly causes several oral diseases such as dental caries, periodontal disease, i.e., gingivitis and periodontitis, and many other oral diseases [1]. Compared with the planktonic microorganism, oral biofilm is masses of bacteria that form structure known as extracellular matrix

(ECM), that allows microorganism to persist under environmental conditions, and able to resist antimicrobial drugs [2]. In biofilm, there is a unique cell-to-cell communication system, namely quorum sensing (QS) that allows bacteria to detect and respond to cell population density mediating gene expression [3, 4]. It has been reported that QS is also responsible in antimicrobial resistance through regulating bacteria multidrug resistance (MDR) efflux pumps, regulating biofilm formation, and regulating bacterial secretion systems [5–9].

For many decades, antimicrobial agents, i.e., chlorhexidine (CHX) have become the best weapon against bacteria in oral cavity. However, CHX is less effective against biofilm bacteria because of the drug resistance properties of biofilm [10, 11]. This condition led researchers to develop another method to fight against biofilm, including use of alternative drugs, such as plant-derived compounds or essential oils. On the other hand, medicinal plants or herbs have been proved empirically and scientifically to have some important biological activities. As antibacterial and antibiofilm, medicinal plant-derived compounds and essential oils could inhibit biofilm formation by inhibiting peptidoglycan synthesis, modulating QS, and damaging bacteria membrane structures [12, 13]. Nowadays the use of natural products and their derivatives in dentistry, especially to prevent dental caries, is receiving large attention [14]. Moreover, many studies have reported the effect of various medicinal plant extracts on inhibiting biofilm formation and inhibiting bacterial adhesion. These suggest that medicinal plant-derived compounds might become promising alternative therapy in dental care.

Curcuma xanthorrhiza Roxb., known as Javanese turmeric or "temulawak," is a native Indonesian medicinal plant, which has been utilized traditionally as an ingredient of *jamu* (Indonesia herbal supplement and medicine) [15]. Most people use the rhizome as they believe it has medicinal effect for stomach illness, liver ailments, constipation, bloody diarrhea, dysentery, arthritis, children's fevers, hypotriglyceridemic, hemorrhoids, vaginal discharge, rheumatism, and skin eruptions empirically [16, 17]. Furthermore, the beneficial medicinal effect of *C. xanthorrhiza* has been proven in scientific studies. *C. xanthorrhiza* has been confirmed to have pharmacology effects such as anti-inflammatory, antibacterial, antioxidative, neuroprotective, nephroprotective, antitumor, and hepatoprotective activities [18–22]. Recently, in dentistry scope, the development of *C. xanthorrhiza*-derived compound as antibacterial drug has been extensively studied especially in East Asia and Southeast Asia countries. Xanthorrhizol (XNT) is the one of main active compound isolated from the essential oil of the rhizomes of *C. xanthorrhizza*, has a variety of pharmacological activities, one of that is antibacterial effects [23]. The bactericidal and bacteriostatic activity of xanthorrhizol against several oral bacteria has been reported using planktonic or biofilm models and showed promising result.

Thus, the use of *C. xanthorrhiza*-derived compounds as antibacterial and antibiofilm agent could be advantageous because natural-based medicines have fewer side effects. In this chapter, we will outline and summarize about inhibition of biofilm formation, mechanism action, and potential roles of *C. xanthorrhiza*-derived compounds as antioralbacterial and antioralbiofilm.

2. Oral biofilm and the most common oral infectious disease

The human oral cavity is a dynamic environment, which houses the most diverse microbiota, inhabited by more than 700 species of bacteria that colonize in the surfaces of both hard and soft tissues [24]. Inside the oral cavity there are two types of bacteria: a single free-living cell known as planktonic bacteria mostly found in saliva, and multicellular-living, where the cells are sessile and live in biofilm. Oral biofilm is a complex community of microorganisms, which are attached on the oral surface and embedded in an extracellular matrix. Thus, the biofilm-associated bacteria differ compared with the planktonic bacteria in many ways, for example, growth rate, gene expression, transcription, and translation because bacteria biofilm lives in different complex microenvironments due to higher cell density of heterogeneous bacteria community [25]. The formation of the three-dimensional structure of biofilm causes the bacteria to be protected from the various environmental stresses, such as antimicrobial drugs.

The development of oral biofilm is a multistep process. The initial stage is pellicle formation on tissue surface, which is composed of a variety of host-derived molecules and source of receptors such as mucins, agglutinins, proline-rich proteins, phosphate-rich proteins, and enzymes such as α-amylase that could be recognized by early colonizer. These receptors allow various planktonic bacteria, which have been classified as early colonizer, such as *Streptococci* species that constitute around 60–90% of the bacteria that first colonize the teeth, and other bacteria include *Actinomyces* sp., *Capnocytophaga* sp., *Eikenella* sp., *Haemophilus* sp., *Prevotella* sp., *Propionibacterium* sp., and *Veillonella* sp. [26]. However, at this stage, the bacteria are still susceptible against antimicrobial drugs, because the biofilm matrix structure is not completely formed.

The interaction between the early-colonizing bacteria has been shown to regulate many gene expression in response to the environment and provide specific direct binding sites (not through salivary glycoprotein for various other bacteria to colonize) and promote the development of biofilm. The bacteria that bind to this initial layer of biofilm are known as known as late colonizers such as *Fusobacterium nucleatum*, *Treponema* sp., *Tannerella forsythensis*, *Porphyromonas gingivalis*, *Aggregatibacter actinomycetemcomitans*, etc. They recognize polysaccharide or protein receptors on the pioneer bacteria cell surface and then attach on them [26]. The presence of late colonizer bacteria causes the change of environment and proportional shift, for example, relative amount of *Streptococci* sp. and *Neisseria* sp. is decreased, while the amount of *Actinomyces* sp., *Corynebacterium* sp., *Fusobacterium* sp., and *Veillonella* sp. increases [27]. The proportional shift occurs due to the interaction between bacteria in the community and the change of environment in biofilm. The competitive and cooperative interaction in biofilm may be essential to develop a successful mixed-species colonization.

During biofilm formation, there's cell-to-cell communication in the biofilm called QS. This phenomenon is mediated through production and release of chemical signals by bacteria termed autoinducer (AI), as response to changes in bacterial density and environment in biofilm. This mechanism initiates modification in gene expression to regulate cell or group behavior. During the maturation biofilm phase, QS also plays an essential role in extracellular matrix (ECM) production [28]. The ECM is a mixture of secreted high-molecular-weight polymers produced by bacteria, consisting of three major components: extracellular polysaccharides (EPS), proteins, and extracellular DNA, which form a cross-linked meshwork that serves as a shield [29]. At this stage, the biofilms show maximum resistance to antimicrobial drugs. The presence of biofilm ECM represents a strong barrier. The molecules of antimicrobial drugs must diffuse through the biofilm matrix to inactivate the bacterial cells. The biofilm ECM contains numerous anionic and cationic molecules that can bind charged molecules of antimicrobial drugs [30]. The resistance provided by ECM may be discouraged by longer exposure and higher concentration of antimicrobial drugs; however, the toxicity for oral application should be the main consideration.

The drug resistance of oral biofilm against antimicrobial drugs becomes the main problem in eliminating oral biofilm. Other mechanisms that have been proposed to explain how bacteria protect itself from the effects of antimicrobials such the ability to adapt to various stress responses; the decrease of growth rate and metabolism; efflux pump mechanism; and QS [7, 10, 31].

Dental caries is the most common oral infectious disease characterized by acidic damage on the tooth surface due to a localized structural demineralization that leads to cavitation [32]. The bacteria that are responsible for the initiation of such cavitation process are the acidogenic, Gram-positive, facultative anaerobic bacteria, *Streptococcus mutans*. *S. mutans* along with other species from the same genus, *S. mitis,* are some of the early colonizers of oral biofilm that provide adherence for other microorganisms promoting the growth and maturation of the biofilm. Recent study by Dongyeop Kim et al. [33] showed that the rotund-shaped biofilm with corona-like cell segregation where *S. mutans* located at the very core created a highly acidic region at the interface between the biofilm and enamel, resulting in the characteristics of localized demineralized surface as commonly seen in clinical setting [33]. Therefore, not only dental caries is a diet-dependent disease but also a biofilm-dependent disease [32]. As the understanding of the nature of dental caries grows, the approach of caries management has been shifted from the previously popular approach that focused more on the symptomatic treatment and removal of carious tissue to be replaced by artificial structure, to the current approach that emphasizes the preventive measures: restriction of dietary sugar consumption, removal of bulk bacterial mass through brushing, and reduction of cariogenic bacteria in dental biofilm through chemotherapeutic methods [34].

While dental caries is a result of a chronic destruction of the tooth hard tissue itself, periodontal disease on the other hand is an inflammatory disease of the surrounding tissue of tooth, which may result in loss of attachment, and induced and maintain by the resident of oral biofilm, especially the biofilm located in the gingival crevices that stay in contact with the gingival epithelium [35, 36]. Different from the microbes of the dental caries-related biofilm located on the tooth surface whose ability is to transform carbohydrate into damaging acidic substrates, the microbes of the biofilm in the gingival crevices gain their source of nutrient mainly from the protein-rich gingival cervicular fluid (GCF) accommodating the growth of Gram-negative bacteria, some of which are responsible for the progression of periodontal diseases [35]. Gram-negative, anaerobic, proteolytic bacteria, namely *P. gingivalis*, *Prevotella intermedia*, and *A. actinomycetemcomitans,* are mostly found in the periodontal biofilm and linked to periodontal diseases due to their ability to release toxins that induce host proinflammatory response, which in turn creates an ecological shift to a dysbiosis and causes damage to the periodontal structure [37, 38].

2.1 Current treatment and challenges using CHX and other antibacterial agents/mouth rinse

The general treatments of periodontal disease are mechanical debridement and ensuring that the proper oral hygiene is maintained by the patient. The use of antibiotics for periodontal disease other than aggressive periodontitis is still controversial to date [36]. Concern has been raised toward drug tolerance and resistance of periodontal bacteria. A study done in Colombia showed that bacterial isolates from subgingival biofilm of patient with aggressive periodontitis (*A. actinomycetemcomitans, P. gingivalis,* and *Tannerella forsythia)* were resistant to amoxicillin, azithromycin, and metronidazole [39]. Considering the nature of periodontal biofilm, mechanical disruption of the biofilm's integrity and reduction of the biofilm mass prior to the administration of antibiotics are considered essential [40].

Although CHX is considered as the gold standard antimicrobial agent in the oral cavity, there are some drawbacks of its usage: the risk of extrinsic staining on tooth surface, alteration in taste perception, and increase in calculus formation [41, 42]. Moreover, the effectiveness of CHX for biofilm eradication is also questioned. Due to the fact that *S. mutans* is the early colonizer of dental biofilm and that it inhibits the lowest strata, administration of CHX results in a concentration gradient from the outermost surface of the biofilm toward its innermost area that in turn exposes the *S. mutans* to only subinhibitory concentration of CHX [33, 43]. This suggestion is supported by another research conducted by spatially mapping the architecture of dental biofilm, which found that the intact corona structure of biofilm that conceals *S. mutans* cells in the core beneath layers of other microbes provides enhanced antimicrobial tolerance against CHX [33]. On the other hand, increasing the concentration of CHX in the aim to eliminate the dental caries-related biofilm is not recommended because the wide spectrum nature of CHX will disturb the balance of the oral environment by perturbing the commensal microbiome. As a prevention of periodontal diseases, several studies found its benefit to prevent bacterial surface adhesion, thus preventing the biofilm formation [44]. However, when the biofilm has formed, Gram-negative bacteria such as *P. gingivalis* are able to secrete outer membrane vesicles to bind CHX and provide protection to the bacteria in the biofilm community [43].

To avoid the aforementioned side effects and concerns, treatment and prevention alternatives from many natural products, herbs, and medicinal plants, in the form of extracts and essential oils, have been developed. Medicinal plant's extract from *Acacia arabica*, *Tamarix aphylla* L., and *Melia azadirachta* L. showed evidence of reducing oral biofilm formation and cleaning the well-developed oral biofilm [45]. Medicinal plant from South East Asia, *C. xanthorrhiza* Roxb., has also been proven through several studies to have eradication and inhibition effects against oral bacteria and candida biofilm [23, 46–53].

3. *Curcuma xanthorrhiza* Roxb

Curcuma xanthorrhiza Roxb., known as Java turmeric or "temulawak," is a native Indonesian medicinal plant that is mainly cultivated in Southeast Asian countries such as Indonesia, Malaysia, Thailand, Vietnam, and Philippines. For a long time, it has been used to enhance the flavor and color of food. Moreover, this plant has been believed and utilized as medication and supplement [15, 17]. In a few decades, turmeric plants including *C. xanthorrhiza* became the main subject of interest in research because many of its biological activities have been confirmed by experimental scientific studies. In addition, *C. xanthorrhiza may* be used as a treatment for COVID-19 because of its ability to inhibit proinflammatory cytokines [54]. However, it's still requiring more evaluation, especially in the clinical trial setting. Thus, recently market demand for *C. xanthorrhiza* rhizome has increased globally.

C. xanthorrhiza is a low-growing plant (2–2.5 m) with a root known as rhizome that looks like ginger. This plant can grow in the lowlands to an altitude of 1500 meters above sea level and has a habitat in tropical forests. The main part of

C. xanthorrhiza that has been proved to have beneficial medicinal activity is rhizome [15]. The rhizome of *C. xanthorrhiza* contains terpenoid and curcuminoid compounds, which reportedly have beneficial properties such as antioxidant, anti-inflammatory, antitumor, and anticancer effects [18, 20–22, 55]. The shape of the rhizome of *C. xanthorrhiza* is oval round shape, 3–4 branched, and reddish brown, dark yellow, or dark green in skin color (**Figure 1**). The rhizome flesh is dark, orange or brown in color, has a sharp pungent aroma and tastes bitter.

3.1 Phytochemical properties of *C. xanthorrhiza* Roxb

The rhizome of *C. xanthorrhiza* contains curcuminoids (1–2%), essential oil (3–12%), xanthorrhizol (44.5%), and camphor (1.39%). Moreover, xanthorrhizol (XNT), a bisabolene-type sesquiterpenoid compound isolated from essential oil of rhizome's *C. xanthorrhiza*, had been well established to possess various medicinal effects XNT is one of the most explored and studied phytochemicals, especially its antibacterial, antifungal, and antibiofilm activity. The major group of secondary metabolites has been identified in the rhizome of *C. xanthorrhiza* and can be seen in **Figure 2** [17]. However, the variation of active metabolite of *C. xanthorrhiza* might be influenced by several external factors, such as climate, sun intensity, altitude, and temperature of cultivation. For example, the high percentage of starch is influenced by the altitude of cultivation. The bioactive compound XNT and curcuminoid also reported higher in low altitude, high temperature, and low rainfall [56]. Thus, these are the challenges for development standardization phytomedicine, because of the vast variation of external factor and the different method of cultivation in each site.

3.1.1 C. xanthorrhiza Roxb. *Extraction preparation*

The *C. xanthorrhiza*-derived products, such as extract or as pure compounds, have provided unlimited opportunities for new drug discovery. However, to take advantage of the beneficial effect of the medicinal plant, an extraction process is carried out to obtain the active secondary metabolite. The extraction solvent selection is very essential because it affects the stability and metabolite profiles that implicate the efficacy of medicinal plant extract. Several commonly used solvents are ethanol, methanol, dichloromethane, acetone, and water [57, 58]. Proper actions must be taken to assure that potential compound is not lost or destroyed during the extraction process.

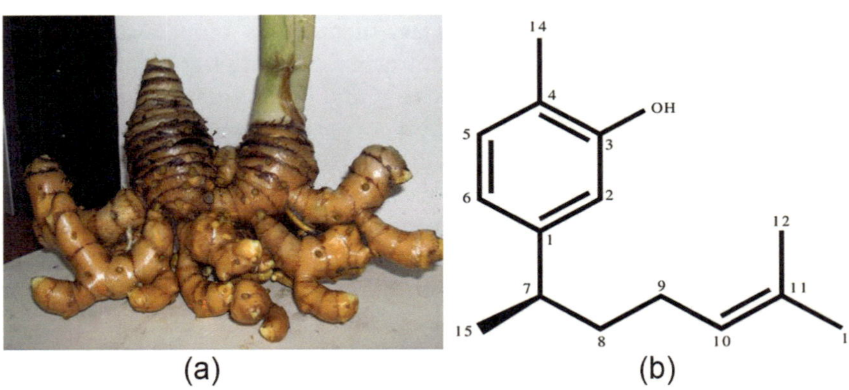

Figure 1.
(a) Rhizome of C. xanthorrhiza *Roxb. (b) Chemical structure of xanthorrhizol.*

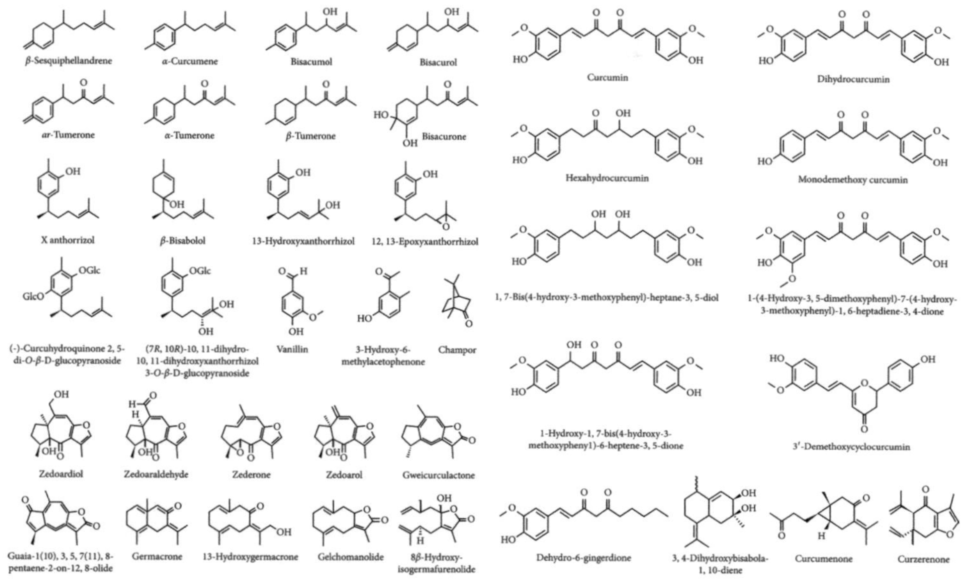

Figure 2.
Secondary metabolite compound of rhizome C. xanthorrhiza *Roxb.*

3.1.2 Xanthorrhizol isolate

XNT is an essential bioactive compound isolated from essential oil of rhizome *C. xanthorrhiza*. There are several methods used to extract the essential oil and XNT, i.e., supercritical fluid carbon dioxide extraction (SCFE-CO_2), Soxhlet extraction, and percolation process [59]. According to Salea et.al (2014), extraction using SCFE-CO_2 method will result in higher XNT compared with Soxhlet or percolation extraction method. Besides that, the conventional method to isolate XNT, which costs less, is still applicable and more efficient, while SCFE-CO_2 method is more applicable in large-scale production in the industry [59].

The interest in XNT as an antibacterial has attracted some researchers to develop as plant-derived drugs. The molecular weight and solubility of XNT are 218.33 g/mol and 28.90 μg/ml, respectively. This makes XNT have lower molecular weight and higher solubility compared with bioactive compound curcumin [60, 61]. Thus, it was expected that XNT might easily penetrate the surface of biofilm. According to the chemical structure, XNT and curcuminoid contain phenolic compounds and hydrocarbons.

4. Antibacterial and antibiofilm activity

4.1 Antibacterial

The antibacterial activities of *C. xanthorrhiza* have been studied using various preparations such as extract or fraction preparation and XNT isolation. *C. xanthorrhiza* extract and XNT have been reported to be effective against a variety of oral bacteria. They have been evaluated by standard in vitro susceptibility tests such as minimum inhibitory concentration (MIC) and minimum bactericidal concentration

(MBC). Our studies have shown that the effectiveness of *C. xanthorrhiza* ethanol extract against Gram-positive bacteria was superior compared with its effect against Gram-negative bacteria. In addition, the efficacy of *C. xanthorrhiza* extract and XNT against Gram-positive bacteria is comparable to CHX [48, 49, 52].

The antibacterial activity of C. *xanthorrhiza* is believed to emerge from XNT and curcuminoid compounds. The mechanism of action of phenol compounds, through interaction between the hydroxyl group (-OH) of the phenol compound with bacterial cells wall to facilitate hydrogen bonds subsequently causes alteration of bacterial membrane permeability. The high concentration of phenol can penetrate into cells subsequently leading to protein coagulation on the cell membrane and cell lysis [62, 63].

The Gram-negative bacteria are more resistant to phenol due to the complexity of their cell wall. Gram-positive bacteria possess thick cell walls containing many layers of peptidoglycan and teichoic acids. In contrast, Gram-negative bacteria possess thinner cell walls, but consist of a few layers of peptidoglycan surrounded by lipid membrane (lipopolysaccharides and lipoprotein). The complex cell wall of Gram-negative bacteria has been predicted to slow down the passage of chemicals. This was supported by a previous study by Inouye et al. [64], which concluded that the antibacterial effect of polyphenols was generally more effective against Gram-positive bacteria than Gram-negative [64].

XNT isolate is more effective against bacteria compared with the extract form. Since the crude extract contains various types of bioactive compounds or phytochemicals, usually unnecessary components are still carried away during the extraction process, for example, starch found in *C. xanthorrhiza* extract. Moreover, that unnecessary component has been suggested can affect the bioactive compounds activity. The XNT has been reported effective against several Gram-negative bacteria such as *Fusobacterium nucleatum* and *Enterococcus faecalis* [65, 66].

In addition, a clinical study evaluated the effectiveness of XNT, neem, cetylpyridinium chloride, and 0.2% CHX to decontaminate 60 children's toothbrushes after being used. Their result showed that the antimicrobial effect of XNT on *S. mutans* (78% reduction in *S. mutans*) was higher compared with CHX, but lesser than neem and cetylpyridinium chloride [67].

4.2 Antibiofilm

The *C. xanthorrhiza* extract and XNT also have been reported to have activity as antibiofilm against several oral bacteria in single species biofilm models. The antibiofilm activity of *C. xanthorrhiza* has been reported in various phases of biofilm formation. Rukayadi study reported that the activity of XNT as an antibiofilm was dependent on the concentration, exposure time, and the phase growth of biofilm. XNT is more effective in the early phase of biofilm formation [68]. Consistent with that, our study also demonstrated that the antibiofilm activity of *C. xanthorrhiza* ethanol extract is more effective in the early phase of biofilm formation. These indicate that the EPS matrix of mature biofilm implicates the resistance [46, 50, 51]. Although high concentration of XNT (1000 μg/mL) reportedly completely killed the biofilm, toxicity should be a major concern.

C. xanthorrhiza extract and XNT have been reported to inhibit several single species biofilm formations in in vitro study. Although not completely eliminated, bacteria were removed in the adhesion phase and early accumulation phase of biofilm development. The mechanism of inhibition biofilm formation is still not clear yet. However, it has been reported that *C. xanthorrhiza* extract has shown to inhibit acid production of *S. mutans* biofilm [53]. Moreover, *C. xanthorrhiza* extract is also reported to have anti-QS or quorum quenching activity [69]. The high level of tannin, phenol, phenolic compound in *C. xanthorrhiza* is suggested to precipitate the proteins that are vital for *rhl* system in *Pseudomonas aeruginosa.* By inhibiting the *rhl* system, the swarming activity of *P. aeruginosa* is inhibited, thus the QS will not take place [69, 70]. Besides that, killing the cells by cell lysis will also degrade and detach the biofilm.

Besides inhibiting the biofilm formation, *C. xanthorrhiza* extract and XNT also reportedly can eradicate the mature biofilm. The in vitro study against single species 72-hour *S. mutans* biofilm model, treated with *C. xanthorrhiza* methanol extract, showed significant fewer colony forming unit (CFU). The TEM and SEM observation showed changes of peptidoglycan layer of *S.mutans* and fewer intact bacteria after treatment [53].

Because the biofilm matrix can limit the penetration of antimicrobial agents, Cho et al. [71] explored the nanoemulsion form of *C. xanthorrhiza* oil in order to facilitate the ease of penetration. The single species *S. mutans* biofilm model, which was treated with nanoemulsion of *C. xanthorrhiza* oil, showed higher dead cells compared with the live cells. Furthermore, quantitative analysis of live/dead biomass and biofilm thickness based on the CLSM images showed that the live/dead ratio with nanoemulsion treatment was 50% less compared with control. It was also reported that nanoemulsions, which were prepared using sonication, are more suitable to be used as antibiofilm materials than emulsions without sonication [71]. These results indicate that *C. xanthorrhiza* extract can penetrate the *S. mutans* biofilm and kill that cell.

Another in vitro study against root canal biofilm *F. nucleatum* presented that XNT at concentrations 1.25% and 1.5% reported similar eradication activity compared with 2.5% NaOCl [72].

The antibiofilm activity of *C. xanthorrhiza* extract and XNT has also been demonstrated in multispecies biofilm models. CLSM analysis demonstrated that biofilm treated with XNT at 2 and 10 μg/ml for 30 min results in reduced bacterial viability in a dose-dependent manner against saliva and multispecies oral biofilm. Moreover, when exposed to 1000 μg/mL XNT, all biofilm cells were completely killed. These results indicate that XNT provides antibiofilm properties by eradicating bacteria viability [73].

Generally, multispecies biofilms were considered to be more resistant to antibiofilm agent compared with single species biofilms. To evaluate this notion, we tested dual species biofilm models (combination Gram-positive and Gram-negative bacteria) treated with *C. xanthorrhiza* ethanol extract, then measured the minimum of biofilm eradication (MBEC) using MTT-assay to assess the viability cell (**Table 1**). Our study demonstrated that *C. xanthorrhiza* ethanol extract was better eradicating dual-species biofilm (for example, *S. sanguinis* with *Porphyromonas gingivalis*; or *S. mutans* with *A. actinomycetemcomitans*), whereas not effective against single-species *P. gingivalis* biofilm nor single-species *A. actinomycetemcomitans* [46, 51]. This result may be possible due to the antagonist interaction between *S. sanguinis* and *P. gingivalis* that causes an incomplete formation of the EPS matrix surrounding the biofilm. It is supported by a clinical study by Stingu et al. [74], who reported that the presence of *S. sanguinis* has an influence on the presence of *P. gingivalis*, where *S. sanguinis* was found more in healthy gingival sulcus [74], while *P. gingivalis* vice versa. *S. sanguinis* also can produce bacteriocin called streptomycin and hydrogen peroxide, which can inhibit the growth of *P. gingivalis* [75].

No	Tested biofilm species	Effect	Reference
1	4–24 hr. *S. sanguinis* ATCC 10556	In the early phase of biofilm formation (4 hr), at concentration 15% shows eradicate biofilm equivalent to CHX. While in 12 hr. and 24 hr. biofilm formation, the $MBEC_{50}$ is 0.5% and 20%, respectively. However, the result at maximum concentration was smaller compared to CHX.	[46]
2	4–24 hr. *S. mutans* ATCC 25175	In the early phase of biofilm formation (4 hr) and 12 hr., at concentration 15–20% shows eradicate biofilm equivalent to CHX. While in 24 hr. biofilm formation, at concentration 20–25%, showed equivalent to CHX	[50]
3	4–24 hr. *Porphyromonas gingivalis* ATCC 33277	In 12 hr. biofilm formation, the $MBEC_{50}$ is 0.5%. However Not effective against 24 hr. biofilm formation. Only reduced <40% bacteria viability	[46]
4	4–24 hr. *Aggregatibacter actinomycetemcomitans* NCTC 9710	In 12 hr. biofilm formation, at concentration 20% the viability still 50%. However Not effective against 24 hr. biofilm formation. Only reduced <30% bacteria viability	[51]
5	4–24 hr. *S. sanguinis–P. gingivalis*	In the early phase of biofilm formation (4 hr), at concentration 15% shows eradicate biofilm equivalent to CHX. While in 12 hr. and 24 hr. biofilm formation, the $MBEC_{50}$ is 0.5%. However, the result at maximum concentration was smaller compared to CHX.	[46]
6	4–24 hr. *S. mutans–P. gingivalis*	In the early phase of biofilm formation (4 hr) and (12 hr), shows can eradicate biofilm. But in the mature phase (24 hr), it is not effective. Maximum concentration only eradicates 50% bacteria viability.	[50]
7	4–24 hr. *S. mutans–A. actinomycetemcomitans*	Only effective in the early phase of biofilm formation (4 hr), at maximum concentration reduce 90% the bacteria viability. While in 12 hr. and 24 hr. only reduce 50% and 20% bacteria viability, respectively	[51]

Table 1.
Effect of C. xanthorrhiza ethanol extract *against single and dual species biofilm.*

5. Conclusion

A fight against oral infectious disease is a fight against an adaptive, highly advanced, multispecies, pathogenic oral microbial community comprising oral biofilm. Inhibition and elimination of oral biofilm by means of preventing and treating oral diseases require pharmacological developments in finding alternative therapies that are able to dodge the defensive nature of oral biofilm and avoid cytotoxicity to the host while maintaining the homeostasis of the oral environment. *Curcuma xanthorrhiza* Roxb.–derived compounds such as XNT have been repeatedly proven to be a promising alternative therapy in dental care for its antimicrobial and antibiofilm activity. The phenolic compound of XNT has been proven to alter the permeability of the bacterial cell wall that leads to cell lysis. It is also proposed to prevent QS by inhibiting the swarming activity of bacteria. Further research to obtain the most effective form of compound and research in clinical settings are still needed to fully harness its potential.

Acknowledgements

The authors would like to thank the Universitas Indonesia for providing grants to our studies.

Author details

Dewi F. Suniarti[1*], Ria Puspitawati[1], Rezon Yanuar[2] and Ranny R. Herdiantoputri[3]

1 Faculty of Dentistry, Department of Oral Biology, Universitas Indonesia, Jakarta, Indonesia

2 Faculty of Dentistry, Department of Pharmacology, Health Sciences University of Hokkaido, Tōbetsu, Japan

3 Department of Oral Pathology, Graduate School of Medical and Dental Sciences, Tokyo Medical and Dental University, Tokyo, Japan

*Address all correspondence to: dewisuniarti17@gmail.com

References

[1] Marsh PD, Zaura E. Dental biofilm: Ecological interactions in health and disease. Journal of Clinical Periodontology. 2017;**44**(S18):S12-S22

[2] Wu H-J, Wang AHJ, Jennings MP. Discovery of virulence factors of pathogenic bacteria. Current Opinion in Chemical Biology. 2008;**12**(1):93-101

[3] Lupp C, Urbanowski M, Greenberg EP, Ruby EG. The Vibrio fischeri quorum-sensing systems Ain and lux sequentially induce luminescence gene expression and are important for persistence in the squid host. Molecular Microbiology. 2003;**50**(1):319-331

[4] Pumbwe L, Skilbeck CA, Wexler HM. Presence of quorum-sensing systems associated with multidrug resistance and biofilm formation in Bacteroides fragilis. Microbial Ecology. 2008;**56**(3):412-419

[5] Subhadra B, Oh MH, Choi CH. RND efflux pump systems in Acinetobacter, with special emphasis on their role in quorum sensing. Journal of Bacteriology and Virology. 2019;**49**(1):1-11

[6] Wang Y, Liu B, Grenier D, Yi L. Regulatory mechanisms of the LuxS/AI-2 system and bacterial resistance. Antimicrobial Agents and Chemotherapy. 2019;**63**(10):e01186-e01119

[7] Zhao X, Yu Z, Ding T. Quorum-sensing regulation of antimicrobial resistance in Bacteria. Microorganisms. 2020;**8**(3):425

[8] Williams P, Cámara M. Quorum sensing and environmental adaptation in Pseudomonas aeruginosa: A tale of regulatory networks and multifunctional signal molecules. Current Opinion in Microbiology. 2009;**12**(2):182-191

[9] Grohmann E, Christie PJ, Waksman G, Backert S. Type IV secretion in gram-negative and gram-positive bacteria. Molecular Microbiology. 2018;**107**(4):455-471

[10] Kouidhi B, Al Qurashi YMA, Chaieb K. Drug resistance of bacterial dental biofilm and the potential use of natural compounds as alternative for prevention and treatment. Microbial Pathogenesis. 2015;**80**:39-49

[11] Bonez PC, dos Santos Alves CF, Dalmolin TV, Agertt VA, Mizdal CR, Flores VC, et al. Chlorhexidine activity against bacterial biofilms. American Journal of Infection Control. 2013;**41**(12):e119-ee22

[12] Cox SD, Mann CM, Markham JL, Bell HC, Gustafson JE, Warmington JR, et al. The mode of antimicrobial action of the essential oil of Melaleuca alternifolia (tea tree oil). Journal of Applied Microbiology. 2000;**88**(1):170-175

[13] Gao M, Teplitski M, Robinson JB, Bauer WD. Production of substances by Medicago truncatula that affect bacterial quorum sensing. Molecular Plant-Microbe Interactions. 2003;**16**(9):827-834

[14] Koo H, Jeon JG. Naturally occurring molecules as alternative therapeutic agents against cariogenic biofilms. Advances in Dental Research. 2009;**21**(1):63-68

[15] Heyne K. In: Kehutanan D, editor. Tumbuhan Berguna Indonesia: Jilid 1. Jakarta: Yayasan Sarana Wana Jaya; 1987

[16] Pertanian BPP. In: Agriculture M, editor. Tanaman Obat - Warisan Tradisi Nusantara untuk Kesejahteraan Rakyat. Bogor: Balai Penelitian Tanaman Rempah dan Obat; 2019. p. 2

[17] Rahmat E, Lee J, Kang Y. Javanese turmeric (*Curcuma xanthorrhiza Roxb.*): Ethnobotany, Phytochemistry, biotechnology, and pharmacological activities. Evidence-based Complementary and Alternative Medicine. 2021;**2021**:9960813

[18] Li S-C, Lin C-C, Lin Y-H, Supriyatna S, Teng C-W. Protective and therapeutic effects of Curcuma xanthorrhiza on Hepatotoxin-induced liver damage. The American Journal of Chinese Medicine. 1995;**23**(03-04):243-254

[19] Ozaki Y et al. Chemical & Pharmaceutical Bulletin. 1990;**38**(4):1045-1048

[20] Lim CS, Jin D-Q, Mok H, Oh SJ, Lee JU, Hwang JK, et al. Antioxidant and antiinflammatory activities of xanthorrhizol in hippocampal neurons and primary cultured microglia. Journal of Neuroscience Research. 2005;**82**(6):831-838

[21] Kim SH, Hong KO, Hwang JK, Park K-K. Xanthorrhizol has a potential to attenuate the high dose cisplatin-induced nephrotoxicity in mice. Food and Chemical Toxicology. 2005;**43**(1):117-122

[22] Park JH, Park KK, Kim MJ, Hwang JK, Park SK, Chung WY. Cancer chemoprotective effects of Curcuma xanthorrhiza. Phytotherapy Research. 2008;**22**(5):695-698

[23] Hwang JK, Shim JS, Pyun YR. Antibacterial activity of xanthorrhizol from Curcuma xanthorrhiza against oral pathogens. Fitoterapia. 2000;**71**(3):321-323

[24] Jenkinson HF, Lamont RJ. Oral microbial communities in sickness and in health. Trends in Microbiology. 2005;**13**(12):589-595

[25] Sharma D, Misba L, Khan AU. Antibiotics versus biofilm: An emerging battleground in microbial communities. Antimicrobial Resistance and Infection Control. 2019;**8**(1):76

[26] Kolenbrander PE, Andersen RN, Blehert DS, Egland PG, Foster JS, Palmer RJ Jr. Communication among oral bacteria. Microbiology and Molecular Biology Reviews. 2002;**66**(3):486-505

[27] Ritz HL. Microbial population shifts in developing human dental plaque. Archives of Oral Biology. 1967;**12**(12):1561-1568

[28] Zhang F, Kwan A, Xu A, Süel GM. A synthetic quorum sensing system reveals a potential private benefit for public good production in a biofilm. PLoS One. 2015;**10**(7):e0132948

[29] Dragoš A, Kovács ÁT. The peculiar functions of the bacterial extracellular matrix. Trends in Microbiology. 2017;**25**(4):257-266

[30] Nadell CD, Drescher K, Wingreen NS, Bassler BL. Extracellular matrix structure governs invasion resistance in bacterial biofilms. The ISME Journal. 2015;**9**(8):1700-1709

[31] Stewart PS. Mechanisms of antibiotic resistance in bacterial biofilms. International Journal of Medical Microbiology. 2002;**292**(2):107-113

[32] Bowen WH, Burne RA, Wu H, Koo H. Oral biofilms: Pathogens, matrix, and Polymicrobial interactions in microenvironments. Trends in Microbiology. 2018;**26**(3):229-242

[33] Kim D, Barraza JP, Arthur RA, Hara A, Lewis K, Liu Y, et al. Spatial mapping of polymicrobial communities reveals a precise biogeography associated with human dental caries. Proceedings

of the National Academy of Sciences. 2020;**117**(22):12375

[34] Yon MJY, Gao SS, Chen KJ, Duangthip D, Lo ECM, Chu CH. Medical Model in Caries Management. Dentistry Journal. 2019;7(2):37

[35] Dahlen G, Fejerskov O, Manji F. Current concepts and an alternative perspective on periodontal disease. BMC Oral Health. 2020;**20**(1):235

[36] Lasserre JF, Brecx MC, Toma S. Oral microbes, biofilms and their role in periodontal and Peri-implant diseases. Materials. 2018;**11**(10):1802

[37] Xu W, Zhou W, Wang H, Liang S. Roles of Porphyromonas gingivalis and its virulence factors in periodontitis. Advances in Protein Chemistry and Structural Biology. 2020;**120**:45-84

[38] Lee HA, Park MH, Song Y, Na HS, Chung J. Role of Aggregatibacter actinomycetemcomitans-induced autophagy in inflammatory response. Journal of Periodontology. 2020;**91**(12):1682-1693

[39] Ardila C-M, Bedoya-García J-A. Antimicrobial resistance of Aggregatibacter actinomycetemcomitans, Porphyromonas gingivalis and Tannerella forsythia in periodontitis patients. Journal of Global Antimicrobial Resistance. 2020;**22**:215-218

[40] Pretzl B, Sälzer S, Ehmke B, Schlagenhauf U, Dannewitz B, Dommisch H, et al. Administration of systemic antibiotics during non-surgical periodontal therapy—A consensus report. Clinical Oral Investigations. 2019;**23**(7):3073-3085

[41] Sakaue Y, Takenaka S, Ohsumi T, Domon H, Terao Y, Noiri Y. The effect of chlorhexidine on dental calculus formation: An in vitro study. BMC Oral Health. 2018;**18**(1):52

[42] Zanatta FB, Antoniazzi RP, Rösing CK. Staining and calculus formation after 0.12% chlorhexidine rinses in plaque-free and plaque covered surfaces: A randomized trial. Journal of Applied Oral Science. 2010;**18**(5):515-521

[43] Cieplik F, Jakubovics NS, Buchalla W, Maisch T, Hellwig E, Al-Ahmad A. Resistance toward chlorhexidine in Oral Bacteria - is there cause for concern? Frontiers in Microbiology. 2019;**10**:587

[44] Chatzigiannidou I, Teughels W, Van de Wiele T, Boon N. Oral biofilms exposure to chlorhexidine results in altered microbial composition and metabolic profile. npj Biofilms and Microbiomes. 2020;**6**(1):13

[45] Khalid M, Hassani D, Bilal M, Butt ZA, Hamayun M, Ahmad A, et al. Identification of oral cavity biofilm forming bacteria and determination of their growth inhibition by Acacia arabica, Tamarix aphylla L. and Melia azedarach L. medicinal plants. Archives of Oral Biology. 2017;**81**:175-185

[46] Yanuar R, Suniarti DF, Djohan W. Efficacy of javanese turmeric ethanol extract in eradicating streptococcus sanguinis and porphyromonas gingivalis biofilm. International Journal of Applied Pharmaceutics. 2019;**11**(1):18-22

[47] Puspitawati R, Lewiyonah R, Herdiantoputri RR, Gultom FP, Suniarti DF. Eradication effect of javanese turmeric (curcuma xanthorrhiza. Roxb.) extract on the early phase of candida albicans biofilm. International Journal of Applied Pharmaceutics. 2018;**9**:117-120

[48] Marpaung LNF, Suniarti DF, Sarwono AT. Effectiveness of identified javanese turmeric ethanol extract for the inhibition of biofilm formation by streptococcus mutans and porphyromonas gingivalis. International Journal of Applied Pharmaceutics. 2019;**11**(1):23-26

[49] Hutauruk R, Suniarti DF, Djohan W. Potential of javanese turmeric ethanol extract in inhibiting streptococcus sanguinis and porphyromonas gingivalis biofilm formation. International Journal of Applied Pharmaceutics. 2019;**11**(1):13-17

[50] Rosyana M, Suniarti DF, Sarwono AT. Effectiveness of a javanese turmeric ethanol extract for eradicating streptococcus mutans and porphyromonas gingivalis biofilms. International Journal of Applied Pharmaceutics. 2019;**11**(1):27-31

[51] Diana R, Joenoes H, Djais AA. The effect of curcuma xanthorrhiza ethanol extract on the viability of streptococcus mutans and aggregatibacter actinomycetemcomitans (dental biofilm research: in vitro study). Asian Journal of Pharmaceutical and Clinical Research. 2017;**10**(17):30-33

[52] Fidhianissa, Joenoes H, Djais AA. The effect of curcuma xanthorrhiza - ethanol extract to biofilm formation of streptococcus mutans and aggregatibacter actinomycetemcomitans (dental biofilm research: in vitro study). International Journal of Applied Pharmaceutics. 2017;**9**(0):20-22

[53] Kim J-E, Kim H-E, Hwang J-K, Lee H-J, Kwon H-K, Kim B-I. Antibacterial characteristics of Curcuma xanthorrhiza extract on Streptococcus mutans biofilm. The Journal of Microbiology. 2008;**46**(2):228

[54] Nugraha RV, Ridwansyah H, Ghozali M, Khairani AF, Atik N. Traditional herbal medicine candidates as complementary treatments for COVID-19: A review of their mechanisms, pros and cons. Evidence-based Complementary and Alternative Medicine. 2020;**2020**:2560645

[55] Maheshwari RK, Singh AK, Gaddipati J, Srimal RC. Multiple biological activities of curcumin: A short review. Life Sciences. 2006;**78**(18):2081-2087

[56] Latifah KD, Mono R, Edy Djauhari P, Waras N. Variasi Bahan Bioaktif dan Bioaktivitas Tiga Nomor Harapan Temulawak pada Lokasi Budidaya Berbeda. Indonesian. Journal of Agronomy. 2012;**40**(2):153-159

[57] Mangunwardoyo RW. Antimicrobial and identification of active compound Curcuma xanthorrhiza Roxb. International Journal of Basic and Applied Sciences. 2012;**12**:9

[58] Diastuti H, Syah YM, Juliawaty LD, Singgih M. Antibacterial Curcuma xanthorrhiza extract and fractions. Journal of Mathematical and Fundamental Sciences. 2014;**46**:11

[59] Salea R, Widjojokusumo E, Veriansyah B, Tjandrawinata RR. Optimizing oil and xanthorrhizol extraction from Curcuma xanthorrhiza Roxb. Rhizome by supercritical carbon dioxide. Journal of Food Science and Technology. 2014;**51**(9):2197-2203

[60] Modasiya MK, Patel VM. Studies on solubility of curcumin. International Journal of Pharmacy and Life Sciences. 2012;**3**(3):8

[61] Lee LY, Shim J-S, Rukayadi Y, Hwang J-K. Antibacterial activity of Xanthorrhizol isolated from Curcuma xanthorrhiza Roxb. against foodborne pathogens. Journal of Food Protection. 2008;**71**(9):1926-1930

[62] Khalid GS, Hamrah MH, Ghafary ES, Hosseini S, Almasi F. Antibacterial and antimicrobial effects of Xanthorrhizol in the prevention of dental caries: A systematic review. Drug Design, Development and Therapy. 2021;**15**:1149-1156

[63] Oon SF, Nallappan M, Tee TT, Shohaimi S, Kassim NK, Sa'ariwijaya MSF, et al. Xanthorrhizol: A review of its pharmacological activities and anticancer properties. Cancer Cell International. 2015;**15**:100

[64] Inouye S, Yamaguchi H, Takizawa T. Screening of the antibacterial effects of a variety of essential oils on respiratory tract pathogens, using a modified dilution assay method. Journal of Infection and Chemotherapy. 2001;**7**(4):251-254

[65] Prijatmoko D, Syafira NL, Lestari PE. Antibacterial activity of essential oil extracts from Curcuma xanthorrhiza roxb. Rhizomes against bacteria causing pulp necrosis. Journal of Dentomaxillofacial Science. 2018;**3**(3):2018. DOI: 1015562/jdmfsv3i3763

[66] Wonyoung Y, Minju S, Si-Mook K, Baek-il K, Tai-Cheol Y, Euiseong K. The antibacterial effect of xanthorrhizol as an endodontic irrigant on enterococcus faecalis. The Journal of the Korean Dental Association. 2016;**54**(3):11

[67] Bhat PK, Badiyani BK, Sarkar S, Chengappa S, Bhaskar NN, Patil S. Effectiveness of antimicrobial solutions on Streptococcus mutans in used toothbrushes. World Journal of Dentistry. 2012;**3**:6-10

[68] Rukayadi Y, Hwang JK. In vitro activity of xanthorrhizol against Streptococcus mutans biofilms. Letters in Applied Microbiology. 2006;**42**(4):400-404

[69] Mustaqim F, Rukayadi Y, Radu S. Inhibition of Pseudomonas aeruginosa quorum sensing by Curcuma xanthorrhiza Roxb. Extract. Journal of Pure and Applied Microbiology. 2019;**13**(3):13

[70] Ugurlu A, Karahasan Yagci A, Ulusoy S, Aksu B, Bosgelmez-Tinaz G. Phenolic compounds affect production of pyocyanin, swarming motility and biofilm formation of Pseudomonas aeruginosa. Asian Pacific Journal of Tropical Biomedicine. 2016;**6**(8):698-701

[71] Cho M-Y, Kang S-M, Lee E-S, Kim B-I. Antimicrobial activity of Curcuma xanthorrhiza nanoemulsions on Streptococcus mutans biofilms. Biofouling. 2020;**36**(7):825-833

[72] Fitri M, Nazar K, Meidyawati R, Azmi R. Antibacterial effect of xanthorrhizol (curcuma xanthorrhiza roxb.) against the biofilm of fusobacterium nucleatum. International Journal of Applied Pharmaceutics. 2020;**12**(2):57-61

[73] Yanti Y, Rukayadi Y, Lee K-H, Han S-H, Hwang J-K. Anti-biofilm activity of Xanthorrhizol isolated from Curcuma xanthorrhiza Roxb. against bacterial biofilms formed by saliva and artificial multi-species Oral strains. Food Science and Biotechnology. 2009;**18**(2):5

[74] Stingu C-S, Eschrich K, Rodloff AC, Schaumann R, Jentsch H. Periodontitis is associated with a loss of colonization by Streptococcus sanguinis. Journal of Medical Microbiology. 2008;**57**(4):495-499

[75] Schlegel RF, Slade HD. Alteration of macromolecular synthesis and membrane permeability by a Streptococcus sanguis bacteriocin. Journal of General Microbiology. 1974;**81**:3

Section 4

Bacterial Biofilms on Livestock, Environment and Pharmaceutical Applications

Chapter 16

Effect of Biofilm on Production of Poultry

Dayamoy Mondal

Abstract

Attachment of bacterial biofilm to the surfaces of farm, fomites and equipments remains chance transmission of infection poultry and human through food chain. Formation of biofilm causes spoilage of poultry products during processing of eggs, meat and distribution. Biofilm may cause many bacterial species in biofilm society. The formation of biofilm deteriorates food quality, water supply system, drugs resistance, and reduces the efficacy of equipments, spread disease and lingering of disease course. Common bacteria cause biofilm in poultry farm and food industries are *Salmonella* sp., *Staphylococcus* spp., *Listeria monocytogenes, Escherichia coli, Klebsiella pneumonae, Campylobacter jejuni, Streptococcus agalactiae*. Formation of biofilm is under stress and regulated by several genes of bacterial. There are several methods of diagnosis of biofilm such as Roll plate method, tube method, microtitre assay, PCR assay, mass spectrometry method and Biological assay of Biofilm. Therapeutic elimination of biofilms for smooth production of poultry is chemical and environmental modifications. Water may be treated with several means, both chemical and physical ways. Food-contaminated biofilm-related treatment is done applying quaternary ammonium compounds, aldehydes, phenolics, alkyl amines, chlorine dioxide, etc. Veterinary medical therapy against biofilms is use of antibiotics with ultrasound, low electric current, phage therapy, nanodrug delivery system, antimicrobial peptides, antiadhesin, antimatrix and chelating substances.

Keywords: biofilm, poultry, diagnosis, therapy, biofilmgene

1. Introduction

Biofilm is a complex structure of microbial populations having different bacterial colonies or monospecies cell type; adhere to the surface of growth. These cells are embedded in extracellular polymeric substances, the matrix substance which is generally composed of extracellular DNA (eDNA), proteins and polysaccharides, showed high resistance to antibiotics and physicochemical tolerance. The formation of biofilm have several impact in the poultry production, dessimination of infection and farm management system. In tropical countries different seasons such as horse summer, dry winter may acts as stress for formation of biofilm. These biofilm may affect the production performance, disease transmission and human health concern. Poultry farm and duckery where there is every chance of formation of biofilm needs special care and intervention against formation of biofilm and proper intervention for effective production and restriction of disease outbreaks.

2. Bacterial biofilming/biofouling conditions

Growth of bacterial population in colony or in a specific area or even in culture containers, the cells are stick to each other as well as with surface of growth container. The adherence materials are extracellular matters that may be composed of wide ranged components of extracellular polymers, these polymers may be with polysaccharides, proteins, lipid, pilli, flagella or even with eDNA). Not all microorganism can produce biofilm, some bacteria (both Gram negative and Gram positive), fungi and protists can produce biofilm. Most common bacteria those can produce biofilm are *Enterococcus faecalis*, *Staphylococcus aureus*, *Staphylococcus epidermidis*, *Streptococcus viridans*, *Escherichia coli*, *Klebsiella pneumoniae*, *Proteus mirabilis, Pseudomonas aeruginosa, Bacillus subtilis*, *Pseudomonus fluorescens*, *Streptococcus mutans, Streptococcus salivaris*, *Acitenobacter baumanni* [1–3].

Ornithobacterium rhinotracheale is a Gram-negative bacillus that causes respiratory disease in birds, and directly affects the poultry industry producing biofilm uncertain conditions [4]. Some common example of a biofilm are dental plaque, heart muscle, pond scum. Biofilms grow in rain forests, in desert as "desert varnish", ocean bottom as deep sea vent, glaciers in Antartic They have been found at the bottom of the ocean as early colonizers of new deep-sea vents and living on glaciers in the Antarctic. The biofilm may grow in normal conditions in industrial infrastructure, hospital, different living tissues and organs of animal and human. Biofilm formation at the air-liquid and solid-liquid interfaces are very common [5]. The origin of biofilm is not just in recent thought; it was present in the primitive earth condition for prokaryotes as defense mechanism. Inside the host, the extra cellular matrix protects biofilm making bacteria from expose to innate immune defenses (phagocytosis, opsonization and antibiotics [6]. Biofilm also helps against desiccation, antibiotics and host body defense immunity.

2.1 Abiotic condition for biofilm formation

Several conditions that may alter the formation of biofilms are temperature (37–40°C), presence of CO_2 (5%), low nutrient supplements in the media, water deprivation/hydrodynamics, osmolality of the medium, concentration metals such as iron and ambient acidity [7]. Several other factors also determine the biofilm formation, presence of toxicants, oxygen concentration, antibiotics and salinity of the environment affects for motion biofilm. Nature of substratum environment of the surface of attachment, glass and stainless steel surface are more hydrophilic for growth of biofilm than hydrophobic rubber, Teflon surface.

2.2 Biotic condition for biofilm formation

The biofilm formation may contain several communities of microbes with different species and class of organisms. The community composition where there may be several microorganisms like bacteria, fungi, algae in a biofilm population. Host stress is another factors growth of biofilm. Microbial genetic factors also a determinant. Several genes are responsible for attachment at the surface and subsequent maturation and dispersion of microbes particularly in *E. coli*. Population of microbes is another determinant that also affects the formation biofilm. Quorum sensing (QS) has big role on production and release of signal molecules called autoinducers. Production of several extracellular proteases that helps in dispersal of biofilm is regulated by QS system in *Staphylococcus aureas* and *B. subtilis*. Production of microbial byproducts such as metabolites like antibiotics, pigments, and siderophores also check the formation of biofilms. Antimicrobial peptides can restrict the development of biofilms. The antimicrobial peptides (bacteriocenes) such as dermicidin, tachyplesins are this kind of antibacterial peptides that may prevent for formation of biofilms and has potential clinical application against drug resistance and against biofilm formation [8].

3. Formation of biofilm

For the formation of biofilm an attachment with a surface is necessary, surface may be biotic or abiotic. Attachment at the surface may be with weak **Van der Wales** force and hydrophobic effect. In case of initial mild attachment is not disturbed, colonies are attached permanently with the cell adhesive structures like pilli, hami (archeal pilli like structure), flagellum [9]. Both motile and nonmotile as well as Gram positive and Gram negative bacteria aggregate together to form biofilm easily. During surface colonization (adhesion) bacterial cells can communicate by quorum sensing (cell to cell communication) traits like virulence factor [10] with the help of products such as N-acetyl homoserin lactone. Once the colonization begins on the settlement surface, the biofilm grows by a combination of cell division and cell recruitment. Besides quorum sensing molecules, several other signals trigger biofilm formation are secondary metabolites of bacteria such as antibiotics, pigments, siderophores. Sub-inhibitory concentration of antibiotic imipenem and tobramycin induce production of biofilm [11].

The composition of the biofilm is mostly with polysaccharides matrix (D-glucose; D-mannose; L-rhamnose [12] which is encloses bacteria forming cocoon like condition. In addition to polysaccharide in biofilm matrix there may be other materials such as protein, eDNA, extracellular enzymes like aminoglycoside modifying enzymes (AMEs), β-lactamase. Gram +ve and Gram –ve bacteria can produce biofilm. Few bacteria are more prone to form biofilm while some are less.

3.1 Stages of biofilm

There are several stages of biofilm formation starting from initial attachment. Five stages are there for complete formation of biofilm. They are stage of initial reversal attachment, irreversible attachment, maturation phase-I, maturation phase-II and dispersion. Other than bacteria protozoa, fungi, algae and archaea can produce biofilm. The common niche where the biofilm produced are slow sand filler, for water purification plant, percholating filler, mammalian intestine, animal and human organs such as urinary tract, endocardium, joint and articulations, heart valve, medical and veterinary tools and devices used may be affected with surface attachment in urinary catheter, prosthetic joints, pacemakers, stomach tube, teat syphone, milking machine etc.

In animal and veterinary medicine biofilm has tremendous impact in livestock industry and animal health that leads to tremendous economic loss. The most challenges posed with biofilm production causes antibiotic resistance which also a big threat to human health through food chain. A wide ranges of bacterial infections in veterinary importance are resistant to antibiotic therapy. Secondly, diseases are not responding to antibiotics when applied on certain disease conditions. Such pathological conditions are mastitis due *Streptococcus* and *Staphylococcus* species infection. Other diseases those also cause less healing response in pasturella pneumonia, enteritis on *E. coli* and *Salmonella* spp., urinogenital tract infection with *E. coli*, periodontal disease (*Staphylococcus* spp.), caseous lymphdenitis (*Corynaebacterium pseudotuberculosis*), wound infection (*Pseudomonus*, *Staphylococcus* spp. etc), pyomettra (*E. coli*) and others [2].

In poultry industries, several bacterial infections such as *Salmonellas* sp., produces biofilm in poultry meat industry that also cross contaminate public health impact [13].

4. Formation of biofilm in poultry industry

In poultry processing-during poultry processing the carcasses may come contact with many solid surfaces and forms biofilms. Bacteria may attach from carcass to the wet equipments. This may acts as continuous cross infection. Poultry plants and equipment's solid surface have different affinity for bacterial attachment and formation of biofilm. An increased extracellular matrix of fibrils and debris are connected with individual bacterial cell. Many bacteria of same species or different species may aligned in side to side pattern. Increase attachment of bacterial population and formation continuous biofilm may act as concern in poultry plant sanitization and pathogen control [14]. Eleven different species of bacteria have been isolated and identified from meat processing unit [15]. This may acts as constant source of infection to other carcasses that lead to public health concern too. Biofilm with slime layers with matrix enclosed bacterial population like population of a metro city. On the same bacterial surfaces similar and different species can adhere each other's side or interfaces. These bacterial population show community homeostatis, primitive circulatory cooperation and exchange of genetic materials as well as metabolic cooperation [16, 17]. Formation of biofilm on equipments and poultry plants cause damage of equipment, product contamination, loss of food energy and dissemination of infection. The microbes those affect the poultry industries are *E. coli*., *Listeria monocytogenes*, *Pseudomonus fluorecens*, *Acenetobactor harbinensis*, *Arthrobactor* sp., *Brochothrix thermosphacta*, *Carnobacterium maltaromaticum*, *Lactococcus piscium*, *Mycobacterium* spp., *Campylobacter jejuni*, *Pseudomonus fragi*, *Psycgrobacter* spp., *Rhodococcus erythropolis*, *Stenotrophomonas* sp. [15, 18]. In food processing environment, bacteria in biofilm as well as suspended forms undergo stresses such as dehydration, temperature variations, antimicrobial agents, therefore, their morphology is changed than their planktonic relatives. As a result they become more resistant (up to 500 times) to antimicrobials [16]. These bacteria also show slow growth not due to nutrient deficiency but due to stress. In the biofilm city/society all the species remain but some of the species contribute to enlarge the size of the biofilm. The formation of biofilm depends on different surface material made up of and nutrients content in the media. It has been reported that glass surface, stainless steel and plastic surface varies. Biofilm can be grown in any surface of stainless steel, glass, rubber, polycarbonate, polyurethane, polystyrene, polypropylene, Teflon, nitrile rubber, titanium, aluminum, ceramic, and wood for developing countries in poultry farms and industries [19]. The formation of biofilm of the above article surfaces remained 96,144 and 240 h with 10^6 cfu/cms for salmonella isolates [20]. Due to contamination of biofilm in poultry and poultry industries several diseases may occur.

5. Advantage and difficulties of biofilm formation in poultry industry

The bacteria show biofilm formation for their survival and to overcome hardship and stress. The extracellular polymeric substances (EPS) of biofilm is negatively charged and hydrophobic in nature helps to keep concentrated ions and dissolved carbon compound from the bulk fluid medium. The advantageous points for bacteria are (i) protection from antibiotics, and antimicrobials (ii) increased availability of nutrition's for their growth (iii) increased capacity of binding water molecules and avoiding dehydration (iv) keep close contacts with progeny, relatives and other bacteria for strategic ecology and transfer of plasmid (v) avoid adverse environments such as temperature, changed pH, antiseptics, disinfectants etc. Biofilm bacteria are more resistant than the planktonic ones, this due to acquisition of resistant genes in plasmid which also transmitted to other species in the biofilm colony.

6. Poultry farm, fomites and water supply system

Biofilm produced by bacterial species and population firmly adhere to the surfaces of attachment with its matrix EPS. These bacterial communities survive for long time and create resistance to various antibiotics; antimicrobials and disinfectants. These being potential contaminants in farm and fomites extend dessimination of infection to other population of birds, animals and human. The contaminants may be at any stage of farm and poultry industry particularly with very common organisms of Salmonella and Campylobactors [21]. The accumulation of biomass of biofilm affect major constrains in water supply in poultry industry. The bacterial biofilm may disturb in area of walls, floors, pipes, watered, drain, feed trough and utensils made up of steel, aluminum, nylon, rubber, plastic, glass and polystyrine [22]. In poultry industry particularly broiler farm, slaughter house, meat processing units, produces large amount of residues mainly proteins and lipids those are accumulated on the surface of containers, drains and waste chambers generate biofilm that eventually target the public health concern. Whatever the top most farm management may be for the poultry farming, there is every chance to be contaminated and formation biofilm with endemic pathogens with *E. coli*, *Pseudomonus* sp., *Salmonella* sp., Coliform bacteria and Enterobacter.

There is also chance of formation of biofilm and transmission of infection with *L. monocytogenes*, *Campylobacer jejuni* associated with poultry industry and diagnostic kit wares.

7. Biofilm potential source of economic loss

Production of biofilm in poultry industries cause huge economic loss, through food spoilage with constant source of potential infection sources and damage of water supply installations, equipments and water supply lines. A wide range of disease conditions causing by food contaminations such as gastroenteritis, abdominal colic, fever, indigestion and several other systemic diseases like respiratory disease, flaccid paralysis in human and veterinary importance. During high summer, the poultry units use cool ventillary system like cooler and wet straw cooling system which may have preexisting biofilm or it may generate biofilm that also spread infection to poultry population and poultry products. The chemicals and supplements used in poultry unit through feed and water may help in the propagation of biofilm.

The biofilm infection also causes certain condition in animals and birds. They are chronic inflammation, impaired wound healing, chronic skin diseases, formation of infectious emboli and antibiotic resistance. Poultry hatchery particularly duck hatchery is also a big sources of biofilm formation epicenter. Several instruments such as incubator, brooder, hover, brooder guards, and humidity chamber may be contaminated with biofilms [23].

Egg cold storage where eggs are stored, packaged and transported, may also be a potential source of biofilm producing concern. Eggs may be kept in trays and basket may be having preformed biofilm that also acts as potential constant source of infection for human through food chain and for next generation chicks/ducklings. Poultry pathogens like *Salmonella enterica*, can cause biofilm formation through feces of chicken and turkey and acts as very possible antimicrobial resistance [24].

8. Poultry drinking water standard

All the water supplied for poultry should be maximum cleaned and hygienic, there should be minimum level of microbe content and mineral composition in water. If the water content for microbial population and minerals are high, there should be option for big correction. More microbes and minerals content induce health hazard [25]. Several microbial loads that affect water quality for poultry farms due to different bacteria such as *E. coli*, *Salmonellas* spp., *Pseudomonas* spp., *Campylobacer jejuni*, coliform bacteria, *Enterobacter* spp., *L. monocytogenes*, *Staphylococcus* spp., are more common contaminants [26–29]. The microbial contents in the water vary with species and their numbers. The minimum and maximum level of bacteria usually occurs and permissible are 0–100 CFU/ml of water (**Table 1**). More bacterial nuclei in the water deviates the standards of health of birds and also the taste of water as well as amount of water used by the birds. There may be restriction in common salts content in water such as sodium (50–150 mg/L), sulphate (15–200 mg/L), nitrate (1–25 mg/L), zinc (0–1.5 mg/L), calcium (60 mg/L), ferrus salt (0.2–2.5 mg/L). The mineral contents in drinking water for poultry needs a standard with restriction of minimum and maximum level [30].

9. Bacterial biofilm and gene regulation in poultry industry

In poultry farming and meat processing industries several bacterial contaminations may be a common sequlae where large numbers of microbial contamination and transmission may occurs through egg, meat, fomites, machinery, water supply system and utensils used in this sector. Very common infections those affect the birds health could be forming biofilms. These may cause various economical and health concern in poultry industries. Salmonella is common pathogens in poultry system. *Salmonella gallinarum, S. typhimurium* and *S. enteritidis* are prevalent. The genes responsible for biofilm formation are *csgD* and *bcsA* adrA, gcpA [20, 31]. Lipopolysaccharide (LPS) producing gene is *rfbA* that helps formation biofilms. Using transposon mutagenesis, several genes such as *metE, ompR, rpoS, rfaG,*

Component of water	Level expected	Lower acceptable level	Higher acceptable level	Correction
Total microbes	0	100	300	Chlorination, sanitizing cleaner for safe water
Total aerobic plate count (cfu/ml)	0	<100	<1000	Water sanitizing
Total coliform (cfu/ml)	0	50	<50	Water sanitizing
Fecal coliform (cfu/ml)	0	0	0–1	Water sanitizing
E. coli (cfu/ml)	0	0	0–1	Water sanitizing
Pseudomonas (cfu/ml)	0	0		Water sanitizing
pH level	5–8	6.5	7.8	pH increase with soda, Na2Co$_3$,NaOH,Ca$(OH)_2$, pH decrease withHPO$_3$, H_2SO_4,HCL,citrate,vinegar.
Total water hardness	0–17	60	150	Acidification and use of polyphosphate to softenwater
Calcium salt mg/L	60	60–80	110	As above
Iron mg/L	0.2	0.3	0.4	Filtration and chlorination
Sodium mg/L	50	100	150	Reverse osmosis, sanitization
Sulphate mg/L	15	40	150	Treatment with oxidizing sanitizers then filtration
Lead mg/L	0.1	0.1	0.1	Water softeners and activated carbon can reduce lead
Nickel mg/L	0.01	0.01	0.05	Water softeners and activated carbon can reduce lead
Cadmium mg/L	0.5	3	5	Zinc oxide (ZnO), manganese oxide (MnO_2), titanium oxides (TiO_2), magnesium oxide (MgO)

Table 1.
Water standards and option for correction for poultry.

rfaJ, rfaK, rfaP, rfbH, rhlE, spiA, and *steB* are found to be associated with biofilm formation of *S. enteritidis* [32]. Similarly, there are several genes in *E. coli* bacterial genomes where many genes controls the formation of biofilms. Several adherence genes such as *luxS, iha, papC, aatA, aggR fimC* have been described [33]. Many other genes also have role in the biofilm formation. They are*fliC. csfA, luxS, adrA, gcpA* [20, 34]. Common poultry contaminate *E. coli* have many genes responsible for biofilm formation. Gene like*fliC, csgA, fimA, luxS, his, papC, aatA, aggR, fimC,* help in the formation and adhesion of bacterial growth on surface [3, 33].

Klebsiella pneumonia causes pneumonia, septicaemia and liver abscess in poultry. They parasitize in respiratory and gastrointestinal system. Formation of biofilm in different organs of poultry and poultry industry is very high by the organism (upto 93.6%). The samples that may transfer the biofilm through surgical wound, feces and other discharges [35, 36]. The genes responsible in *Klebsiella pneumonae* for formation of biofilm in poultry are *treC, sugE* which produce more capsular saccharides (cps) that helps in biofilm formation [37]. The *Enterococcus* sp. procuses biofilm frequently. The quorum sensing peptide pheromones (*cpd, cob, fsr, ccf*) are secreted by the cell to induce conjugate apparatus of doner cell. The bacteria transfer the pherome responsive plasmid which carry virilence genes promotes biofilm formation. *Enterococcus fecalis, E. faecium, E.durans, E. hirae,* and *E. cecorum* show biofilm formation in poultry. The most genes responsible for biofilm formation are *ebpB, ebpC* and *srt*. *Acenobacter baumanni* have some gene that cause biofilm. Serotypes have several gene that regulate biofilm are *omp*A, *bap*, *bla*PER-1, *csu*E, *csg*A, and *fim*H. *Proteus mirabilis* cause several diseases in poultry such as cellulitis, digestive disorder, urinary infection and hydronephrosis [31]. Several biofilm producing genes in poultry due to *Proteus* sp. are *mrpA*, *pmfA*, *ucaA*, *atfA*, *zapA*, *ptA*, *hpmA*, and *ireA*, ureC, zapA, rsmA, hmpA, mrpA, atfA and pmfA (**Table 2**) [45, 48]. *Pseudomonas aerugenosa* is very common poultry pathogens causes diarrhea, septicaemia and respiratory diseases. The bacteria may transmit from animals and inanimate objects where they form biofilm. Several virulent genes have been isolated responsible for disease production. Ggenes responsible for biofilm formation are *katA* and *kpsM.*

Campylobacter is also a pathogenic bacterium in poultry flock and several genes responsible for production of biofilm in surfaces of stainless steel and polystyrine articles at different temperatures and oxygen concentration. The genes responsible for production of biofilm are *bhpC, cadF, clpP, dnaJ, docA., flaA, flaB, katA, kspM, luxS, racR* and *sodB* [49].

Ornithobacterium rhinotracheale is a Gram positive bacterium, causes respiratory disease in poultry and other birds that affect the productivity. All serovar A-E can produce biofilm at optimal condition of 40°C after 72 hours of incubation in elevated CO_2 concentration [4].

Listeria monocytogenes is an important poultry bacterium that causes septicemic condition in poultry. The bacteria has significant role in the public health concern

Name of organism	Causes disease in poultry	References
Escherichia coli	Avian colibacillosis	Grakh et al. [38]
Riemerella anatipestifer	Epizootic infectious disease	Sun et al. [39]
Salmonella enteritides *S.typhimurium*	Salmonellosis	Afshari et al. [40]
Klebsiella pneumonae	Respiratory disease	Ammar et al. [41]
Listeria monocytogenes	Septicemia, encephalitis	Ossaili et al. [42]
Campylobacter jejuni	Transient diarrhea in chicks	Shanes [43]
Pseudomonas auruginosa *P. fragi*	Gastroenteritis and zoonotic infection	Wafaa and Ghany [44]
Proteus mirabilis	Cellulitis, GI disorder	Sanches et al. [45]
Staphylococcus aureus, *S. intermedius*, *S. schleiferi*, *S. pseudointermedius*, *S. lutrae*	Arthritis, synovitis, and osteomyelitis	Marek et al. [46]
Staphylococcus aureus, Staphylococcus hyicus, Streptococcus agalactiae	Septicaemia, peritonitis, salpingitis, tropical infection and endocarditis	Olson et al. [47]

Table 2.
Bacteria cause poultry diseases and have biofilm forming capacity.

through egg and meat food chain. Samples collected from different poultry outlets revealed biofilm forming capacity [50]. The high capability for biofilm formation in this organism derived out of several genes such as *lux*S and *fla*A [51]. The ability of *L. monocytogenes* have adaptability in refrigerated environment in poultry slaughter houses and industry, food processing unit, fish processing unit as well as in vegetable processing industries [52]. It has been found *hly*A gene may have role in the formation of biofilm in stainless steel and polypropyline surface [53]. Different *Mycobacterium* sp. are also have role in the biofilm formation process in poultry farm and meat food industries. Many species other than *Mycobacterium tuberculosis* are involved in the formation of biofilm. *Mycobacterium avium*, *M. fortuitum*, *M.smegmatis* produce biofilm and transmission of diseases to new hosts. The *Mycobacterium* spp. may produce biofilm in the variable temperature and conditions (**Table 2**).

10. Diagnosis of biofilm formation

Formation of biofilm in veterinary and medical related instrument, tools and in different tissues and in vitro structures may be due to various methods. Several methods of direct and indirect methods are there for detecting the biofilm formation. In direct methods, observing the microbial colonization with several techniques such as contact plates, enzymatic reaction, electron transmission (transmission electron microscopy, TEM), scanning electron microscopy, (SEM), laser scanning confocal, epifluorescence microscopy. Indirect methods of detection of biofilm where it may be done based on detaching the microorganism from the surface before counting them.

For detection of biofilm formation several instruments and devices have been developed for clinical microbiological investigation. Some of the instruments are modified Robins device, Calgary biofilm device, flow well disc reactor, profusion biofilm fermenter, model blade etc. The substratums of the tools cited above are mainly made up of sialic (silicon), plastic, teflon stuff and cellulose derivatives. Biofilm in urinary catheter can be detected directly by Scanning electron microscopy or transmission electron microscopy (SEM/TEM). The rate of biofilm formation on model system i.e. in different tools may be altered with the composition of medium used such as amount of glucose, iron, antimicrobial agents, cation of Ca^{++}, Mg^{++} present [54].

Several methods of studies have been used to detect and determination of biofilms. The methods are tube method, Congo red agar method, microtitre plate assay, plate counting of biofilm covered bacteria (Sessile bacteria), PCR study, mass spectrometry etc. Some of the methods used for detection of biofilm are as follow.

10.1 Microscopic observation

Both light and electron microscopic studies can be made for direct observation of biofilm. The confocal laser scanning microscopy (CLSM), scanning (SEM) and transmission electron microscopy (TEM) are done for observation of microorganism adhere on surface, fluorescent dye can be used for clarity of organism and biofilm materials and their thickness. Indirect observation of **biofilm** of bacterial origin can be observed by various methods, they are roll plate method, Congo red agar method, tube method, microscopic assay etc.

10.2 Roll plate method

In Roll plate method, where the development of biofilm on the surface of cylindrical device and tools such as urinary catheter and vascular graft. It is not considered the growth of microorganisms inside the tubular device. The Congo red agar method is a qualitative test for detection of biofilm producing bacteria, the colony color is changed in the medium. Blackish crystalline colonies are produced by the biofilm forming sessile organism while the planktonic bacterial cells produced red in the medium [55].

The **tube method** of qualitative assay of detection of biofilm formation. In this assay a visible film is developed around the glass tube of culture of bacteria with tryptic soy broth. The sessile bacteria form biofilm on the wall of the polystyrene test tube which may be stained with Safranine for 1 h dye exposure. The plankotonic cells are discharged by waiting twice with Phosphate-buffered saline (PBS). The sessile bacterial test tube showing visible stained at the bottom while the Planktonic cells contain bacterial culture tube become clear after washing with PBS.

10.3 Biofilm assay by microtitre assay

Microtitre plate assay is quantitative test to determine biofilm production by microplate reader. Bacterial broth suspension is prepared in Muller Hintone broth (MHB) with 1% glucose solution. An amount of 20 μL of bacterial isolate is in 180 μL MHB. Microplate with 96 well polystyrene stuff is incubated at 37°C for 24 h. The sessile bacterial form biofilm on the wall of the wells those can be stained with Safranine for 15 min. The planktonic cells well are rinsed with PBS (pH 7.2) and air dried at 60°C for an hour. Biofilm of well can be fixed with 150 μL methanol for 20 min. Air dry of micropipette is resolubilized by 150 μL of 95% ethanol, or 33% of glacial acetic acid. The study is repeated in triplicates. Microplates are measured photometrically at 570 nm filter in spectrophotometer by microreader. Uninoculated well with MHB medium is considered negative control as blank [56]. The cut off value (ODc) can be categorized of the isolates by biofilm producer or not.

$$\mathrm{ODc} = \mathrm{OD\,of\,negative} + \left(3 \times \mathrm{SD\,of\,negative\,control}\right)$$

$$\mathrm{OD}_{\mathrm{isolate}} = \mathrm{average\,OD\,of\,isolate} - \mathrm{ODc}$$

Interpretation of Result:
OD $\leq$ ODc no production of biofilm.
ODc < OD $\leq$ 2× ODC production of weak biofilm.
2× ODc < OD $\leq$ 4OD is moderate production of biofilm.
4× OD < OD is indication of strong production of biofilm.

10.4 PCR based biofilm detection

Amplification of target gene helps in species diagnosis for microbiological studies; similarly genes responsible for biofilm formation can be identified using gene specific primers. Biofilm related genes are amplified by PCR machine as qualitative real time PCR. Several species specific gene of different microbial species have different gene segment that express the biofilm formation. Several genes in different bacterial species have been discussed in the text.

10.5 Mass spectrometry method

The extracellular polymeric substances (EPS) composed of polysaccharides and proteins (extracellular enzymes) are produced in biofilm The proteins in biofilm matrix can be detected and characterized by mass spectrometry (MS), complex biological structures like EPS can be characterized by MS. The matrix assisted laser desorption ionization (MALDI) and Electrospray ionization (ESI) is similar to that of massspectrometry. The time of flight mass spectometer (TOF) with which mass is analyzed by ion desorped in cacuum chamber. If these two techniques (MALDI and TOF) are combined called MALDI-TOF) can help in the analysis of biofilm mass. In recent years matrix assisted laser desorption ionization-time of flight mass spectrometry (MALDI-TOF MS) has emerged as a potential tool for microbial identification and diagnosis [57]. Here, the matrix mass is ionized and vapourized by laser beam, depending on mass/charge ratio of the samples molecules are measured by TOF. Bacteria are identified by expressing of proteins like surface proteins, Co-enzymes (β-lactamase) response to antimicrobial can be monitored.

10.6 Biological assay of biofilm

Biofilm colony may produce numbers of bacterial species and wide range of biological products. Estimation of biofilm embedded products and characterization of the products. The planktonic and sessile bacterial producers are very similar. Standardization of curves of each microorganism tested needs to be formed. Estimation of total protein at 550 nm or 950 nm ansorbance. Estimation of tryptophan fluorescence, urease, formazan and endotoxins are also assayed.

11. Treatment of biofilm intrigue

The biofilm causes several inefficacy of equipments and infrastructures as well as spread of infection and resistant to antimicrobial therapy. The biofilm cause equipment inefficient and corrosion that reduces the efficacy any equipment. Biofilm in industry causes better insulator which is scale type, this insulator increases energy cost. It also changes the water passing capacity in the water supply in poultry industry. Biofilm hamper the water distribution system with disinfectant residual, increase bacterial level, reduction of O_2 level in water, reduce water taste and produce bad odor. Red and black water problem due to iron and sulphate reducing bacteria.

Chemical and environmental modification is the main tools to prevent biofilm formation. Several antibiotics, biocides, and ion coating are commonly used against biofilm in veterinary and human medicines. Biofilm prevention is two types; prevention of growth and prevention of surface attachment. Microbial growth can be preventing giving antimicrobial coating in indwelling, medical device etc. Several antibiotic, biocides and ion coating are used. All these coating remains effective for few days to week, later they disperse. Silver ions have antibacterial property for water purification in reverse osmosis process. Other way of purification of water is electric deionization, exposure of UV light and application of ozone.

11.1 Therapeutic intervention for poultry production

Several microorganisms affect poultry production both egg and meat through infection and diseases production. Besides getting infection, other risk factors for biofilm production in poultry farm and meat industry are scarcity of quality water, negligence of biosecurity standard, co-existence of other animals in vicinity of poultry premises, inadequate infrastructures and their condition. A scarcity of water is lethal for growth of biofilm, interrupts water supply through drinking fountain and drips are sufficient source of biofilm bacteria. Control of water distribution system reduces the microbial load and infection. Several chemicals such as chlorine, chlorine dioxide, organic acid, hydrogen peroxide may be used but they are used in some occasion. Intermittent used of such antibacterials and unhygienic used of water supply invites biofilm formation.

11.2 Water purification

The equipment and water supply system will be such that the coating of the equipment and water supply pipes will be free from corners, cracks, valve, joint and pores. A mechanical sensor system have been developed to monitor biofilm formation in the system where production of gas due bacterial fermentation will be alarmed by the device. Once biofilm is established it may be dismantled through cleansing by physical and chemical means and disinfection of tools and fomites are to be done regularly. Water can be purified applying Ozone exposure (1.0–2.0 mg/L). It disintegrate bacterial cell into fragments. Chlorine and chloramine are highly effective method of water disinfection, but in the pipes it produces small amounts of chemicals dirt if the water contains much impurities and the taste of water is also changed. The amount of chlorine used is 4 milligrams per liter (mg/L i.e. 4 ppm). Different chlorines used are chlorine gas, sodium hypochlorite, and calcium hypochlorite. The biofilm polymeric surface charged can be modified by electrostatic charged particle that will repeal other particle of same charge. The electrostatic charge and biofilm polymeric charge are negative so, they dispel each other.

11.3 Food industries

In food industries, most disinfectants used are quaternary ammonium compounds (amphoteric compounds, hyperchlorides, peroxides (H_2O_2, peracetic acid), aldehydes (formaldehyde, glutaraldehide), phenolics, alkyl amines, chlorine dioxide etc. [58].

11.4 Veterinary medical therapy and biofilm

Antimicrobials can be on the medical devices surface using long flexible polymeric chain. The chain forms a covalent bonds with device surface killing microbial organisms. Several such antibacterial materials used are N-alkylpyridimidinium bromide can act against *E. coli*, *Streptococcus epidermidis*, *Pseudomonas aurugenosa*. The dispersion force dispel the organism on the surface of the device to prevent adhesion and biofilm formation. For effective result with biofilm infected patients, combination of antibiotics and antibiofilm can be used in poultry and veterinary therapy. Usually, quorum sensing mechanism binds the whole biofilm population of the society through a complex cascade of events which unit the biofilm population. So antibiotic and use of ultrasound device that enhance the antibiotic activities. The ultrasound helps to pass energy weave through the cell of biofilm particularly in tropical infection. Several antibiotics along with application of different antibiofilm agent and their use are presented (**Table 3**).

Use of ultrasound can destruct the bacterial cell penetrating the biofilm and the antibiotic can pass through the biofilm to reach the bacterial cell and act upon it

Low electric current-Passing of low level of electric with antibiotic can provide effective response in biofilm Society that may be situated in tissues. Electromagnetic pulse may increase the antimicrobial response of cationic antibiotic against biofilm. Gentamicin with mild electric current cans synergistic effect against *Staphylococcus aureas*.

11.5 Phage therapy

The phage virus may act on biofilm bacteria penetrating the biofilm through diffusion and even propagation of phage into biofilm environment. The function of phage virus also depends on nature of biofilm matrix, species of bacteria etc. Usually phage generates EPS degradating enzymes (depolymerases) that may digest the matrix. Another function of phage is that in biofilm the bacteria remains under several stress condition, this stress can enhance the phage to disintegrate the biofilm Community, particularly in *Pseudomonas aurugenosa* and *Fusebacterium nucleatum*, *Streptococcus* sp., *Proteus mirabilis*., *Listeria* sp., *E. coli* etc. The phage can be used are pyobacteriophage, phage PB-1, T4 etc. The antibiotic and phage combination acts suitably in complicated cases of infection [59].

Therapy	Action	Usefulness	Response against bacteria
Ultrasound	Destroy bacterial cell penetrating biofilm	Helps to penetrate antibiotics in deep seated bacteria	*Staphylococcus aureas*
Phage therapy	Destroy bacteria penetrating biofilm	Generates EPS degradating enzymes that destroy bacteria	*Pseudomonas auguginosa Streptococcus sp, Listeria sp, Proteas sp*
Drug delivery system	Nano carrier drug delivery system prolongs the drug stability	Different nanoparticles like silverAg, Zn,Ti, Au with antibiotics	*Fusebacterium nucleatum*
Antibacterial peptides	Some peptides have antibacterial properties	Peptides can be used to check bacterial infection	*Staphylococcus sp, Pseudomonas auruginosa*
Antiadhesin agents	Inhibits formation of biofilm preventing surface attachment by pilli, flagella etc	Prevent adhesion on surface by bacteria	*E. coli, Salmonella sp Proteus spp etc*
Antimatrix agents	Disintegration of biofilm by enzymes	Digest the matrix of biofilm and remove it	*E. coli, Pseudomonas sp Klensiella pneumonae, Staphylococcus epidermides, Enterococcus fecalis*
Chelating agent	Chelates out of the matrix metallic ions	Helps in disintegration of biofilm matrics	*Staphylococcus aureus, Staphylococcus epidemidis Pseudomonas aeruginosa*

Table 3.
Therapeutic intervention against biofilm in contaminants.

11.6 Chelating agents

Several metalic ions such as Ca^{++}, Mg^{++} and Fe^{++} are abundant in the biofilm matrix for their integrity. Chelating agents such as Sodium citrate, trisodium citrate, Na-EDTA can be used to chelate out the cations from the biofilms matrix and this helps in disintegration of biofilm society.

11.7 Drug delivery system

Encapsulated nano carrier drug delivery system that prolongs the activity of active molecules against a *Fusebacterium nucleatum* bacteria. Such combination of antibiotic such as gentamicin, ciprofloxacin, ampicillin, along with nano carriers of phosphotydylcholine, polyethylene glycerol, polyamidoamine are used. Silver nanoparticle has also antibacterial property. This is due to positive charge of Ag- and –ve charge of biofilm attract and a strong bacteriocidal action of nano silver (Ag) provides antibacricidal function. Other nanoparticles used are zinc (Zn), titanium (Ti), gold (au) nano particle.

11.8 Antimicrobial peptides

Antimicrobial peptides (AMPs) are small peptides that widely exist in nature and they are an important part of the innate immune system of different organisms. The AMPs have a various inhibitory effects against microorganisms. The emergence of antibiotic-resistant concern and the increasing of concerns about the use of antibiotics resulted in the development of AMPs, which have a good application prospect in veterinary medicine, food Science, agriculture, aquaculture and human medicine. It could be novel types of antibacterial in the regime of antibiotic resistance. Antibacterial peptides must be assayed before use about their spectrum and mechanism. Several peptides such as SMAP-29 (Sheep myoloid antimicrobial peptide), BAMP-28(bovine antimicrobial peptide), BAMP-27 have property to reduce significant biofilm reduction property against multidrug resistant *Pseudomonas aurugenosa*. These peptides kill the microorganism in the beginning of biofilm formation [60]. High efficacy of α-helical cecropin/melitin hybrid peptide CEME reported against *Staphylococcus aureas*. Due to increasing concern of AMR with different antibiotics, the use of antibacterial peptides in poultry have been tried and found that 2 truncated cathelicidins and 4 avian β-defensins are potent peptides against bacterial infection and immunomodulatory effect [61].

11.9 Antiadhesin agents

Several antiadhesion agents could be used against biofilm in-vivo and in-vitro. Use of mannocides, pilicides and culicides. Mannocides are small molecules of drug that contains mannose sugar group. The bacterial fimbri bound to mannose. Mannocide fits the FimH mannose binding pockets and completely inhibits FimH site to the host receptor. Similarly pillicides are those chemical that inhibits the formation of the pilli of bacteria. Pillicides are designed such that interfare the process of pilli formation through inhibition of export of pillin subunits. The curli is a protenaciuos fiber that produced by certain bacteria like *E. coli, Salmonella* spp. It helps in the formation biofilm. Curlicides are those chemicals which inhibits formation of curli. All these three forms of antiadhesins agents are used in upper urinary tract infection with *E.coli, Proteas* sp. etc.

11.10 Antimatrix agents

Bacterial matrix aggregation in the biofilm colony with extracellular matrix is a hardle for therapy and elimination of bacterial propagation. Several natural and engineered enzymes and used of bacteriophage that can disintegrate the biofilm society and matrix. The N-acetyle-D- glucosamine-1 phosphate acetyle transferase (GlmU) can be used against *E.coli, Pseudomonas aurogenosa*, *Klensiella pneumonae, Staphylococcus epidermides, Enterococcus fecalis*. Other enzymes have potential use are Dnase, dispersinB etc.

11.11 Chelating agents

Several metalic ions such as Ca^{++}, Mg^{++} and Fe^{++} are abundant in the biofilm matrix for their integrity. Chelating agents such as Sodium citrate, trisodium citrate, Na-EDTA can be used to chelate out the cations from the biofilms matrix and this helps in disintegration of biofilm society.

12. Conclusion

Biofilm formation is a real problem in the therapeutic and poultry management. In poultry a large numbers of bacteria that form biofilm have several direct and indirect effects on disease transmission and resistance to antibiotic therapy. Several infectious diseases whose course remains longer might be due to biofilm formation. Besides therapeutic difficulties poultry industries and water supply system also hampered. To avoid biofilm formation and treatment with different areas of biofilm have been discussed. Regular investigation for biofilm formation and therapeutic interventions as deem fit should be taken regularly.

Acknowledgements

The author is very thankful to the director, Indian Veterinary Research Institute for providing facilities to write this chapter.

Author details

Dayamoy Mondal
Eastern Regional Station, Indian Veterinary Research Institute, Kolkata, India

*Address all correspondence to: dayamoy21@gmail.com

References

[1] Daniel L, Vlamakis H, Kolter R. Biofilms. Cold Spring Harbour Perspective Biology. 2010;**2**(7):a000398. DOI: 10.1101/cshperspect.a000398

[2] Umar FA, Igwenagu E, Mu'azu A, Aliyu S, Umar MI. Intrigues of biofilm: A perspective in veterinary medicine. Veterinary World. 2016;**9**(1):12-18

[3] Minh TTT, Wibowo D, Bernd HAR. *Pseudomonas aeruginosa* Biofilms. International Journal of Molecular Sciences. 2020;**21**(22)1-25. Article ID: 8671. DOI: 10.3390/ijms21228671

[4] Miguel A, Rosa-Ramos D, Rodríguez-Cruz M, López-Villegas EO, Castro-Escarpulli G, Guerra-Infante FM. Conditions that induce biofilm production by Ornithobacterium rhinotracheale. Avian Pathology. 2015;**44**(5):366-369

[5] Martí S, Rodríguez-Baño J, Catel-Ferreira M, JouenneT VJ, Seifert H, Dé E. Biofilm formation at the solid-liquid and air-liquid interfaces by *Acinetobacter* species. BMC Research Notes. 2011;**4**:5. DOI: 10.1186/1756-0500-4-5

[6] Cerca N, Maira-Litran T, Jefferson KK, Grout M, Goldmann DA, Pier GB. Protection against *Escherichia coli* infection by antibody to the *Staphylococcus aureus* poly-N-acetylglucosamine surface polysaccharide. Proceedings of the National Academy of Sciences. 2007;**104**:7528-7533

[7] Ponomareva AL, Buzoleva LS, Bogatyrenko EA. Abiotic environmental factors affecting the formation of microbial biofilms. Biological Bulletin of Russian Academy Sciences. 2018;**45**:490-496. DOI: 10.1134/S106235901805014X

[8] Lei J, Sun L, Huang S, Zhu C, Li P, He J, et al. The antimicrobial peptides and their potential clinical applications. American Journal of Translational Research. 2019;**11**(7):3919-3931

[9] Hannah HT, Weibel DB. Bacteria-surface interactions. Soft Matter. 2013;**9**(18):4368-4380

[10] Rutherford ST, Bassler BL. Bacterial Quorum sensing: Its role in virulence and possibilities for its control. Cold Spring Harbor Perspectives in Medicine. 2012;**2**(11):a012427. DOI: 10.1101/cshperspect.a012427

[11] Hoffman LR, D'Argenio DA, MacCoss MJ, Zhang Z, Jones RA, Miller SI. Aminoglycoside antibiotics induce bacterial biofilm formation. Nature. 2005;**436**:1171-1175

[12] Ma L, Conover M, Lu H, Parsek MR, Bayles K, Wozniak DJ. Assembly and development of the *Pseudomonas aeruginosa* biofilm matrix. PLoS Pathogens. 2009. DOI: 10.1371/journal.ppat.1000354

[13] Lina M, Francisco P, Fernando M, Trejo DJ, Bueno M, Golowczyc A. Biofilm formation by *Salmonella* sp. in the poultry industry: Detection, control and eradication strategies. Food Research International. 2019;**119**:530-540

[14] Galié S, García-Gutiérrez C, Miguélez EM, Villar CJ, Lombó F. Biofilms in the food industry: Health aspects and control methods. Frontiers in Microbiology. 2018;**9**(898). DOI: 10.3389/fmicb.2018.00898

[15] Wagner EM, Fischel K, Rammer N, Beer C, Palmetzhofer AL, Conrady B, et al. Bacteria of eleven different species isolated from biofilms in a meat processing environment have diverse biofilm forming abilities. International Journal of Food Microbiology. 2021;**349**:109232

[16] Costerton JW, Lewandowski Z, Caldwell DE, Korber DR, Lappin-Scott HM. Microbial biofilms. Annual Review of Microbiology. 1999;**49**:711-745

[17] Trachoo N. Biofilms and the food industry. Songklanakarin Journal of Science and Technology. 2003;**25**(6):807-815

[18] Takeuchi K, Frank JF. Penetration of *Escherichia coli* O157:H7 into lettuce tissues as affected by inoculum size and temperature and the effect of chlorine treatment on cell viability. Journal of Food Protection. 2000;**63**:434-440

[19] Chia TW, Goulter RM, McMeekin T, Dykes GA, Fegan N. Attachment of different *Salmonella serovars* to materials commonly used in a poultry processing plant. Food Microbiology. 2009;**26**:853-859. DOI: 10.1016/j.fm.2009.05.012 PMid:19835771

[20] Jayaweera TSP, Ruwandeepika HAD, Deekshit VK, Kodithuwakku SP, Cyril HW, Karunasagar I, et al. Biofilm forming ability of broiler chicken meat associated *Salmonella* spp on food contact surfaces. Tropical Agricultural Research. 2021;**32**(1):17-26

[21] Rossi DA, MeloRT MEP, MoteiroGP. Biofilm of *Salmonella* and Campylobactor in the poultry Industry. IntachOpenScience. 2017. DOI: 10.5772/65254

[22] Vivian RC. The evaluation of biofilm formation and sensitivity to peracetic acid of *Salmonella* spp. isolated from poultry abattoir [dissertation]. São Paulo, Brasil: Universidade Estadual Paulista; 2017

[23] Osman KM, Kappell AD, Elhadidy M, ElMought F, El-Ghany WAA, Orabi A, et al. Poultry hatcheries as potential reservoirs for antimicrobial-resistant *Escherichia coli*: A risk to public health and food safety. Scientific Reports. 2018;**8**:5859. DOI: 10.1038/s41598-018-23962-7

[24] Cwiek K, KorzekwaK TA, Bania J, Bugla-Płosko G, Wieliczko A. Antimicrobial resistance and biofilm formation capacity of *Salmonella enterica* serovar enteritidis strains isolated from poultry and humans in Poland. Pathogens. 2020;**9**(643)

[25] King AJ. Water quality and poultry production. Poultry Science. 1996;**75**:852-853. DOI: 10.3382/japr/pfw010

[26] Robbins JB, Fisher CW, Moltz AG, Martin SE. Elimination of Listeriamonocytogenes biofilms by ozone, chlorine, and hydrogen peroxide. Journal of Food Protection. 2005;**68**:494-498

[27] Sparks N. The role of the water supply system in the infection and control of Campylobacter in chicken. World's Poultry Science Journal. 2009;**65**:459-474

[28] Dou X, Jiansen G, Xiangan H, MingX HS, Di Z, Linlin Z, et al. Characterization of avian pathogenic *Escherichia coli* isolated in eastern China. Gene. 2016;**576**:244-248

[29] Maharjan P, Clark CT, Kuenzel C, Foy MK, Watkins S. On farm monitoring of the impact of water system sanitation on microbial levels in broiler house water

supplies. Journal of Applied Poultry Research. 2016;**25**(2):266-271

[30] Watkins SW. Identifying and correcting challenges. Avian Advice. 2008;**10**:10-15

[31] Chen CY, Hsu Chen Y, Lu PL, Ru Lin W, Chen TC, Lin CY. *Proteus mirabilis* urinary tract infection and bacteremia: Risk factors, clinical presentation, and outcomes. Journal of Microbiology, Immunology, and Infection. 2012;**45**(3):228-236

[32] Peng D. Biofim formation of *Salmonella*. Intech Open Book Series. 2016

[33] Wang YYL, Wang Y, Wang Y, Cai Y, Zhao W, Ding D. Isolation, phylogenetic group, drug resistance, biofilm formation, and adherence genes of *Escherichia coli* from poultry in central China. Poultry Science. 2016;**95**(12):2895-2901

[34] Silva VO, Soares LO, SilvaJúnior A, Mantovani HC, Chang Y, MAS M. Biofilm formation on biotic and abiotic surfaces in the presence of antimicrobials by *Escherichia coli* isolates from cases of bovine mastitis. Applied and Environmental Microbiology. 2014;**80**(19):6136-6145

[35] Franklin-Alning FV, Kaspersen H, MAK H, Bakksjø RJ, Nesse LL, Leangapichart T, et al. Exploring *Klebsiella pneumoniae* in healthy poultry reveals high genetic diversity, good biofilm-forming abilities and higher prevalence in Turkeys than broilers. Frontiers in Microbiology. 2021;**7**(12):725414

[36] Saife K, Kezemian H, Heidari H, Rezagholizadeh F, Saee Y, Shirvani F, et al. Evaluation of biofilm formation among *Klebsiella pneumonia*e isolates and molecular characterization by ERIC-PCR. Jundishapur Journal of Microbiology. 2016;**9**(1):e30682. DOI: 10.5812/jjm.30682

[37] Wu MC, Lin TL, Hsieh PF, Yang HC, Wang JT. Isolation of genes involved in biofilm formation of a *Klebsiella pneumoniae* strain causing pyogenic liver abscess. PLoS One. 2011;**6**(8):e23500. DOI: 10.1371/journal.pone.0023500

[38] Grakh K. Epidemiology and characterization of biofilm producing Avian Pathogenic *Escherichia coli* in broiler chickens [thesis for: MVSc in Veterinary Public Health and Epidemiology]. Hisar, Hariya, India: Lalalajpat Rai University of Veterinary and Animal Science; 2020

[39] Sun B, Xue Y, Du X, He Z, Zou Z, Tian X, et al. Identification of genetic determinants of hemolytic activity of Riemerella anatipestifer using random transposon mutagenesis. Veterinary Research. 2021;**52**(1):19. DOI: 10.1186/s13567-021-00900-6

[40] Afshari A, Baaratpour A, Khanzade S, Jamshidi A. *Salmonella* Enteritidis and *Salmonella* Typhimorium identification in poultry carcasses. Iran Journal of Microbiology. 2018;**10**(1):45-50

[41] Ammar AM, Abd El-Aziz NK, Mohamed SS. Biofilm formation and its correlation with antimicrobial resistance in Klebsiella pneumoniae. Zagazig Veterinary Journal. 2020;**48**(4):366-377. DOI: 10.21608/zvjz.2020.37640.1115

[42] Osaili TM, Alaboudi AR, Nesiar EA. Prevalence of Listeria spp. and antibiotic susceptibility of Listeria monocytogenes isolated from raw chicken and ready-to-eat chicken products in Jordan. Food Control. 2011;**22**(3-4):586-590

[43] Shane SM. The significance of *Campylobacter jejuni* infection in poultry: A review. Avian Pathology. 1992;**21**(2):189-213. DOI: 10.1080/03079459208418836

[44] El-Ghany E. *Pseudomonas aeruginosa* infection of avian origin: Zoonosis and one health implications. Veterenary World. 2021;**14**(8):2155-2159

[45] Sanches MS, Batista AAS, Suza M, Menk-Costa MF. *Proteus mirabilis* causing cellulitis in broiler chickens. Brazilian Journal of Microbiology. 2020;**51**(3):1353-1362. DOI: 10.1007/s42770-020-00240-1

[46] Marek A, Pyzik E, Stępień-śniak D, Dec M, Jarosz LS, Nowaczek A, et al. Biofilm-formation ability and the presence of adhesion genes in coagulase-negative *Staphylococci* isolates from chicken broilers. Animals (Basel). 2021;**11**(3):728

[47] Olson ME, Ceri H, Morck DW, Buret AG, Reed RR. Biofilm bacteria: Formation and comparative susceptibility to antibiotics. Canadian Journal of Veterinary Research. 2002;**66**(2):86-92

[48] Sun Y, Wen S, Zhao L, Xia Q, Pan Y, Liu H, et al. Association among biofilm formation, virulence gene expression, and antibiotic resistance in *Proteus mirabilis* isolates from diarrhetic animals in Northeast China. BMC Veterinary Research. 2020;**16**:176

[49] Scheik LK, Maia DSV, Würfel SFR, Ramires T, Kleinubing NR, Lopes GV, et al. Biofilm-forming ability of poultry *Campylobacter jejuni* strains in the presence and absence of *Pseudomonas aeruginosa*. Canadian Journal of Microbiology. 2021;**67**(4):301-309. DOI: 10.1139/cjm-2020-0256

[50] Alonso-Calleja C, Gómez-Fernández S, Carballo J, Capita R. Prevalence, molecular typing, and determination of the biofilm-forming ability of *Listeria monocytogenes* serotypes from poultry meat and poultry preparations in Spain. Microorganisms. 2019;**7**(11):529. DOI: 10.3390/microorganisms7110529

[51] WarkeSR IVC, Kurkure NV, Tembhurne PA, Prasad M, Chaudhari SP, Barbuddhe SB. Biofilm formation and associated genes in Listeria monocytogenes. The Indian Journal of Veterinary Sciences & Biotechnology. 2017;**12**(3):07-12

[52] Mazaheri T, BRH C-H, Bermúdez-Capdevila M, Ripolles-Avila C, Rodríguez-Jerez JJ. Listeria monocytogenes biofilms in the food industry: Is the current hygiene program sufficient to combat the persistence of the pathogen. Microorganisms. 2021;**9**:181. DOI: 10.3390/microorganisms9010181

[53] Grudlewska-Buda K, Skowron K, Gospodarek-Komkowska E. Comparison of the intensity of biofilm formation by *Listeria monocytogenes* using classical culture-based method and digital droplet PCR. AMB Express. 2020;**10**:75. DOI: 10.1186/s13568-020-01007-5

[54] Kirmusaoglu S. The methods for detection of biofim and screening antibiofilm activity of agents. In: Antimicrobials, Antibiotic Resistance, Antibiofilm Strategies and Activity Methods. Intech Open Book Series; 2019. DOI: 105772/intechopen.84411

[55] Kaiser TDL, Pereira EM, Nettodos Santos KR, Noia Maciel EL, Schuenck RP, Ferreira Nunes AP. Modification of the Congo red agar method to detect biofilm production by *Staphylococcus epidermidis*.

Diagnostic Microbiology and Infectious Disease. 2013;**75**(3):235-239

[56] Bendouah Z, Barbeau J, Hamad WA, Desrosiers M. Use of an in vitro assay for determination of biofilm-forming capacity of bacteria in chronic rhinosinusitis. American Journal of Rhinology. 2016;**20**(5):434-438

[57] Singhal N, Kumar M, Kanaujia PK, Virdi JS. MALDI-TOF mass spectrometry: An emerging technology for microbial identification and diagnosis. Frontiers in Microbiology. 2015;**6**:1-16. Article ID: 791. Online published

[58] Carrascosa C, Raheem D, Ramos F, Saraiva A, Raposo A. Microbial biofilms in the food industry: A comprehensive review. International Journal of Environmental Research Public Health. 2021;**18**(4). DOI: 10.3390/ijerph18042014

[59] Gonzalez CF, Chalap PD. Phages for biofilm removal. Antibiotics (Basel). 2020;**9**(5):268

[60] Pompilio A, Scocchi M, Pomponio S, Guida F, Di Primio A, Fiscarelli E, et al. Antibacterial and anti-biofilm effects of cathelicidin peptides against pathogens isolated from cystic fibrosis patients. Peptides. 2011;**32**:1807-1814

[61] Nguyen TTT, Allean B, Wheler C, Koster W, Gerdts V and Dar A. Avian antimicrobial peptides: In vitro and in ovo characterization and protection from early chick mortality caused by yolk sac infection. Scientific Reports. 2021;**11**:2132

Chapter 17

Chronic Intraocular Leptospiral Infection Relying on Biofilm Formation inside the Vitreous Cavity Leads to Recurrent Uveitis in Horses

Bettina Wollanke and Hartmut Gerhards

Abstract

Equine recurrent uveitis (ERU) is a disease known and feared for centuries, as it almost always leads to blindness even with careful and meticulous conservative treatment of the individual episodes of uveitis. In about one-third of horses, both eyes are affected, often necessitating euthanasia. A link between ERU and leptospiral infection has been suspected for nearly 80 years. Vitreous lavage (vitrectomy) can preserve vision in affected eyes. After surgery, no further episodes of uveitis occur in up to more than 95% of operated eyes. With routine performance of vitrectomies, numerous vitreous samples could be used for further investigations. Intraocular anti-Leptospira antibody production was proven, leptospires could be cultured from the vitreous samples, and the LipL32 gene could be detected in the vitreous samples by PCR. Thus, there was convincing evidence of a chronic intraocular leptospiral infection, which can be eliminated most reliably by vitrectomy. Recently, it has been shown that the intraocular leptospires produce biofilm in the equine vitreous. Biofilm formation explains not only the success of vitrectomy, but also the survival of leptospires in the vitreous cavity for many years despite the presence of high intraocular antibody titers and immunocompetent cells, as well as the high tolerance to antibiotics.

Keywords: equine recurrent uveitis (ERU), pathogenic *Leptospira* spp., biofilm formation, vitreous cavity, intraocular specimens, intraocular antibody production, leptospiral culture, real-time PCR targeting LipL32

1. Introduction

Equine recurrent uveitis (ERU) occurs in mules and horses and is a disease that has been known for a long time. From about the beginning of time-counting, ancient writings have described symptoms that are consistent with today's definition of ERU. Since the end of the nineteenth century and the beginning of the twentieth century, more and more detailed descriptions of this disease have been published [1]. In earlier times, the working power of the horse was quite crucial for the survival of men [2]. Not only during war, but also for the cultivation of the fields, for the transport of people and freight as well as for serving as living motors in preindustrial times, people were dependent on horses and mules.

Without horses, the development of mankind would not have been possible to the extent that has been achieved in the past centuries. All the more the health maintenance of the horses was of paramount importance [3]. The recurrent and painful episodes of uveitis led to reduced performance and not infrequently to blindness and thus often to unserviceability of the affected horses. For this reason, equine recurrent uveitis has preoccupied many generations of owners and veterinarians [3, 4]. There are the most diverse historical treatment approaches and theories about the causes of this disease [3, 5].

Among many causes that had not been confirmed, wet pastures and flooding as well as heritability were discussed [5–7]. An infectious etiology has been suspected for over 100 years, although *Leptospira* spp. were not known at that time [3]. Since the first description of Weil's disease in humans, "eye complications" were known to be associated with this disease [8]. A first description of leptospires was given in 1915 [9]. At that time, the identification of *Leptospira* spp. was made in Japan and Germany at approximately the same time [10].

After a link between leptospiral infections and uveitis had been established in human medicine, the Swiss ophthalmologist Gsell and coworkers studied aqueous humor from equine ERU eyes and described for the first time a link between ERU (then called "moon blindness" or "periodic ophthalmia") and leptospiral infection [11]. Since then, there have been numerous investigations addressing the leptospiral etiology of ERU.

Because antibody detection in intraocular fluids was relatively common [11–17], but uveitis bouts typically do not become apparent until months or even years after the acute systemic infection [18–22], it was assumed that the infection was a trigger of ERU, but the bacteria were no longer present when the uveitis attacks started [18, 23]. In addition, a culture of *Leptospira* spp. from equine intraocular samples failed many times [12, 16, 22, 24–28]. For this reason, ERU has also been considered by some authors to be an "autoimmune" disease [29–31].

Different causes of uveitis can occur in horses just like in other species [30, 32, 33]. However, in equine uveitis associated with painful recurrent episodes causing the typical ocular changes, chronic intraocular leptospiral infection has been found to be the cause [34, 35]. Therefore, the term "ERU" will be used hereafter to refer to leptospiral-induced recurrent uveitis.

It was not until the routine use of vitrectomy (irrigation of the vitreous chamber) in horses [36–38] and the resulting ability to obtain intraocular specimens from eyes affected with ERU [39], that the importance of leptospiral etiology in ERU was confirmed [34, 35, 40–44].

Only recently it was recognized that recurrent uveitis in horses is a biofilm-mediated disease [45]. The ERU has many aspects that had raised questions and been incomprehensible before the discovery of biofilm formation of pathogenic *Leptospira* spp. in the vitreous chamber. However, knowing the characteristics of chronic and biofilm-associated infections, the pathogenesis of ERU can now be better understood [33].

2. Incidence and clinical course of ERU

Leptospiral-induced uveitis is not only in horses a late consequence of systemic infection [18, 34, 46], but also human leptospiral uveitis often occurs a long time after the acute infection [11, 46–51]. A causal relationship between uveitis and a previous leptospiral infection is often difficult to recognize when uveitis occurs, because systemic leptospirosis is predominantly inapparent in horses [19, 52] and can also be inapparent in humans [53].

ERU affects quite a lot of horses. In the United States, where there are many leopard coat pattern horses (Appaloosas), it has been reported that up to 25% of horses are affected and lose vision in one or both eyes during the course of the disease [30]. However, in that study, leopard coat pattern uveitis (Section 4.), which accounts for a large proportion of affected horses in the United States, was also classified as ERU. In other studies, the percentage of horses affected with ERU ranges from 7 to 10% [54–56], with up to one-third of the horses suffering from the disease on both sides [3, 34, 57]. The attacks of uveitis in both eyes often do not start at the same time, but with a time delay of several months up to about 2 years [34].

The first episodes of uveitis are usually noticed in younger horses between 4 and 6 years of age [34]. More rarely, however, horses can still develop ERU up to over 20 years of age. Foals up to 6 months of age typically do not develop ERU. When uveitis occurs in foals younger than 6 months, it is typically septicemia-associated and bilateral, e.g., in the course of rhodococcosis [58–61].

In ERU, recurrent episodes of uveitis occur in unpredictable intervals and oftentimes not, as the former term "periodic ophthalmia" suggests, periodically. The interval between episodes of uveitis can be less than 2 weeks and up to more than a year. In most cases, ERU episodes are associated with blepharospasm, epiphora, and photophobia, so the owner notices the eye disease and seeks veterinary advice. The severity of uveitis also varies greatly from horse to horse. Sometimes very mild episodes occur, which subside after 1–2 days. Other ERU attacks are so severe that after one or two attacks, the eye may already show significant and irreversible changes and in the worst case may even lose vision. In most horses, the clinically quiescent intervals between episodes of uveitis become shorter over time, and at the same time the uveitis bouts become more severe.

3. Clinical signs of ERU

Descriptions of the ophthalmologic findings in ERU have been given repeatedly and in broad agreement [3, 30, 32, 34, 62–64]. Acute attacks are usually painful or even very painful. Affected horses are depressed, show decreased appetite, can have a moderate rise in body temperature, severe blepharospasm, serous and later sero-mucous lacrimation, and more or less swollen eyelids. These symptoms, although typical, are not pathognomonic and can also occur with other eye lesions.

Ocular examination in horses is the easiest and most informative when a simple handheld (direct) ophthalmoscope with bright light source is used. The handheld ophthalmoscope can be used as a focal light source, magnifying glass, and slit lamp, and is most crucial for examining the posterior segment of the eye (posterior lens surface, vitreous cavity, and fundus). Since serum in horses is yellowish in color, aqueous humor and vitreous humor in acute uveitis ("leakage") are also jaundiced.

Ophthalmic examination typically reveals the following findings during an acute ERU episode:

- Reddening of the conjunctiva
- Low-grade diffuse corneal haziness
- Incipient circular vascularization of the cornea
- Jaundiced aqueous humor with positive Tyndall effect, usually also fibrin in the anterior chamber of the eye
- Miosis and only delayed and often incompletely medically achievable mydriasis
- Diffuse haziness of the vitreous humor
- Ocular hypotension (intraocular pressure often <10 mmHg)

In the inflammation-free interval, after mild ERU episodes and meticulous conservative treatment, sometimes no definite changes can be detected in early stages of the disease. However, when multiple ERU attacks have occurred, pathologic changes become increasingly apparent that are also evident during the clinically quiescent phase of the disease:

- Gradually increasing atrophy of the globe (if necessary, the inner anterior-posterior diameter can be measured by ultrasound; the difference is definite as from ≥ 2 mm side-to-side difference)
- Delayed pupillary response to light, drug-induced mydriasis also only achievable with delay
- In mydriasis, otherwise hidden posterior synechiae or iris residuals may be detected on the anterior lens capsule
- Diffuse vitreous opacification may still be recognizable in the inflammation-free interval (in some cases only evident by comparison with the other eye, and if the fundus on the diseased side is less clear compared with the other side)
- Vesicular cataract, typically in the periphery of the posterior lens capsule
- Dense vitreous deposits, initially visible only in mydriasis and typically located high dorsally close to the ciliary body; in the course of the disease, these deposits become more pronounced and can eventually also be seen in the center of the vitreous cavity, many times in combination with a murky yellowish discoloration of the liquefied vitreous
- In more advanced stages of the disease, moderate to severe bulbar atrophy or even phthisis, cataract, lens luxations, and retinal detachment may occur

In 3% of ERU cases, the inflammation occurs primarily in the posterior segment of the eye [65]. Hardly any pain is evident in these horses, and this form of ERU is sometimes detected only as an incidental finding during routine examinations or purchase examinations of horses. Only rarely do very observant owners notice a change in the fundus reflex of the diseased eye and call a veterinarian. In most cases, however, iritis occurs in the course of the disease, which then leads to the typical and easily recognizable pain symptoms. Depending on the changes that have already occurred in the posterior segment of the eye, the prognosis for preservation of vision is often guarded at this point. Sometimes these horses are not presented to the veterinarian until "sudden" blindness due to cataract formation or retinal detachment has occurred.

4. Differential diagnosis

A significant and strinkly common type of uveitis not caused by leptospires occurs in leopard coat pattern horses [65, 66]. This type of uveitis is strikingly common in leopard coat pattern horses. In contrast to ERU, leopard coat pattern uveitis progresses insidiously and does not present as recurrent painful episodes of uveitis. In the literature, it is therefore often referred to as "insidious uveitis," but not distinguished from ERU. Other forms of uveitis may be phacogenic, traumatic, tumor-associated, septicemia-associated, or triggered by other infectious causes such as parasites (*Micronema* (syn. *Halicephalobus*) *deletrix* or *Sertaria* spp.) or, e.g., staphylococci [33]. In addition, a chronic iritis similar to Fuchs' heterochromia iritis in humans occurs in horses [33, 67]. In most cases, all these forms of uveitis can be relatively clearly differentiated from ERU based on the clinical picture and/or the course of the disease (**Table 1**) [33].

Sometimes recurrent keratitis is misinterpreted as ERU, as some types of keratitis can also cause painful with miosis and responds to the same conservative therapy as ERU. However, in keratitis cases, medical dilation of the miotic pupil results usually more rapidly and completely than in ERU. In recurrent keratitis, however, the changes that almost always are evident in ERU after several episodes of uveitis, even in the inflammation-free interval, are absent.

If an ocular disease is clinically not clearly assignable to an etiology (e.g., "recurrent keratitis" or "uveitis of unknown cause"), it is possible to take aqueous humor during the inflammation-free interval [33]. In horses, approximately 1 ml of aqueous humor can be safely collected and then used for laboratory tests [33, 68–70]. To investigate for the presence of ERU, testing for both anti-*Leptospira* antibodies and by PCR for, e.g., LipL32 is advisable [35, 71–73]. For scientific questions, a leptospiral culture can additionally be performed [34, 35]. Depending on the laboratory findings, a decision can then be made on the further course of action. In case of positive leptospiral findings, vitrectomy is indicated. With negative leptospiral laboratory findings, a leptospiral infection of the vitreous cavity can be excluded with a high probability. These horses would not benefit from vitrectomy—except to remove vitreous opacities that impair vision. In this case, however, a preoperative aqueous humor examination would be superfluous—just as in the case of unequivocal findings in terms of ERU.

For the detection of intraocular anti-*Leptospira* antibodies, the microscopic agglutination test (MAT) is used in most cases. The MAT is highly sensitive and specific when examining aqueous humor or vitreous samples [34, 35, 40]. In addition, other antibody tests can be used, which are either commercially available or available as in-house ELISA tests. Specific anti-*Leptospira* immunoglobulin class A (IgA) antibodies are particularly reliable for detecting intraocular leptospiral infection [72, 74]. Another well-suited test is the SNAP Lepto, which detects anti-LipL32

	History	Typical ophthalmologic findings	Therapy	Prognosis
ERU	most times recurring episodes of painful uveitis attacks	see Section 3.	acute uveitis: topical atropine and dexamethasone, systemic nonsteroidal anti-inflammatory drugs, see Section 6. quiet interval: see Section 7. and 8.	unfavorable, if exclusively conservative therapy is given, usually increasing cloudiness of the transparent media; good for prevention of recurrences with vitrectomy; good for permanent preservation of vision if vitrectomy is performed before ERU has caused irreversible intraocular damage
Leopard coat pattern uveitis	no obvious painful uveitis attacks, "suddenly" noticed impaired vision or cloudiness of the eye	lens pathology (initially or very early in the course of the disease), cataract formation, lens (sub-) luxation, sometimes glaucoma, sometimes atrophy of the globe	so far no etiologic therapy possible, only symptomatic treatment (anti-inflammatory, cyclosporine devices, combatting an elevated intraocular pressure)	unfavorable, often enucleation is required; if both eyes are affected, euthanasia may be indicated
Phacogenic uveitis	often mild uveitis, rarely severe uveitis attacks, sometimes presentation only because of cataract formation	lesions of the lens capsule, sometimes lens fragments in the anterior chamber or in the vitreous chamber, mild or moderate amount of fibrin in the anterior chamber, posterior synechiae (sometimes sealing defects in the anterior lens capsule)	conservative therapy see section 6. phacoemulsification might be considered in selected cases	depending on the degree of leakage of lens proteins and the course of the disease good to unfavorable; phacoemulsification: guarded (risk of retinal detachment)
Traumatic uveitis	sudden onset of eyelid swelling, no previous uveitis attacks observed; in some cases: observed head and/or globe trauma	sometimes accompanying lesions of the eyelids or the cornea, often sero-hemorrhagic uveitis, sometimes hyphema or even hemophthalmus	if no corneal defects: see section 6., in case of corneal defects: no corticosteroids until the cornea is fluorescein-negative; if the fibrin and especially blood are not decreasing within about 10 days: injection of fibrinolytics (e.g., urokinase) or mechanical removal of the inflammatory products and the blood coagulum (e.g. using a small vitrectomy-cutter)	good if there has been no damage to lens and retina and no severe damage to cornea and sclera
Chronic iritis, similar to Fuchs' heterochromic iritis in humans	no painful attacks, most often presented because of corneal cloudiness, rarely because of pigment loss of the iris	often endothelial precipitates and circumscribed depigmentations in the iris, increasing corneal edema, usually slow progression of the disease over years, glaucoma might finally occur	no etiologic, but only symptomatic therapy possible	guarded concerning long-term vision, sometimes enucleation required

	History	Typical ophthalmologic findings	Therapy	Prognosis
Uveitis caused by severe keratitis or corneal infections (kerato-uveitis)	often very painful, increasing corneal opacification, no satisfying response to treatment	deep corneal ulcer or outlined round dense opacities, sometimes annular opacities (ulcus serpens)	corneal debridement, in case of severe weakening of the cornea and imminent rupture: suturing of a conjunctival flap into the lesion; in case of corneal rupture: corneal transplant	depending on the stage of the disease, the extent of the infection and the pretreatment (topical or systemic nonsteroidal drugs and especially corticosteroids worsen the prognosis) good to unfavorable
Uveitis accompanying septicemia	uveitis, most often in foals <6 months, rarely adult horses	typical acute uveitis, comparable to an acute ERU-attack, but often both eyes involved, presence of a severe general infection (e.g. rhodococcosis)	treatment of acute uveitis (Section 6.) and treatment of the septicemia	good if septicemia can be treated successfully and if conservative treatment achieves mydriasis; usually no further uveitis attacks occur
Uveitis caused by intraocular parasitic infections	chronic severe (kerato-) uveitis, not responding to therapy	cloudy cornea, depending on the causative parasite parasites might be visible in the anterior chamber or seen in the ultrasound examination	treatment of acute uveitis (section 6.), but usually uveitis is not responding; if parasites are visible in the anterior chamber: surgical removal may be discussed; enucleation might be indicated	for preserving vision: unfavorable for preserving the globe: guarded to unfavorable
Uveitis caused by intraocular tumors	presentation because of (mild) uveitis, impaired vision and / or corneal or intraocular opacities	most often iris-melanoma, causing keratitis and sometimes glaucoma, other tumors occur less frequently (e.g. medulloepithelioma or malignant lymphoma); uveitis can be caused mechanically and immunologically	symptomatic treatment as long as possible, if the intraocular tumor causes chronic pain or if the tumor is going to spread enucleation is required	unfavorable for preserving vision and for preserving the globe
Endophthalmitis	disturbed general condition, fever, severe eyelid swelling	phlegmon of the eyelids, purulent epiphora, circular deep corneal vascularization, severe yellow-green corneal opacity	in case of very early stages: lavage of the vitreous cavity; otherwise enucleation	for vision: unfavorable for preserving the globe: guarded to unfavorable

Table 1.
Different types of uveitis in horses, symptoms, therapy, and prognosis.

antibodies and is neither immunoglobulin-specific nor serovar-specific. It can be used for samples from different species. With its easy handling and the result visible within 10 minutes, this is a very useful test with a sensitivity and specificity comparable to MAT for intraocular specimens [73, 75]. In contrast to MAT, which is too unspecific for serum testing, SNAP Lepto is well qualified as a screening method even when serum is tested [71].

Antibody detections are equally reliable in vitreous and aqueous humor samples [70, 76, 77]. Both PCR and leptospiral culture are somewhat more reliable when testing vitreous humor samples compared with testing aqueous humor samples [34, 35, 78]. However, the collection of a vitreous sample is disproportionately risky and should be rejected for a preoperative diagnosis, because the aqueous humor analysis is overall very informative [33]. In rare cases, e.g., no anti-*Leptospira* antibodies are detectable in the aqueous humor, but at the same time the PCR yields a positive result. In routine diagnostics, culture has been largely replaced by the much faster and less expensive PCR.

If time is not an issue, but economic reasons have to be taken into account, a reasonable approach for the examination of aqueous humor samples is to first perform an on-site rapid test for the detection of anti-*Leptospira* antibodies. If this test is negative, the MAT can be commissioned externally if necessary. If the MAT is also negative, further antibody tests (e.g., specific in-house ELISA tests) and a PCR can be performed. The more laboratory tests are performed, the fewer "false negatives" can be expected, but the higher the costs for laboratory diagnostics will be.

5. Interpretation of intraocular antibodies

In eyes with a history of recurrent inflammations, but without clear evidence of ERU, and thus without aqueous or vitreous humor opacities, protein levels are typically not elevated. If protein levels in intraocular fluids are not elevated, leakage from the blood can be excluded. In these cases, even very low MAT titers are indicative of intraocular antibody production. The authors consider a MAT result of 1:50 as sufficient indication for vitrectomy in these cases. In eyes with obvious aqueous humor and vitreous opacities, however, the diagnosis of ERU is usually unambiguous even without aqueous humor examination. In cases of doubt, the Goldmann-Witmer coefficient can be used to differentiate leakage from intraocular antibody production [79]. It is crucial that not only the intraocular and the serum titer are evaluated, as it often could be read lately [80–86], but that—as described by Goldmann and Witmer—a reference value is determined both in the aqueous humor and in the serum. Any other antibody titer (e.g., tetanus) can be used as a reference value, provided that antibodies are present in the serum. Alternatively, the total IgG content or, if necessary, even the total protein content can be used as a reference value [33, 34].

6. Therapy of acute uveitis

Acute ERU is treated in the same way as any other equine uveitis [30, 32, 57, 58]. First of all, it is important to achieve mydriasis to avoid posterior synechiae and resulting cataract formation. Atropine is the drug of choice for this purpose and can be used as of 1–2% eye drops or eye ointment. Since the ophthalmic ointment adheres slightly better and acts more protracted, ointment is preferable, if available.

Atropine should initially be given several times daily or even hourly until the pupil dilates. Thereafter, the intervals can be adjusted to the pupil width and often considerably prolonged. Systemic side effects associated with the topical use of 1–2% atropine in horses do not play a significant role in the authors' experience and after having treated thousands of horses over a 30-year period. Colic, e.g., due to an impaction or a meteorism, can occur in any hospitalized horse, not just ophthalmic patients. By feeding mash and monitoring the fecal consistency, an impaction can be detected early and countermeasures (e.g., administration of laxatives) can be taken to avoid more serious colic.

Apart from mydriasis, anti-inflammatory treatment is important. Topical application of ophthalmic ointments containing dexamethasone is particularly effective, provided the corneal epithelium is intact. If corneal defects are present, topical corticosteroids must not be given.

In addition, the administration of a nonsteroidal anti-inflammatory drug (NSAID) orally is indicated. Only in exceptional situations and in case of very significant diffuse vitreous opacification, systemic administration of prednisolone (1 mg/kg per os) for several days may be considered additionally. In these particularly severe cases with significant diffuse vitreous opacification, adjunctive therapy with a systemically given antibiotic, e.g., enrofloxacin [87], can also be performed, to eliminate at least part of the intraocular bacteria—even if this does not completely eliminate the infection [88].

Other measures accompanying the therapy are keeping the horse in a dark place and resting in the stall or just light exercise until the acute inflammation has subsided. If it is not possible to keep the horse in the dark, wearing a light absorbing mask can be considered.

6.1 Brief historical overview of the development of conservative treatment of uveitis in horses valid today (without treatment proposals that did not prove successful or were even questionable from an animal welfare point of view)

- Topical atropine has been recognized as an essential therapeutic mydriatic for equine uveitis as early as 1821 and has been considered a standard treatment for ERU in textbooks since 1842 [89]

- Topical cocaine has been recommended for the control of pain since the beginning of the last century [90]

- Salicylic acid preparations have been included among the treatment options for uveitis since 1922 [91]

- Corticosteroids have been used both parenterally [92] and topically [93] to treat uveitis since the middle of the last century

- In addition to eye drops and ointments, subconjunctival injections with corticosteroids [94, 95], later also with cocaine and atropine, were suggested to intensify the local effect

- Systemic administration of nonsteroidal anti-inflammatory drugs (NASIDs) has also been part of the standard treatment of acute uveitis in horses since their approval for veterinary use in the late 1970s (flunixin meglumine and phenylbutazone) [96, 97].

7. Vitrectomy during the quiet intervals

The most effective treatment for ERU is vitrectomy (removal of diseased vitreous and irrigation of the vitreous cavity). This surgery is performed exclusively in intervals without acute inflammation. Mechanical removal of the vitreous opacities caused by inflammation and accessible vitreous parts very reliably and permanently eliminates the leptospires in the biofilm. Postoperatively, up to 98% of eyes remain free of recurrences when surgery is performed properly [98]. If, exceptionally, further episodes of inflammation occur after surgery, a second vitrectomy can, if necessary, permanently eliminate the infection and prevent further episodes.

Vitrectomy as a vision-preserving procedure is a demanding surgery, having a relatively long learning curve. Prerequisites for successful performing vitrectomies are solid training, availability of for equine ophthalmo-surgery optimized, custom-made instrumentation and equipment as well as careful and intensive perioperative examination and conservative treatment. Any complication may have devastating consequences and can lead to blindness or even enucleation. Only rarely eyes that are already blind undergo surgery in order to prevent both future painful uveitis attacks and removal of the globe, which is cosmetically unsightly.

In order to perform vitrectomy with minimal complications, an experienced team (surgeon, sterile and nonsterile assistant, skilled anesthesiologist) is required, as well as expensive equipment and instruments specially adapted to the dimensions of the horse's eye. For this reason, only a few specialized equine clinics perform vitrectomies to date. In clinics in which vitrectomy is performed as a routine procedure, it is a quick (total anesthesia time is about 40 minutes, the surgical instrument is in the eye <10 minutes) and relatively safe procedure with a very good prognosis [38, 39, 89, 99].

8. Other treatment options for ERU

Apart from vitrectomy, other treatment options have been described, of which two in particular are favored in recent publications. One consists of an intravitreal gentamicin injection. However, the recommended dosage for this purpose (4–6 mg) [80, 100, 101] is 3–4 times the drug concentration that was found to be "safe" with regard to retinal toxicity in experimental studies [102]. So far, there are no long-term results after these injections and the number of horses treated in this way is still limited. Surprisingly, gentamicin injection is not recommended exclusively for equine eyes with intraocular leptospiral infection; other forms of uveitis are also treated with this injection. Improvement after the intravitreal injection is also thought to result from the antibiotic gentamicin having immunomodulatory effects [103].

The second therapeutic option described since the turn of the millennium is the deep intra- or subscleral implantation of a cyclosporine device [104–107]. These implants lead to less frequent and milder episodes of uveitis over a period of up to about 2 years. However, the uveitis does not stop completely, and if the effect wears off, a new implant may have to be inserted. Like gentamicin injection, implantation of cyclosporine devices is performed independently of leptospiral infection in the vitreous cavity. Only individual authors differentiate and use the implants exclusively when no leptospiral infection is detectable [86]. Attention should also be paid to the drug law in its current version, which currently prohibits the import of cyclosporine devices, at least in the EU [108].

However, neither gentamicin injection nor implantation of cyclosporine devices can remove the dense vitreous floaters that often lead to impaired vision. Over time, these deposits also often adhere to the posterior capsule of the lens and, just like extensive posterior synechiae, can lead to a cataract formation.

9. Results of the examination of intraocular specimens

In the literature, before the introduction of vitrectomy to the therapeutic measures against ERU, there were only very sporadic reports of cultural detection of leptospires in intraocular specimens from eyes affected with ERU [109, 110]. Numerous investigators failed to obtain cultural evidence of leptospires, casting doubt on chronic intraocular leptospiral infection. It was rather assumed that although leptospires somehow trigger ERU, the inflammations are not subsequently maintained by the presence of the pathogen [22, 24, 63, 68, 111, 112].

Vitrectomy was initially performed to remove vitreous opacities. The aim was to improve vision in the eyes affected by ERU [36, 37]. However, it soon became apparent that vitrectomy was surprisingly effective in preventing further episodes of uveitis. Therefore, more and more horses were sent to the clinic for vitrectomy.

It was only with the routine performance of vitrectomy that it had become possible to examine numerous vitreous samples from horses suffering from ERU. The peculiarity was that the samples were predominantly from eyes still able to see at an early stage of the disease. By collecting the first milliliters aspirated from the vitreous cavity before opening the intraocular infusion line, it was possible to use undiluted vitreous material for investigations. The results of these examinations, in turn, provided insights into which ocular findings were associated with leptospiral infection and which were not. It was also shown that the prognosis in terms of postoperative absence of recurrences was best when eyes with an intraocular leptospiral infection were treated by vitrectomy [98]. In this way, on the other hand, the indication for vitrectomy was optimized.

With careful assessment of the indication for vitrectomy and examination of undiluted vitreous specimens, MAT titers of 1:100 or higher were detected in 382 of 426 vitreous samples (90%) examined [34, 35]. In some MAT-negative specimens, specific anti-*Leptospira* antibodies (especially immunoglobulin class A) could be detected by an in-house ELISA [74]. Leptospires were culturally detected in 189 of the undiluted vitreous samples from 358 eyes (53%) affected with ERU [34, 35]. The positive cultures had been obtained only after optimization of the sampling technique and immediate sterile inoculation into a transport medium for mailing to a laboratory. The sensitivity of PCR is in between culture and antibody detection. In 70–77% of vitreous samples from eyes affected by ERU, the PCR result was positive [35, 73, 75, 113].

In Germany and neighboring countries, infections with leptospires of the serogroup Grippotyphosa are dominating, accounting for about 80% of intraocular infections in horses suffering from ERU. Infections with leptospires of the Australis serogroup account for about 13–14% of intraocular infections. Less frequently, leptospires of the serogroups Pomona, Sejroe, and Javanica were also detected in the vitreous samples from ERU eyes [34, 35, 114].

Vitreous samples obtained during vitrectomies from eyes affected by ERU were also used for histological and ultrastructural studies. It has been shown that the leptospires in the vitreous of eyes affected with ERU are surrounded by a homogeneous layer, which is lacking the leptospires from culture [115]. This homogeneous layer surrounding the leptospires could be extracellular matrix. In another study, in addition to phagocytosed leptospires, dense roundish structures were detected in vitreous material from eyes affected with ERU [116]. Some of these roundish structures had been phagocytosed, but others of these structures were so large that phagocytosis was impossible. These dense round structures could represent mature leptospiral biofilm constructs.

In 1971, Williams reported on immunologically mediated tissue damage in cases of equine uveitis [22]. However, autoimmune reactions that can be detected at the same time as the leptospiral infection [117–121] must be autoimmune phenomena accompanying the infection, since they cease as soon as the infection has been eliminated [33, 35]. Thus, there is no evidence of autoimmune disease following ERU.

10. Pathogenic *Leptospira* spp. and biofilm

Since many chronic infections are associated with biofilm formation, it has long been suspected that leptospires also form biofilm in vivo. In in vitro studies, biofilm formation of pathogenic *Leptospira* spp. was observed [122], and a detailed description of the three-dimensional structure of these biofilms was given [123]. The main focus with regard to in vivo biofilm formation was on small rodents, which are considered the main vectors of pathogenic leptospires and are chronic shedders. Following experimental infections, evidence of biofilm formation in the proximal renal tubules had been observed [124, 125]. Recently, there was also a description of in vivo biofilm formation in naturally infected rats [126]. At about the same time, biofilm formation of leptospires in vitreous samples from eyes affected with ERU could be demonstrated by immunohistochemistry (IHC) [45].

11. Characteristics of biofilm infections in ERU

Recurrent episodes of uveitis and the concomitant intraocular persistence of leptospiral infection over a long period of time meet the criteria of a biofilm infection [127, 128] very well [129].

The infection primarily affects the vitreous cavity. Possibly following the vitreous clearance, leptospires (more rarely) can also enter the anterior chamber of the eye and be detected there [33–35, 130, 131]. However, the infection obviously remains limited to the eye, there is no evidence of further spreading. As with other local biofilm infections, IgA antibodies are of particular importance in diagnostics [72, 132–135].

One criterion of biofilm infections is the difficult cultural detection of the causative pathogen and ERU meets this criterion. Despite urgent suspicion of leptospiral infection in ERU (high intraocular antibody titers, intraocular antibody production), however, cultural detection of leptospires is demanding and often failed [24–26, 136].

In the vitreous of horses suffering from ERU, there are not only high antibody titers, but also immunocompetent cells (besides lymphocytes, especially plasma cells, macrophages, and granulocytes) [116, 137, 138]. The epithelium of the ciliary body shows many plasma cells in eyes affected by ERU [139]. In the area of the ciliary body and the iris root, even lymph follicles develop during the course of ERU, which contain B lymphocytes in the center [30, 140, 141]. Nevertheless, the immune system fails to eliminate the infection from the large vitreous chamber of the horse.

Leptospira spp. could be visualized in vitreous samples from eyes affected with ERU by immunohistochemistry (IHC). The infecting bacteria are bound to each other, and extracellular matrix could be demonstrated around and in between the bacteria [33, 45, 115, 142]. *Leptospira* spp. could be demonstrated in planktonic forms as well as in smaller and larger cell aggregates and in larger biofilm structures [33, 45].

Leptospires localized in the vitreous chamber show high tolerance to antibiotics. The first cultures were performed with samples from the entire lavage fluid collected during vitrectomy [41, 42, 44]. In the lavage fluid, the vitreous material was diluted about 10-fold and the lavage fluid contained 0.08 mg gentamicin/ml. This concentration had been shown to be 100 times the minimum inhibitory concentration (MIC) for WHO strains of pathogenic leptospires in vitro [143]. Cultures with these vitreous samples were less frequently positive than in later studies performed with undiluted vitreous samples [34, 35, 88], but nevertheless several culture sets eventually became positive after further inoculations and thus dilution of the antibiotic concentration [41, 42, 44].

Similar results were found in a study in which horses had been treated preoperatively intravenously with enrofloxacin. In the undiluted vitreous samples obtained at vitrectomy, the enrofloxacin content was above the MIC. Compared with the control group, in which more than 50% of the cultures were positive for pathogenic *Leptospira* spp., only 30% of the cultures in the group treated with enrofloxacin were positive. Thus, although the probability of a positive culture had been reduced with antibiotic treatment, reliable elimination of the infection was not achieved.

12. Discussion

ERU with persistent intraocular leptospiral infection over a long period of time meets all criteria of an infection associated with biofilm formation. The most likely route by which leptospires enter the vitreous cavity during acute systemic infection is by the fenestrated capillaries in the Pars plicata region of the ciliary body [30, 33]. In the healthy vitreous with its collagen fibers and viscosity, there are ideal conditions for the formation of leptospiral biofilm (**Figure 1**) [129].

The vitreous body is 98% water and contains a collagen fiber scaffold. It has been shown that plant fibers in rice fields are important sites for biofilms [144]. The vitreous fibers [138] might also serve as "surfaces" to which *Leptospira* may adhere and start biofilm production. Furthermore, viscous media promote biofilm production of *Leptospira* spp. [145], and healthy vitreous humor is such a viscous substance. With the collagen fiber scaffold and viscous consistency, the vitreous thus represents an ideal medium for biofilm formation of *Leptospira* spp. [33, 45, 129, 145].

Another factor to consider is that the vitreous cavity of the horse has a volume of approximately 28 ml, making it a large immunologic niche [34]. In addition, there is the immune privilege of the eye [146, 147], which effectively suppresses the immune defense. In this way, pathogenic *Leptospira* spp. can remain clinically unnoticed in the eye for a long time. The latency period can be many months or several years. It probably varies with individual factors of the host, the amount of *Leptospira* spp. in the vitreous, and possibly the leptospiral serovar involved.

Only after months or years, when a threshold is exceeded due to gradual multiplication of the leptospires and increase of immune reactions despite the ocular immune privilege, a uveitis attack with disturbance of the blood-aqueous barrier or blood-ocular barrier becomes apparent [33, 34]. The immune response that occurs in conjunction with the

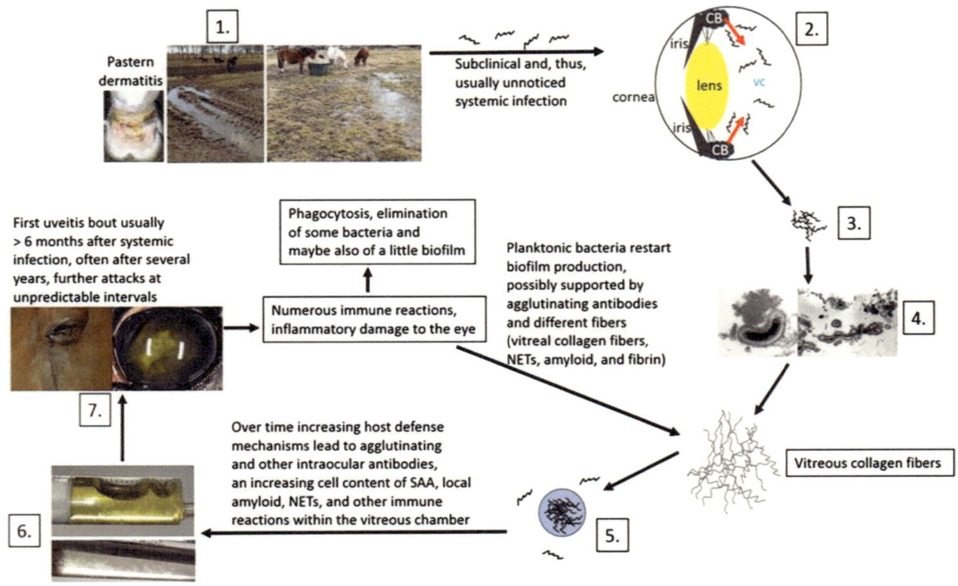

Figure 1.
Schematic illustration of the discussed pathogenesis of equine recurrent uveitis (ERU) caused by a leptospiral biofilm infection in the vitreous chamber. Each uveitis bout leads to increasing damage to the intraocular structures. 1. Infection of horses with Leptospira spp. may occur on humid and muddy pastures or by drinking from standing waters. The bacteria can enter the blood stream via intact mucous membranes (e.g., oral cavity) or small skin lesions (e.g., on the legs). 2. Leptospira spp. most probably enter the vitreous chamber (VC) via the fenestrated capillaries of the pars plicata of the ciliary body (CB). 3. Leptospira spp. within the vitreous chamber attach to each other and to vitreous fibers, starting biofilm production. 4. Transmission electron microscopy using a vitreous sample from an ERU eye: Leptospira spp. are surrounded by extracellular matrix (reprint of [115] courtesy of Schluetersche specialized media GmbH, Hanover, Germany). 5. Most Leptospira spp. are protected within the biofilm, single planktonic bacteria are in the vitreous chamber. 6. Vitreous samples from ERU eyes, containing visible inflammatory products ("vitreous floaters"); the yellow color indicates increased permeability of the blood-ocular barrier. 7. Threshold exceeded, immune privilege of the eye temporarily suspended, clinically apparent uveitis bout (left: Epiphora and blepharospasm; right: Much fibrin in the anterior chamber).

inflammation likely results in the elimination of some planktonic bacteria. Other bacteria in the biofilm outlast the inflammatory bout. After the inflammation subsides under antiphlogistic treatment and with the help of intraocular immunosuppressive mechanisms, a clinically apparently inflammation-free interval occurs, which, however, does not represent a totally quiescent phase immunologically [137].

There are reports, and some own experiences seem to support this, that episodes of uveitis can be triggered by exposure to stressful situations (e.g., competitions, long-distance transport, change of stables, general anesthesia and major surgery). It is conceivable that endogenous cortisol release in stressful situations further reduces the immune defense in the eye (in addition to the ocular immune privilege). This in turn might increase the number of planktonic *Leptospira* spp. in the vitreous cavity after a stress situation and lead to contact with the uvea—which then causes an exaggerated immune reaction resulting in a uveitis attack.

A gradual spread of biofilm structures in the vitreous cavity could explain that ERU episodes occur at shorter intervals and become more severe over time. In addition, there are immune reactions that fail to eliminate the leptospires but may result in damage to the ocular structures adjacent to the vitreous chamber. One example is neutrophil extracellular traps (NETs), which have been detected in vitreous samples from eyes affected by ERU [148]. These NETs are formed by granulocytes to remove pathogens too large for phagocytosis [149]. A disadvantage of the formation of NETs is that tissue-damaging substances are also secreted, which in turn promote an inflammatory reaction of the surrounding tissue [150, 151], which in ERU cases is the uvea.

The high MAT titers in eyes affected by ERU certainly also play a crucial role in the course of the disease, as they promote agglutination of planktonic leptospires. However, since complete elimination of the bacteria is usually not possible, this agglutination can also be the starting point for new biofilm formation. During agglutination, leptospiral aggregates are formed, extracellular matrix is produced after surface contact of bacteria with each other, and thus new biofilm structures can be built. In this way, the agglutinating antibodies could accelerate the biofilm formation of pathogenic *Leptospira* spp. [33].

High levels of serum amyloid A (SAA) [152] and the formation of AA amyloid [153, 154] were detected in intraocular samples from eyes affected with ERU. The formation of amyloid is a good explanation for the fact that the dense vitreous floaters in ERU fail to resolve, but instead increase as the disease progresses. Besides the collagen fibers of the vitreous scaffold, the NETs and the amyloid fibers provide additional fiber structures that could be used for biofilm formation. The formation of NETs and biofilm promote each other [155, 156]. Similar to what has been described for otitis media [157], these numerous fibers could be incorporated into the biofilm and help to reinforce the biofilm scaffold, so that therapeutically only mechanical removal is promising.

With knowledge of the successful cultivation of leptospires from vitreous specimens that contained an active level of gentamicin or enrofloxacin above the MIC, it is questionable whether intraocular gentamicin injections, which are performed therapeutically by some veterinarians, provide lasting success. Biofilms can increase tolerance to antibiotics up to 1000-fold compared with planktonic bacteria [158, 159]. The described improvement of eyes suffering from ERU after gentamicin injection could be due to the fact that planktonic bacteria are eliminated. However, it is questionable whether the bacteria in the biofilm can really be eliminated by the injection. It could also be that the structure and composition of the biofilm change accordingly, so that the bacteria survive protected in the biofilm and then lead to ERU relapses again after some time. With the therapeutically used cyclosporin-devices, spread of the leptospiral biofilm in the vitreous cavity could even be favored, since immune reactions of the host, including those directed against the bacterial pathogen, are suppressed.

In vivo biofilm formation has also been described for other spirochetes. In human medicine, for example, chronic Lyme disease with its various organ manifestations plays an important role [160, 161]. In patients with Lyme disease, in vivo biofilm formation was shown to be associated with the long-term persistence of *Borrelia* spp. [162], and biofilms were found to contain both *Borrelia* spp. and Chlamydiae [163]. For example, alginates have been found in biofilms of *Borrelia* [164]. Alginates induce a distinct immune response [165] and result in the biofilm being more pathogenic than the planktonic bacteria. For lymphocytoma [166] and Alzheimer's disease [167, 168], there are detailed descriptions of biofilm formation and indications for improved treatment options. In Alzheimer's disease, *Borrelia* bacteria in planktonic form do not appear to cause noticeable harm. Here, too, it is the biofilms that create the pathology [167]. Biofilm formation and approaches to improve therapy have also been demonstrated following experimental *Borrelia* infections of mice as a model for Lyme disease [169].

The composition of leptospiral biofilms in the vitreous cavity in ERU is still largely unknown. Neither alginates nor curli fibers (bacterial amyloid) could be detected in the in vitro *Leptospira* biofilms [123]. The in vitro biofilms of *Leptospira* spp. consisted predominantly of extracellular DNA. However, the composition of in vivo biofilms of leptospires could be quite different [170]. It is possible that further analysis of the leptospiral biofilms in the vitreous cavity of horses suffering from ERU may provide further information on how to disperse these biofilms in a manner that is as tissue (retina, lens capsule) compatible as possible. This could provide new insights for the treatment of other biofilm-associated infections that are also relevant to human medicine.

13. Conclusions

ERU is a spontaneously occurring intraocular leptospiral biofilm infection. For centuries, only symptomatic conservative treatment was possible, which has become increasingly effective with the availability of modern anti-inflammatory drugs. However, even the most potent anti-inflammatory treatment could not prevent recurrences of uveitis, which led to gradual damage and even destruction of the affected globe. It was not until the introduction of vitrectomy in equine ophthalmology that causative therapy had become possible. Samples containing leptospiral biofilm can easily be collected in the course of therapeutic vitrectomy. Not only can these samples be used for laboratory diagnostics regarding intraocular leptospiral infection, but further studies can be performed on the composition of the biofilm. There could be significant differences between the composition of the biofilm formed in vitro and that formed in vivo, as host tissues (here: vitreous material and collagen fibrils) and interactions with the host immune system (e.g., agglutinating antibodies, macrophages, granulocytes, NETs, fibrin, and amyloid) influence the composition of the biofilm. ERU provides possibilities for investigation of an in vivo biofilm infection without the need for animal experiments and, thus, could serve as a naturally occurring entity for further research.

Author details

Bettina Wollanke* and Hartmut Gerhards
Ludwig-Maximilians-University, Munich, Germany

*Address all correspondence to: b.wollanke@lmu.de

References

[1] Braun D. Die Geschichte der Erforschung und Behandlung der "periodischen Augenentzündung" des Pferdes im deutschsprachigen Raum von 1750-1950 [Doctoral Thesis]. Munich, Germany: Ludwig-Maximilians-University (LMU); 1994

[2] Paglia DT, Miller PE, Dubielzig RR. James Wardrop and equine recurrent uveitis. Archives of Ophthalmology. 2004;**122**:1218-1223. DOI: 10.1001/archopht.122.8.1218

[3] Bayer J. Mondblindheit oder periodische Augenentzündung. In: Bayer J, editor. *Handbuch der Tierärztlichen Chirurgie und Geburtshilfe*. 2nd ed. Vol. V. Wien, Oesterreich & Leipzig, Germany: Wilhelm Braumüller; 1906. pp. 427-484

[4] Zündel A. Die Mondblindheit der Pferde. Der Gesundheitszustand der Hausthiere in Elsass-Lothringen vom 1. April 1880 bis 1. April 1881. Zeitschrift für vergleichende Augenheilkunde. 1882;**I**:153

[5] Fröhner E. Die Mondblindheit oder periodische Augenentzündung. In: Fröhner E, editor. *Kompendium der speziellen Chirurgie für Tierärzte*. 4th ed. Stuttgart, Germany: Ferdinand Enke Verlag; 1910. pp. 31-35

[6] Kellner R. Ein Beitrag zu Erbfehlerstudien. Deutsche Landwirtschaftliche Tierzucht. 1934;**38**:209-211

[7] Schwarznecker G. Die periodische Augenentzündung im Saargebiet. Zeitschrift für Veterinärkunde 1892;**IV**:1-10

[8] Weil A. Über eine eigenthümliche mit Milztumor, Icterus und Nephritis einhergehende acute Infectionskrankheit. Deutsches Archiv für Klinische Medizin. 1886;**39**:209-232

[9] Köhler W. Weilsche Krankheit. In: Gerabek WE, Haage BD, Keil G, Wegner W, editors. *Enzyklopädie Medizingeschichte*. Berlin, Germany: De Gruyter; 2005. p. 1470

[10] Uhlenhut P, Fromme W. Experimentelle Untersuchungen über die sogenannte Weilsche Krankheit. Medizinische Klinik. 1915;**46**

[11] Gsell O, Rehsteiner K, Verrey F. Iridocyclitis als Spätfolge von leptospirosis Pomona (Schweinehüterkrankheit). Agglutinin- und Lymphozytose-Befund im Kammerwasser [Iridocyclitis as a late consequence of leptospirosis Pomona (porter's disease): Agglutinin and lymphocytosis in the aqueous humor]. Ophthalmologica. 1946;**112**:320-334. DOI: 10.1159/000300399

[12] Heusser H. Zur Ätiologie der periodischen Augenentzündung. Schweizer Archiv für Tierheilkunde. 1952;**94**:296-306

[13] Witmer R, Löhrer J, Wiesmann E. Zur Ätiologie, Diagnose und Therapie der periodischen Augenentzündung (p. A.) des Pferdes. Schweizer Archiv für Tierheilkunde. 1953;**95**:419-439

[14] Rossi P, Kolochine-Erber B. Iridocyclites des équidés et leptospiroses. Revue de Pathologie Générale et de Physiologie Clinique. 1954;**54**:432-477

[15] Witmer R. Periodic ophthalmia in horses. American Journal of Ophthalmology. 1954;**37**:243-253. DOI: 10.1016/0002-9394(54)91570-6

[16] Kemenes F, Surjan J, Vizy L. Leptospira as the cause of periodic

ophthalmia in horses. The Veterinary Bulletin. 1961;**31**:12

[17] Bürki F, Egli P, Wiesmann E. Experimentelle Infektion von Pferden mit Leptospira pomona. Berliner und Munchener Tierarztliche Wochenschrift. 1963;**76**:265-269

[18] Lowe RC. Equine uveitis: a UK perspective. Equine Veterinary Journal. Supplement. 2010;**37**:46-49. DOI: 10.1111/j.2042-3306.2010.tb05634.x

[19] Zwierzchowski J. Klinik und Therapie der Leptospirosen der Haus- und Nutztiere. In: Kathe J, Mochmann H, editors. *Leptospiren und Leptospirosen*. Vol. I. Jena, Germany: Gustav Fischer Verlag; 1967. pp. 79-137

[20] Roberts SJ. Sequelae of leptospirosis in horses on a small farm. Journal of the American Veterinary Medical Association. 1958;**133**:189-194

[21] Bolte HF. Uveitis, a Sequela to Experimentally Induced Leptospira Pomona Infection in the Shetland Pony. West Lafayette, IN, USA: Purdue University; 1966

[22] Williams RD, Morter RL, Freeman MJ, Lavignette AM. Experimental chronic uveitis. Ophthalmic signs following equine leptospirosis. Investigative Ophthalmology. 1971;**10**:948-954

[23] Sillerud CL, Bey RF, Ball M, Bistner SI. Serologic correlation of suspected Leptospira interrogans serovar Pomona-induced uveitis in a group of horses. Journal of the American Veterinary Medical Association. 1987;**191**:1576-1578

[24] Bryans JT. Studies on equine leptospirosis. The Cornell Veterinarian. 1955;**45**:16-50

[25] Williams RD. The presence and duration of persistence of Leptospira Pomona in Equine Ocular Tissues following experimentally induced systemic infection [Master of Science Thesis]. West Lafayette, IN, USA: Purdue University; 1968

[26] Bohl E, Ferguson L. Leptospirosis in domestic animals. Journal of the American Veterinary Medical Association. 1952;**121**:421-428

[27] Tomasek V. Etiology and treatment of equine periodic ophthalmia. Veterinarski Glasnik. 1954;**8**:243

[28] Morter RL, Herschler RC, Fessler JF, Lavignette A. Experimental equine leptospirosis (Leptospira pomona). Proceedings, Annual Meeting of the United States Animal Health Association. 1964;**68**:147-152

[29] Deeg CA, Ehrenhofer M, Thurau SR, Reese S, Wildner G, Kaspers B. Immunopathology of recurrent uveitis in spontaneously diseased horses. Experimental Eye Research. 2002;**75**: 127-133. DOI: 10.1006/exer.2002.2011

[30] Gilger BC, Hollingsworth SR. Diseases of the uvea, uveitis, and recurrent uveitis. In: Gilger BC, editor. *Equine Ophthalmology*. 3rd ed. Wiley Blackwell: Ames, IA, USA; 2017. pp. 369-415

[31] Williams RD. Equine Uveitis: A model system for study of immunologically mediated tissue injury [PhD Thesis]. West Lafayette, IN, USA: Purdue University; 1973

[32] Gerhards H, Wollanke B. Uveitis bei Pferden-Diagnose und Therapie. Pferdeheilkunde. 2001;**17**:319-329. DOI: 10.21836/PEM20010402

[33] Wollanke B, Gerhards H, Ackermann K. Infectious uveitis in horses and new insights in its leptospiral biofilm-related pathogenesis. Microorganisms. 2022;**10**(2):387. DOI: 10.3390/microorganisms10020387

[34] Wollanke B. Die Equine Rezidivierende Uveitis (ERU) Als Intraokulare Leptospirose. Habilitation [Thesis]. Munich, Germany: Ludwig-Maximilians-University (LMU); 2002. DOI: 10.13140/RG.2.2.16979.50720

[35] Wollanke B, Gerhards H, Brem S, Meyer P, Kopp H. Ätiologie der equinen rezidivierenden Uveitis (ERU): Autoimmunkrankheit oder intraokulare Leptospireninfektion [Etiology of equine recurrent uveitis (ERU): autoimmune disease or intraocular leptospiral infection]. Pferdeheilkunde. 2004;**20**: 327-340. DOI: 10.21836/PEM20040403

[36] Werry H, Gerhards H. The surgical therapy of equine recurrent uveitis. Tierärztliche Praxis. Ausgabe G, Grosstiere/Nutztiere. 1992;**20**:178-186

[37] Werry H, Gerhards H. Technique and indications for surgical treatment of equine recurrent uveitis. Pferdeheilkunde. 1991;7:321-331. DOI: 10.21836/PEM19910602

[38] Gerhards H, Wollanke B. Surgical treatment of equine recurrent uveitis: Trans-pars-plana vitrectomy in horses. In: Gilger BC, editor. *Equine Ophthalmology*. 1st ed. Elsevier Saunders: St. Louis, MO, USA; 2005. pp. 314-319

[39] Gerhards H, Wollanke B, Brem S. Vitrectomy as a diagnostic and therapeutic approach for equine recurrent uveitis (ERU). In: Proceedings of the 45th Annual Convention of the American Association of Equine Practitioners (AAEP). Albuquerque, NM, USA; 1999. pp. 89-93

[40] Wollanke B, Rohrbach BW, Gerhards H. Serum and vitreous humor antibody titers in and isolation of Leptospira interrogans from horses with recurrent uveitis. Journal of the American Veterinary Medical Association. 2001;**219**:795-800. DOI: 10.2460/javma.2001.219.795

[41] Wollanke B, Gerhards H, Brem S, Wolf E, Kopp H, Meyer P. Leptospiral aetiology of equine recurrent uveitis (ERU). Results of studies on vitreous and serum samples. Tierärztliche Praxis. Ausgabe G, Grosstiere/Nutztiere. 2000;**28**:153-158

[42] Brem S, Gerhards H, Wollanke B, Meyer P, Kopp H. 35 leptospira isolated from the vitreous body of 32 horses with recurrent uveitis (ERU). Berliner und Münchener Tierärztliche Wochenschrift. 1999;**112**:390-393

[43] Wollanke B, Gerhards H, Brem S, Kopp H, Meyer P. Intraocular and serum antibody titers to Leptospira in 150 horses with equine recurrent uveitis (ERU) subjected to vitrectomy. Berliner und Münchener Tierärztliche Wochenschrift. 1998;**111**:134-139

[44] Brem S, Gerhards H, Wollanke B, Meyer P, Kopp H. Demonstration of intraocular leptospira in 4 horses suffering from equine recurrent uveitis (ERU). Berliner und Münchener Tierärztliche Wochenschrift. 1998;**111**:415-417

[45] Ackermann K, Kenngott R, Settles M, Gerhards H, Maierl J, Wollanke B. In vivo biofilm formation of pathogenic Leptospira spp. in the vitreous humor of horses with recurrent uveitis. Microorganisms. 2021;**9**:1915. DOI: 10.3390/microorganisms9091915

[46] Verma A, Stevenson B. Leptospiral uveitis - there is more to it than meets the eye! Zoonoses and Public Health. 2012;**59**(*Suppl. 2*):132-141. DOI: 10.1111/j.1863-2378.2011.01445.x

[47] Moro F. Les Uvéites leptospirosiques. Documenta Ophthalmologica. 1960;**14**: 383-398

[48] Shukla D, Rathinam SR, Cunningham ET Jr. Leptospiral uveitis in the developing world. International Ophthalmology Clinics. 2010;**50**:113-124. DOI: 10.1097/IIO.0b013e3181d2df58

[49] Rathinam SR. Ocular leptospirosis. Current Opinion in Ophthalmology. 2002;**13**:381-386. DOI: 10.1097/00055735-200212000-00007

[50] Rathinam SR, Rathnam S, Selvaraj S, Dean D, Nozik RA, Namperumalsamy P. Uveitis associated with an epidemic outbreak of leptospirosis. American Journal of Ophthalmology. 1997;**124**:71-79. DOI: 10.1016/s0002-9394(14)71646-0

[51] Pappachan JM, Mathew S, Thomas B, Renjini K, Scaria CK, Shukla J. The incidence and clinical characteristics of the immune phase eye disease in treated cases of human leptospirosis. Indian Journal of Medical Sciences. 2007;**61**:441-447. DOI: 10.4103/0019-5359.32929

[52] Ellis WA. Animal leptospirosis. Current Topics in Microbiology and Immunology. 2015;**387**:99-137. DOI: 10.1007/978-3-662-45059-8_6

[53] Sivakumar R, Balakrishnan V, Gowri P, Visalakshi J. Leptospiral uveitis: Usefulness of clinical signs as diagnostic predictors. Ocular Immunology and Inflammation. 2018;**26**:569-576. DOI: 10.1080/09273948.2016.1217341

[54] Henriksen ML, Dwyer AE, Krarup Nielsen R, Bäcklund S, Dahlmann Christensen N, Holberg Pihl T. Ocular abnormalities in the Icelandic horse with a focus on equine recurrent uveitis: 112 Icelandic horses living in Denmark and 26 Icelandic horses living in the United States. Veterinary Ophthalmology. 2021. DOI: 10.1111/vop.12961 [Online ahead of print]

[55] Szemes P, Gerhards H. Study on the prevalence of equine recurrent uveitis in the Cologne-Bonn area. Praktische Tierarzt. 2000;**81**:408-420

[56] Alexander CS, Keller H. Ätiologie und Vorkommen der periodischen Augenentzündung des Pferdes im Raum Berlin [Etiology and occurrence of periodic eye inflammation of horses in the area of Berlin]. Tierärztliche Praxis. 1990;**18**:623-627

[57] Tóth J, Hollerieder J, Sótoni P. *Augenheilkunde beim Pferd - Lehrbuch und Atlas*. Stuttgart, Germany: Schattauer GmbH; 2010

[58] Barnett KC. Uveitis. In: Barnett KC, Crispin SM, Lavach JD, Matthews AG, editors. *Equine Ophthalmology - an Atlas and Text*. 2nd ed. Edinburgh, UK: Saunders; 2004. pp. 191-197

[59] Blogg J, Barton M, Graydon R, Cust R. Blindness caused by Rhodococcus equi infection in a foal. Equine Veterinary Journal. 1983;**15**:25-26

[60] Tarancón I, Leiva M, Jose-Cunilleras E, Ríos J, Peña T. Ophthalmologic findings associated with Rhodococcus equi bronchopneumonia in foals. Veterinary Ophthalmology. 2019;**22**:660-665. DOI: 10.1111/vop.12637

[61] Barr BS. Rhodococcus equi pneumonia in a foal. The Veterinary Clinics of North America. Equine Practice. 2006;**22**:239-246. DOI: 10.1016/j.cveq.2005.12.022

[62] Cross RS. Equine periodic ophthalmia. The Veterinary Record. 1966;**78**:8-13. DOI: 10.1136/vr.78.1.8

[63] Gelatt KN, Peiffer RL Jr, Gwin RM, Williams LW. The status of equine ophthalmology. Journal of Equine Medicine and Surgery. 1977;**1**:13-19

[64] Heusser H. Die Periodische Augenentzündung, Eine Leptospirose?

Schweizer Archiv für Tierheilkunde. 1948;**90**:287-312

[65] Baumgart A, Gerhards H. Characteristics of uveitis in horses with leopard coat color and potential use of cyclosporine a in its therapy in Germany. Pferdeheilkunde. 2014;**30**:626-632

[66] Baumgart A. Cyclosporin a Und Dessen möglicher Einsatz Bei der Tigerschecken-Uveitis [Cyclosporine a and its Potential Use in Leopard Coat Pattern Uveitis] [Doctoral Thesis]. Munich, Germany: Ludwig-Maximilians-University (LMU); 2014. Available from: https://edoc.ub.uni-muenchen.de/16899/1/Baumgart_Alessandra.pdf

[67] Pinto NI, McMullen RJ Jr, Linder KE, Cullen JM, Gilger BC. Clinical, histopathological and immunohistochemical characterization of a novel equine ocular disorder: Heterochromic iridocyclitis with secondary keratitis in adult horses. Veterinary Ophthalmology. 2015;**18**:443-456

[68] Halliwell RE, Hines MT. Studies on equine recurrent uveitis. I: Levels of immunoglobulin and albumin in the aqueous humor of horses with and without intraocular disease. Current Eye Research. 1985;**4**:1023-1031. DOI: 10.3109/02713688509003347

[69] Gelatt KN. Ophthalmic examination and diagnostic procedures. In: Gelatt KN, editor. *Veterinary Ophthalmology*. Philadelphia, PA, USA: Lea and Febiger; 1981. pp. 219-221

[70] Wollanke B, Gerhards H. Differential diagnosis of equine recurrent uveitis: The importance of a paracentesis of the anterior chamber and aqueous analysis. In: Proceedings of the International Society of Veterinary Ophthalmology (IVO) Pre-Congress Programme of the 30th WSAVA Congress Mexico City, Mexico, 9th – 10th May 2005. 2005

[71] Geiger T, Gerhards H, Wollanke B. Detection of anti-LipL32 antibodies in serum samples from horses with chronic intraocular infection with Leptospira spp. Pathogens. 2021;**10**:1325. DOI: 10.3390/pathogens10101325

[72] Loibl JK, Gerhards H, Brem S, Wollanke B. Improving the laboratory diagnosis of leptospiral uveitis in horses by using an indirect ELISA for the detection of antibodies against Leptospira spp. in intraocular samples. Pferdeheilkunde Equine Medicine. 2018;**34**:267-277. DOI: 10.21836/PEM20180308

[73] Wollanke B, Geiger T, Gerhards H. Evaluation of "SNAP® Lepto"-ELISA and comparison with MAT and PCR results for diagnosis of leptospiral uveitis in horses using intraocular samples. Pferdeheilkunde Equine Medicine. 2018;**34**:508-516. DOI: 10.21836/PEM20180601

[74] Loibl JK. Immunologische und mikrobiologische Untersuchungen zur intraokular persistierenden Leptospireninfektion bei Pferden mit rezidivierender Uveitis [Doctoral Thesis]. Munich, Germany: Ludwig-Maximilians-University (LMU); 2009. Available from: https://edoc.ub.uni-muenchen.de/10508/1/Loibl_Julia.pdf

[75] Geiger T. Evaluation of a Rapid ELISA Test (SNAP-Lepto®) for the Detection of Antibodies against LipL32 in Serum and Intraocular Specimen Material from Horses as a Diagnostic Tool for Equine Recurrent Uveitis [Doctoral Thesis]. Munich, Germany: Ludwig-Maximilians-University (LMU); 2019. Available from: https://edoc.ub.uni-muenchen.de/24938/1/Geiger_Tobias_Xaver.pdf

[76] Gesell S, Wollanke B, Brem S, Gerhards H. Vergleich der Antikoerpertiter

gegen Leptospiren in Kammerwasser- und Glaskoerperproben bei Pferden mit rezidivierender Uveitis. In: Proceedings of the 19th DVG-Fachtagung Pferdekrankheiten. Hanover, Germany; 2006. pp. 239-241

[77] Eule JC, Wagner B, Leibold W, Deegen E. Occurrence of various immunoglobulin isotopes in horses with equine recurrent uveitis (ERU). Berliner und Münchener Tierärztliche Wochenschrift. 2000;**113**:253-257

[78] Sauvage AC, Monclin SJ, Elansary M, Hansen P, Grauwels MF. Detection of intraocular Leptospira spp. by real-time polymerase chain reaction in horses with recurrent uveitis in Belgium. Equine Veterinary Journal. 2019;**51**:299-303. DOI: 10.1111/evj.13012

[79] Goldmann H, Witmer R. Antibodies in the aqueous humor. Ophthalmologica. 1954;**127**:323-330. DOI: 10.1159/000301976

[80] Fischer BM, McMullen RJ Jr, Reese S, Brehm W. Intravitreal injection of low-dose gentamicin for the treatment of recurrent or persistent uveitis in horses: Preliminary results. BMC Veterinary Research. 2019;**15**:29. DOI: 10.1186/s12917-018-1722-7

[81] Himebaugh N, Gilger B. Role of Leptospira spp. testing and ocular examination in horses with equine recurrent uveitis: A retrospective study of 63 horses. Equine Veterinary Education. 2021:1-10. DOI: 10.1111/eve.13543

[82] Gilger BC. Association of acute leptospirosis with systemic disease and uveitis in horses. Equine Veterinary Education. 2018;**30**:137-138. DOI: 10.1111/eve.12693

[83] Gerding JC, Gilger BC. Prognosis and impact of equine recurrent uveitis. Equine Veterinary Journal. 2016;**48**: 290-288. DOI: 10.1111/evj.12451

[84] Gilger BC, Salmon JH, Yi NY, Barden CA, Chandler HL, Wendt JA, et al. Role of bacteria in the pathogenesis of recurrent uveitis in horses from the southeastern United States. American Journal of Veterinary Research. 2008;**69**:1329-1335. DOI: 10.2460/ajvr.69.10.1329

[85] Allbaugh R. Equine recurrent uveitis: A review of clinical assessment and management. Equine Veterinary Education. 2017;**29**:279-288. DOI: 10.1111/eve.12548

[86] Voelter K, Vial Z, Pot SA, Spiess BM. Leptospiral antibody prevalence and surgical treatment outcome in horses with equine recurrent uveitis (ERU) in Switzerland. Veterinary Ophthalmology. 2020;**23**:648-658. DOI: 10.1111/vop.12767

[87] Divers TJ, Irby NL, Mohammed HO, Schwark WS. Ocular penetration of intravenously administered enrofloxacin in the horse. Equine Veterinary Journal. 2008;**40**:167-170. DOI: 10.2746/042516408x255972

[88] Popp MK, Gerhards H, Wollanke B. Enrofloxacinkonzentrationen im Glaskörper und im serum an equiner rezidivierender uveitis (ERU) erkrankter Pferde nach wiederholter intravenöser Verabreichung [Enrofloxacin concentrations in the vitreous of horses with equine recurrent uveitis (ERU) after repeated intravenous administration]. Pferdeheilkunde. 2013;**29**:574-580. DOI: 10.21836/PEM20130501

[89] Braun D. [History of the research and treatment of periodic ophtalmia (ERU) in horses in the German speaking area between 1750 and 1950] Die Geschichte der Erforschung und Behandlung der "Periodischen Augenentzündung" des

Pferdes im deutschsprachigen Raum von 1750-1950. Pferdeheilkunde. 1995;**11**: 43-49. DOI: 10.21836/PEM19950104

[90] Jakob H. *Tierärztliche Augenheilkunde*. Berlin, Germany: Verlagsbuchhandlung Richard Schoetz; 1920

[91] Schleich G. §73 Die periodische, intermittierende Augenentzündung, die Mondblindheit des Pferdes, Iridochorioiditis recidiva equi, Moonblindness. In: Graefe A, Saemisch T, editors. *Handbuch der gesamten Augenheilkunde*. 2nd ed. Berlin, Germany: Tier-Augenheilkunde; Julius Springer Verlag; 1922. pp. 141-151

[92] Unger HA. Allergie Und Herdinfekt in Ihren Beziehungen Zur Periodischen Augenentzündung der Pferde [Anaphylaxis, Allergy, and Herd Infection in their Relationships to Equine Periodic Ophthalmia] [Doctoral Thesis]. Zurich, Switzerland: University of Zurich; 1955

[93] Eikmeier H. [corticosteroids in ophthalmology] das Cortison in der Augenheilkunde. Berliner und Münchener Tierärztliche Wochenschrift. 1955;**68**(83-85):110-112

[94] Komar G, Szutter L. Die innere periodische Augenentzündung ("Mondblindheit") der Pferde. In: Komar G, Szutter L, editors. *Tieraerztliche Augenheilkunde*. Berlin, Germany: Paul Parey; 1968. pp. 231-237

[95] Schmidt V. Periodische Augenentzündung des Pferdes. In: Schmidt V, editor. *Augenkrankheiten der Haustiere*. 2nd ed. Jena, Germany: VEB Gustav Fischer Verlag; 1988. pp. 174-176

[96] Bistner S, Shaw D. Uveitis in the horse. Compendium: Continuing Education for Veterinarians. 1980;**2**:535-543

[97] Tobin T, Chay S, Kamerling S, Woods WE, Weckman TJ, Blake JW, et al. Phenylbutazone in the horse: A review. Journal of Veterinary Pharmacology and Therapeutics. 1986;**9**:1-25. DOI: 10.1111/j.1365-2885.1986.tb00008.x

[98] Wollanke B, Gerhards H, Schinagl C. Results of 654 trans-pars plana vitrectomies of equine eyes with recurrent uveitis - follow-up until 18 years after surgery. Pferdeheilkunde Equine Medicine. 2021;**37**:204-214. DOI: 10.21836/PEM20210301

[99] Winterberg A, Gerhards H. Longterm-results of pars-plana-vitrectomy in equine recurrent uveitis. Pferdeheilkunde. 1997;**13**:377-383. DOI: 10.21836/PEM19970409

[100] Kleinpeter A, Göpfert A, Köhler E, Brehm W. Intravitreal injection of low-dose gentamicin for the treatment of ERU-affected horses. Tierärztliche Praxis. Ausgabe G, Grosstiere/Nutztiere. 2019;**47**:25-34. DOI: 10.1055/a-0816-7156

[101] Launois T, González Hilarión LM, Barbe F, Leurquin C, Bihin B, Hontoir F, et al. Use of Intravitreal injection of gentamicin in 71 horses with equine recurrent uveitis. Journal of Equine Veterinary Science. 2019;**77**:93-97. DOI: 10.1016/j.jevs.2019.02.018

[102] Zachary IG, Forster RK. Experimental intravitreal gentamicin. American Journal of Ophthalmology. 1976;**82**:604-611. DOI: 10.1016/0002-9394(76)90549-3

[103] Neumann M, Ohnesorge B. The intravitreal gentamicin-injection for treatment of the equine recurrent uveitis (ERU) – Status quo. Pferdeheilkunde Equine Medicine. 2021;**37**:302-311. DOI: 10.21836/PEM20210312

[104] Gilger BC, Wilkie DA, Clode AB, McMullen RJ Jr, Utter ME, Komaromy AM, et al. Long-term

outcome after implantation of a suprachoroidal cyclosporine drug delivery device in horses with recurrent uveitis. Veterinary Ophthalmology. 2010;**13**:294-300. DOI: 10.1111/j.1463-5224.2010.00807.x

[105] Gilger BC, Salmon JH, Wilkie DA, Cruysberg LP, Kim J, Hayat M, et al. A novel bioerodible deep scleral lamellar cyclosporine implant for uveitis. Investigative Ophthalmology & Visual Science. 2006;**47**:2596-2605. DOI: 10.1167/iovs.05-1540

[106] Gilger BC, Wilkie DA, Davidson MG, Allen JB. Use of an intravitreal sustained-release cyclosporine delivery device for treatment of equine recurrent uveitis. American Journal of Veterinary Research. 2001;**62**:1892-1896. DOI: 10.2460/ajvr.2001.62.1892

[107] Gilger BC, Malok E, Stewart T, Horohov D, Ashton P, Smith T, et al. Effect of an intravitreal cyclosporine implant on experimental uveitis in horses. Veterinary Immunology and Immunopathology. 2000;**76**:239-255. DOI: 10.1016/s0165-2427(00)00219-1

[108] Wollanke B, Gerhards H. Cyclosporine a (CsA) concentrations in aqueous and vitreous humour samples and clinical and ophthalmological findings in 16 equine eyes after implantation of a sustainedrelease CsA delivery device. Pferdeheilkunde Equine Medicine. 2021;**37**:234-242. DOI: 10.21836/PEM20210304

[109] Hartwigk H, Stoebbe E. Kultureller Nachweis von Leptospiren bei Hund und Pferd. Berliner und Munchener Tierarztliche Wochenschrift. 1952;**65**:212-214

[110] Gelatt K. Ophthalmoscopic studies in the normal and diseased ocular fundi of horses. Journal of the American Animal Hospital Association. 1971;**7**:158-167

[111] Kalsow CM, Hainworth SA, Russell NS, Dwyer AE. Retinal autoimmunity in leptospiral associated uveitis. Investigative Ophthalmology & Visual Science. 1998;**39**:S782

[112] Kemenes F, Tamas L. Ist die fibrinöse Iridozyklitis der Pferde eine Leptospirose. Acta Veterinaria. 1952;**2**:327-336

[113] Wollanke B, Gerhards H, Kaufmann S. Untersuchungen zur Beteiligung von Borrelien an der Ätiologie der equinen rezidivierenden Uveitis (ERU). Pferdeheilkunde Equine Medicine. 2017;**33**:447-451. DOI: 10.21836/PEM20170504

[114] Hartskeerl RA, Goris MG, Brem S, Meyer P, Kopp H, Gerhards H, et al. Classification of leptospira from the eyes of horses suffering from recurrent uveitis. Journal of Veterinary Medicine. B, Infectious Diseases and Veterinary Public Health. 2004;**51**:110-115. DOI: 10.1111/j.1439-0450.2004.00740.x

[115] Niedermaier G, Wollanke B, Hoffmann R, Brem S, Gerhards H. Detection of leptospira in the vitreous body of horses without ocular diseases and of horses with equine recurrent uveitis (ERU) using transmission-electron microscopy. DTW. Deutsche Tierarztliche Wochenschrift. 2006;**113**:418-422

[116] Brandes K, Wollanke B, Niedermaier G, Brem S, Gerhards H. Recurrent uveitis in horses: Vitreal examinations with ultrastructural detection of leptospires. Journal of Veterinary Medicine. A, Physiology, Pathology, Clinical Medicine. 2007;**54**:270-275. DOI: 10.1111/j.1439-0442.2007.00921.x

[117] Degroote RL, Deeg CA. Immunological insights in equine recurrent uveitis. Frontiers in Immunology. 2020;**11**:609855. DOI: 10.3389/fimmu.2020.609855

[118] Deeg CA, Hauck SM, Amann B, Pompetzki D, Altmann F, Raith A, et al. Equine recurrent uveitis--a spontaneous horse model of uveitis. Ophthalmic Research. 2008;**40**:151-153. DOI: 10.1159/000119867

[119] Deeg CA, Hauck SM, Amann B, Kremmer E, Stangassinger M, Ueffing M. Major retinal autoantigens remain stably expressed during all stages of spontaneous uveitis. Molecular Immunology. 2007;**44**:3291-3296. DOI: 10.1016/j.molimm.2007.02.027

[120] Deeg CA, Marti E, Gaillard C, Kaspers B. Equine recurrent uveitis is strongly associated with the MHC class I haplotype ELA-A9. Equine Veterinary Journal. 2004;**36**:73-75. DOI: 10.2746/0425164044864651

[121] Deeg CA, Kaspers B, Gerhards H, Thurau SR, Wollanke B, Wildner G. Immune responses to retinal autoantigens and peptides in equine recurrent uveitis. Investigative Ophthalmology & Visual Science. 2001;**42**:393-398

[122] Ristow P, Bourhy P, Kerneis S, Schmitt C, Prevost MC, Lilenbaum W, et al. Biofilm formation by saprophytic and pathogenic leptospires. Microbiology. 2008;**154**:1309-1317. DOI: 10.1099/mic.0.2007/014746-0

[123] Thibeaux R, Soupé-Gilbert ME, Kainiu M, Girault D, Bierque E, Fernandes J, et al. The zoonotic pathogen Leptospira interrogans mitigates environmental stress through cyclic-di-GMP-controlled biofilm production. npj Biofilms and Microbiomes. 2020;**6**:24. DOI: 10.1038/s41522-020-0134-1

[124] Brihuega B, Samartino L, Auteri C, Venzano A, Caimi K. In vivo cell aggregations of a recent swine biofilm-forming isolate of Leptospira interrogans strain from Argentina. Revista Argentina de Microbiología. 2012;**44**:138-143

[125] Yamaguchi T, Higa N, Okura N, Matsumoto A, Hermawan I, Yamashiro T, et al. Characterizing interactions of Leptospira interrogans with proximal renal tubule epithelial cells. BMC Microbiology. 2018;**18**:64. DOI: 10.1186/s12866-018-1206-8

[126] Santos AAN, Ribeiro PDS, da França GV, Souza FN, Ramos EAG, Figueira CP, et al. Leptospira interrogans biofilm formation in Rattus norvegicus (Norway rats) natural reservoirs. PLoS Neglected Tropical Diseases. 2021;**15**:e0009736. DOI: 10.1371/journal.pntd.0009736

[127] Parsek MR, Singh PK. Bacterial biofilms: An emerging link to disease pathogenesis. Annual Review of Microbiology. 2003;**57**:677-701. DOI: 10.1146/annurev.micro.57.030502.090720

[128] Hall-Stoodley L, Stoodley P. Evolving concepts in biofilm infections. Cellular Microbiology. 2009;**11**:1034-1043. DOI: 10.1111/j.1462-5822.2009.01323.x

[129] Geißler P, Wollanke B. Biofilm formation in persistent infections and its role in the pathogenesis of equine recurrent uveitis (ERU) – A literature review. Pferdeheilkunde Equine Medicine. 2021;**37**:225-233. DOI: 10.21836/PEM20210303

[130] Faber NA, Crawford M, LeFebvre RB, Buyukmihci NC, Madigan JE, Willits NH. Detection of Leptospira spp. in the aqueous humor of horses with naturally acquired recurrent uveitis.

Journal of Clinical Microbiology. 2000;**38**:2731-2733. DOI: 10.1128/jcm.38.7.2731-2733.2000

[131] Polle F, Storey E, Eades S, Alt D, Hornsby R, Zuerner R, et al. Role of intraocular Leptospira infections in the pathogenesis of equine recurrent uveitis in the southern United States. Journal of Equine Veterinary Science. 2014;**34**:1300-1306

[132] Bjarnsholt T, Jensen P, Fiandaca MJ, Pedersen J, Hansen CR, Andersen CB, et al. Pseudomonas aeruginosa biofilms in the respiratory tract of cystic fibrosis patients. Pediatric Pulmonology. 2009;**44**:547-558. DOI: 10.1002/ppul.21011

[133] Høiby N, Bjarnsholt T, Moser C, Bassi GL, Coenye T, Donelli G, et al. ESCMID guideline for the diagnosis and treatment of biofilm infections 2014. Clinical Microbiology and Infection. 2015;**21**(*Suppl. 1*):1-25. DOI: 10.1016/j.cmi.2014.10.024

[134] Pedersen SS, Shand GH, Hansen BL, Hansen GN. Induction of experimental chronic Pseudomonas aeruginosa lung infection with P. aeruginosa entrapped in alginate microspheres. APMIS. 1990;**98**:203-211

[135] Aanaes K, Johansen HK, Poulsen SS, Pressler T, Buchwald C, Høiby N. Secretory IgA as a diagnostic tool for Pseudomonas aeruginosa respiratory colonization. Journal of Cystic Fibrosis. 2013;**12**:81-87. DOI: 10.1016/j.jcf.2012.07.001

[136] Halliwell RE, Brim TA, Hines MT, Wolf D, White FH. Studies on equine recurrent uveitis. II: The role of infection with Leptospira interrogans serovar Pomona. Current Eye Research. 1985;**4**:1033-1040. DOI: 10.3109/02713688509003348

[137] Roth T, Brandes K, Gerhards H, Giving E, Wollanke B. Histologische Untersuchungen des Glaskörpers bei Pferden mit equiner rezidivierender Uveitis. Pferdeheilkunde. 2014;**30**:512-520. DOI: 10.21836/PEM20140501

[138] Niedermaier G, Wollanke B, Hoffmann R, Matiasek K, Gerhards H. Depiction of the structure of the vitreous body in horses without ocular diseases and in horses with equine recurrent uveitis (ERU) using transmission electron microscopy. DTW. Deutsche Tierarztliche Wochenschrift. 2006;**113**:211-217

[139] Dubielzig R, Render J, Morreale R. Distinctive morphologic features of the ciliary body in equine recurrent uveitis. Veterinary and comparative ophthalmology. 1997;**7**:163-167

[140] Gilger BC, Deeg CA. Equine recurrent uveitis. In: Gilger BC, editor. *Equine Ophthalmology*. 2nd ed. Maryland Heights, MO, USA: Elsevier Saunders; 2011. pp. 317-349

[141] Romeike A, Brügmann M, Drommer W. Immunohistochemical studies in equine recurrent uveitis (ERU). Veterinary Pathology. 1998;**35**:515-526. DOI: 10.1177/030098589803500606

[142] Ackermann K. Untersuchungen von Glaskörperproben aus an equiner rezidivierender Uveitis (ERU) erkrankten Augen im Hinblick auf Leptospiren und deren Biofilmbildung [Doctoral Thesis]. Munich, Germany: Ludwig-Maximilians-University (LMU); 2021. Available from: https://edoc.ub.uni-muenchen.de/28756/7/Ackermann_Kerstin.pdf

[143] Brem S. Personal information, State Office for Health and Food Safety, 85764 Oberschleissheim, Germany. Unpublished data. 1996

[144] Vinod Kumar K, Lall C, Vimal Raj R, Vedhagiri K, Vijayachari P. Molecular detection of pathogenic leptospiral protein encoding gene (lipL32) in environmental aquatic biofilms. Letters in Applied Microbiology. 2016;**62**:311-315. DOI: 10.1111/lam.12533

[145] Trueba G, Zapata S, Madrid K, Cullen P, Haake D. Cell aggregation: A mechanism of pathogenic Leptospira to survive in fresh water. International Microbiology. 2004;**7**:35-40

[146] Grisanti S. Immune privilege of the eye. Der Ophthalmologe. 1998;**95**: 124-135. DOI: 10.1007/s003470050250

[147] Zhou R, Caspi RR. Ocular immune privilege. F1000 Biology Reports. 2010;**2**(1):3. DOI: 10.3410/b2-3

[148] Fingerhut L, Ohnesorge B, von Borstel M, Schumski A, Strutzberg-Minder K, Mörgelin M, et al. Neutrophil extracellular traps in the pathogenesis of equine recurrent uveitis (ERU). Cell. 2019;**8**:1528. DOI: 10.3390/cells8121528

[149] Branzk N, Lubojemska A, Hardison SE, Wang Q, Gutierrez MG, Brown GD, et al. Neutrophils sense microbe size and selectively release neutrophil extracellular traps in response to large pathogens. Nature Immunology. 2014;**15**:1017-1025. DOI: 10.1038/ni.2987

[150] Tan C, Aziz M, Wang P. The vitals of NETs. Journal of Leukocyte Biology. 2021;**110**:797-808. DOI: 10.1002/jlb.3ru0620-375r

[151] Hänsch GM. Host defence against bacterial biofilms:"Mission impossible"? International Scholarly Research Notices. 2012;**2012**:853123. DOI: 10.5402/2012/853123

[152] Waldner J, Gerhards H, Wollanke B. Investigations into the occurrence of serum amyloid a in the equine eye. Pferdeheilkunde Equine Medicine. 2018;**34**:461-467. DOI: 10.21836/PEM20180508

[153] Linke RP, Brandes K, Cielewicz MB, Gerhards H, Wollanke B. Ocular leptospiral infection leads to ciliary induction and local AA-amyloidosis in horses. Amyloid. 2019;**26**:127-128. DOI: 10.1080/13506129.2019.1584100

[154] Ostevik L, de Souza GA, Wien TN, Gunnes G, Sørby R. Characterization of amyloid in equine recurrent uveitis as AA amyloid. Journal of Comparative Pathology. 2014;**151**:228-233. DOI: 10.1016/j.jcpa.2014.04.007

[155] Thanabalasuriar A, Scott BNV, Peiseler M, Willson ME, Zeng Z, Warrener P, et al. Neutrophil extracellular traps confine Pseudomonas aeruginosa ocular biofilms and restrict brain invasion. Cell Host & Microbe. 2019;**25**:526-536.e524. DOI: 10.1016/j.chom.2019.02.007

[156] Papayannopoulos V. Neutrophils facing biofilms: The Battle of the barriers. Cell Host & Microbe. 2019;**25**:477-479. DOI: 10.1016/j.chom.2019.03.014

[157] Tsuprun V, Shibata D, Paparella MM, Cureoglu S. Formations of host Fibers and Bacteria in human temporal bones with otitis media. Otology & Neurotology. 2021;**42**:e949-e957. DOI: 10.1097/mao.0000000000003126

[158] Donlan RM, Costerton JW. Biofilms: Survival mechanisms of clinically relevant microorganisms. Clinical Microbiology Reviews. 2002;**15**:167-193. DOI: 10.1128/cmr.15.2.167-193.2002

[159] Nielsen SM, Nørskov-Lauritsen N, Bjarnsholt T, Meyer RL. Achromobacter species isolated from cystic fibrosis patients reveal distinctly different biofilm Morphotypes. Microorganisms. 2016;**4**:33. DOI: 10.3390/microorganisms4030033

[160] Di Domenico EG, Cavallo I, Bordignon V, D'Agosto G, Pontone M, Trento E, et al. The emerging role of microbial biofilm in Lyme Neuroborreliosis. Frontiers in Neurology. 2018;**9**:1048. DOI: 10.3389/fneur.2018.01048

[161] Rudenko N, Golovchenko M, Kybicova K, Vancova M. Metamorphoses of Lyme disease spirochetes: Phenomenon of Borrelia persisters. Parasites & Vectors. 2019;**12**:237. DOI: 10.1186/s13071-019-3495-7

[162] Sapi E, Kasliwala RS, Ismail H, Torres JP, Oldakowski M, Markland S, et al. The long-term persistence of Borrelia burgdorferi antigens and DNA in the tissues of a patient with Lyme disease. Antibiotics. 2019;**8**(4):183. DOI: 10.3390/antibiotics8040183

[163] Sapi E, Gupta K, Wawrzeniak K, Gaur G, Torres J, Filush K, et al. Borrelia and Chlamydia can form mixed biofilms in infected human skin tissues. European Journal of Microbiology and Immunology. 2019;**9**:46-55. DOI: 10.1556/1886.2019.00003

[164] Sapi E, Theophilus PA, Pham TV, Burugu D, Luecke DF. Effect of RpoN, RpoS and LuxS pathways on the biofilm formation and antibiotic sensitivity of Borrelia Burgdorferi. European Journal of Microbiology & Immunology. 2016;**6**:272-286. DOI: 10.1556/1886.2016.00026

[165] Høiby N, Ciofu O, Bjarnsholt T. Pseudomonas aeruginosa biofilms in cystic fibrosis. Future Microbiology. 2010;**5**:1663-1674. DOI: 10.2217/fmb.10.125

[166] Sapi E, Balasubramanian K, Poruri A, Maghsoudlou JS, Socarras KM, Timmaraju AV, et al. Evidence of In vivo existence of Borrelia biofilm in Borrelial Lymphocytomas. European Journal of Microbiology & Immunology. 2016;**6**: 9-24. DOI: 10.1556/1886.2015.00049

[167] Allen HB. A novel approach to the treatment and prevention of Alzheimer's disease based on the pathology and microbiology. Journal of Alzheimer's Disease. 2021;**84**:61-67. DOI: 10.3233/jad-210429

[168] Senejani AG, Maghsoudlou J, El-Zohiry D, Gaur G, Wawrzeniak K, Caravaglia C, et al. Borrelia burgdorferi Co-localizing with amyloid markers in Alzheimer's disease brain tissues. Journal of Alzheimer's Disease. 2022;**85**(2):889-903 DOI: 10.3233/jad-215398

[169] Feng J, Li T, Yee R, Yuan Y, Bai C, Cai M, et al. Stationary phase persister/biofilm microcolony of Borrelia burgdorferi causes more severe disease in a mouse model of Lyme arthritis: Implications for understanding persistence, post-treatment Lyme disease syndrome (PTLDS), and treatment failure. Discovery Medicine. 2019;**27**:125-138

[170] Bjarnsholt T, Whiteley M, Rumbaugh KP, Stewart PS, Jensen P, Frimodt-Møller N. The importance of understanding the infectious microenvironment. The Lancet Infectious Diseases. 2022;**22**(3):e88-e92. DOI: 10.1016/s1473-3099(21)00122-5

Chapter 18

Sub-Aerial Cyanobacteria: A Survey of Research with Antimicrobial Properties for Pharmaceutical Approaches

Lakshmi Singh

Abstract

Cyanobacteria also known as Blue Green Algae (BGA) are widely distributed in environments. Cyanobacteria or BGA commonly being aquatic are also reported from terrestrial ecosystems like sub-aerial surface of temples, monuments and building facades etc., represent their versatile habitats and extremophilic nature. These organisms are the excellent material for primary and secondary metabolites has been investigated by ecologists, physiologists, biochemists and molecular biologists. Scientists and young researchers require knowledge of the potential cyanobacteria and their exploitation in order to formulate effective natural compound or drug remedies. A large number of reports in literature stress have acknowledged the use of Cyanobacteria in pharmaceutical and industries, due to the production of different secondary metabolites with diverse bioactivities. However, very less study is being carried out with respect to exploitation of these sub-aerial Cyanobacteria group for production of different secondary metabolites with biological activities. Since many cyanobacteria are also able to survive most type of stress/and or extreme, they may become even more important as antimicrobial agents of pharmaceuticals in the future. Hence, special attention is paid to these groups of organisms.

Keywords: sub-aerial cyanobacteria, extreme environment, antimicrobial agents, pharmaceutical sector

1. Introduction

The appearances of multi drug resistance among pathogens growing day by day. This could be attributable to prolonged and indiscriminate use of antibiotics and chemotherapeutic agents, over and/or under use of drugs, use of antibiotics without prior knowledge of antibiotic sensitivity pattern of the pathogens, non-completion of dose. In addition, prolonged use of antibiotics and chemotherapeutics results in many side effects too. So there has been a growing demand in search of some new source group of alternative antibiotics. Most of the academicians and researchers all over the world, starting from the ancient age, exploited medicinal and aromatic plants, to a great extent for treatment of diseases and discovery of new antimicrobials or compounds with bioactivities. Based on the complexity in composition, extractions of compounds from microorganisms now are studies again for new antimicrobial compounds. A greater interest has been raised in the field of research towards bioactive compounds from algae. Secondary or primary metabolites of algae consist of diverse groups of chemical compounds. The antibiotic activity of algae has been reported since 1944 [1]. More than 164,784 algae species and infraspecific taxa are reported from all over the world in AlgaeBase whereas, regarding Cyanobacteria, 5152 species have been reported [2]. Among them few have been identified or tested for their efficiency. Algae are sources of amino acid, terpenoids, phlorotannins, steroids, phenolic compounds, halogenated ketones and alkanes and cyclic polysulphides [3]. The natural products from a wide variety of taxa have been isolated and tested for their potential biological activities [4]. Sub-aerial cyanobacteria are one of the important taxa of prokaryotic algae; distributed in extreme habitats need to be explored for their efficiency with respect to bioactivities, as prior research in this area has been inconclusive.

1.1 Cyanobacteria distribution in diverse habitats

Cyanobacteria (BGA) are gram negative photoautotrophic bacteria found in almost all ecological habitats, of aquatic and terrestrial origin. Aquatic forms are abundantly found in both marine and fresh water ecosystems including stagnant water bodies, under running water bodies, lagoons etc. Brackish water bodies also harbor a large number of Cyanobacterial species. Terrestrial habitats, including extreme environments also reflects the tolerance of Cyanobacteria have been reported as biofilms/ or crusts on the exposed surfaces of solid substrata in almost all climatic zones [5]. These organisms grow as epiphytes on tree bark, as epiliths on rocks and stones, and also on anthropogenic surfaces such as facades, concrete floors of roofs and other artificial surfaces of buildings where they cause esthetically unacceptable discolouration of the structures [6]. Such growths are common in humid places on uneven surfaces such as holes, crevices and also on damp building walls due to leaking, roof guttering, inadequate drainage of flat areas or from adjacent water courses. Their adaptation on surfaces of both modern and ancient buildings as well as old monuments represents them as sub-aerial Cyanobacteria/extremophiles since enduring extreme environments. They are particularly abundant in tropics as compared to temperate regions due to their capacity to resist very harsh conditions such as very high temperature, prolonged dry periods, extreme light intensity and UV radiation as they are the prolific producers of secondary metabolites, extracellular glycans, heat shock proteins and, UV pigments such as Mycosporine like amino acids (MAAs) and Scytonemins [7–9]. This population has been reported to have a characteristics appearance and develop a large number of photosynthetic pigments (chlorophyll, carotenoids and phycobiliproteins) including UV - absorbing compounds and pigments which play a key role in their protection and adaptability [10, 11]. These are certain attributes for their colonization and also have of great importance implications in scientific research and for human welfare.

According to literature stresses, the organisms those occur on such substrata mainly consists of coccoid forms of the order Chroococcales (*Chroococcidiopsis*, *Chroococcus*, *Gloeocapsa*, *Myxosarcina*), filamentous forms of the order Oscillatoriales (*Plectonema*, *Leptolyngbya*, *Lyngbya*, *Microcoleus*, *Oscillatoria*, *Phormidium*, *Pseudophormidium*, *Schizothrix*) and Nostocales (*Calothrix*, *Nostoc*, *Scytonema*,

Tolypothrix) etc. Several researchers have studied and reported this type of forms from almost all climatic zones. Examples are building facades in Greece [12], buildings in South Eastern, Spain [13–18], building in American countries [19], building facades in France [20, 21], stone monument and building facades in Italy [22–25], monuments in Portugal [26, 27], monuments in Slovakia [28, 29] and modern/old monuments, India [30–39]. In general, the knowledge on sub-aerial Cyanobacteria diversity colonizing building facades and their exploitation in different applications is still limited. To our knowledge, few reports have been published that deal specifically with the presence of secondary metabolites and pigments of sub-aerial Cyanobacteria and algae isolated from facades of buildings, on structural of cultural heritage and on rock surfaces of different monuments. No systematic scientific approach has been taken yet in India including other countries on this subject. A few research workers have worked on the microorganisms from facades of buildings, cultural heritages and monuments and other material in different parts of the globe [40]. However, no effective chemical or compound which can be employed as an antimicrobial agent from sub-aerial species in Pharmaceutical and Nutraceuticals industries has not yet been reported for which search is on.

2. Bioactive compounds from cyanobacteria

Literature stresses isolation and identification of Cyanobacteria from a diverse environment with bioactivities, but only few research has focused on a variety of bioactive compounds produced by Cyanobacteria after analysis of a great number of marines [41–43], freshwater [44–46], terrestrial [47, 48], and hot spring [49, 50]. Cyanobacterial natural products still seem to prevail followed at much lesser proportions by alkaloids, aromatic compounds, cyclic depsipeptides, cyclic peptides, cyclic peptide, cyclophane, fatty acids, linear peptides, lipopeptides, nucleosides, phenols, macrolides, polyketides, polyphenyl ethers, porphinoids and terpenoids [51]. These interesting and biochemically active compounds possess biological activity covering a wide range of antibacterial [52–55], antifungal [56], antialgal [56], antiviral [57], anticancer effectiveness [58–60], and immunosuppressive [61] activities. Some bioactive lead compound are bastadin, bis-x-butyrolactone, hapalindole, didehydromirazole, kawaguchipeptin B, muscoride, noscomin, nostocine A, scytophytin, and lipids [62] exhibited with antibacterial activity and, ambiguines, calothrixin, cyanobacterin, fischerindole A, hapalindole, hassallidin, phytoalexin, scytophycin, tjipanazole and *Y*-lactone [63, 64] with antifungal activity and few compound such as 4,4′-dihydroxybiphenyl, norhamane pyrido (3,4-*b*)indole, beta-glucan, bacteriocin, ambiguines, parsiguine, scytoscalarol, hapalindole [65] which have been reported to show antimicrobial activity. However, only few of them have been investigated in details [66, 67] are described under this subpoint 2.1. Some known bioactivities as per reported are listed below (**Table 1**). Thus, screening efforts aimed to identify antimicrobial agents in sub-aerial Cyanobacteria which might reveal promising compounds.

2.1 Bioactive compounds and its inhibitory activity with actions

Earlier reports indicate that bioactive compounds contradict synthetic drugs in their composition and their arrangement of radicals and atoms. However, their inhibitory activities are much more depends on the nature of interaction between donor and target organisms. They may inhibit growth or photosynthesis, kill the competitor or exclude

Cyanobacteria	Bioactive compound	Properties	Bioactivity	References
Order: Chroococcales *Microcystis aeruginosa*	Kawaguchipeptin B	It is an antibacterial cyclic undecapeptide	Antibacterial	[71]
Chrococcus turgidus	Beta-glucan	It is a group of beta-D-glucose polysaccharides, form a linear backbone with 1–3 β-glycosidic bonds	Antimicrobial	[94]
Order: Nostocales *Scytonema hofmanni* PCC7110	Scytophytin	It is a congeneric macrolide	Antibacterial	[63]
Anabaena sp. BIR JV1 and *Anabaena* sp. HAN7/1 KP701033	Hassallidin	It is a glycosylated lipopeptides, an esterified eight residue cyclic peptide bonded with a carbohydrate and fatty acid chain. It is composed of non proteingenic amino acids	Antifungal	[63]
Anabaena cf. cylindrica PH133 AJ293110 and *Anabaena* sp. HAN21/1 KP701032	Scytophycin	It is a congeneric macrolide	Antifungal	[63]
Nostoc sp. CENA 219 KP701037 and *Nostoc calcicula* 6sf Cale KP701034	Hassallidin	It is a cyclic glycosylated lipopeptides, contains structural variation of dihydroxy fatty acids and complex glycosylated pattern of monosaccharides	Antifungal	[63, 98]
Fischerella ambigua, *Haplosiphon hibernicus*	Ambiguines	It is isonitrile containing alkaloids	Antibacterial Antifungal	[72, 73]
Scytonema sp.	Scytoscalarol	It is a sesquiterpene	Antibacterial Antifungal	[74]
Tolypothrix sp.	Hassallidin	It is a Glycosylated lipopeptides, an esterified eight residue cyclic peptide bonded with a carbohydrate and fatty acid chain	Antifungal	[75]

Cyanobacteria	Bioactive compound	Properties	Bioactivity	References
Scytonema hofmanni	*Y* – Lactone	They are cyclic esters of hydroxycarboxylic acids, which contain 1-oxacycloalkan-2-one structure, or analogues having heteroatom replacing one or more carbon atoms of the ring	Antifungal	[76]
Scytonema mirabile	Didehydromirabazole	It is an alkaloid otherwise known as 4-methylthiazoline	Antibacterial	[77]
Anabaena basta	Bastadin	It is available in many forms but Bastadin 5 is the bioactive compound	Antibacterial	[78]
Anabaena variabilis	Bis - (x -butyrolactone)	It acts as an intermediate compound for the synthesis of other chemicals like methyl-2-pyrrolidone and as prodrug for gamma hydroxybutyric acid. Used as recreational drug for humans	Antibacterial	[79]
Tolypothrix tjipanasensis	Tjipanazole	It is compound having the pyrrolo [3, 4 - c] ring of indolo [2, 3 - a] carbazoles	Antifungal	[80]
Nostoc sp.	Muscoride	It is a peptide alkaloid possessing *N*-(2-methyl-3-buten- 2- yl) valine and two contiguous methyloxazoles	Antibacterial	[81]
Fischerella muscicola	Fischerellin A	It is an allelochemicals compound	Antibacterial Antifungal	[82]
Nostoc commune	Nostofungicidin	It is a novel lipopeptide fungicide and contains β-amino acid, 3-amino-6-hydroxy stearic acid	Antifungal	[83]
N. commune	Noscomin	It is a diterpenoid skeleton, 8 - [(5-carboxy-2 - hydroxy) benzyl]-2-hydroxy - 1, 1, 4a, 7, 8 –pentamethyl - 1, 2, 3, 4, 4a, 6, 7, 8, 8a, 9, 10, 10a dodecahydrophenanthrene	Antibacterial	[84]
Nostoc spongigaeforme TISTR 8169	Nostocine A	It is a violet color nitrogen rich alkaloid with rare heterocyclic structure.	Antibacterial	[85]

Cyanobacteria	Bioactive compound	Properties	Bioactivity	References
Fischerella muscicola	Fischerindole L	It is an isonitrile tetracyclic allelochemicals	Antibacterial Antifungal	[86, 88]
Fischerella ambigua	Parsiguine	It is a cyclic polymer, oxygenated ethylenic compound	Antibacterial Antifungal	[87]
Scytonema ocellatum	Scytophycin	It is a phenolic compound, macrolide polyketides	Antibacterial Antifungal	[89]
Scytonema pseudohofmanni	Scytophytin	It is a congeneric macrolide	Antibacterial	[90]
Scytonema hofmanni	Cyanobacterin	It is a phenolic compound	Antifungal	[91]
Nostoc sp.	Cyanobacterin LU-1 and LU- 2	It contains a nitrous heterocycle with sugar and phenolic substituents	Antifungal	[92]
Calothrix sp.	Calothrixine A	It is an alkaloid compound	Antifungal	[92]
Fischerella sp.	12-epi-Hapalindole E isonitrile	These are polycyclic bioactive compounds	Antifungal	[92]
Fischerella sp.	Hapalindole	It is polycyclic, isothiocyanate bioactive compounds	Antibacterial	[93]
Nostoc CCC537	Hapalindole	It is a polycyclic ring bioactive compound	Antibacterial	[93]
Nostoc sp. 78–11 A-E	Bacteriocin	It is a proteinaceous or peptide toxins	Antimicrobial	[96]
Nostoc insulare 54.79	4,4′-dihydroxybiphenyl	It is a phenolic compound	Antibacterial Antifungal	[97]
Nodularia harveyana	Norhamane pyrido (3,4-*b*) indole	It is a prototype of β -carboline and nitrogen heterocycle	Antibacterial Antifungal	[97]
Anabaena oryza	Pentadecane	It is a constitution of diacetone alcohol and mesityl oxide as major components	Antibacterial Antifungal	[99]
Stigeonema ocellatum	(3E)-3-Icosene, (Z)-14-Tricosenyl formate, 6- Octen-1-ol 3,7-dimethyl- acetate and 9-Hexadecenoic acid octadecyl ester	It is a constitution of diacetone alcohol, mesityl oxide and heptadecane as major components	Antibacterial Antifungal	[99]

Cyanobacteria	Bioactive compound	Properties	Bioactivity	References
Order: Oscillatoriales *Oscillatoria* sp.	m-Xylene, 2,6,10,14-Tetramethylheptadecane, 2-Ethoxy2-methylbutane, propanedioic acid dimethyl ester, hexaethylene glycol dimethyl ether, propylene glycol trimer 3 and phthalic acid mono-(2-ethylhexyl) ester	It is a constitution of diacetone alcohol, acetic acid butyl ester and mesityl oxide as major components	Antibacterial Antifungal	[99]
Synechococcus sp., strain GFB01	6-pentadecanol and octadecyl acetate	It is a volatiles organic compound	Antibacterial Antifungal	[100]

Table 1.
Bioactive molecules or compound produced by various cyanobacteria on database.

it from the donor vicinity, may be potent in inhibiting protein–protein interactions resulting in effective immune response, signal transduction; mitosis and ultimately apoptosis without causing much harm to living organisms [68, 69]. A large number of novel antimicrobial agents have been identified with antimicrobial, antibacterial and antifungal activities globally represented in (**Table 1**). However, few compounds like ambiguines, calothrixine A, cyanobacterin, fischerindole L, hapalindole, hassallidin, muscoride, noscomin, nostocine, phytoalexin, scytophycin, scytoscalorol and tjipanazole etc., either synthetized by ribosomal pathways or by non-ribosomal pathways [70] have attained importance for their antimicrobial activity in the field of pharmaceutical sector. Most of the cyanobacteria bioactive compound reported here are generally soluble in organic solvents and with low molecular weight. With respect to their mode of action, a relatively limited number of compounds have been studied or identified based on growth inhibition against target organisms. Kawaguchipeptin B, an antibacterial cyclic undecapeptide isolated from the cultured cyanobacterium *Microcystis aeruginosa* (NIES-88) showed antibacterial activity by growth inhibition towards gram positive bacterium *Staphylococcus aureus* at a concentration of 1 μg/mL (MIC) [71]. Ambiguines reported from *Fischerella ambigua* and *Haplosiphon hibernicus* was found to inhibit bacteria like *Mycobacterium tuberculosis* and *Bacillus anthracis*, and fungi such as *Aspergillus oryzae*, *Candida albicans*, *Penicillium notatum*, *Saccharomyces cerevisiae* and *Trichophyton mentagrophytes* [72–74]. Hassallidin reported with various types (hassallidin A, hassallidin B, hassallidin D, hassallidin 12, hassallidin 14 and hassallidin 15) from three different species, *Tolypothrix*, *Anabaena* strain (BIR JV1 and HAN7/1) and *Nostoc* strain (6sf Calc and CENA 219) showed as a potent antifungal agent against *Aspergillus fumigatus* and *C. albicans* [75] through inhibiting growth. Similarly, many other compounds such as gamma lactone from *Scytonema hofmanni* [76], didehydromirabazole from *Scytonema mirabile* [77], bastadin and Bis-x-butyrolactones from *Anabaena basta* and *A. variabilis* [78, 79], tjipanazole from *T. tjipanasensis* [80], muscoride from *Nostoc muscorum* [81], fischerellin A produced by *Fischerella muscicola* [82], nostofungicidin, noscomin and nostacine A from *Nostoc commune* and *Nostoc spongigaeforme* TISTR 8169 against *Bacillus cereus*, *Staphylococcus epidermidis* and *Escherichia coli* [83–85], fischerindole L and Parsiguine from *Fischerella muscicola* and *Fischerella ambigua* [86–88], and Scytophycins from *Scytonema pseudohofmanni*, *S. hofmanni* PCC7110, *Nostoc* sp. HAN11/1 and *Anabaena cf. cylindrica* (BIR JV1 and HAN7/1) [89, 90] are demonstrated with antibacterial /or antifungal activity based on growth inhibition but the type of target organisms and mode of action is unclear. However, few compounds have been shown to exhibit their mode of action through inhibition of photosystem - II, or enzyme or nucleic acid synthesis and/ or cellular paralysis. Phytoalexin from *Scytonema Ocellatum* exhibited inhibition of fungal enzymes and mycelial growth including cytoplasmic granulation, disorganization of the cellular contents and rupture of the plasma membrane of fungi like *Aspergillus oryzae*, *C. albicans*, *Penicillium notatum* and *S. cerevisiae* [89]. Cyanobacterin from *Scytonema hofmanni* and *Nostoc* sp., both found to inhibit the photosystem II-mediated photosynthetic electron transfer [91, 92]. Calothrixine A from *Calothrix* sp., as antifungal activity leads to growth inhibition because of RNA synthesis inhibition [92]. Two alkaloids, hapalindole a polycyclic isothiocyanate and 12-*epi*-hapalindole E isonitrile from *Fisherella* sp., and *Nostoc CCC537* have pointed to inhibition towards bacteria (*Bacillus subtilis*, *M. tuberculosis* H37Rv, *S. aureus* ATCC25923, *Salminella typhi* MTCC3216, *Pseudomonas aeruginosa* ATCC27853, *E. coli* ATCC25992 and *Enterobacter aerogenes* MTCC2822) and fungi (*C. albicans*) based on RNA polymerase, DNA and protein synthesis [92, 93]. β-glucans, a beta-D-glucose polysaccharides from *Chrococcus turgidis* exhibited phagocytic activity and resistance towards *B. subtilis*, *E. coli*, *P. aeruginosa* and *S. aureus* and have shown chronic wound healing activity either directly or indirectly by modulating the activity of diverse cells and growth factors to reparative process [94, 95]. Bacteriocin, an antimicrobial protein/or peptide toxin isolated from *Nostoc* sp. 78–11 A-E found to be inhibits protein and its actions against bacteria and cyanobacteria [96]. Other bioactive molecules like 4–4′-dihydroxybiphenyl (*Nostoc insulare* 54, 79), Norhamane pyrido (3,4-b) indole (*Nodularia harveyana*), Pentadecane (*Anabaena oryzae*), 6-pentadecanol and octadecyl acetate (*Synechococcus* strain), m-Xylene, 2,6,10,14-Tetramethylheptadecane, 2-Ethoxy2-methylbutane, propanedioic acid dimethyl ester (*Oscillatora* sp.), hexaethylene glycol dimethyl ether, propylene glycol trimer 3 and phthalic acid mono-(2-ethylhexyl) ester, (3E)-3-Icosene, (Z)-14-Tricosenyl formate (*Stigonema ocellatum*), 6- Octen-1-ol 3,7-dimethyl- acetate and 9-Hexadecenoic acid octadecyl ester [97–100] are reported with activity, although the mode of that action is still unknown.

3. Bioactivity of sub-aerial cyanobacteria

Many sub-aerial Cyanobacteria are known to tolerate environmental extremes as they possess a great capacity for producing biologically active compounds. Researchers are in believe that more harsh and extreme conditions lead to a wider production of a diverse range of more or less, specific substances thus pointing towards these organisms as brilliant candidates for antimicrobial properties. A few numbers of sub-aerial cyanobacteria compounds are found to inhibit the target organisms, making them an attractive source of antimicrobial agents. Some known bioactivities from ten sub-aerial cyanobacteria as per reported are listed below (**Table 2**). The chloroform fraction of *Scytonema* br1 isolated from wall and Terrace, Konark Temple, Puri, Odisha showed significant anticyanobacterial activity against *Anabaena* BT2 and *Nostoc* pbr01 and antialgal activity against a green alga *Bracteacoccus* [55]. The lipids extract from *Toxopsis calypsus* and *Phormidium melanochroun* isolated from caves established good antibacterial activity against *Enterococcus faecium* (VRE), *Enterococcus faecalis* (ATCC) and *S. aureus* (MRSA) by disrupting cellular membranes [101]. Another study reported the chloroform extracts of *Scytonema hofman* isolated from building facades showed antibacterial activity against *E. coli*, followed by *Klebsiella pneumonia* and *P. aeruginosa*, *S. aureus* [102]. There is a report that acetone extract of sub-aerial species, *Scytonema ocellatum* isolated from sub-aerial habitats exhibits antibacterial activity towards *E. coli*, *B. subtilis* and *S. aureus* and GC analysis showed 98% and 95.6% purity antibiosis [103]. The sub-aerial Cyanobacteria *Anabaena* sp. (VBCCA 052002) as dominant species on terracotta monuments of Bishnupur showed highest antibacterial activity against *S. aureus*, *Salmonella typhimurium* and *E. coli* with a MIC value of 100 μg/ml against *S. aureus* and 150 μg/ml against *S. typhimurium* [104]. In another study reported three different type of bioactive compounds such as 2, 4-Bis (2-methyl-2-propanyl) phenol - phosphorous acid ($C_{42}H_{69}O_6P$:Mw- 700 g/mol) as phenolic, and other two compound Ergost-5-en-3-ol ($C_{28}H_{38}O_4$: Mw-704 g/mol) and 7, 11-dihydroxysolasodine ($C_{27}H_{43}NO_4$: Mw-413 g/mol) as steroidal alkaloid from three sub-aerial cyanobacteria species, *Tolypothrix rechingeri*, *Scytonema hyalinum* and *Scytonema ocellatum* respectively which exhibiting antimicrobial activity against *E. coli*, *P. aeruginosa*, *S. aureus*, *C. albicans* and *Epidermaphyton flocossum* etc. [105]. Out of ten, one of the sub-aerial cyanobacteria, *Fischerella* sp. (NCBI Accession number MN593556) reported with most potent active compound with R_f value 0.96 of acetone fraction showed complete growth inhibition against *E. coli* and moderate

Building materials	Cyanobacteria	Bioactive compounds	Properties	References
Terrace wall, Temple	*Scytonema* br1	—	Anticyanobacterial Antialgal	[55]
Rock (Cave)	*Toxopsis calypsus Phormidium melanochroun*	Lipid	It is constitutions of different class of lipids like glycolipids, sphingolipids, sterol lipids, glycerolipids etc.	[101]
Limewashed wall (Building)	*Scytonema hofman*	—	Antibacterial	[102]
Cement wall (Building)	*Scytonema ocellatum*	—	Antibacterial	[103]
Terracotta wall (Monument)	*Anabaena* sp. (VBCCA 052002)	—	Antibacterial	[104]
Stone carving (Temple)	*Tolypothrix rechingeri*	2, 4-Bis (2-methyl-2-propanyl) phenol - phosphorous acid	It is phenolic compound	[105]
Rock (Cave)	*Scytonema hyalinum*	Ergost-5-en-3-ol and 7	It is a steroidal alkaloid	[105]
Cement wall (Building)	*S. ocellatum*	11-dihydroxysolasodine	It is a steroidal alkaloid	[105]
Cement wall (Building)	*Fischerella* sp. (Accession Number -MN593556)	Pentaphenyl ferrocene carboxamide	It is a heterocyclic alkaloid	[106]

Table 2.
Bioactive molecules or compound produced by various sub-aerial cyanobacteria on database.

activity to *C. albicans*, was identified as, Iron (2+) amino (cyclopenta 2,4 diene-1-ylidine) methanolate 1,2,3,4,5-pentaphenycyclopenta-2.4, dien-1-ide (Pentaphenyl ferrocene carboxamide), $C_{41}H_{31}FeNO$: Mw-610 g/mol and was found to be non-toxic against cells lines of *Catla thymus macrophage* and osteoblast precursor cell line of *Mus musculus* up to 72 hours, with a concentration range of 0.875 - 4 mg/ml indicated their potentiality for development of new antimicrobial compounds [106].

4. Sub-aerial Cyanobacteria: as a source of antimicrobial compounds towards pharmaceutical approaches.

In modern research, a number of significant advancements have been made in Cyanobacterial pharmacologically active compounds from natural resources like marine, freshwater, and very few terrestrial etc., and has received ever increasing interest. A large number of antibiotic compounds, many with novel structures, have been isolated and characterized, but few compounds such as dolastatins, soblidotin, Tasidotin, cryptophycin, curacin D and micropeptins exhibited very interesting results and successfully reached Phase II and Phase III of clinical trials [107–111]. Isolation of these compounds from cyanobacteria species like *Symploca* sp., *Nostoc* sp., and *Lyngbya majuscule* offers great opportunity and a platform for the discovery of anticancer and antitumor agents. Furthermore, a few have focused on baseline information for promoting the use of cyanobacterial bioactive compounds as drugs using the computational approach. They can be profitable to mankind in multidirectional ways and probably they constitute a principal group of organisms for biotechnological exploitation, especially for valuable products, processes and services, with significant impact in food and pharmaceutical industries as well as in public health. However, still the active principles and their mode of action are yet unknown in most cases. Since there is a direct need for an alternate antimicrobial drug due to the emergence of multi drug resistant pathogens throughout the Globe, as one of the major concerns. Literature stresses the study of emerald compound of algae including Cyanobacteria having antimicrobial property. The search of new active substances with antimicrobial activity from Sub-aerial Cyanobacteria (BGA) of extreme environments, form a major group among algae too are the potential and promising candidates. It is of its kind to mention here that, In the past [33] a number of sub-aerial Cyanobacteria from old temples, monuments, caves, building facades were isolated to accelerate their survival strategies and control mechanisms; only few made an effort for their bioactivity [55]. Few are proved to be antiviral drug, anticancer drug, antibacterial drug and or antifungal drug too [112, 113]. In this review, ten major activities of sub-aerial cyanobacteria have been listed from the literature (anticyanobacterial, antialgal, antibacterial and antimicrobial activities) as describe in **Table 2**. However, to the best of knowledge these sub-aerial Cyanobacteria of unique environment are not explored for their biotechnological applications in terms of bioactivities and/or antimicrobial activities to find out their possible use in pharmaceuticals for development of new antimicrobial compounds which need to be further analyzed.

5. Conclusion

Nowadays, the production of secondary metabolites from extreme enduring cyanobacteria has catapulted this group of organisms into the midst of intense research. The survival strategies of cyanobacteria to various stress fixed secondary metabolites sources in term of growth, physiology and different metabolic processes are of great interest as they able to secrete different metabolites with environmental stress and ability for their adaptation to extreme environments. No systematic scientific approach has been taken yet on secondary metabolite with their antimicrobial properties from sub-aerial cyanobacteria in India or other countries on this subject. A few research workers have worked on the bioactive compound and their approaches in pharmaceutical sectors of these sub-aerial cyanobacteria to represent as a new source of biologically active compounds in the form of secondary metabolites with production of different antimicrobial compounds, further more studies are desired to find its way for use in pharmaceutical industries, for development of newer antimicrobials, against costly harmful antibiotics and chemotherapeutics, in order to enjoy the benefits and/or the fruits of this investigation for future uses. However, this knowledge may be important in developing strains of sub-aerial cyanobacteria with higher efficiency for antimicrobial properties.

Funding

No funding was received for this contribution.

Conflicts of interest

The author declares no conflict of interest.

Author details

Lakshmi Singh
Department of Botany, College of Basic Science and Humanities, Odisha University of Agriculture and Technology, Bhubaneswar, Odisha, India

*Address all correspondence to: lakshmisinghouat@gmail.com

References

[1] Pratt R, Daniels TC, Eiler JB, Gunnison JB, Kumler WD. Chlorellin, an antibacterial substance from *chlorella*. Science. 1944;**99**(2754):351-352

[2] Guiry MD, Guiry GM. AlgaeBase. World-wide electronic publication. Galway: National University of Ireland; 2020. https://www.algaebase.org. 2020

[3] Singh R, Parihar P, Singh M, Bajguz A, Kumar J, Singh S, et al. Uncovering potential applications of cyanobacteria and algal metabolites in biology, agriculture and medicine: Current status and future prospects. Frontiers Microbiology. 2017;**8**:515

[4] Dixit RB, Suseela MR. Cyanobacteria: Potential candidates for drug discovery. Antonie Van Leeuwenhoek. 2013;**103**(5):947-961

[5] Gaysina LA, Saraf A, Singh P. Cyanobacteria in diverse habitats. In: Mishra AK, Tiwari DN, Rai AN, editors. Cyanobacteria from Basic Science to Application. Academic Press; 2019. pp. 1-28

[6] Ferrari C, Santunione G, Libbra A, Muscio A, Sgarbi E, Siligardi C, et al. Review on the influence of biological deterioration on the surface properties of building materials: Organisms, materials and methods. International Journal of Design & Nature and Ecodynamics. 2015;**10**(1):21-39

[7] Drovac-cik AV, Dulic TI, Stojanovic DB, Svircev ZB. The importance of extremophiles cyanobacteria in the production of biologically active compounds. Natural Product Sciences. 2007;**112**:57-66

[8] Rastogi RP, Incharoensakdi A. Characterization of UV-screening compounds, mycosporine-like amino acids, and scytonemin in the cyanobacterium *Lyngbya* sp. CU2555. FEMS Microbiology Ecology. 2013;**8**(1):254-256

[9] Rezayian M, Niknam V, Ebrahimzadeh H. Oxidative damage and antioxidative system in algae. Toxicology Reports. 2019;**6**:1309-1313

[10] Böhm GA, Pfleiderer W, Böger P, Scherer S. Structure of a novel oligosaccharide mycosporine amino acid ultraviolet a/B sunscreen pigment from the terrestrial cyanobacterium *Nostoc commune*. Journal of Biological Chemistry. 1995;**270**:8536-8539

[11] Sinha RP. Klisch M, Helbling EW, H€ader DP. Induction of mycosporine-like amino acids (MAAs) in cyanobacteria by solar ultraviolet radiation. Journal of Photochemistry and Photobiology B: Biology. 2001;**6**:129-135

[12] Anagnostidis K, Economou-Amilli A, Roussomoustakaki M. Epilithic and chasmolithic microflora (Cyanophyta, Bacillariophyta) from marbles of the Parthenon-acropolis, Athens, Greece. Nova Hedwigia. 1983;**38**:227-288

[13] Saiz-Jimenez C, Hermosin B, Ortega-Calvo JJ, GómezAlarcón G. Applications of analytical pyrolysis to the study of stony cultural properties. Journal of Analytical and Applied Pyrolysis. 1991;**20**:239-521

[14] Ortega-Calvo JJ, Sanchez-Castillo PM, Hernandez-Marine M, Saiz-Jimenez C. Isolation and characterization of epilithic chlorophyta and cyanobacteria from two Spanish cathedrals (Salamanca and Toledo). Nova Hedwigia. 1993a;**57**:239-253

[15] Ortega-Calvo JJ, Hernandez-Marine M, Saiz-Jimenez C. Cyanobacteria and algae on historic buildings and monuments. In: Garg KL, Garg N, Mukerji NG. editors. Recent Advances in Biodeterioration and Biodegradation Calcutta: Naya Prokash; 1993b. pp. 173-203

[16] Gómez-Alarcón G, Muñoz M, Ariño X, Ortega-Calvo JJ. Microbial communities in weathered sandstones: The case of Carrascosa del campo church. Spain. Science of the Total Environment. 1995;**167**:249-254

[17] Zurita YP, Cultrone G, Castillo PS, Sebastián E, Bolívar FC. Microalgae associated with deteriorated stonework of the fountain of Bibatauín in Granada. Spain. International Biodeterioration and Biodegradation. 2005;**55**:55-61

[18] Uher B, Aboal M, Kovacik L. Epilithic and chasmoendolithic phycoflora of monuments and buildings in South-Eastern Spain. Cryptogamie Algologie. 2005;**26**:275-358

[19] Gaylarde PM, Gaylarde CC. Algae and cyanobacteria on painted buildings in Latin America. International Biodeterioration and Biodegradation. 2000;**46**:93-97

[20] Crispim CA, Gaylarde PM, Gaylarde CC. Algal and cyanobacterial biofilms on calcareous historic buildings. Current Microbiology. 2003;**46**:79-82

[21] Barberousse H, Tell G, Yéprémian C, Couté A. Diversity of algae and cyanobacteria growing on buildings facades in France. Algological Studies. 2006;**120**:81-105

[22] Tomaselli L, Lamenti G, Bosco M, Tiano P. Biodiversity of photosynthetic micro-organisms dwelling on stone monuments. International Biodeterioration and Biodegradation. 2000a;**46**:251-258

[23] Tomaselli L, Tiano P, Lamenti G. Occurrence and fluctuation in photosynthetic biocoenoses dwelling on stone monuments. In: Ciferri O, Tiano P, Mastromei G. editors. The Role of Microbial Communities in the Degradation and Protection of Cultural Heritage. New York: Kluwer; 2000b. pp. 63-76

[24] Lamenti G, Tiano P, Tomaselli L. Biodeterioration of ornamental marble statues in the Boboli gardens (Florence, Italy). Journal of Applied Phycology. 2000;**12**:427-433

[25] Bellinzoni AM, Caneva G, Ricci S. Ecological trends in travertine colonization by pioneer algae and plant communities. International Biodeterioration and Biodegradation. 2003;**51**:203-210

[26] Ortega-Calvo JJ, Ariño X, Hernandez-Marine M, Saiz-Jimenez C. Factors affecting the weathering and colonization of monuments by phototrophic microorganisms. Science of the Total Environment. 1995;**167**:329-341

[27] Oliveira BP, Rosa JM, Miller AZ, Saiz-Jimenez C. An integrated approach to assess the origins of black films on a granite monument. Environmental Earth Science. 2011;**63**:1677-1690

[28] Uher B, Kováčik L. Epilithic cyanobacteria and algae in subterrean mausoleum Chatam Sófer. Bulletin of Slovakia Botanical Spoločnkia, Bratislava. 2004;**10**:83-86

[29] Uher B. Spatial distribution of cyanobacteria and algae from the tombstone in a historic cemetery in Bratislava, Slovakia. Fottea. 2008;**9**:81-92

[30] Tripathy P, Roy A, Adhikary SP. Survey of epilithic blue green algae (cyanobacteria) from temples of India and Nepal. Archives für Hydrobiology Supplement Algological Studies. 1997;**87**:43-57

[31] Tripathy P, Roy A, Anand N, Adhikary SP. Blue-green algae flora of rock surface of temples and monuments of India. Feddes Repertorium. 1999;**110**:133-144

[32] Pattanaik B, Adhikary SP. Blue-green algal flora of some archeological site and monuments of India. Feddes Repertorium. 2002;**113**:289-300

[33] Samad LK, Adhikary SP. Diversity of micro-algae and cyanobacteria on building facades and monuments in India. Algae. 2008;**23**(2):91-114

[34] Sahu JK, Nayak B, Pradhan P, Bhattacharyya S. Epilithic cyanobacteria on temples of Western Odisha. A Scientific Journal of Biological Sciences. 2011;**2**(1&2):47-50

[35] Khan AB, Kulathuran G. Composition of microorganisms in deterioration of stone structures of monuments. The Bioscan. 2010;**1**:57-67

[36] Karande VC, Uttekar GV, Kamble KB, Karande CT. Diversity of cyanobacteria in biofilms on buiding facades of Western Maharashtra. Phykos. 2012;**42**(2):54-58

[37] Keshari N, Adhikary SP. Diversity of cyanobacteria on stone monuments and building facades of India and their phylogenetic analysis. International Biodeterioration and Biodegradation. 2014;**90**:45-51

[38] Pradhan P, Bhattacharyya S, Deep PR, Sahu JK, Nayak B. Biodiversity of Cyanoprokaryota from monuments of Western Odisha, India-I (Chroococales and Stigonematales). Phykos: Journal of the Phycological Society. 2018;**48**(1):58-66

[39] Pattnaik S, Singh L. Cyanobacteria bioactive compound, their production and extraction with pharmaceutical application - a review. International Journal of Current Microbiology and Applied Science. 2020;**9**(6):3394-3405

[40] Pandey VD. Cyanobacterial natural products as antimicrobial agents. International Journal of Current Microbiology and Applied Science. 2015;**4**(1):310-317

[41] Sethubathi GVB, Prabu VA. Antibacterial activity of cyanobacterial species from Adirampattinam coast, southeast coast of palk bay. Current Research Journal of Biological Sciences. 2010;**2**(1):24-26

[42] Engene N, Paul VJ, Byrum T, Gerwick WH, Thor A, Ellisman MH. Five chemically rich species of tropical marine cyanobacteria of the genus *Okeania* gen. Nov. (Oscillatoriales, Cyanoprokaryota). Journal of Phycology. 2013;**49**(6):1095-1106

[43] Skočibušić M, Lacić S, Rašić Z. Evaluation of antimicrobial potential of the Marine cyanobacterium. *Rivularia mesenterica*. Journal of Advances in Microbiology. 2019;**16**(4):1-11

[44] Madhumathi V, Deepa P, Jeyachandran S, Manoharan C, Vijayakumar S. Antimicrobial activity of cyanobacteria isolated from freshwater lake. International Journal of Microbiological Research. 2011;**2**(3):213-216

[45] Mugilan V, Sivakami R. Antimicrobial activity of microalgae isolated from fresh water pond, Tamil

Nadu, India. International Journal of Current Microbiology and Applied Sciences. 2016;**5**(6):588-595

[46] Yadav S, Agrawal M, Raipuria N, Agrawal MK. Antimicrobial activity of *Nostoc calcicola* (cyanobacteria) isolated from Central India against human pathogens. Asian Journal of Pharmaceutics. 2016;**10**(6):554-558

[47] Nagaranjan M, Maruthanayagam V, Sundararaman M. SAR analysis and bioactive potentials of freshwater and terrestrial cyanobacterial compounds: A review. Journal of Applied Toxicology. 2013;**33**(5):313-349

[48] Srikong W, Mittraparp-arthorn O, Rattanaporn P, Bovornreungroj N. Antimicrobial activity of seaweed extracts from Pattani, southeast coast of Thailand. Food and Applied Bioscience Journal. 2015;**3**(1):39-49

[49] Fish SA, Codd GA. Bioactive compound production by thermophilic and thermotolerant cyanobacteria (blue-green algae). World Journal of Microbiology and Biotechnology. 1994;**10**:338-341

[50] Sherwani N, Abed Raeid MM, Dobretsov S, Mary S. Antibacterial and antifungal activities of cyanobacterial strains isolated from hot springs in Oman. Journal of Sciences. 2015;**20**(1):11-19

[51] Swain SS, Paidesetty SK, Padhy RN. Antibacterial, antifungal and antimycobacterial compounds from cyanobacteria. Biomedicine and Pharmacotherapy. 2017;**90**:760-776

[52] Ramya VP, Muralitharan G. Evaluation of antibacterial activity of cyanobacteria isolated from freshwater ecosystem of Tiruchirappalli district, Tamil nadu, India. Research Journal of Life Sciences, Bioinformatics, Pharmaceutical and Chemical Sciences. 2019;**5**(2):331-344

[53] Kumar S, Marketo M, Andreja K, Katrina D, Jan H, Hrouzek P. Antimicrobial activity and bioactive profiling of heterocystous cyanobacterial strains using MS/MS-based molecular networking. Folia Microbiologia. 2019;**64**:645-654

[54] Shaieb FA, Issa AAS, Meragaa A. Antimicrobial activity of crude extracts of cyanobacteria *Nostoc commune* and *Spirulina platensis*. Archives of Biomedical Sciences. 2014;**2**(2):34-41

[55] Yadav S, Agrawal M, Raipuria N, Agrawal MK. Potential of cyanobacteria collected from Central India against human pathogenic bacteria- *In vitro*. International Journal of Pharmaceutical Sciences and Research. 2015;**6**(9):4047-4052

[56] Volk RB, Furkert FH. Antialgal, antibacterial and antifungal activity of two metabolites produced and excreted by cyanobacteria during growth. Microbiological Research. 2006;**161**:180-186

[57] Mazur-Marzec H, Ceglowska M, Konel R, Pyrć K. Antiviral cyanometabolites - a reviews. Biomolecules. 2021;**11**:474

[58] Semary NA, Fouda M. Anticancer activity of *Cyanothece* sp. strain extracts from Egypt: First record. Asian Pacific Journal of Tropical Biomedicine. 2015;**5**(12):992-995

[59] Karan T, Aydin A. Anticancer potential and cytotoxic effect of some freshwater cyanobacteria. Tropical Journal of Pharmaceutical Research. 2018;**17**(11):2183-2188

[60] Silva MR, Silva GM, Silva A, Lima L, Bezerra R, Marques D. Bioactive

compounds of *Arthrospira* spp. (*spirulina*) with potential anticancer activities: A systematic review. ACS Chemical Biology. 2021;**16**(11): 2057-2067

[61] Vijaykumar S, Menakha M. Pharmaceutical applications of cyanobacteria- a review. Journal of Acute Medicine. 2015;**5**(1):15-23

[62] Singh RK, Tiwari SP, Rai AK, Mohapatra TM. Cyanobacteria: An emerging source for drug discovery. The Journal of Antibiotics. 2011;**64**(6):401-412

[63] Shishido TK, Humisto A, Jokela J, Liu L, Wahlsten M, Tamrakar A, et al. Antifungal compounds from cyanobacteria. Marine Drugs. 2015;**13**(4):2124-2214

[64] Marrez DA, Sultan YY. Antifungal activity of the cyanobacterium *Microcystis aeruginosa* against mycotoxigenic fungi. Journal of Applied Pharmaceutical Science. 2016;**6**(11):191-198

[65] Rath B, Priyadarshini I. Antibacterial and antifungal activity of marine cyanobacteria from Odisha coast. International Journal of Current Treasure Research. 2013;**2**(1):248-251

[66] Saad MH, El-Fakharany EM, Salem MS, Sidkey NM. The use of cyanobacterial metabolites as natural medical and biotechnological tools: Review article. Journal of Biolmolecular Structure and Dynamics. 2020;**8**:1-23

[67] Zahra Z, Choo DH, Lee H, Parveen A. Cyanobacteria: Review of current potentials and applications. Environments. 2020;**7**(13):1-17

[68] Koehn FE, Carter GT. The evolving role of natural products in drug discovery. Nature Reviews Drug Discovery. 2005;**4**(3):206-220

[69] Guaadaoui A, Benaicha S, Elmajdoub N, Bellaoui M, Hamal A. What is a bioactive compound? A combined definition for a preliminary consensus. International Journal of Nutrition and Food Sciences. 2014;**3**(3):174-179

[70] Walton K, Berry JP. Indole alkaloids of the Stigonematales (Cyanophyta): Chemical diversity, biosynthesis and biological activity. Marine Drugs. 2016;**14**(73):1-28

[71] Ishida K, Matsuda H, Murakami M, Yamaguchi K. Kawaguchipeptin B, an antibacterial cyclic undecapeptide from the cyanobacterium Microcystis aeruginosa. Journal of Natural Product. 1997;**60**(7):724-726

[72] Smitka TA, Bonjouklian R, Doolin L, Jones ND, Deeter JB, Yoshido WY, et al. Ambiguine isonitriles, fungicidal hapalindole-type alkaloids from three genera of blue green algae belonging to the Stigonemataceae. Journal of Organic Chemistry. 1992;**57**:857-861

[73] Raveh A, Carmeli S. Antimicrobial ambiguines from the cyanobacterium *Fischerella* sp. collected in Israel. Journal of Natural Products. 2007;**70**(2):196-201

[74] Mo S, Krunic A, Pegan SD, Franzblau SG, Orijala J. An antimicrobial guanidine-bearing sesterterpene from the cultured cyanobacterium *Scytonema* sp. Journal of Natural Products. 2009;**72**(11):2043-2045

[75] Neuhof T, Schmieder P, Preussel K, Dieckmann R, Pham H, Bartl F, et al. Hassallidin a, a glycosylated lipopeptide with antifungal activity from the cyanobacterium *Hassallia* sp. Journal of Natural Products. 2005;**68**(5):695-700

[76] Pignatello JJ, Porwoll J, Carlson RE, Xavier A, Gleason FK, Wood JM. Structure of the antibiotic cyanobacterin, a chlorine-containing gamma lactone from the freshwater cyanobacterium *Scytonema hofmanni*. Journal of Organic Chemistry. 1983;**48**(22):4035-4038

[77] Stewart JB, Bornemann V, Chen JL, Moore RE, Caplan FR, Karuso H, et al. Cytotoxic, fungicidal nucleosides from blue green algae belonging to the Scytonemataceae. Journal of Antibiotics. 1988;**41**:1048-1056

[78] Miao SC, Anderson RJ, Allen TM. Cytotoxic metabolites from the sponge *Ianthella basta* collected in Papua New Guinea. Journal of Natural Products. 1990;**53**(6):1441-1446

[79] Ma LX, Led JJ. Determination by high field NMR spectroscopy of the longitudinal electron relaxation rate in Cu (II) plastocyaninform *Anabaena variabilis*. Journal of American Chemical Society. 2000;**122**:7823-7824

[80] Bonjouklian R, Smitka TA, Doolin LE, Molloy RM, Debono M, Shaffer SA, et al. Tjipanazoles, new antifungal agents from the blue green alga *Tolypothrix tjipanasensis*. Tetrahedron. 1991;**47**:7739-7750

[81] Nagatsu A, Kajitani H, Sakakibara J. Muscoride a - a new oxazole peptide alkaloid from freshwater cyanobacterium *Nostoc muscorum*. Tetrahedron Letters. 1995;**36**(23):4097-4100

[82] Hagmann L. Jüttner F, Fischerellin A. a novel photosystem-II inhibiting allelochemical of the cyanobacterium *Fiseherella museieola* with antifungal and herbicidal activity. Tetrahedron Letters. 1996;**37**:6539

[83] Kajiyama S, Kanazaki H, Kawazu K, Kobayashi A. Nostifungicide, an antifungal lipopeptide from the field grown terrestrial blue-green alga. Tetrahedron Letters. 1998;**39**:37-40

[84] Jaki B, Orjala J, Heilmann J, Linden A, Vogler B, Sticher O. Novel extracellular diterpenoids with biological activity from the cyanobacterium *Nostoc commune*. Journal of Natural Products. 2000;**63**(3):339-343

[85] Hirata K, Yoshitomi S, Dwi S, Iwabe O, Mahakhant A, Polchai J, et al. Bioactivities of nostocine a produced by a fresh water cyanobacterium nostoc spongiaefrome TISTR 8169. Journal of Bioscience and Bioengineering. 2003;**95**:512-517

[86] Park A, Moore RE, Patterson GML. Fischerindole L, a new isonitrile from the terrestrial blue-green alga *Fischerella muscicola*. Tetrahedron Letters. 1992;**33**:3257-3260

[87] Ghasemi Y, Yazdi MT, Shafiee A, Amini M, Shokravi S, Zarrini G. Parsiguine, A Novel Antimicrobial Substance from *Fischerella ambigua*. Pharmaceutical Biology. 2004;**42**(4-5):318-322

[88] Kim H, Krunie A, Lantvit D, Shen Q, Kroll DJ, Swanson SM, et al. Nitrile containing fischerindoles from the cultured cyanobacterium *Fischerella* sp. Tetrahedron. 2012;**68**:3205-3209

[89] Patterson GML, Bolis CM. Scytophycin production by axenic cultures of the cyanobacterium *Scytonema oscellatum*. Natural Toxins (USA). 1997;**2**:280-285

[90] Ishibashi M, Moore RE, Patterson GML, Xu C, Clardy J. Scytophycins, cytotoxic and animycotic agents from cyanophyte *Scytonema pseudohofmanii*. Journal of Organic Chemistry. 1986;**51**:5300-5306

[91] Gleason FK, Baxa CA. Activity of the natural algicides, cyanobacterin, on eukaryotic microorganisms. FEMS Microbiology Letters. 1986;**33**:85-88

[92] Doan NT, Rickards RW, Rothschild JM, Smith GD. Allelopathic actions of the alkaloid 12-epi-hapalindole E isonitrile and calothrixin a from cyanobacteria of the genera *Fischerella* and *Calothrix*. Journal of Applied Phycology. 2000;**12**:409-416

[93] Asthana RK, Deepali S, Tripathi MK, Srivastava A, Singh AP. Isolation and identification of a new antibacterial entity from the Antarctic cyanobacterium *nostoc* CCC 537. Journal of Applied Phycology. 2009;**21**:81-88

[94] Chinnu K, Muthukumaran M, Mukund S, Sivasubramanian V. Antimicrobial and antifungal activity of isolated betaglucan from *Chroococcus turgidus*. Indian Journal of Pharmaceutical Science & Research. 2014;**4**(4):217-220

[95] Shin MS, Lee S, Lee KY, Lee HG. Structural and biological characterization of aminated-derivatized oat β-glucan. Journal of Agriculture and Food Chemistry. 2005;**53**(14):5554-5558

[96] Flores E, Wolk CP. Production, by filamentous, nitrogen- fixing cyanobacteria, of a bacteriocin and of other antibiotics that kill related strains. Archives of Microbiology. 1986;**145**(3):215-219

[97] Volk RB. Screening of microalgal culture media for the presence of algicidal compounds and isolation and identification of two bioactive metabolites excreted by the cyanobacteria *Nostoc insulare* and *Nodularia harveyana*. Journal of Applied Phycology. 2005;**17**(4):339-347

[98] Vestola J, Shishido TK, Jokela J, Fewer DP, Aitio O, Permi P, et al. Hassallidins, antifungal glycolipopeptides, are wide spread among cyanobacteria and are the end-product of a non-ribosomal pathway. Proceedings of the National Academy of Sciences of the United States of America. 2014;**111**:E1909-E1917

[99] Seddek NH, Fawzy MA, El-Said WA, Ragaey MM. Evaluation of antimicrobial, antioxidant and cytotoxic activities and characterization of bioactive substances from freshwater blue-green algae. Global NEST Journal. 2019;**21**(3):329-337

[100] Amaral R, Fawley KP, Němcová Y, Ševčíková T, Lukešová A, Fawley MW, et al. Toward modern classification of Eustigmatophytes, including the description of Neomonodaceae fam. Nov. and three new genera. Journal of Phycology. 2020;**56**(3):630-648

[101] Lamprinou V, Tryfinopoulou K, Velonakis EN, Vatopoulos A, Antonopoulou S, Fragopoulou E, et al. Cave cyanobacteria showing antibacterial activity. International Journal of Speleology. 2015;**44**(3):1-9

[102] Panigrahi S, Sethi AK, Samad LK. Antibacterial activities of *Scytonema hofman* extracts against human pathogenic bacteria. International Journal of Pharmacy and Pharmaceutical Science. 2015;**7**(5):123-126

[103] Pattnaik S, Samad LK. Spectral analysis of biologically active compounds and antibacterial activity of *Scytonema ocellatum* isolated from sub-aerial habitats. International Journal of Current Research. 2018;**10**(9):73218-73224

[104] Bhakat S, Saha S, Mandal S, Rath J. Antibacterial activity of desiccated cyanobacterium *Anabaena* sp. isolated from Terracotta Monuments of

Bishnupur, West Bengal. International Journal of Pharmaceutical Science and Drug Research. 2020;**12**(2):1333

[105] Pattnaik S, Rath CC, Singh L. Isolation, identification and screening for bioactive compounds with antimicrobial activities from sub-aerial cyanobacteria of eastern regions, Odisha. International Journal of Pharmaceutical Sciences and Research. 2021;**12**(7):3716-3730

[106] Pattnaik S, Mishra BB, Singh L. Isolation and characterization of a novel antimicrobial compound from sub-aerial cyanobacterium *Fischerella* sp., isolated from building facades. Applied Biological Research. 2021;**23**(1):1-12

[107] Luesch H, Moore RE, Paul VJ, Mooberry SL, Corbett TH. Isolationof dolastatin 10 from the marine cyanobacterium Symploca species VP642 and total stereochemistry and biological evaluation of its analogue symplostatin 1. Journal of Natural Product. 2001a;**64**:907-910

[108] Edelman MJ, Gandara DR, Hausner P, Israel V, Thornton D, DeSanto J, et al. Phase 2 study of cryptophycin 52 (LY355703) in patients previously treated with platinum-based chemotherapy for advanced non-small cell lung cancer. Lung Cancer. 2003;**39**:197-199

[109] D' Agostino G, Del Campo J, Mellado B, Izquierdo MA, Minarik T, Cirri L, et al. A multicancer phase in study of the cryptophycin analog LY355703 in patients with platinum-resistant ovarian cancer. International Journal of Gynecology Cancer. 2006; **16**:71-76

[110] Sammet B, Bogner T, Nahrwold M, Weiss C, Sewald N. Appeoaches for the synthesis of functionalized cryptophycins. Journal of Organic Chemistry. 2010;**75**:6953-6960

[111] Weiβ C, Bogner T, Sammet B, Sewald N. Total synthesis and biological evaluation of fluorinated cryptophycins. Beilstein Journal of Organic Chemistry. 2012;**8**:2060-2066

[112] Lingington RG, Clark BR, Trimble EE, Almanza A, Urina LD, Kyle DE, et al. Antimalarial peptides from marine cyanobacteria: Isolation and structural elucidation of gallianamide a. Journal of Natural Products. 2009;**72**(1):14-17

[113] Abdo SM, Hetta MH, El-Senousy WM, El Din RAS, Ali GH. Antiviral activity of freshwater algae. Journal of Applied Pharmaceutical Science. 2012;**2**(2):21-25

Chapter 19

Growing Environmental Bacterium Biofilms in PEO Cryogels for Environmental Biotechnology Application

Galina Satchanska

Abstract

This Chapter discusses the entrapment, growing and biofilm formation by an environmental bacterium immobilized in polyethyleneoxide cryogel to be applied in environmental biotechnology. The KCM-R5 bacterium was isolated from the heavy metal-polluted environment near a large Pb-Zn smelter, also producing precious metals in Bulgaria. Molecular-genetic analysis revealed affiliation with *Pseudomonas rhodesiae.* The strain is capable of growing in high concentrations of phenol and different phenol derivatives. Polyethylene oxide was found to be friendly and nontoxic to bacteria polymer enabling bacteria easy to penetrate in it and fast to grow. KCM-R5 biofilms were grown for 30 days in batch culture with phenol (300-1000 mg L^{-1}) dissolved in the mineral medium. The bacterium was able to involve phenol in its metabolism and use it as a single carbon supplier. The results obtained in the study showed 98% phenol biodegradation using the biotech installation described. The proposed PEO cryogel-*P. rhodesiae* KCM-R5 bacterium biotech biofilter can be used for environmental biotechnology application in industrial wastewater detoxification.

Keywords: PEO cryogels, environmental bacterium, biodegradation, phenol derivatives, biofilms

1. Introduction

In recent years, the growing amount of polymer-encapsulated bacteria and engineered bacterial biofilms have enhanced both wastewater management and biodegradation of industrial pollutants. Amidst the aromatic substances, monocyclic phenol and its nitro- and chlorophenol derivatives represent one of the most harmful environmental pollutants. Phenol (**Figure 1**) is a by-product of benzene production and widely exploited in the chemical industry.

The continuous application of phenol and its derivatives such as *ortho*-nitrophenol (*o*-NP), 2,4- dinitrophenol (2,4-dNP), 2,5-dinitrophenol (2,5-dNP), penthachlorophenol (PCP) and 2,4- dichlorophenoxyacetic acid (2,4-D) (**Figure 2**) in the

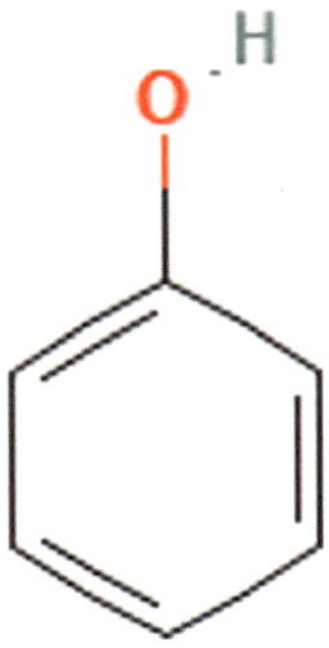

Figure 1.
Structural formula of phenol.

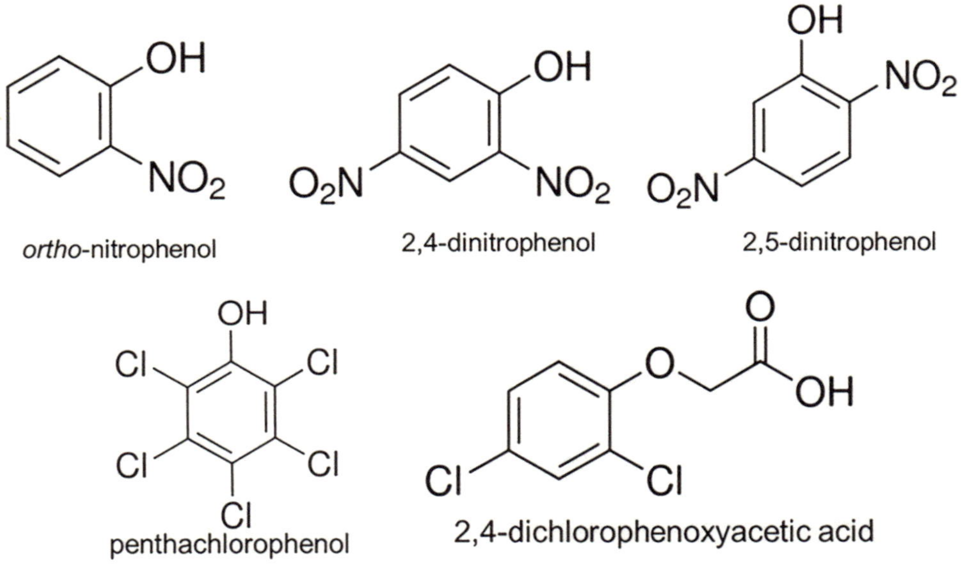

Figure 2.
Structural formulas of nitro- and chlorophenol derivatives.

chemical, agricultural, woodworking and oil processing industries has resulted in their persistent presence in the environment.

Worldwide, high concentrations of phenol and phenol derivatives were detected in industrial wastewaters, which further flow into rivers, seas and oceans. Bisphenol A (BPA) as phenol derivative is amid he most prominent plasticizers and is omnipresent in surface and ground water. This toxic substance is detected in many aquatic organisms. Mathieu-Denoncourt *et al.* [1] reported that BPA was the most toxic (96 h LC50s) to aquatic invertebrates (0.96-2.70 mg/L) and less toxic to fish (6.8-17.9 mg/L). It plays toxic effect on amphibians being more noxious to embryos than to juveniles. It plays neuro-toxic and reproductive effect reported by Santoro *et al.* [2].

Phenol is harmful to human causing blood pressure increase, leukemia, skin necrosis and pores creation, damages of the phospholipid bilayer, heart arrhythmia, tight junctions disruption, liver and kidney injury, earlier child birth and gastro-intestinal perforations [3, 4]. Phenol did not demonstrate a carcinogenic effect (**Figure 3**) [4].

It is estimated that the median *lethal dose* of phenol in humans is 14-214 mg kg^{-1}or 1-15 g [4].

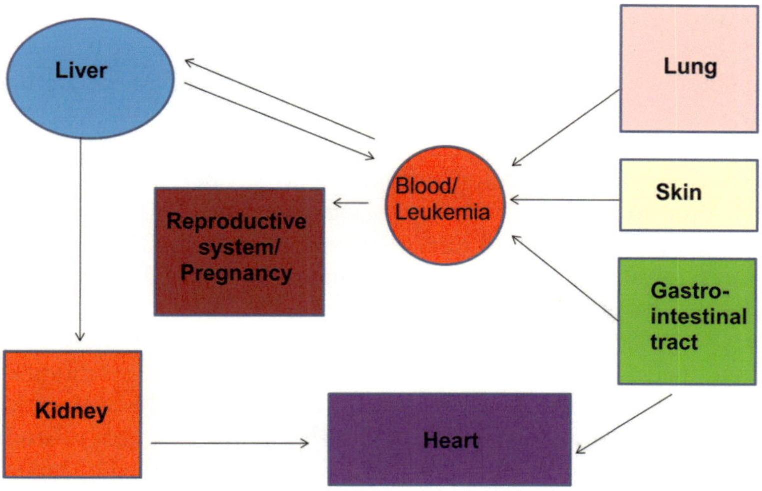

Figure 3.
Harmful effect of phenol on human organs.

Among bacteria, bacterial species like *Pseudomonas* [5–7], *Bacillus* [8] and *Geobacter* [9] are capable of effective phenol biodegradation. Besides bacteria, fungi of the genera *Aspergillus* [10], *Trichosporon* can also successfully degrade phenol. Most authors describe mainly the degradation by free planktonic cells, but data about degradation by encapsulated bacteria are scarce. Both natural or synthetic polymers can be used as bacteria carriers. Biodegradation of phenol is accomplished *via ortho-* or *metha* cleavage of the aromatic ring. First step is conversion of phenol to catechol by attachment of additional hydroxyl group (**Figure 4**).

Further the catechol is degraded either *via* the *metha*-mechanism, a process catalyzed by the enzyme catechol 2,3-dioxigenase or *via* the *ortho*-mechanism using catechol 1,2-dioxigenase [11].

The current chapter discusses the variety of natural and synthetic polymers used for bacterial entrapment; the content, development and structure of bacterial biofilms, and encapsulation of the xenobiotic degrading bacterium *Pseudomonas rhodesiae* KCM-R5 in PEO cryogels, creating a biofilter, bacterial biofilm formation and phenol degradation by said polymer-bacterium biofilter.

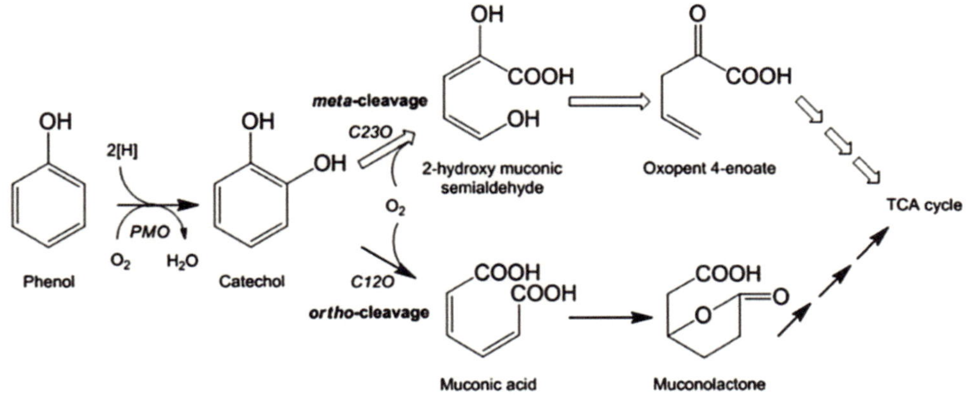

Figure 4.
Mechanism of phenol degradation [11].

2. Natural and synthetic polymers used for entrapment of bacteria

Natural polymers most commonly used are: (i) Alginate: alginate, alginate/soy protein isolate (SPI), algnate/cashew gum, (ii) Cellulose derivatives: cellulose acetate, ethyl cellulose, cellulose fibers, (iii) Chitosan: chitosan, the binary system beta-cyclodextrin modified chitosan, chitosan/synthetic poly(ethylene oxide), (iv) Starch and maltodextrin: gum acacia/maltodextrin, Arabic gum/maltodextrin/starch, (v) Whey protein, (vi) Fibroin: fibroin/poly-caprolactone, (vii) Gelatine. The main advantages of natural polymers are their biocompatibility and nontoxicity to living cells and biological structures, e.g. essential oils [12].

Synthetic polymers used for bacterial encapsulation are polyvinylchloride, polylactic acid, polycaprolactone, polycaprolactone/hydroxiapatite composites, poly(methil methacrilate), poly(vinyliden fluoride), poly(ethileneoxide), poly(ethylene brassilate-co-squaric acid) [12–14]. A limited number of studies have been reported for phenol degradation by bacterial biofilms formed by immobilized bacteria. Immoblilization of bacteria was conducted in polyacrylamide [15], polyurethane [16], polyamide [17], polyacrylonitrile [18, 19] or polyvinyl alcohol [20].

In the last 20 years, different organic carriers for bacterial immobilization were investigated [15, 16]. Among synthetic polymers, poly(ethylene oxide) hydrogels are excellent candidates because they are nontoxic and biocompatible materials which meet all of the requirements for strength, absorbency, flexibility and adhesiveness [17]. Hydrogels of poly(ethylene oxide) have been synthesized *in situ* by applying a facile optimized protocol, which will be further described.

3. Structure and development of bacterial biofilms

Biofilms are an excellent strategy for bacterial survival in a sessile way and 40-80% of bacteria on earth can form biofilms [21–26]. The first to observe under a microscope microbes living on the surfaces of teeth was the Dutch merchant Antony van Leuwehoek. He can also be considered the first discoverer of bacterial biofilms. The invention of the electronic microscope in the 1930-ies provided an insight into the structure and organization of biofilms. Biofilms colonize different surfaces like plant and animal tissues, medical devices, potable water pipes, and natural lakes and rivers. In the early 1970-ies, the ambiguous role of disinfectants in the disruption of bacterial biofilms was proved, a finding published by [24]. The authors discussed bacterial resistance to chlorine, one of the most widely used disinfectants, due to bacterial biofilms.

Bacterial biofilms are complex living communities composed of a wide range of components and molecules such as bacterial cells, their polysaccharides, proteins, lipids, DNA and RNA. The external DNA (eDNA) in particular plays an important role in the early phase of biofilm arrangement [27].

Several factors can influence biofilm generation [28, 29]. The main factors are related to the bacterial surface and its charge. Hydrophobicity is a main factor influencing the adsorption and change in the surface tension of bacteria. Biofilm formation involves all flagellar and non-flagellar bacterial structures - fimbriae, pilli and flagella [30]. Investigations on the structure of fimbriae show that they contain predominantly residues of hydrophobic amino acids, such as valine, leucine, isoleucine, methionine, phenylalanine, tryptophan, and cystein [23]. Fimbirae also contain adhesion molecules [29] which attach to substrates and thus bacteria can deliver nutrients for their metabolism. Temperature and substrate availability also impact biofilm formation.

It is important to note that bacterial adhesion [29] and biofilm formation increase on rough surfaces compared to smooth surfaces. The larger surface area and the weaker shear forces facilitate biofilm formation. As Donlan [29] described, the physicochemical properties of the surface is of great importance in biofilm build-up. Bacteria attach more easily to hydrophobic, nonpolar, rough surfaces like Teflon or plastics than to hydrophilic surfaces like glass or steel [31–33]. Bacterial biofilms develop on tooth enamel in the oral cavity. The pellicle contains albumin, lipids, glicoproteins, gingvinal fissure liquid, lysozime and bacteria dwelling in the oral cavity. Mittelman [34] discussed in his publication that the host produces complex bacterial biofilms as saliva, respiratory secretion, tears, urine and blood, which strongly influence bacterial attachment. The development of bacterial biofilm is shown in **Figure 5**.

As shown in **Figure 5**, the stages of bacterial biofilm development include the crucial initial steps of finding, interacting with, and adhering of planktonic bacteria to a surface [35, 36]. Once irreversibly attached to a surface, bacteria form microcolonies. Biofilm matures and when it has completely matured it is affected by shear forces and undergoes rupture resulting in free planktonic cells. The liberated planktonic cells fall on new surfaces and colonize them, forming new biofilms [22, 37].

Both pH and the high amount of nutrients increase the concentration of ferric, sodium and calcium cations. These cations affect the adhesion of *Pseudomonas fluorescens* reducing the chemical forces between the negatively charged bacterial cells and the glass surface [29]. Several studies reported that mycolic acid-containing bacteria like *Mycobacterium* [38], *Corynebacterium* [39] and *Nocardia* [40] attach more intensively than non-mycolic ones. The longer chain length of mycolic acid correlates with high and rapid bacterial adhesion. Silva and de Ataujo [41] discussed the inhibitory role of lectins on biofilm formation. Lectins are proteins which bind to carbohydrates

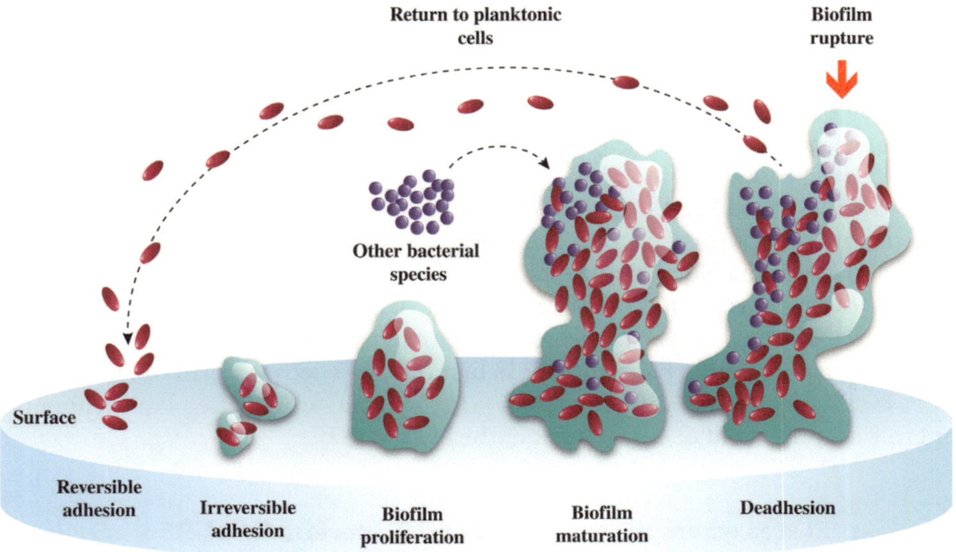

Figure 5.
Bacterial biofilm development.

and polysaccharides of the outer membrane of bacteria. Lectins are ubiquitous in nature and can be found in large amounts in cereals and legumes.

Depending on the affinity of motile and nonmotile bacteria to adhesion, motile bacteria are capable of more active attachment. Nonmotile bacteria are slower in forming biofilms. The flagella of motile bacteria are crucial for the early stages of biofilm formation [30, 42].

Pseudomonas aeruginosa, one of human opportunistic pathogens was used as model organism to study bacterial biofilm formation cells [43]. Authors show that three extracellular polysaccharides (EPS) - alginate, Psl, and Pel are mainly responsible for the biofilm formation. EPS can represent between 50% and 90% of the total organic carbon in the biofilm [44]. EPS composed of polysaccharides are neutral biopolymers [45]. When EPS contain uronic acids such as D-glucoronic or D-galacturonic acid, they contribute to their anionic nature. The anionic property is important for the association with calcium and magnesium bivalent cations, which cross-link and provide greater strength to the bacterial biofilm. EPS can be either hydrophobic or hydrophilic but are generally highly hydrated due to water accumulation *via* the hydrogen bonding. This is the reason why natural biofilms can hardly be desiccated. In addition to divalent cations, EPS can bind to metal ions, proteins, DNA or lipids. Some EPS can even bind to humic acids [46].

The main extracellular polymeric substances which bacteria produces when exposed to phenol are PN (exopolymeric protein) and PS (lower polysaccharides) as described by Gao *et al.* [47]. During the biotransformation of heavy metals synthesize as EPS both homopolysaccharides and heteropolysaccharides [48]. Among the homopolysaccharides are identified dextrane, mutane, alternant, reuteran, gurdlan, levan and inulin. Gupta *et al.* reported the most abundant amidst heteropolysaccharides - alginate, xanthan, hyaluronan and sphingans [48]. *P. aeruginosa* responds to chlorine-based disinfectants by synthesis of alginate-based EPS as described by Xue *et al.* [49].

Undoubtedly, the architecture of each bacterial biofilm is unique. They can be mono-, double or multi-layer thick. When consisting of several layers, a network of many water channels can be observed inside the biofilm. According to the bacterial diversity, biofilms can consist of one bacterial strain but most often they contain mixed bacterial cultures. Different bacteria form thicker or thinner biofilms. Sometimes, when the biofilm is formed in the human body, it can also include non-bacterial compartments like erythrocytes or fibrin. Such types of biofilms form on heart valves. Bacterial biofilms formed on urinary catheters are known to consist of bacteria capable of urease-catalyzed degradation of urea, resulting in the release of ammonia. Ammonia induces precipitation of the calcium and magnesium inside the biofilm, leading to encrustation and catheter blockage [50].

Bacterial biofilms are perfect structures for plasmid DNA horizontal transfer, which occurs more easily between cells in biofilms than between planktonic cells because of the tighter cell-to-cell contact [51]. Quorum sensing also plays an important role in attachment or detachment of the biofilm [52].

4. Industrial area where the environmental bacterium KCM-R5 was isolated

KCM-R5 is an environmental bacterial isolate collected from a Pb-Zn smelter area and successfully entrapped in a synthetic polymer – poly(ethylene oxide) hydrogels (PEO) [53]. PEO hydrogels are macroporous polymers with high molecular weight and appropriate for bacterial immobilization due to their biocompatibility, strength and adhesiveness [54]. Additionally, they demonstrate nontoxicity, flexibility and durability. Initially, PEO hydrogels were obtained *in situ* by γ-irradiation of aqueous solutions [55], and two decades later, *via* methods based on chemical crosslinking [56]. UV crosslinking at cryogenic temperatures contributes to an important feature of the PEO hydrogels, namely the formation of macroporous structure. This macroporous structure is highly compatible with bacteria and enable their easy penetration, movement, hence, biofilm generation inside the hydrogels. The second main advantage of poly(ethylene oxide) hydrogel synthesis under cryogenic conditions than at room temperature is the extraordinarily high yield of gel fraction and better crosslinking [55, 56].

The environmetal bacterial isolate KCM-R5 was isolated from a soil sample collected at the industrial area of KCM Pb-Zn smelter (plant for production of non-ferrous metals), located in Central Bulgaria, near the town of Plovdiv. This plant is the biggest smelter on the Balkan Peninsula and producer of Pb, Zn, Au, Ag and Pt and their alloys since 1962. At approximately 1 km away is the pesticide factory AGRIA Ltd., founded in 1932. Both plants have been polluting the environment with heavy metals and hydrocarbons for years. Recently, the new wastewater treatment plant operating at KCM has reduced the outflow of polluted water. The produce of both plants is sold on the local market but is mainly exported worldwide. After the isolation, the bacterium was successfully cultivated in nutrient broth and nutrient agar and on selective media containing various heavy metals and 2,4-Dichlorophenoxyacetic acid (2,4-D).

5. Molecular-genetic analysis of KCM R5 bacterial isolate

A molecular-genetic analysis of the bacterial DNA was conducted aiming the identification of the bacterium. 16S rDNA of the KCM R5 strain was amplified, restricted with the frequently cutting endonuclases *Msp*I, *Hae*III and *Rsa*I (New England BioLabs, UK) and sequenced. PCR amplification was performed using the primers 8F (forward) (5'-AGAGTTTGATCCTGGCTCAG-3′) and 1513R (reverse) (5'-GGTTACCTTGTTACGACTT-3′). This primer pair is preferable because it generates the longest amplicon ofapproximately 1400 bp. The amplification protocol consisted of one cycle of initial denaturation at 95°C for 3 min, 35 cycles of DNA denaturation at 94°C for 90 sec, primer annealing at 55°C for 40 sec, and primer extension at 72°C for 1.5 min, ending with a final extension step at 72°C for 20 min. Sequencing was accomplished with an automated sequencer 310 ABI-PRISM (Applied Biosystems, USA). The sequences obtained were analyzed using BLAST program and the bioinformatic analysis showed that the 16S rDNA sequence of KCM-R5 is affiliated with *Pseudomonas rhodesiae* with 99.9% identity. The 16S rDNA sequence of the strain was submitted to the Gene Bank-EMBL Database under the accession number AJ 830707. **Figure 6** presents the dendrogram of the strain *P. rhodesiae* KCM-R5 (Gamma- Proteobacteria) with its closely related relatives.

Members of the genus *Pseudomonas* are heterotrophs, rod-shaped, psychrotrophic and motile. According to Gram staining, pseudomonads are Gram-negative. Gram staining of the bacterial isolate KCM R5 shown in **Figure** 7 demonstrated that it is a Gram-negative bacterium.

Ubiquitous in nature, the size of the bacteria of genus *Pseudomonas* varies between 1 and 5 micrometers in length and 0.5-1.0 micrometers in width. Bacterial flagella and pilli are important for the adhesion process. Pseudomonads are known to produce a vast amount of extracellular polysaccharides (EPS) [57]. They are able to produce biofilms even on smooth stainless steel surfaces, multiplying alone in the biofilm or co-existing with other bacterial species [58]. The biodegradation of phenol in wastewater by immobilized cells of *Pseudomonas putida* was described by [7, 59, 60].

When pseudomanads exist in mixed biofilms, they are more stable. In such biofilms *P. aeruginosa* or *P. fluorescens* synthesize a blue toxic substance called pyocianin (**Figure 8**) able to kill bacteria competing pseudomonads [27]. Norman et al. [61] demonstrated that

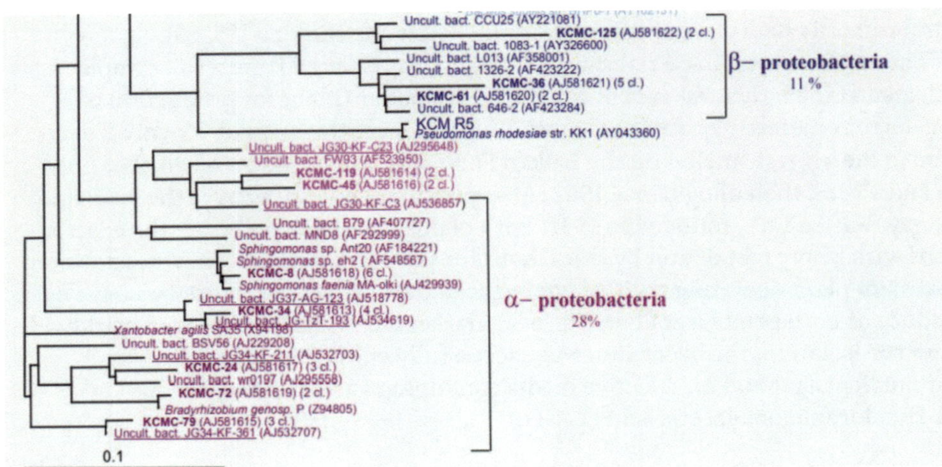

Figure 6.
Dendrogram of the strain Pseudomonas rhodesiae KCM R5.

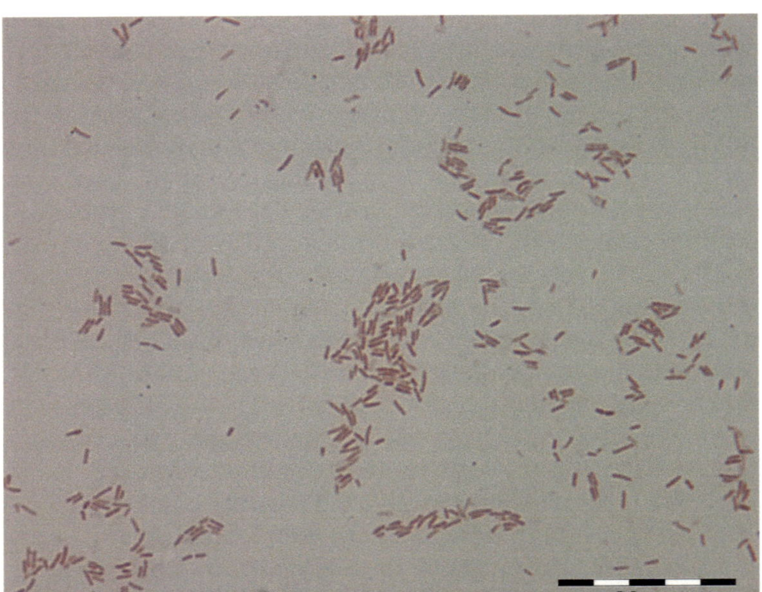

Figure 7.
Gram staining of Pseudomonas rhodesiae KCM R5.

DOI: http://dx.doi.org/10.5772/intechopen.104813

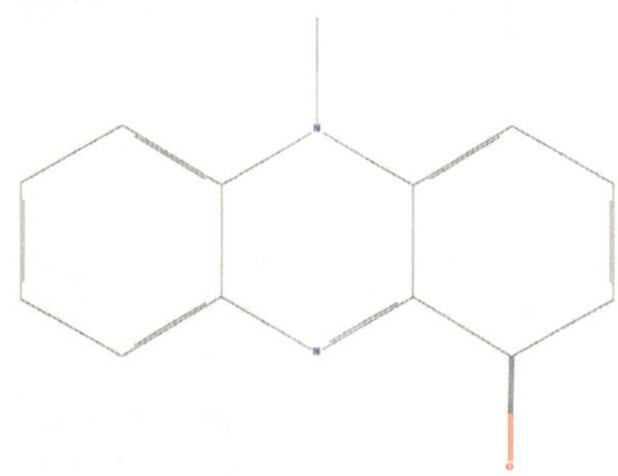

Figure 8.
Structural formula of pyocyanin.

pyocyanin influenced the functional diversity of a crude oil-degrading culture containing *P. aeruginosa and* affected the overall degradation of the crude oil.

6. Heavy metal tolerance and growth of Pseudomonas rhodesiae KCM R5 on phenol and phenol derivatives as planktonic cells

The tolerance of planktonic cells of *P. rhodesiae* KCM-R5 to phenol and phenol derivatives was studied by cultivation of the strain on phenol, *o*-nitrophenol, pentachlorophenol, 2,4-dinitrophenol, 2,5-dinitrophenol and 2,4-dichlorophenoxiacetic acid (2.4-D) added to mineral media of Furukawa and Chakrabarty [62]. The medium contained per liter 5.6 g $K_2HPO_4x3H_2O$, 3.4 g KH_2PO_4, 2 g $(NH_4)_2SO_4$, 0.34 g $MgCl_2x6H_2O$, 0.001 g $MnCl_2x4H_2O$, 0.0006 g $FeSO_4x7H_2O$, 0.026 g $CaCl_2x2H_2O$ and 0.002 g $Na_2MoO_4x2H_2O$. Phenol was applied at a concentration of 100 mg L^{-1} while its five derivatives were added at a lower concentration of 20 mg L^{-1} due to their higher toxicity and carcinogenicity, which may cause bacterial cells death. Xenobiotics were metabolized as a sole carbon source with no glucose or other carbohydrate addition. The investigation was performed for 144 h at 28°C. **Figure 9** shows

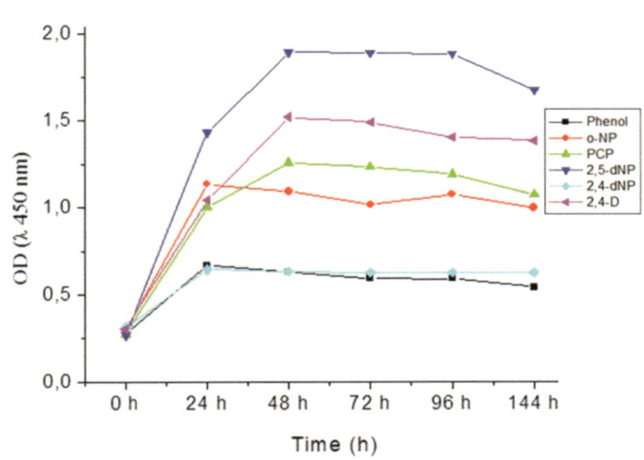

Figure 9.
Growth of P. rhodesiae KCM-R5 on phenol and nitro- and chlorophenol derivatives as sole carbon sources.

the growth of *P. rhodesiae* KCM R5. The strain demostrated the most intensive growth on 2,5 - dinitrophenol, 2.4-D, and pentachlorophenol (**Figure 9**).

7. Cryogels preparation

PEO cryogels necessary for bacteria entrapment were kindly supplied by Prof. Petar Petrov, DSc, Institute of Polymers, Bulgarian Academy of Sciences. Polyethylene oxide was dissolved in distilled water and polymerized by adding a photo initiator (4-benzoylbenzyl) trimethylammonium chloride. The obtained solution was poured into Teflon dishes forming layers 50 mm in diameter. The layers were further placed at -20°C for 2 h and irradiated with UV–VIS light for 2 min. PEO cryogels were extracted in distilled water for 7 days and freeze dried at -55°C, adopted from Doycheva *et al.* [55]; Petrov *et al.*, [56], Satchanska *et al.*, [63], Berillo *et al.*, [54].

8. Entrapment of the bacteria into the PEO cryogels

The dried PEO cryogels were swelled by soaking without shaking in Furukawa and Chakrabarty medium for 24 h. The strain *P. rhodesiae* KCM R_5 was prepared for entrapment in the PEO cryogels by cultivation in Furukawa and Chakrabarty mineral medium with added 0.1% sterile glucose and 100 mg/L phenol until reaching OD 0.550. Then the bacterial culture was mixed with the pre-swollen PEO cryogels and shaken mildly at 100 rpm for 48 h. The resulting PEO-KCM R_5 unit consisting of cryogel and immobilized inside bacteria was gently placed inside the sterile Top Filter 45 mm, 500 ml system (Nalgene, Rochester, USA) and the locking rings were softly screwed up in order to avoid cutting of the cryogels, adopted from Satchanska *et al.*, [63]; Donelli *et al.*, [64] and Berillo *et al.*, [65]. In the control swelled but empty (without immobilized bacteria inside) PEO cryogel was used.

9. Phenol biodegradation by PEO cryogel-*P. rhodesiae* KCM R5 biofilter and biofilm formation

The phenol biodegradation by the PEO cryogel-*P. rhodesiae* KCM R5 biofilm occurred *via* a sequencing batch process [66, 67]. The cycle of feeding *via* the upper container (phenol inflow) was 24 h and phenol concentrations was increased from 300 to 1000 mg L^{-1}. Volume of phenol inflow was 250 mL. The experiment was conducted 28°C, in triplicate. Every 24 hours 250 mL sterile medium that contained increasing phenol concentrations on the following scheme: 7 days with 300 mg L^{-1}, 5 days with 400 mg L^{-1}, 4 days with 600 mg L^{-1} and 12 days with 1000 mg L^{-1} phenol was poured into the upper funnel. The experiment lasted 28 days. No pressure was applied to the phenol-containing liquid and it run through the PEO cryogel-*P. rhodesiae* KCM R5 biofilm by only its gravity force, adopted by Satchanska *et al.*, [63].

Inside the PEO cryogel-*P. rhodesiae* KCM R5 biofilm phenol degradation occurred and the solution of degraded phenol flowed out into the lower container (phenol outflow) [51, 52]. Phenol concentration in both phenol inflow and outflow was measured in succession at every 24 hours for a period of 28 days. Assessment of the phenol concentration in both inflow and outflow was carried out by colorimetric method using pyramidone. The protocol can be briefly described as follows: 0.125 ml phenol outflow liquid, 0.250 ml ammonium chloride buffer pH 9,3, 0.125 ml 3.5% pyramidone and 0.375 ml ammonium persulfate pH 7.0 were added to 12.375 ml distilled water to obtain 13 ml total volume. The reaction was incubated at room T °C for 45 min and its absorption was measured with a UV/VIS spectrophotometer at

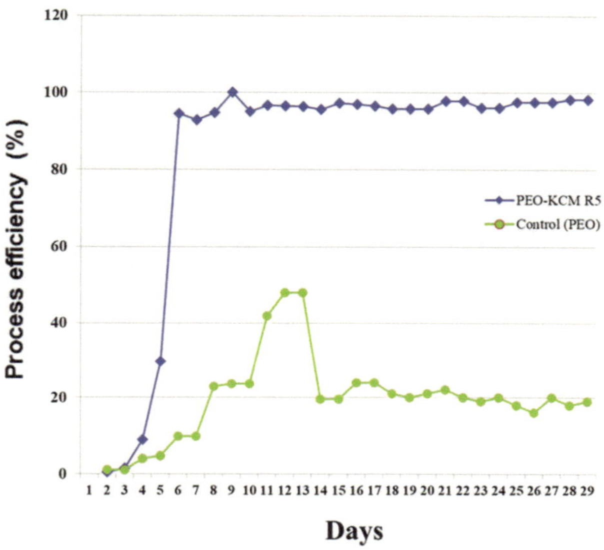

Figure 10.
Phenol degradation by PEO cryogel-P. rhodesiae KCM R_5 biofilter.

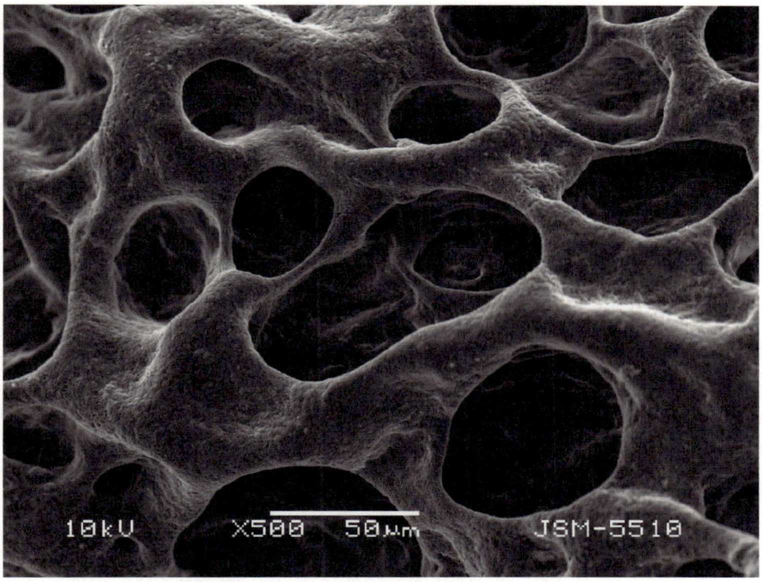

Figure 11.
Macrostructure of swelled PEO cryogel without bacteria.

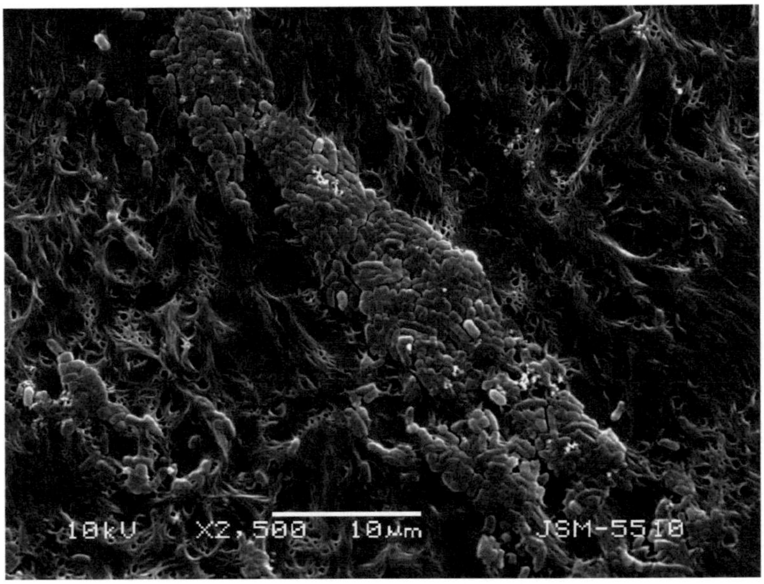

Figure 12.
Bacterial P. rhodesiae KCM R5 biofilmo engeneered inside the biofilter.

540 nm. In the control, instead of the phenol outflow liquid 0.125 ml distilled water was added adopted by Satchanska *et al.*, [63].

The phenol amount and biodegradation was calculated according to a standard curve and phenol biodegradation was calculated according the equation:

$$\text{Efficiency in}\,(\text{X})\,\text{hour}\,(\%) = Ci - Cf\,/\,Ci \times 100 \qquad (1)$$

Data about phenol biodegradation [54–57] by the PEOcryogel-*P. rhodesiae* KCM R_5 biofilm is presented in **Figure 10**.

After 28 days of biodegradation, the PEO-KCM R5 biofilter was disassembled and the cryogel with bacteria degrading phenol inside was taken out and subjected to Scanning Electron Microscopy analysis (SEM). The biofilter sample was covered with an Au microlayer and observed at JSM-5510 Scanning Electron Microscope (Jeol, Japan) in vacuum at 10000 V voltage and under different magnifications ranging from x500 to x20 000 (**Figures 11** and **12**).

10. Conclusions

Our molecular-genetic analysis showed that the environmental bacterium KCM-R5 is affiliated to *Pseudomonas rhodesiae.* The strain is tolerant to xenobiotics and can grow as planktonic cells on phenol and nitro- and chlorophenol derivatives as sole carbon sources. The constructed PEO cryogel-*P. rhodesiae* KCM R5 biofilm is capable of phenol degradation at a concentration of 1000 mg L^{-1}/24 h. Phenol biodegradation is due to the biofilm formed by *P. rhodesiae* KCM R_5 inside the PEOcryogel approved by observation using Scanning Electron Microscope. The so engineered PEO cryogel-*P. rhodesiae* KCM R5 biofilm can be used for environmental biotechnology application in industrial wastewater detoxification.

Acknowledgements

The author is grateful to Prof. P. Petrov, Institute of Polymers, Bulgarian Academy of Sciences, for supplying the PEO-cryogels and to Prof. Maria Angelova, Stephan Angeloff Institute of Microbiology, Bulgarian Academy of Sciences, for the critical reading of the manuscript.

The author thanks RIDACOM Ltd., Bulgaria for the financial support in publishing the chapter.

Conflict of interest

The author declares no conflict of interest.

Author details

Galina Satchanska
Department of Natural Sciences, New Bulgarian University, Sofia, Bulgaria

*Address all correspondence to: gsatchanska@nbu.bg

References

[1] Mathieu-Denoncourt J, Wallace J, de Solla S, Langlois V. Influence of lipophilicity on the toxicity of bisphenol A and phthalates to aquatic organisms-A review. Bulletin of Environmental Contamination and Toxicology. 2016;**97**: 4-10. DOI: 10.1007/s00128-016-1812-9

[2] Santoro A, Chianese R, Troisi J, Richards S, Nori S, Fasano S, et al. Neuro-toxic and reproductive effects of BPA – A review. Current Neuropharmacology. 2019;**17**:1109-1132. DOI: 10.2174/1570159X17666190726112101

[3] McCall I, Betanzos I, Weber D, Nava P, Miller G, Parkos C. Effects of phenol on barrier function of a human intestinal epithelial cell line correlate with altered tight junction protein localization. Toxicology and Applied Pharmacology. 2009;**241**:61-70. DOI: 10.1016/j.taap.2009.08.002 (TJ Junction)

[4] Public Health England. Phenol Toxicological Overview. 2016. https://assets.publishing.service.gov.uk/government/uploads/system/uploads/attachment_data/file/500822/Phenol_PHE_TO_120216.pdf. [Accessed 14.03.2022]

[5] Neumann G, Tera R, Monson L, Kavisaar M, Schauer F, Heipieper H. Simultaneous degradation of atrazine and phenol by *pseudomonas* sp. strain ADP: Effect and adaptation. Applied and Environmental Microbiology. 2004;**70**:1907-1912

[6] Polymenakou P, Stephanou E. Effect of temperature and additional carbon sources on phenol degradation by an indigenous soil pseudomonad. Biodegradation. 2005;**16**:403-413

[7] Tsai S, Juang R. Biodegradation of phenol and sodium salicylate mixtures by suspended *pseudomonas putida* CCRC 14365. Journal of Hazardous Materials. 2006;**138**:125-132

[8] Tallur P, Megadi V, Kamanavalli C. Ninnekar H biodegradation of p-cresol by *bacillus* sp. strain PHN 1. Current Microbiology. 2006;**53**:529-533

[9] Kunapuli U, Jahn M, Lueders T, Geyer R, Heipieper HJ, Meckenstock R. *Desulfitobacterium aromaticivorans* sp. nov. and *Geobacter toluenoxydans* sp. nov., iron-reducing bacteria capable of anaerobic degradation of monoaromatic hydrocarbons. International Journal of Systematic and Evolutionary Microbiology. 2010;**60**:686-695

[10] Jones K, Trudgill P. Hopper D. evidence of two pathways for the metabolism of phenol by *aspergillus fumigatus*. Archives of Microbiology. 1995;**163**:176-181

[11] Hasan S, Jabeen S. Degradation kinetics and pathway of phenol by *pseudomonas* and *bacillus* species. Biotechnology & Biotechnological Equipment. 2015;**29**:45-53

[12] Chiriac A, Rusu A, Nita L, Chiriac v, Neamtu J, Sandu A. Polymeric carriers designed for encapsulation of essential oils with biological activity – A review. Pharmaceutics. 2021;**13**:631. DOI: 10.3390/pharmaceutics13050631

[13] Rodrigo-Navarro A, Sankaran S, Dalby M, del Campo A. Manuel Salmeron-Sanchez. Engineered living biomaterials. Nature Reviews Materials. 2021;**6**:1175-1190

[14] Lu Y, Aimetti A, Langer R, Gu Z. Bioresponsive materials. Nature Reviews Materials. 2017;**2**:16075

[15] Lusta K, Starostina N, Gorkina N, Fikhte B, Lozinski V. Immobilization of *E. coli* cells in macroporous polyacrylamide cryogels. Applied Biochemistry and Microbiology. 1988;**24**:418-426

[16] Stormo K, Crawford R. Preparation of encapsulated microbial cells for environmental applications. Applied and Environmental Microbiology. 1992;**58**:727-730

[17] Godjevargova T, Aleksieva Z, Ivanova D, Shivarova N. Biodegradation of phenol by *Trichosporon cutaneum* cells covalently bound to polyamide granules. Process Biochemistry. 1998;**33**:831-835

[18] Godjevargova T, Ivanova D, Aleksieva Z, Dimova N. Biodegradation of toxic organic components from industrial phenol production waste waters by free and immobilized *Trichosporon cutaneum* R57. Process Biochemistry. 2003;**38**:915-920

[19] Yordanova G, Ivanova D, Godjevargova T, Krastanov A. Biodegradation of phenol by immobilized *aspergillus awamori* NRRL 3112 on modified polyacrylonitrile membrane. Biodegradation. 2009;**20**:717-726

[20] El-Naas M, Al-Muhtaseb S. Biodegradation of phenol by *pseudomonas putida* immobilized in polyvinyl alcohol (PVA) gel. Journal of Hazardous Materials. 2009;**164**:720-725

[21] Alvarez-Paino M, Munoz-Bonila A, Fernandez-Garcia M. Antimicrobial polymers in the nanoworld. Nanomaterials. 2017;7:48. DOI: 10.3390/nano7020048

[22] Muhammad M, Idris A, Fan X, Guo Y, Yu Y, Jin X, et al. Beyond risk: Bacterial biofilms and their regulating approaches-a review. Frontiers in Microbiology. 2020;**11**:1-20. DOI: 10.3389/fmicb.2020.00928

[23] Carrascosa C, Raheem D, Ramos F, Ariana Saraiva A, Raposo A. Microbial biofilms in the food industry - A comprehensive review. International Journal of Environmental Research and Public Health. 2021;**18**(4):2-31. DOI: 10.3390/ijerph18042014

[24] Characklis W, Cooksey K. Biofilms and microbial fouling. Advances in Applied Microbiology. 1983;**29**:93-138

[25] Aumeran C, Balestrino D, Forestier C. Bacterial Biofilms and Health. 2020; 1-7 https://www.encyclopedie-environnement.org/en/health/bacterial-biofilms. [Accessed 21.01.2021]. Encyclopédie de l'environnement. University of Grenoble

[26] Zhang W, Ding W, Li Y-X, Tam C, Bougouffa S, Wang R, et al. Marine biofilms constitute a bank of hidden microbial diversity and functional potential. Nature Communications. 2019;**10**:517

[27] Das T, Ibugo A, Klare W, Manefield M. Microbial Biofilms - Importance and Applications. Role of Pyocyanin and Extracellular DNA in Facilitating Pseudomonas aeruginosa Biofilm Formation. London: IntechOpen; 2016. ISBN: 978-953-51-5439-6

[28] An Y, Friedman R. Handbook of Bacterial Adhesion: Principles, Methods, and Applications. 2000th ed. Totowa, NJ, USA: Humana Press Inc. p. 629. ISBN-13: 978-0896037946

[29] Donlan R. Biofimls: Microbial life on surfaces. Emerging Infectious Diseases. 2002;**8**:881-890

[30] Liaqat I, Mirza S, Iqbal R, Ali N, Saleem G, Majid S, et al. Flagellar

motility plays important role in biofilm formation of *Bacillus cereus* and *yersinia enterocolitica*. Pakistan Journal of Pharmaceutical Sciences. 2018;**3**(Supplementary):2047-2052

[31] Tallawi M, Opitz M, Lieleg O. Modulation of the mechanical properties of bacterial biofilms in response to environmental challenges. Biomeatreial Science. 2017;**5**:887-900. DOI: 10.1039/c6bm00832a

[32] Flemming H, Wingender J, Szewzyk U, Steinberg P, Rice S, Kjelleberg S. Biofilms: An emergent form of bacterial life. Nature Reviews Microbiology. 2016;**14**:563-575

[33] Rumbaugh K, Sauer K. Biofilm dispersion. Nature Reviews Microbiology. 2020;**18**:571-586

[34] Mittelman M. Bacterial biofilms in pharmaceutical water systems. In: Filtration and Purification in the Biopharmaceutical Industry. 3rd ed. Boca Raton: Imprint CRC Press; 2019. p. 21. eBook ISBN 9781315164953

[35] Pinto A, Cerqueira M, Bañobre-Lópes M, Pastrana L, Sillankorva S. Bacteriophages for chronic wound treatment: From traditional to novel delivery systems. Viruses. 2020;**12**:235. DOI: 10.3390/v12020235

[36] Simoes M, Borges A, Simoes L. Recent Trends in Biofilm Science and Technology. 1st ed. UK: Academic Press, Elsevier; 2020. p. 416. ISBN-10: 0128194979

[37] Mosselhy D, Assad M, Sironen T, Elbahri M. Nanotheranostics: A possible solution for drug-resistant *Staphylococcus aureus* and their biofilms. Nanomaterials. 2021;**11**:82. DOI: 10.3390/nano11010082

[38] Esteban E, García-Coca. *Mycobacterium* biofilms. Frontiers in Microbiology. 2018;**8**:2651. DOI: 10.3389/fmicb.2017.02651

[39] Yaacob M, Murata A, Nor N, Jesse F, Yahya M. Biochemical composition, morphology and antimicrobial susceptibility pattern of *Corynebacterium pseudotuberculosis* biofilm. Journal of King Saud University-Science. 2021;**33**:101225. DOI: 10.1016/j.jksus.2020.10.022

[40] Mehtaa H, Wenga J, Pratera A, Elworthb R, Hanc X, Shamooa Y. Pathogenic *nocardia cyriacigeorgica* and *Nocardia nova* evolve to resist trimethoprim-sulfamethoxazole by both expected and unexpected pathways. Antimicrobial Agents and Chemotherapy. 2018;**62**:00364-00318. DOI: 10.1128/AAC.00364-18

[41] Silva N, de Araújo F. Antibacterial activity of plant lectins: A review Brazilian archives of biology and technology. Brazilian Archives of Biology & Technology. 2021;**64**:e21200631. DOI: 10.1590/1678-4324-2021200631

[42] Mishra R, Panda A, De Mandal S, Shakeel M, Bisht S, Khan J. Natural anti-biofilm agents: Strategies to control biofilm-forming pathogens – A review. Frontiers in Microbiology. 2020;**11**:1-23. DOI: 10.3389/fmicb.2020.566325

[43] Ghafoor A, Hay I, Rehm B. Role of exopolysaccharides in *Pseudomonas aeruginosa* biofilm formation and architecture. Applied and Environmental Microbiology. 2011;**77**:5238-5246. DOI: 10.1128/AEM.00637-11

[44] Singh S, Datta S, Narayanan B, Rajnish N. Bacterial exo-polysaccharides in biofilms: Role in antimicrobial resistance and treatments. Journal of Genetic Engineering and Biotechnology. 2021;**19**:140-158

[45] Di Martino P. Extracellular polymeric substances, a key element in understanding biofilm phenotype. AIMS Microbiology. 2018;**4**:274-288

[46] Cohen E, Merzendorfer H. Extracellular Sugar-Based Biopolymers Matrices (Biologically-Inspired Systems Book 12). 1st ed. Switzerland: Springer Link; 2019. p. 693. ISBN-13: 978-3030129187. DOI: 10.1007/978-3-030-12919-4

[47] Gao M, Diao M-H, Yuan S, Wang Y-K, Xu H, Wang X-H. Effects of phenol on physicochemical properties and treatment performances of partial nitrifying granules in sequencing batch reactors biotechnology reports. Biotechnology Reports. 2017;**13**:13-18

[48] Gupta P, Diwan B. Bacterial exopolysaccharide mediated heavy metal removal: A review on biosynthesis, mechanism and remediation strategies. Biotechnology Reports. 2017;**13**:58-71

[49] Xue Z, Hessler C, Panmanee W, Hassett D, Seo Y. *Pseudomonas aeruginosa* inactivation mechanism is affected by capsular extracellular polymeric substances reactivity with chlorine and monochloramine. FEMS Microbiology Ecology. 2013;**83**:101-111. DOI: 10.1111/j.1574-6941.2012.01453.x

[50] Krajewska B. Urease-aided calcium carbonate mineralization for engineering applications: A review. Journal of Advanced Research. 2018;**13**:59-67. DOI: 10.3934/microbiol.2018.2.274

[51] Flemming H-C, Wingender J. The biofilm matrix. Nature Reviews Microbiology. 2010;**8**:623-633

[52] Carradori S, Di Giacomo N, Lobefalo M, Luisi G, Campestre C, Sisto F. Biofilm and quorum sensing inhibitors: The road so far – A review. Expert Opinion on Therapeutic Patents. 2020;**30**:917-930. DOI: doi.org/10.1080/13543776.2020.1830059

[53] Caroline L, Araújo A, Melo de Oliveira M. Effect of heavy metals on the biofilm formed by microorganisms from impacted aquatic environments. In: Bacterial Biofilms. London: IntechOpen; 2019. ISBN:978-1-78985-900-3. DOI: 10.5772/intechopen.82929

[54] Berillo D, Al-Jwaid A, Capl J. Polymeric materials used for immobilisation of bacteria for the bioremediation of contaminants in water– A review. Polymers. 2021;**13**:1073. DOI: doi.org/10.3390/polym1307107

[55] Doycheva M, Petrova E, Stamenova R, Tsvetanov C, Riess G. UV-induced cross-linking of poly(ethylene oxide) in aqueous solution. Macromolecular Materials and Engineering. 2004;**289**:676-680

[56] Petrov P, Petrova E, Stamenova R, Tsvetanov C, Reiss G. Cryogels of cellulose derivatives prepared via UV irradiation of moderately frozen systems. Polymer. 2006;**47**:6481-6484

[57] El-Naas M, Al-Muhtaseb S. Biodegradation of phenol by *pseudomonas putida* immobilized in polyvinyl alcohol (PVA) gel. Journal of Hazardous Materials. 2009;**164**:720-725

[58] Baumgarten T, Vazquez J, Bastisch C, Veron W, Feuilloley M, Nietzsche S, et al. Alkanols and chlorophenols cause different physiological adaptive responses on the level of cell surface properties and membrane vesicle formation in *pseudomonas putida* DOT-T1E. Applied Microbiology and Biotechnology. 2012;**93**:837-845

[59] Neumann G, Tera R, Monson L, Kavisaar M, Schauer F, Heipieper H.

Simultaneous degradation of atrazine and phenol by *pseudomonas* sp. strain ADP: Effect and adaptation. Applied and Environmental Microbiology. 2004;**70**:1907-1912

[60] Polymenakou P, Stephanou E. Effect of temperature and additional carbon sources on phenol degradation by an indigenous soil pseudomonad. Biodegradation. 2005;**16**:403-413

[61] Norman R, Moeller P, McDonald T, Morris P. Effect of pyocyanin on a crude-oil-degrading microbial community. Applied and Environmental Microbiology. 2004;**70**:4004-4011. DOI: 10.1128/AEM.70.7.4004-4011.2004

[62] Furukawa R, Chakrabarty S. Common induction and regulation of biphenyl, xylene, toluene and salycilate in *pseudomonas paucimobilis*. Journal of Bacteriology. 1983;**154**:1356-1363

[63] Satchanska G, Topalova Y, Ivanov I, Golovinsky E. Xenobiotic biotransformation potential of *Pseudomonas rhodesiae* KCM-R_5 and *Bacillus subtilis* KCM-RG_5 tolerant to heavy metals and phenol derivatives. Biotechnology & Biotechnological Equipment. 2014;**20**:97-102

[64] Donelli G. Microbial Biofilms: Methods and Protocols (Methods in Molecular Biology, 1147). New York, USA: Humana Press, Springer Science+Business Media; 2014. p. 379. ISBN-13: 978-1493904662

[65] Berillo B, Caplin J, Cundy A, Savina I. A cryogel-based bioreactor for water treatment applications. Water Research. 2019;**153**:324-334

[66] Edwards S, Kjellerup B. Applications of biofilms in bioremediation and biotransformation of persistent organic pollutants, pharmaceuticals/personal care products, and heavy metals. A Mini-Review. Applied Microbiology and Biotechnology. 2013;**97**:9909-9921

[67] Sehar S, Naz I. Role of the biofilms in wastewater treatment. In: Microbial Biofilms - Importance and Applications. London. ISBN: 978-953-51-5439-6: IntechOpen; 2016. DOI: 10.5772/63499